T0135309

Lecture Notes in Networks and Systems

Volume 489

The series "Lecture Notes in Networks and Systems" publishes the latest developments in Networks and Systems—quickly, informally and with high quality. Original research reported in proceedings and post-proceedings represents the core of LNNS.

Volumes published in LNNS embrace all aspects and subfields of, as well as new challenges in, Networks and Systems.

The series contains proceedings and edited volumes in systems and networks, spanning the areas of Cyber-Physical Systems, Autonomous Systems, Sensor Networks, Control Systems, Energy Systems, Automotive Systems, Biological Systems, Vehicular Networking and Connected Vehicles, Aerospace Systems, Automation, Manufacturing, Smart Grids, Nonlinear Systems, Power Systems, Robotics, Social Systems, Economic Systems and other. Of particular value to both the contributors and the readership are the short publication timeframe and the world-wide distribution and exposure which enable both a wide and rapid dissemination of research output.

The series covers the theory, applications, and perspectives on the state of the art and future developments relevant to systems and networks, decision making, control, complex processes and related areas, as embedded in the fields of interdisciplinary and applied sciences, engineering, computer science, physics, economics, social, and life sciences, as well as the paradigms and methodologies behind them.

Indexed by SCOPUS, INSPEC, WTI Frankfurt eG, zbMATH, SCImago.

All books published in the series are submitted for consideration in Web of Science.

For proposals from Asia please contact Aninda Bose (aninda.bose@springer.com).

More information about this series at https://link.springer.com/bookseries/15179

Mohamed Lazaar · Claude Duvallet ·
Abdellah Touhafi · Mohammed Al Achhab
Editors

Proceedings of the 5th International Conference on Big Data and Internet of Things

Editors
Mohamed Lazaar
ENSIAS
Mohammed V University
Rabat, Morocco

Claude Duvallet
UNILEHAVRE, UNIROUEN
Normandie Université
Le Havre, France

Abdellah Touhafi
Vrije Universiteit Brussel
Brussels, Belgium

Mohammed Al Achhab
ENSA
Abdelmalek Essaâdi University
Tetuan, Morocco

ISSN 2367-3370 ISSN 2367-3389 (electronic)
Lecture Notes in Networks and Systems
ISBN 978-3-031-07968-9 ISBN 978-3-031-07969-6 (eBook)
https://doi.org/10.1007/978-3-031-07969-6

This Springer imprint is published by the registered company Springer Nature Switzerland AG
The registered company address is: Gewerbestrasse 11, 6330 Cham, Switzerland

Contents

Big Data and Cloud Computing

Toward an Automatic Assistance Framework for the Selection and Configuration of Machine Learning Based Data Analytics Solutions in Industry 4.0

Moncef Garouani[1,2,3](✉), Mohamed Hamlich[2], Adeel Ahmad[1], Mourad Bouneffa[1], Gregory Bourguin[1], and Arnaud Lewandowski[1]

[1] Univ. Littoral Côte d'Opale, UR 4491, LISIC, Laboratoire d'Informatique Signal et Image de la Côte d'Opale, 62100 Calais, France
moncef.garouani@etu.univ-littoral.fr,
{adeel.ahmad,mourad.bouneffa,gregory.bourguin,
arnaud.lewandowski}@univ-littoral.fr

[2] CCPS Laboratory, ENSAM, University of Hassan II, Casablanca, Morocco
moha.hamlich@gmail.com

[3] Study and Research Center for Engineering and Management, HESTIM, Casablanca, Morocco

Abstract. Machine Learning (ML) based data analytics provide methods to cope with the massive data amounts, generated by the various manufacturing processes. In this respect, the maintenance is among the most critical activities concerned by the industrial data analytics in the contexts of industry 4.0. We believe that the maintenance activities can be managed by the predictive processes dwelling on ML methods with the help of meta-learning based data analytics solutions. The challenge is then to facilitate the industry 4.0 actors, who are supposedly not AI specialists, with the application of machine learning. The automated machine learning seems to be the area dealing with this challenge. In this paper, we first show the problematic of assisting industry 4.0 actors to implement ML algorithms in the context of predictive maintenance. We then present a novel AutoML based framework. It aims to enable industry 4.0 actors and researchers, who presumably have limited competencies in machine learning, to generate ML-based data analytics solutions and their deployment in the manufacturing workflows. The framework implements primarily the approaches based on the meta-learning for this purpose. In the context of Industry 4.0 such approaches lead to the implementation of the smart factory concepts. It makes the factory processes more proactive on the basis of predictive knowledge extracted from the various manufacturing devices, sensors, and business processes.

Keywords: Machine learning · AutoML · Data analytics · Meta-learning · Industry 4.0

M. Lazaar et al. (Eds.): BDIoT 2021, LNNS 489, pp. 3–15, 2022.
https://doi.org/10.1007/978-3-031-07969-6_1

1 Introduction

The recent research in the automation, data exchange, and interoperability in manufacturing technologies is intended towards the concrete implementation of the "smart factory" which is among the core concepts of the Industry 4.0 [1]. It is generally achieved by combining the use of hardware and software technologies including Cyber Physical Systems and frameworks, which effectively deal with the knowledge and data management including Artificial Intelligence (AI) based algorithms and models [2]. The studies in this paper deal with such implementations concerning the maintenance activities, applied using in general the AI and more specifically machine learning algorithms and models.

The current manufacturing market is more competitive with improved availability, sustainability, and quality of manufacturing services in smart factories, which has become a crucial concern. The actual scenarios, in this regard, urge the requirement of using AI solutions on production lines. It is more decisive to addresse the maintenance activities in order to prevent the critical failure occurrences. It also reassures the availability and the safety of the concerned system. The AI, particularly machine learning, supports the knowledge extraction process using algorithms, methods, and tools to build models from diverse data representations. These data representations emerge from the different activities of a company and its environment. The careful literature study reveals that the machine learning has demonstrated its usefulness and benefits in many fields. Its successful applications in the context of manufacturing require a lot of effort from human experts since there is not one fit all algorithm to perform well on all possible problems.

The industrial researchers yet lack the machine learning expertise required to handle the massive industrial heterogeneous data sources even though being familiar with manufacturing data [31]. The industrial researchers may collaborate with data science experts to confront such a lack. This collaboration may involve the actors from various expertise, for instance the domain experts can be the production planners, engineers in assembly lines, or the ones who devise strategies to pursue the goals. The existing data analytical approaches do not sufficiently enable these domain experts to specify the problem formulation and devising strategies to achieve the reflected goals. Similarly, the data collection in industrial processes involves a series of activities and continuous evolution of manufacturing data from heterogeneous data sources such that the hardware (digital sensors, machine data, etc.) and software (e-services, databases, etc.). The collaboration between domain experts and data scientists risks to be enacted in a continuous to and fro input among them without yielding useful results. Hence, a successful collaboration among industrial experts and data scientists must prevail in a systematic methodology to guide industry 4.0 actors for proper business data analytical solutions for their personalized problems.

Thereby, the automated assistance for the required human expertise can allow the stakeholders of smart factories to rapidly build, validate, and deploy ML solutions. It may also improve their quality of service, productivity, and more importantly, reduce the need for ML human experts. Motivated by this goal, the

AutoML [3] has emerged across smart factories as a new research field that aims to automatically select, compose, and parameterize machine learning models which are able to achieve an optimal performance on a given task.

In this paper, we present the novel architecture of automated ML based framework that assists domain experts in selecting and configuring ML-based data analytics solutions and automatically explain the provided solutions. The framework is based on the meta-learning [32] approaches to implement the automatic assistance tool, meaning that the system learns from its past experiences to improve the automation of further arriving business data analytics.

The rest of the paper is organized as follows: The Sect. 2 discusses the closely related works in respect of ML-based data analytics in industry 4.0. The Sect. 3 derives the goals that proposed framework is expected to achieve in the industry 4.0 domain. The Sect. 4 introduces the main components of the proposed framework and discusses how these components collaborate to achieve the pursued goals. We later on show how to implement the framework for representative application scenarios, in the Sect. 5, and we discuss its prototypical implementation. Finally, the Sect. 6 concludes the paper and outlines future perspectives.

2 Related Works

The Predictive modeling is crucial to transform large manufacturing data or "big industrial data" into actionable knowledge for various industrial applications. Predictive models can guide manufacturing decision making as well as personalized manufacturing. For example, by predicting the time of occurrences of machinery failures and the criticality of the failures, we can avoid shutdowns in the production processes.

As a major approach to predictive modeling, machine learning is aided with computer algorithms, such that Decision trees, Random forest, Neural networks, and Support Vector Machine, that use past experiences to improve both predictions and decision-making. A recent survey shows that 15% of hospitals use ML predictive models for industrial purposes, and many more are planning to use it [4]. It is important to note that, historically, most of the machine learning algorithms have been criticized for providing no explanations for their prediction results. Nevertheless, new methods have been recently developed to automatically explain prediction results of machine learning models without losing accuracy [5]. Nowadays, providing explanations is considered as a critical requirement of applying AI in any human area. This leads to the development of a new discipline called Explainable AI [6]. Nevertheless, machine learning still presents two major challenges to use them in industrial area, both inefficiently supported by existing solutions such as AutoWeka [7], TPOT [8], and RapidMiner [9]. Both of these two challenges concern the feature engineering (dealing with the inputs of the ML algorithms) and the automatic selection/parametrization of the adequate model, as detailed in the following:

Challenge 1: Efficiently Performing Features Engineering
Features engineering is the process of generating and selecting features from a given data set for the subsequent modeling step. As the overall model performance highly depends on the available features. Feature engineering is a crucial process in the life cycle of a ML pipeline construction. The performance of an ML pipeline can be increased many times over by building good features. In many cases, the original features from the data may not be good enough, e.g., their dimensionality may be too high or samples may not be discriminable in the feature space [10]. Consequently, it is necessary to perform some prerocessing on these features to improve the learning performance.

Feature engineering involves the application of some transformation functions such as arithmetic and aggregate operators on given features to build new ones and remove data errors such as missing values in an input data entry, invalid values or broken links between entries of multiple data sets. Given a predictive modeling problem, an analyst manually examines the quality of the available data, performs adequate transformations and then builds the model which is evaluated later on. If the model accuracy is insufficient, the analyst changes the applied transformation functions and operators for some attributes and rebuilds the model. This labor-intensive process requires interactions between industrial professionals and computer scientists. Moreover, it is often repeated many times before converging, causing a time and human resource bot-tleneck especially in fields that do not tolerate delays such as the manufacturing industry.

Challenge 2: Efficiently and Automatically Selecting Algorithms and Hyperparameters Values
All machine-learning algorithms have two kinds of model parameters: (1) the ordinary parameters that are automatically learned and optimized during the training of the model; (2) the hyperparameters (categorical and continuous) which are manually set by the user of a machine-learning tool before beginning the training of the model (as shown in Table 1). Given a modeling problem like, predicting whether an equipment failure will occur, an analyst builds a model manually and iteratively. Initially, the analyst selects an algorithm among the many other applicable algorithms like Logistic Regression, SVM, Random Forest or Naïve-Bayes. Subsequently, (s)he sets the hyperparameter values for the selected algorithm. Later on, (s)he trains the model to automatically optimize the ordinary parameters. If the model accuracy is insufficient, the analyst changes the hyperparameters values and/or the algorithm and rebuilds the model. This process is iterated until (s)he obtains a model with sufficient accuracy, or (s)he no longer has time to optimize it or the model accuracy cannot be improved anymore.

Numerous combinations of algorithms and hyperparameter values result in hundreds or thousands of labor-intensive manual iterations to build a model, which can be difficult even for experts in machine learning [5]. It is largely observed in the available literature and empirically proved that the algorithms and the used hyperparameter values, affect the model accuracy. Thornton et al. [7] have shown in their study that for the 39 ML algorithms in Weka that the effect on model accuracy, which in average is equal to 46% on 21 data sets and it is of 94% on one data set. Even when considering only a few common algorithms such as support vector machine and random forest; the effect is still greater than 20% on 14 out of 21 data sets. Furthermore, the effective combination of an algorithm and the hyperparameter values varies with respect to the problem we attempt to model. In the literature, some authors explore the automatic search of algorithms and hyperparameter values [11]. Evidently, it shows that automatic search methods can obtain equivalent or even better results than those resulting from the manual tuning done by machine-learning experts [5]. However, when the set of possible algorithms becomes large, previous efforts such as Hyperopt-sklearn [12] and AutoWEKA [7] established limited usefulness as they cannot quickly find the good algorithms as well as their corresponding hyperparameter values and especially, when operating on large datasets.

3 Automated Machine Learning

Automated Machine Learning or AutoML [3] is among the rapidly emerging subfields of ML that attempts to address the theoretical and algorithmic challenges that fully automate the ML process. Its main goal is to automatically generate and configure data analytics solutions through empirical testing under predefined conditions on a target dataset. Furthermore, it leverages human expertise by allowing users to define the conditions that restrict the algorithms to be used, as well as the performance metrics to be used while evaluating candidate algorithms. The core problem considered by AutoML can be formulated as follows: given a dataset, a machine learning task and a performance criterion; solve the task with respect to the dataset while optimizing the performance [13]. Finding an optimal solution is especially challenging due to the growing amount of available machine learning models and their hyperparameters configurations. This can adversely affect the performance of the model [14].

Multiple approaches have been proposed to tackle the above problem. These approaches range from automatic features engineering [15,16] to automatic model selection [17,18] and hyperparameters tuning [19]. Some approaches attempt to automatically and simultaneously choose a learning algorithm and optimize its hyperparameters. These approaches are also known as Combined Algorithm Selection and Hyperparameters optimization problem (CASH) [5,8,15]. A solver for the CASH problem aims to pick an algorithm from a list of options and then tune it to give the highest validation performance amongst all the possible combinations of algorithms and hyperparameters.

Automation of analytics workflows or their parts have been studied and attempted actively over the past decade. As a result, several open source tools have been developed that provide partial or complete ML automation, such as Autosklearn [12], AutoWEKA [7], TPOT [8] as well as commercial systems such as RapidMiner [9], H2O.ai [20] and google tables [21]. However, the available literature lacks systematic methodologies to guide the selection and configuration of algorithms for concrete real-world scenarios [17,22].

3.1 Meta-learning for Automatic Algorithms Selection and Configuration

Meta-learning or learn to learn is a general process used for predicting the performance of an algorithm on a given dataset. It is a method that aims at finding relationships between dataset characteristics and data mining algorithms [17]. Given the characteristics of a dataset, a predictive meta-model can be used to forcast the performance of a given data mining algorithm. For instance, in a classification task, meta-learning can be used to predict the accuracy of a classification algorithm on a given dataset and hence provide user support in the mining step.

The process of ranking consists of three phases (see Fig. 1). First, a *meta-learning space* is established using meta-data. The meta-data consist of dataset characteristics along with some performance measures for data mining algorithms on those particular datasets. Then, the *meta-learning phase* generates a model (i.e., predictive meta-model) which defines the area of competence of the data mining algorithm [23]. Finally, when a dataset arrives, the dataset characteristics are extracted and fed to the predictive meta-model, which predicts the potential well performing ML pipelines on the considered dataset. At this point, by comparing the obtained predictions for the different pipelines on similar tasks, we are able to rank the pipelines depending on their predicted impact on the given dataset. This concludes the *recommending* phase.

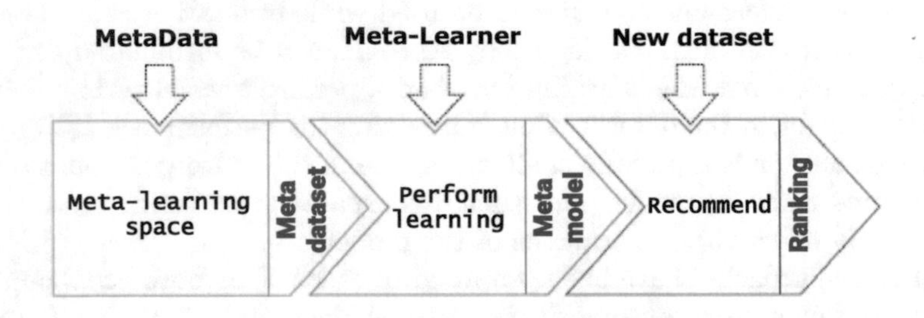

Fig. 1. The meta-learning process.

For the sake of concreteness, let us assume that, the user wants to perform predictive analytics to a dataset using the *Accuracy* performance measure, to deal with a classification problem at hand. The meta-learning based system, first, extracts the necessary meta-features (characteristics) from the dataset and uses them as input to the predictive meta-model which is specifically built for the Accuracy performance measure. The meta-model is built by training a meta-learner (e.g., KNN or any other classification algorithm) on existing/historical meta-data consisting of dataset characteristics and a performance measure (e.g., Accuracy) of multiple ML pipelines on the datasets. This meta-model is used to produce a prediction for provided dataset.

4 Automated Machine Learning in Industry 4.0

Despite the advances achieved in the field of AutoML, few works have been conducted to apply these techniques in the manufacturing field and hence, the industrial needs are yet to be fulfilled. As we mentioned before, there are several challenges that tackle the application of machine learning in the manufacturing space. One of the main challenges is the construction of a high quality and representative dataset. Ideally, a ML model should be trained with data that reflects as close as possible the original data. This is not easy as in a manufacturing unit, data are heterogeneous and each individual process generates different amounts of data with various formats (text, images, sensors data, etc.) and with different quality levels.

Besides the difficulty of constructing a high-quality dataset, a much bigger issue arises, consisting of a lack of transparency regarding the decisions made by AutoML systems making them as black boxes [6]. The lack of transparency leads machine learning experts and novices alike to question the results that were automatically obtained. If users cannot interpret the obtained results, they will not trust the AutoML system they are attempting to use and hence, they will hesitate to implement the model in critical applications, especially in manufacturing fields where interpretation and transparency of algorithms are a must for a system to be adopted into a workflow [24]. Another reason that justifies the low adoption rate of AutoML solutions in the industrial space is that the current methods for the ML pipeline optimization are inefficient on the large datasets originating from the manufacturing environment.

In order to address some of these issues, Villanueva Zacarias et al. [25] proposed a framework that automatically recommends suitable analytics techniques with respect to a domain-specific problem at hand. Similarly, Lechevalier et al. [26] presented a framework for the semi-automatic generation of analytical models in manufacturing and a proof-of concept prototype that allows practitioners to generate artificial neural networks for prediction tasks through a user interface. Both frameworks have represented promising approaches to tackle the problem of automated analytics technique configuration in the manufacturing domain. However, these frameworks do not achieve the goal of identifying the promising combinations of analytics and the application areas in the first place. Therefore, they are not suitable for decision making at the managerial level [2].

All of these findings point to one take-away message: intelligent systems are able to automatically design the whole or parts of machine learning pipelines, which can save practitioners considerable amounts of time by automating one of the most laborious parts of machine learning pipelines. To our knowledge, the currently available literature lacks the attempts to address the idea of automating the optimization of the entire ML pipelines for industrial big data analytics. Thus, the work presented in this paper establishes a blueprint for future research on the automation of machine learning pipeline design. The proposed framework focuses on three major tasks of AutoML. Firstly, the algorithm selection, i.e., which type of algorithm fits well the problem at hand. Secondly, the hyperparameters tuning which deals mainly with the question of how parameters should be set to get satisfactory results and finally, providing a rational explanation of the provided solution combining the aforementioned approaches.

5 Framework and Methodology

The overall architecture of the proposed framework is depicted in Fig. 2. The pipeline suggestions are provided by the *Suggestion Engine* through the use of a knowledge base (KB). This knowledge base is a collection of inductive meta-features that describe the datasets, the pipelines and their interdependencies. Whenever a new dataset is presented to the system, the suggestion engine provides a suggestion of the most appropriate classifiers by combining the pipelines of the knowledge base with the morphological characteristics provided by the meta-model which maps the characteristics of the given dataset to a label that describes similar seen datasets in the meta-learning space. The proposed framework has been developed on the meta-learning concept, consequently, it consists of two main phases which are the *learning* phase and the *recommendation* one.

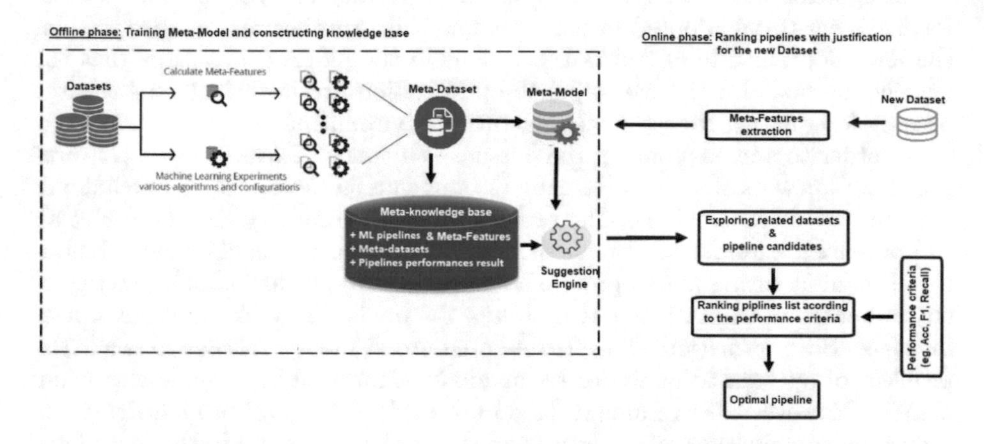

Fig. 2. The functional architecture of the proposed meta-learning based framework.

5.1 Learning Phase

Two important activities are performed in the learning phase. First, a meta-knowledge base (i.e., set of meta-datasets) is generated for all the performance measures considered (see Algorithm 1), and then on top of it, a learning algorithm is applied (see Algorithm 2). As a result, a statistical model (meta-model) is generated for every considered perfomance measure. The inputs required to construct the meta-knowledge base are datasets, hyperparameters values that are likely to improve the performance of the considered classification algorithms.

Algorithm 1. Establish the meta-knowledge base.

```
1: Input: ClassificationAlgs[..],                  ▷ available classification algorithms
   HpSpace[..],                                    ▷ set of HP configurations to be applied
   PerfMeasures[..]                                ▷ set of performance measures to acquire
2: Output: meta_KB[#measure][#metadata]            ▷ meta-knowledge base
3: function CreateMetaKB(datasets[])
4:     metadata[] = ∅
5:     for each measure in PerfMeasures do
6:         for each dataset DS in datasets do
7:             ds_mf = ComputeMetaFeatures(DS);
8:             for each algorithm Alg in ClassAlgs do
9:                 for each hyperparameters_configuration Hp in HpSpace do
10:                    ds_pm = GetPerformanceWith5FoldCV(Alg, Hp, DS);
11:                    metadata[] ← ds_mf ∪ ds_pm;
12:        meta_ds[measure] ← metadata[];
13:    return meta_KB
```

For the sake of simplicity, let us consider that we want to create the meta-dataset for a predictive metric (e.i. accuracy). In line 7 of Algorithm 1, we first extract the dataset characteristics (i.e., meta-features from the datasets). Next, we apply the classification algorithms with all possible and resonable HP configurations and then take the corresponding performance measures (e.g., predictive accuracy)—line 10. The latter is the meta-response, which together with the meta-features of the dataset, and the performance measure of the dataset compile the complete set of metadata—see line 11.

Once a meta-dataset for each performance measure is obtained, next, a learning algorithm (i.e., meta-learner) is applied on top - line 5 of Algorithm 2, and as a result, a meta-model (i.e., statistical model) for each of the performance measures is obtained. The proposed framework uses the KNN algorithm as meta-learner.

Algorithm 2. Create meta-models.

```
1: Input: meta_KB[..][..],                         ▷ See Algorithm 1
   PerfMeasures[..]                                ▷ set of available performance measures
2: Output: meta_models[..]                         ▷ meta-model for each performance measure
3: function PerformMetaLearning( )
4:     meta_models[] = ∅
5:     meta_model = KNN();
6:     for each measure in PerfMeasures do
7:         meta_models[measure] ← ApplyMetaModel(meta_KB[measure][]);
8:     return meta_models
```

5.2 Recommendation Phase

The recommendation phase is initiated when a new dataset to be analyzed arrives. At this point, the user selects a performance measure to be used for the analysis and the system automatically recommends the ML algorithms along with related HP configurations to be applied, such that the final result is optimal. This phase is described in Algorithm 3, where, first, the meta-features are extracted from the dataset in lines 5. Next, the extracted features are then fed to the predictor (meta-model) in line 6. The predictor in line 6 apply the meta-model to the extracted features, to find the optimal or near optimal ML pipelines for the given dataset. After, the ranked list of the potential well performing pipelines are obtained for the given dataset and desired performance measure in line 7.

Algorithm 3. Recommend ML pipelines.

```
1: Input: Dataset[..],                      ▷ new dataset chosen by the user
   meta_models[..]                          ▷ meta-model for each performance measure
2: Output: MLpipelines[..]                  ▷ ML pipelines ranked according to the PM
3: function RecommandPipelines(Dataset[])
4:     recommendations[] = ∅
5:     ds_mf = ComputeMetaFeatures(Dataset);
6:     PotentialPipelines[] ← ApplyMetaModel(ds_mf, meta_models[measure])
7:     recommendations[] ← Rank(PotentialPipelines, desc = True)
8:     return recommendations
```

5.3 Prototypical Implementation

The learning phase is performed offline and consists of two main steps. In the first one, meta-datasets are established. For each classification algorithm that is considered by the system for application, i.e., for the time being the proposed system supports 08 classification algorithms: *SVM, Stochastic Gradient Descent (SGD), AdaBoost, Random Forest, Extra Trees, Logistic Regression, Gradient Boosting, and Decision Tree.* The meta-dataset is constructed by extracting 42 datasets characteristics (Meta-features), and by generating different measures on the performance of the classification algorithms on the datasets (e.g., predictive accuracy, precision, recall). Dataset characteristics and performance measures altogether are referred to as meta-data. The Meta-knowledge base is provided as a data source for the suggestion engine and an application for metadata management is built on top of it to implement the system and the graphical user interface. In the second step, meta-learning is performed on top of the Meta-Knowledge base.

We implement a prototype of the proposed framework using the Python programming language. The used ML algorithms implementations are provided by the Sklearn data-mining library. We performed an experimental study to evaluate the performance that can be achieved by using the proposed framework on various datasets. These data are gathered from state-of-the-art papers dealing with industry 4.0 related problems using ML solutions. It covers binary and multiclass classification problems from different industrial levels to ensure a meaningful evaluation of the prototype's capabilities.

Table 1. Performances of the proposed AutoML framework on the benchmark datasets.

Dataset	Task	Recommended config. result	Original paper result
[27]	Pipeline networks Failure risk analysis	93.74	85
[28]	RUL prediction	99.41	98.95
[29]	APS system failure prediction	99.10	92.56
[30]	Chatter prediction	97.06	95

As shown in Table 1, it is obvious, that the results of some machine learning solutions, oriented from manufacturing industry, can be improved simply through the use of better ML models and related hyperparameters configuration.

6 Conclusion

Advanced data analytics is expected to have a significant impact on manufacturing. However, practical implementation is still in its early stages. Among others, one reason is that decision-makers lack tools that allow the choice of suitable techniques from a wide range of advanced data analytics methods devoted to problem solving in the manufacturing field. In traditional approaches, the selection of algorithms is a tedious task because of trial-and-error strategy. It is therefore, to support automation in algorithm selection and hyperparameters optimization, the data science community finds a way in which the selection of the classification/regression algorithms is possible with the help of past experiences. In this paper, we described the design of an AutoML based framework aimed at enabling the industry 4.0 actors and researchers, who have limited computing expertise to develop ML predictive models. The proposed framework supports the whole process of iterative machine learning on big manufacturing data. This extracts manufacturing parameters, and then assists to build and evaluate the appropriate predictive models. The proposed approach provide means to discover novel uses of big manufacturing data for many Industry 4.0 researchers and increase the ability to foster the smart factory discovery and improve their safety. We are currently in the process of building the system.

In the future, we plan to further extend and evaluate the scope of our framework by applying it to different kinds of use cases with a special attention for the predictive maintenance activity. This may also require adding support for further data formats and ML algorithms, as needed by the respective use cases.

Acknowledgements. This work has been supported, in part, by Hestim, CNRST Morocco, and University of the Littoral Cote d'Opale, Calais France.

References

1. Usuga Cadavid, J.P., et al.: Machine learning applied in production planning and control: a state-of-the-art in the era of industry 4.0. J. Intell. Manuf. **31**(6), 1531–1558 (2020). https://doi.org/10.1007/s10845-019-01531-7
2. Wolf, H., et al.: Bringing advanced analytics to manufacturing: a systematic mapping. In: Ameri, F., Stecke, K.E., von Cieminski, G., Kiritsis, D. (eds.) APMS 2019. IAICT, vol. 566, pp. 333–340. Springer, Cham (2019). https://doi.org/10.1007/978-3-030-30000-5_42
3. Hutter, F., Kotthoff, L., Vanschoren, J. (eds.): Automated Machine Learning. TSSCML, Springer, Cham (2019). https://doi.org/10.1007/978-3-030-05318-5
4. Mustafa, A., Rahimi Azghadi, M.: Automated machine learning for healthcare and clinical notes analysis. Computers **10**(2), 24 (2021). https://doi.org/10.3390/computers10020024
5. Garouani, M., et al.: AMLBID: an auto-explained automated machine learning tool for big industrial data. SoftwareX **17**, 100919 (2022). https://doi.org/10.1016/j.softx.2021.100919
6. Garouani, M., et al.: Towards big industrial data mining through explainable automated machine learning. Int. J. Adv. Manuf. Technol. **120**, 1169–1188 (2022). https://doi.org/10.1007/s00170-022-08761-9
7. Thornton, C., et al.: Auto-WEKA: combined selection and hyperparameter optimization of classification algorithms. In: Proceedings of the 19th ACM SIGKDD International Conference on Knowledge Discovery and Data Mining, KDD 2013, pp. 847–855. Association for Computing Machinery, New York (2013). https://doi.org/10.1145/2487575.2487629
8. Olson, R.S., Moore, J.H.: TPOT: a tree-based pipeline optimization tool for automating machine learning. In: Hutter, F., Kotthoff, L., Vanschoren, J. (eds.) Automated Machine Learning. TSSCML, pp. 151–160. Springer, Cham (2019). https://doi.org/10.1007/978-3-030-05318-5_8
9. RapidMiner—Best Data Science & Machine Learning Platform. Rapid-Miner. https://rapidminer.com/
10. Yang, C., et al.: OBOE: collaborative filtering for AutoML model selection. In: Proceedings of the 25th ACM SIGKDD International Conference on Knowledge Discovery & Data Mining, pp. 1173–1183 (2019). https://doi.org/10.1145/3292500.3330909
11. Guyon, I., et al.: Analysis of the AutoML challenge series 2015–2018. In: Hutter, F., Kotthoff, L., Vanschoren, J. (eds.) Automated Machine Learning. TSSCML, pp. 177–219. Springer, Cham (2019). https://doi.org/10.1007/978-3-030-05318-5_10
12. Feurer, M., et al.: Auto-sklearn: efficient and robust automated machine learning. In: Hutter, F., Kotthoff, L., Vanschoren, J. (eds.) Automated Machine Learning. TSSCML, pp. 113–134. Springer, Cham (2019). https://doi.org/10.1007/978-3-030-05318-5_6
13. Drori, I., et al.: AlphaD3M machine learning pipeline synthesis (2018)
14. Luo, G.: PredicT-ML: a tool for automating machine learning model building with big clinical data. Health Inf. Sci. Syst. **4**(1), 5 (2016). https://doi.org/10.1186/s13755-016-0018-1
15. Katz, G., Shin, E.C., Song, D.: ExploreKit: automatic feature generation and selection. In: 2016 IEEE 16th International Conference on Data Mining (ICDM) (2016). https://doi.org/10.1109/ICDM.2016.0123

16. Nargesian, F., et al.: Learning feature engineering for classification, pp. 2529–2535 (2017)
17. Garouani, M., et al.: Towards the automation of industrial data science: a meta-learning based approach. In: 23rd International Conference on Enterprise Information Systems, pp. 709–716 (2021). https://doi.org/10.5220/0010457107090716
18. Reif, M., et al.: Automatic classifier selection for non-experts. Pattern Anal. Appl. **17**(1), 83–96 (2012). https://doi.org/10.1007/s10044-012-0280-z
19. Wang, Q., et al.: ATMSeer: increasing transparency and controllability in automated machine learning. In: Proceedings of the 2019 CHI Conference on Human Factors in Computing Systems, CHI 2019, pp. 1–12. Association for Computing Machinery, New York (2019). https://doi.org/10.1145/3290605.3300911
20. H2O.Ai—AI Cloud Platform. H2O.ai. https://www.h2o.ai/
21. AutoML Tables. Google Cloud. https://cloud.google.com/automl-tables/docs
22. Muñoz, M.A., et al.: Algorithm selection for black-box continuous optimization problems: a survey on methods and challenges. Inf. Sci. **317**, 224–245 (2015). https://doi.org/10.1016/j.ins.2015.05.010
23. Bilalli, B.: Learning the Impact of Data Pre-processing in Data Analysis. Universitat Politècnica de Catalunya, Barcelona (2018)
24. Samek, W., Müller, K.-R.: Towards explainable artificial intelligence. In: Samek, W., Montavon, G., Vedaldi, A., Hansen, L.K., Müller, K.-R. (eds.) Explainable AI: Interpreting, Explaining and Visualizing Deep Learning. LNCS (LNAI), vol. 11700, pp. 5–22. Springer, Cham (2019). https://doi.org/10.1007/978-3-030-28954-6_1
25. Villanueva Zacarias, A.G., Reimann, P., Mitschang, B.: A framework to guide the selection and configuration of machine-learning-based data analytics solutions in manufacturing. In: Procedia CIRP, 51st CIRP Conference on Manufacturing Systems, vol. 72, pp. 153–158 (2018). https://doi.org/10.1016/j.procir.2018.03.215
26. Lechevalier, D., et al.: A methodology for the semi-automatic generation of analytical models in manufacturing. Comput. Ind. **95**, 54–67 (2018). https://doi.org/10.1016/j.compind.2017.12.005
27. Mazumder, R.K., Salman, A.M., Li, Y.: Failure risk analysis of pipelines using data-driven machine learning algorithms. Struct. Saf. **89**, 102047 (2021). https://doi.org/10.1016/j.strusafe.2020.102047
28. Benkedjouh, T., Medjaher, K., Zerhouni, N., Rechak, S.: Health assessment and life prediction of cutting tools based on support vector regression. J. Intell. Manuf. **26**(2), 213–223 (2013). https://doi.org/10.1007/s10845-013-0774-6
29. Costa, C.F., Nascimento, M.A.: IDA 2016 industrial challenge: using machine learning for predicting failures. In: Boström, H., Knobbe, A., Soares, C., Papapetrou, P. (eds.) IDA 2016. LNCS, vol. 9897, pp. 381–386. Springer, Cham (2016). https://doi.org/10.1007/978-3-319-46349-0_33
30. Saravanamurugan, S., et al.: Chatter prediction in boring process using machine learning technique. Int. J. Manuf. Res. (2017). https://doi.org/10.1504/IJMR.2017.10007082
31. Garouani, M., Ahmad, A., Bouneffa, M., Hamlich, M., Bourguin, G., Lewandowski, A.: Towards meta-learning based data analytics to better assist the domain experts in industry 4.0. In: Dang, N.H.T., Zhang, Y.D., Tavares, J.M.R.S., Chen, B.H. (eds.) ICABDE 2021. LNDECT, vol. 124, pp. 265–277. Springer, Cham (2022). https://doi.org/10.1007/978-3-030-97610-1_22
32. Garouani, M., et al.: Using meta-learning for automated algorithms selection and configuration: an experimental framework for industrial big data. J. Big Data **9**, 57 (2022). https://doi.org/10.1186/s40537-022-00612-4

New Deep Learning Architecture for Improving the Accuracy and the Inference Time of Traffic Signs Classification in Intelligent Vehicles

Btissam Bousarhane(✉) and Driss Bouzidi

Smart Systems Laboratory (SSL), National School of Computer Science and Systems Analysis ENSIAS, Mohammed V University, Rabat, Morocco
{ibtissam_bousarhane,driss.bouzidi}@um5.ac.ma

Abstract. Vehicular Ad-hoc Network (VANET) is a new technology on which are based Intelligent Transportation Systems (ITS). The goal of this technology is to improve the vehicular environment, and to provide more safety for both vehicles and drivers. In this global context, characterized by the use of IoT, vehicles are integrating more intelligent tools that help to insure vehicle-to-vehicle, vehicle-to-infrastructure and vehicle-to-driver communications. Hence, IoT involves the necessity to handle and manage a huge amount of data, including processing and analyzing traffic signs and road scene images. In fact, dealing with these aspects of Big Data (volume, variety, etc.) presents a big challenge, for which many approaches are proposed, including Deep Learning (DL). Although their high performances, this type of approaches still faces many difficulties, which are related essentially to the computational load and the hardware requirements. From this perspective, we have adopted a new Deep architecture to ensure traffic signs classification. The objective of the proposed architecture is to speed up the training and the inference stages, and that without affecting classification's performances. The obtained results show that the adopted approach accelerates the training & the inference speed, and reaches high accuracies using a limited number of parameters.

Keywords: Big Data · Deep Learning · Neural networks · Traffic signs recognition

1 Introduction

Our digital society is characterized by the presence of millions or even billions of devices and objects that are connected together. That represents in fact the main objective of Internet of Objects or Things (IoT). This concept was proposed in the 1990s and consists more precisely on combining existing resources or objects over the Internet. Hence, IoT is related to many fields of application [1], and it is essentially characterized by three main aspects, which are more exactly: Things, Internet & Human.

One of the most important application fields of IoT is Intelligent Transportation Systems (ITS). In effect, with the proliferation of vehicles and the rapid increase of population, traffic congestion becomes a more complex issue. Furthermore, every year

M. Lazaar et al. (Eds.): BDIoT 2021, LNNS 489, pp. 16–31, 2022.
https://doi.org/10.1007/978-3-031-07969-6_2

traffic accidents cause the death of approximately 1.25 million people worldwide [2]. In addition to that, congestion's problems cause pollution, wasted fuel, delays, stress, etc. Where comes the role of Vehicular Ad-hoc Networks (VANETs) to ameliorate this congestion and improve road safety.

The main objective of VANETs is to create distributed networks that enable the exchange of information among devices, and also among vehicles and people [3]. On other words, these networks consist on implementing Vehicle-to-Vehicle (V2V), Vehicle-to-Infrastructure (V2I) [4], and more generally Vehicle-to-Everything (V2E) communications [5].

In fact, these vehicular networks are characterized by a high mobility, strong dynamic topologies and distributed operations [6]. The main categories of VANETs' applications are more precisely: safety, public service, driving improvement and convenience service [7].

In the near future, and with the development of new technologies, vehicles will integrate more intelligent and advanced equipment that generate a huge amount of data. This massive exchange of information, through VANETs, will help to improve vehicular safety by ensuring a more accurate perception of the environment [8].

From this perspective, Big Data will present an opportunity but also a very big challenge for VANETs. On the one hand, VANETs' Big Data facilitate many applications related to road safety, such as vehicle collision avoidance, security distance warning, road condition warnings etc. In addition to that, this exchange of information will facilitate also other applications, like traffic management, self-driving systems, etc.

However, managing traffic congestion leads, on the other hand, to the generation of a large amount of data, that come from heterogeneous sources. Where comes the need for managing data collection, storage and computation in VANETs. This data management includes, necessarily, processing and analyzing traffic signs and road scene images in general.

In this context, many approaches are proposed by researchers to face these Big Data challenges [9], including Deep Learning approaches. However, this type of methods still presents a certain number of limitations although their high performances (very expensive in terms of computational load and hardware requirements).

From this perspective, we have adopted a new Deep Learning approach for traffic signs classification. The objective of the adopted approach is to face some of these challenges, and that by reducing the number of used parameters in order to accelerate the training and the inference stages.

Hence, our paper is organized as follow, Sect. 2 discusses the challenges that still face Big Data using Machine and Deep Learning approaches. Section 3 presents some works relating to traffic signs recognition. Section 4 discusses the proposed approach. Finally, the last section shows the obtained results.

2 Big Data Challenges

Internet of Things consists on connecting things and smart technologies to enable person-to-object and object-to-object communications [10]. With this explosion of information, managing a huge amount of data is getting more and more complicated, and that although the great improvement realized in terms of hardware optimization (computational load, storage capabilities, etc.).

This situation is due to a certain number of challenges that still face Big Bata. These difficulties concern more specifically [9] the Volume, the Variety, the Velocity, the Veracity and the Volatility of data:

- Volume: Huge amount of generated data;
- Variety: Different types of files and formats (text, image, audio, video, pdf, etc.);
- Velocity: Data generated at a fast speed, especially in real time applications;
- Veracity: Usefulness, validity and quality of data;
- Volatility: Needed time for data to remain in the system

To deal with these five aspects of Big Data, the classical approaches have shown their limitations, where comes the role of Artificial Intelligence (AI), especially Machine Learning approaches [11] to bring some relevant answers to these big challenges.

2.1 Machine Learning Approaches

Machine Learning (ML) represents a subfield of Artificial Intelligence. The objective of ML techniques is to give "computers the ability to learn without being explicitly programmed" [12]. Which means teaching systems to make smart, fast and efficient decisions using unknown data [2].

Generally, Machine Learning methods are based on supervised, unsupervised, semi-supervised and reinforcement learning. In this context, different techniques are used, like Support Vector Machines (SVM), Random Forest, AdaBoost, Discriminative Analysis, Naive Bayes Classifier, Decision Trees, Neural Networks, etc.

Due the availability of massive amounts of data and systems with high computational and storage capabilities, ML is used in many domains and fields of research. These domains include biotechnology, arts, business, robotics, Intelligent Transportation Systems, etc. [2]. ML is also used for VANETs Big Data, and we find that Decision Tree, for example, gives more reliable results compared to other methods like K Nearest Neighbors and Naïve Bays [7].

In effect, ML methods consists on manually extracting features or Regions of Interest (ROIs) from data (using HOG features, Dense_Sift, LBP, Gabor, etc.). These techniques are characterized by their high accuracy, quick response time and speed, which makes them very suitable for real time applications.

However, they show best results just on relatively small datasets, while their performances decrease considerably when dealing with huge amount of data. Where comes the need for using more developed methods to face these limitations.

From this perspective, Deep Learning (DL) based approaches are proposed by some researchers to handle these challenges faced by Big Data [13].

2.2 Deep Learning Approaches

Unlike classical ML methods, Deep Learning approaches consist instead on automatically extracting features from data. Furthermore, they deal directly with original datasets, and use less classifiers.

This type of approaches is based on learning data representations inspired by information processing in the human brain [2]. The application fields of DL are multiple, including natural language processing, audio recognition, computer vision, etc.

DL approaches include Convolutional Neural Networks (CNNs), Recurrent Neural Networks (RNNs), Deep Belief Networks (DBNs), etc. However, the most common and widely used DL approaches are based on Convolutional Neural Networks [14–16]. Compared to traditional Machine Learning approaches, CNNs have good classification performances.

In effect, a CNN is a multi-layered network, where each layer includes multiple neurons, and each of these neurons receives an input to perform a specific task. The outputs of each neuron represent the input of the next neurons [17, 18].

This type of networks has shown high performances in many fields of research [19, 20], including speech recognition, text mining, language processing, and also in computer vision, more precisely for the detection and the classification of different types of objects [21–23]. However, although their outstanding performances, this type of approaches still face a certain number of challenges, especially when dealing with Big Data.

In fact, Deep Learning plays an important role in Big Data [9, 24]. Table 1 presents the DL solutions proposed for the principal challenges faced by Big Data.

Table 1. Deep Learning solutions for Big Data challenges.

Big Data challenges	Deep Learning solutions
1.Volume	Parallel training based on modules and blocks, using Deep Stacking Networks [25], Disbelief [26], etc. GPUs for parallel training [27]
2.Variety	Concatenate extracted features from different types of sources, using Multi-Modal Deep Learning models [28]
3.Velocity	The knowledge of the model is continuously extended by input data, using Incremental Deep Learning models [29]
4.Veracity	Data quality improvement via Deep Learning [30]

Table 1 shows that, there are many proposed solutions to the faced challenges, especially for the "Volume" aspect of Big Data. In effect, DL approaches help to extract and learn high level features from a huge amount of data. However, the computational load of this type of approaches still presents a big challenge.

In fact, the number of used parameters overpass generally millions of parameters, which makes the training and the validation process a hard and very time-consuming task. Furthermore, deeper architectures are based on multiple number of layers, and each layer contains hundreds and even thousands of nodes. Hence, to accelerate the training and the inference process, hardware optimization is required. This optimization includes the use of multiple processing cores, parallelism [31], Graphic Processing Units (GPUs), FPGA, etc.

Certainly, hardware optimization accelerates enormously the training and the inference speed, especially for real time applications. However, it limits on the other hand the use of Deep Learning approaches in low resources environment (as mobile devices, etc.).

From this perspective, our work aims to face some of these challenges in order to ensure the classification of traffic signs using a Deep Learning based approach. Hence, Sect. 3 presents some works relating to traffic signs recognition, and Sect. 4 discusses the proposed approach.

3 Traffic Signs Recognition

Traffic signs recognition plays an important role in Intelligent Transportation Systems. It presents also a key issue in traffic safety [2]. Generally, road signs are divided into three main categories, which are more precisely prohibitory, mandatory and danger signs [32]. These signs are characterized by their specific colors (red, blue, green and yellow), and unique shapes (circular, rectangular, triangular, etc.).

Although the fact that traffic signs have standard shapes and colors, their recognition still presents however a very challenging subject, and that due to many external factors that affect their appearance, especially weather conditions, illumination changes, occlusions, etc.

Generally, a traffic signs recognition system includes two stages, which are more precisely traffic signs detection and traffic signs classification.

3.1 Traffic Signs Detection

Traffic signs detection consists on locating and extracting traffic signs from road scene images. This detection is realized using classical and Deep Learning approaches.

For the classical approaches, they are essentially based on extracting basic features from images, such as colors and contours. In this context, we can mention the work of Islam [33], which consists of detecting signs with a background color of red and blue, and with circular and triangular shapes. To select the ROIs, the images are blurred to reduce the noise using Gaussian blur, then they are converted from RGB (Red, Green and Blue) to HSV (Hue, Saturation and Value). In effect, RGB defines colors in terms of three primary colors, while HSV separates the intensity from the color, which makes it more robust to illumination changes. This color space is more similar to the human eye perception of objects. After HSV conversion, image masking is used to hide and reveal some portions of the images. Then, morphological processes have been realized to filter out the unnecessary parts. Finally, the ROIs that contain traffic signs are selected by extracting the region bounded by the bounding boxes.

Another approach adopted by Sun, Ge & Liu [17] is also based on HSV color space and image enhancement to detect circular signs. To ensure this detection, the color space is transformed and RGB images are converted into HSV. This conversion consists of extracting H and S components. In the process of segmentation, hue plays an important role because it shows more invariance to illumination changes and color saturation. To ensure a more accurate detection, morphological operations are also used to keep the basic shape of images, and eliminate irrelevant and redundant information. Finally, Hough Transform is used for detecting and locating the circular signs.

For Vennelakanti, *et al.* [34], their approach relies on obtaining the ROIs using traffic signs color and shapes (red triangular and circular signs). For the detection, the captured images are preprocessed by converting the RGB images into HSV color space. Hence, once the red threshold is verified, the number of edges is calculated using the contours of detected red objects. This shapes' validation is realized using the Douglas-Peucker algorithm. According to that, objects with 3 edges are considered to be triangles, while ROIs with more than 5 edges are considered to be circles. Finally, the bounding boxes that represent the outer triangle or circle separate the ROIs from the rest of the environment.

From the proposed approaches above, we find that the majority of these methods are based on signs' colors and shapes. In fact, that makes the detection process more sensitive to illumination and weather changes especially at night. Occlusions and signs' deformations is another problem for shape-based techniques. Where comes the role of Deep Learning approaches to overcome these difficulties.

Compared to classical ones, Deep Learning approaches show outstanding performances for traffic signs detection [35]. This type of methods is more robust to occlusions and illumination changes, because they use large training datasets with a diversity of samples.

In this context, we can mention the work of William, *et al.* [36]. The adopted approach is based on transfer learning to ensure the real-time recognition taking into consideration various challenges (weather, illumination and visibility challenges). This detection is realized using Faster Recurrent Convolutional Neural Networks (F-RCNN), and Single Shot Multi-Box Detector (SSD) combined with various feature extractors (MobileNet v1, Inception v2 and Tiny-YOLOv2). The obtained results show that F-RCNN Inception v2 and Tiny YOLO v2 achieve the best results. YOLO v2 (You Only Look Once) is a state-of-the-art real-time object detection system. While TINY-Yolov2 is developed to run on portable devices such as cellphones, etc. Hence, to ensure the detection, the images are converted to HSV color space. Each image is passed after that to the network for training. Finally, the ROIs are extracted and Non-Maximum Suppression is realized to choose only the ROIs with the highest confidence.

Another Deep Learning approach is proposed by Nguyen [35]. This paper proposes a Deep Learning-based approach to improve the performance of small traffic signs detection. For that, a lightweight and efficient architecture is adopted to handle the issue of inference speed. To enhance the performance, a deconvolution module is adopted to generate an enhanced feature map. Then, two improved Region Proposal Networks are used to generate the ROIs. The proposed framework uses ESPNetv2 network for increasing the processing speed. The improved Region Proposal Network is designed based on the original one for fast and efficient proposal generation. The proposed approach is evaluated on German Traffic Signs Detection Benchmark and Tsinghua-Tencent 100K dataset. The results show that the proposed approach achieves competitive performance compared to current state-of-the-art approaches while being faster and simpler. The proposed network reduces the computational costs of the whole framework for proposal generation. Hence, the proposed framework can be applied in real-time systems.

In the same context, we find the work of Xu & Srivastava [32]. In their approach, the histogram equalization method is used to pre-process traffic signs images. That includes images enhancement and contrast improvement. Then, traffic signs images are recognized using a Convolution Neural Network to automatically extract the feature maps. Next, the areas of interest are extracted by using the hierarchical significance model.

As already presented, finding the right region of interest represents the first step in traffic signs recognition. This detection process consists on locating and extracting the ROIs, while the classification stage consists on identifying the meaning of each detected sign.

3.2 Traffic Signs Classification

For classification purposes, different Machine Learning based approaches are used. Among these algorithms, we find Support Vector Machines (SVM), which is the most popular one. Neural Networks are also used in this field.

In this context, Islam [33] propose an approach based on two separate Neural Networks. The first one for classifying object's shapes (triangular, circular, irregular or random), while the second one is for signs' classification. The results obtained from the two classifiers decide if the image contains a traffic sign or not. Convolutional Neural Networks have been chosen for this purpose as it performs better than SVM. The CNN used for signs' classification is based on LeNet architecture, which consists of two convolutional layers and two fully connected ones. Whereas the CNN used for shapes' classification consists of one convolutional layer and two fully connected ones. Categorical Cross Entropy method was used to determine the loss. Adadelta optimizer is also used to achieve the convergence as quickly as possible. For training purposes, the images were taken from the British traffic signs image database, while the images used to measure the performance were taking from Ukrainian and Bangladeshi streets. The signs chosen have a background color of red and blue with circular and triangular shapes.

Sun, Ge & Liu [17] use also Convolutional Neural Networks (CNNs). In the classification stage, the extracted traffic signs are used as input, and a Convolutional Neural Network is used to classify the detected signs. The lightweight CNN adopted consists of two convolutional layers, two pooling layers and two full connected ones. The kernel size is set as 5×5, and the size of feature maps is set to 32×32 & 16×16 respectively. The hidden nodes of the fully connected layers are 512 and 128. Furthermore, Rectified Linear Unit (ReLU) is used as the activation function. For training and testing the network, GTSRB and a self-designed dataset were used. In the German dataset, the approach was able to identify the circular signs with more than 98.2%, in terms of accuracy.

For Vennelakanti, *et al.* [34], they have used instead a CNN ensemble (3 CNNs). Their approach represents a feed-forward network with six convolutional layers. All the layers of the proposed CNN have ReLU as activation function. The output of the last convolutional layer is fed to a fully connected layer which uses Softmax function.

The optimizer used is Stochastic Gradient Descent with Nesterov momentum. Belgium dataset and German Traffic Signs Benchmark were used for training and testing the proposed model. The overall training set for triangles had 10582 images and the testing set had 3456 images. For circular signs, the training set had 16106 images and the testing set had 5277 images. The three used CNNs are later aggregated to form a single model by averaging the outputs of each of these CNNs. The proposed approach has achieved an accuracy of 98.11% for triangles and 99.18% for circles.

Convolutional Neural Networks are also used by Xu & Srivastava [32] for traffic signs classification, and the Softmax classifier is selected to classify the generated feature maps.

Although the fact that Deep Learning methods perform better in difficult driving environments, processing real time Big Data, generated at a high speed, still presents real difficulties for DL approaches due to the high computational load of this type of techniques.

In VANET networks the time for delivering safety messages is a very challenging task, and the latency of this exchange should not exceed few milliseconds [2]. Hence, ensuring the QoS (Quality of Services) for safety messages is a very crucial issue, especially in these networks characterized by high-speed mobility and random traffic environments.

From this perspective, we have adopted a new Deep Learning architecture for traffic signs classification. The objective of the adopted approach is to reduce the computational complexity in order to speed up the training and the inference stages, and that without affecting the performances of classification accuracy.

4 Proposed Architecture

The adopted architecture proposed in our work is shown in Fig. 1.

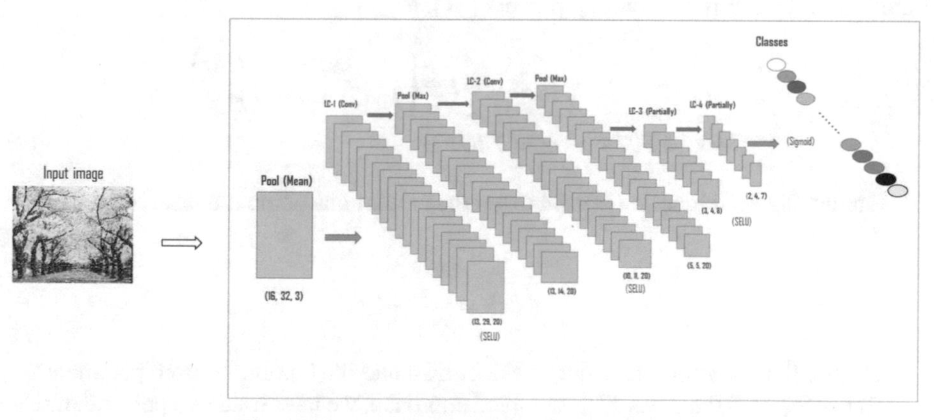

Fig. 1. The proposed architecture (Mean-LC4).

Our adopted model (Mean-LC4) is a composite Deep Neural Network for images' classification. The adopted approach includes 4 layers based on Local Connections.

Unlike other types of Deep Learning architectures, the first layer in our approach is a subsampling layer. The objective of this layer is to reduce the dimension of input images, and that before features' extraction. Hence in the adopted approach, features' maps are extracted from the representative values calculated in this layer. These values are set to the average (Mean) of the (2 × 1) matrix.

For features' extraction, it is totally based on Local Connections, using 4 Locally Connected layers (LC).

Sharing weights are adopted for the first two layers (convolutional ones) (1). These two layers has 20 × (4 × 4) kernels. Furthermore, each of these layers is followed by a Batch Normalization layer, and a Max pooling layer of (1 × 2) & (2 × 2) sizes, respectively.

$$f_l^k(p,q) = \sum_c \sum_{x,y} i_c(x,y) e_l^k(u,v) \quad (1)$$

$i_c(x, y)$: element of the input.
$e_l^k(u, v)$: element kernel of a layer.
$f_l^k(p, q)$: element of the feature map.
Concerning the third & fourth layers, their weights are not shared [37] (contrary to the two first layers). That is realized by using instead two partially connected layers in this stage. These two layers has 8 × (3 × 2) & 7 × (2 × 1) kernels, respectively.

This type of layers is characterized by applying different sets of filters to each location in the maps, which helps to extract more representative features in a deeper way.

As for the first two layers, a Batch Normalization [38] layer is added after each of these layers to ensure the generalizability of the model regardless of input data.

To guarantee the performance of the model for unseen data, we have added Dropout [39, 40] to deactivate randomly selected nodes from each of the four layers during the training process.

Furthermore, Scaled Exponential Linear Units (SELU) is used as the activation function to accelerate the training process [41].

$$Selu(x) = \begin{Bmatrix} \lambda x, & x < 0 \\ \lambda\alpha(e^x - 1), & x \leq 0 \end{Bmatrix}$$

$$(\lambda = 1.0507 \textit{ and } \alpha = 1.6733) \quad (2)$$

Finally, Sigmoid function is used to get the predictions of the model at the last layer.

$$g(x) = \frac{1}{1+e^{-x}}$$

Table 2 Illustrates the structure of Mean-LC4 and the number of used parameters.

To test the performances of the adopted approach, we have used two public datasets, which are more precisely German Traffic Signs Recognition Benchmark (GTSRB) [42] and Belgium Traffic Sign Classification Database (BTSCD) [43], as presented in the next section.

Table 2. Structure of Mean-LC4 and used parameters.

Layer	Output	Parameters
Mean_1	(16, 32, 3)	0
Conv_1	(13, 29, 20)	980
Batch_1	(13, 29, 20)	116
Max_1	(13, 14, 20)	0
Drop_1	(13, 14, 20)	0
Conv_2	(10, 11, 20)	6 420
Batch_2	(10, 11, 20)	80
Max_2	(5, 5, 20)	0
Drop_2	(5, 5, 20)	0
Partially_1	(3, 4, 8)	11 616
Batch_3	(3, 4, 8)	32
Drop_3	(3, 4, 8)	0
Partially_2	(2, 4, 7)	952
Batch_4	(2, 4, 7)	28
Drop_4	(2, 4, 7)	0
Flatten	(56)	0
Output	(62)	3 534
Total parameters: 23 758		

5 Obtained Results

5.1 Training Results

The public datasets GTSRB & BTSCD have been created in order to facilitate the benchmark of traffic igns classification algorithms. These two datasets respect the Vienna Convention on Road Signs and Signals [44], and are widely used by researchers in this field of research.

For this specific reason, we have used these datasets for training & testing our approach. Table 3 shows the number of classes, training samples and testing images in each of these datasets.

Table 3. Structure of GTSRB & BTSCD.

Dataset	Classes	Training images	Testing images
GTSRB	43 classes	39 209 images	12 630 images
BTSCD	62 classes	4 575 images	2 520 images

Figure 2 & Fig. 3 illustrate the different types of signs in each of these datasets, and Fig. 4 & Fig. 5 show the balance between the different classes in GTSRB & BTSCD.

Fig. 2. Types of signs in BTSCD

Fig. 3. Types of signs in GTSRB

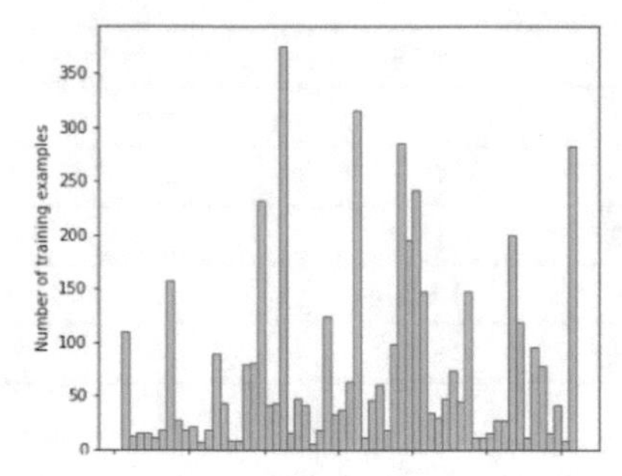

Fig. 4. Balance between the 62 classes of BTSCD.

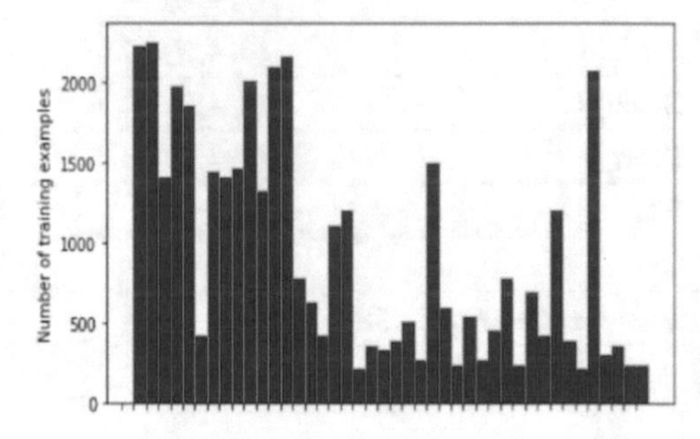

Fig. 5. Balance between the 43 classes of GTSRB.

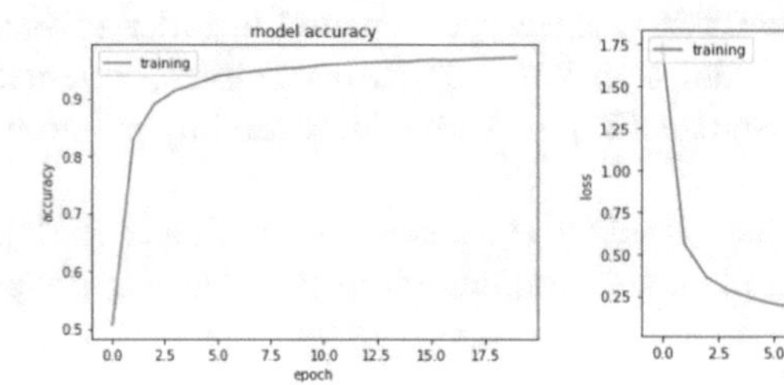

Fig. 6. Accuracy and loss obtained for training GTSRB.

For the preprocessing stage, we have scaled the training and testing data to the size of 32 × 32, because the images in these two datasets have different sizes. Hence, to train our model, we have used 39 209 training images from GTSRD. The obtained results on training samples are presented in Fig. 6.

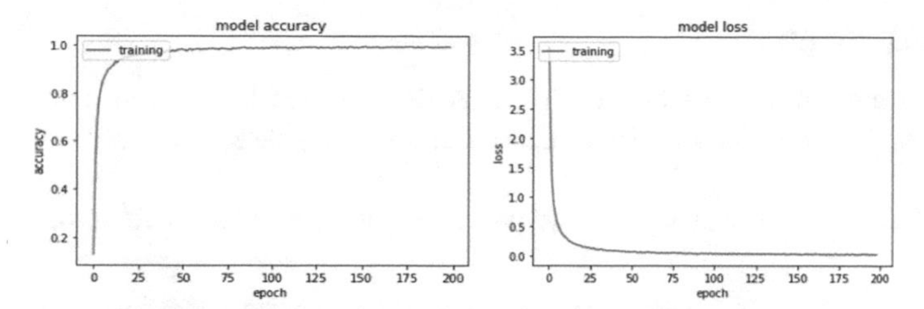

Fig. 7. Accuracy and loss obtained for training BTSCD.

We have also used the 4 575 training images of BTSRD to train our model. The obtained results are shown in Fig. 7.

The obtained results in Fig. 6 & Fig. 7 show that, the model reaches a high accuracy for the training samples of the two datasets. The figures show also that, the model takes only 20 epochs to reach 97% as training accuracy.

To test the performances of the model to unseen data, we have trained our model again using a validation dataset (90% for training and 10% for validation). Figure 8 & Fig. 9 show the obtained results after using the validation datasets.

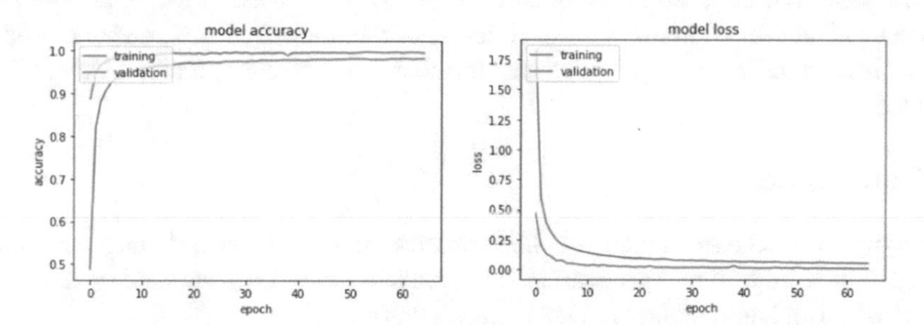

Fig. 8. Accuracy and loss obtained for the training & validation datasets in GTSRB.

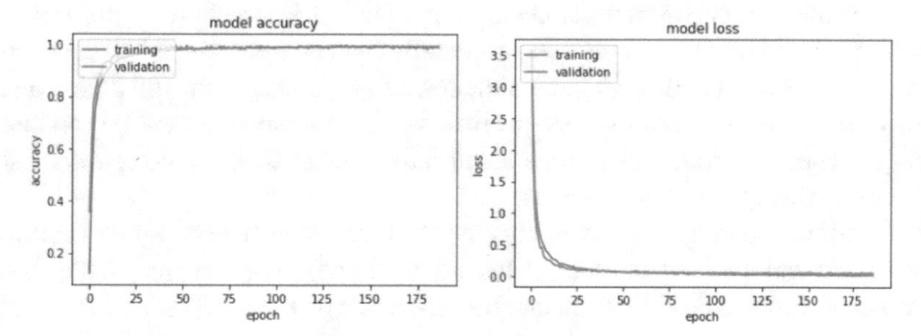

Fig. 9. Accuracy and loss obtained for the training & validation datasets in BTSCD.

Figure 8 & Fig. 9 show that the adopted approach gets high accuracies for the training samples and also for the validation or unseen data.

After the validation process, we have tested our approach using the testing datasets of GTSRB & BTSCD as presented in the next section.

5.2 Testing Results

The testing dataset of German Traffic Signs Recognition Benchmark contains 12 630 images. For Belgium Traffic Signs Classification Database, it includes 2 520 testing images.

The obtained results using these two datasets are presented in Table 4.

Table 4. Testing results obtained using GTSRB & BTSCD.

Dataset	Classes	Testing images	Accuracy	Parameters	Time	Configuration
GTSRB	43	12 630	97.29%	22 675	2 ms	2.0 GHz, CPU
BTSCD	62	2 520	98.05%	23 758	467 μs	

Table 4 shows that the proposed approach get high cassification accuracies with 97.29% & 98.05% for GTSRB and BTSCD respectively. Furthermore, the adopted approach has a very quick response time, that not exceeds few milliseconds per image (using 2.0 GHz, CPU).

Knowing that state-of-the art methodes [45–47] are based on a huge number of parameters that exceed in many cases millions of parameters [21, 34, 48], the proposed approach involves instead a very reduced number of parameters (that not exceed 24000 parameters).

6 Conclusion

IoT involves the need to efficiently collect, process, store and manage a huge amount and variety of data. To achieve this goal, many approaches have been adopted by researchers to deal with Big Data, including Deep Learning approaches.

In fact, Big Data provide a big opportunity for DL, and that by making available a huge amount of data for the training process, which contributes enormously on improving and ameliorating DL performances. On the other hand, Deep Learning approaches help to handle many of Big Data challenges, especially those related to the "Volume" aspect.

However, although their high performances, DL approaches are still facing a certain number of difficulties. From this perspective, we have adopted a new DL architecture for images' classification, which aims essentially to deal with the computational load complexity of this types of approaches.

The presented method achieves high accuracies using a very limited number of parameters (in comparison to state-of-the art methods). The adopted model helps to speed up the training, the validation and the inference process.

7 Further Work

Our future work will focus on optimizing our approach to improve the accuracy and the inference speed of the proposed model, in addition to the application of Mean-LC4 for traffic signs detection. We intend also to accelerate the calculation during the training stage using the adopted approach.

References

1. Khanna, A., Kaur, S.: Internet of Things (IoT), applications and challenges: a comprehensive review. Wireless Pers. Commun. **114**, 1687–1762 (2020)
2. Hossain, M.A., Noor, R.M., Yau, K.A., Azzuhri, S.R., Z'aba, M.R., Ahmedy, I.: Comprehensive survey of machine Learning approaches in cognitive radio-based vehicular ad hoc networks. IEEE Access **8**, 78054–78108. IEEE (2020)
3. Cheng, N., et al.: Big data driven vehicular networks. IEEE Netw. **32**(6), 160–167 (2018)
4. Anadu, D. et al.: Internet of things: vehicle collision detection and avoidance in a VANET environment. International Instrumentation and Measurement Technology Conference (I2MTC), pp. 1–6. IEEE (2018)
5. Liang, L., Ye, H., Li, G.Y.: Toward intelligent vehicular networks: a machine learning framework. IEEE Internet Things J. **6**(1), 124–135 (2019)
6. Tangade, S., Manvi, S., Hassan, S.: A deep Learning based driver classification and trust computation in VANETs. In: 90th Vehicular Technology Conference (VTC2019-Fall), pp. 1–6, IEEE (2019)
7. Najat, B., Salah, E.H.: Comparative study of classification algorithms for Big Data in VANET. In: International Conference on Advances in Computing and Communication Engineering (ICACCE), pp. 327–330, IEEE (2018)
8. Wang, Y., Menkovski, V., Ho, I.W., Pechenizkiy, M.: VANET meets deep learning: the effect of packet loss on the object detection performance. In: 89th Vehicular Technology Conference (VTC2019-Spring), pp. 1–5, IEEE (2019)
9. Zhang, Q., Yang, L., Chen, Z., Li, P.: A survey on deep learning for big data. Inf. Fusion. **42**,146–157 (2018)
10. Kuyoro, S., Osisanwo, F., Akinsowon, O.: Internet of Things (IoT): an overview. In: 3rd International Conference on Advances in Engineering Sciences & Applied Mathematics, pp. 53–58 (2015)
11. Ongsulee, P.: Artificial intelligence, machine learning and deep learning. In: 15th International Conference on ICT and Knowledge Engineering (ICT&KE), pp. 1–6 (2017)
12. Awad, M., Khanna, R.: Machine Learning. Efficient Learning Machines, pp. 1-18 (2015)
13. Chauhan, N., Singh, K.: A review on conventional machine learning vs deep learning. In: International Conference on Computing, Power and Communication Technologies (GUCON), IEEE, pp. 347–352 (2018)
14. Strumberger, I., Tuba, E., Bacanin, N., Jovanovic, R., Tuba, M.: Convolutional neural network architecture design by the tree growth algorithm framework. In: International Joint Conference on Neural Networks (IJCNN), pp. 1–8, IEEE (2019)
15. ElAdel, A., Ejbali, R., Zaied, M., Ben Amar, C.: Deep learning with shallow architecture for image classification. In: International Conference on High Performance Computing & Simulation (HPCS), pp. 408–412, IEEE (2015)
16. Gubbi, J., Varghese, A., Balamuralidhar, P.: A new deep Learning architecture for detection of long linear infrastructure. In: Fifteenth IAPR International Conference on Machine Vision Applications (MVA), pp. 207–210, IEEE (2017)
17. Sun, Y., Ge, P., Liu, D.: Traffic sign detection and recognition based on convolutional neural network. In: Chinese Automation Congress (CAC), pp. 2851–2854, IEEE (2019)
18. Santos, D.C., et al.: Real-time traffic sign detection and recognition using CNN. IEEE Latin America Trans. **18**(3), 522–529 (2020)
19. Aloysius, N., Geetha, M.: A review on deep convolutional neural networks. In: International Conference on Communication and Signal Processing (ICCSP), pp. 588–592 (2017)
20. Wick, C.: Deep learning. Informatik-Spektrum **40**(1), 103–107 (2016). https://doi.org/10.1007/s00287-016-1013-2

21. Ciregan, D., Meier, U., Schmidhuber, J.: Multi-column deep neural networks for image classification. In: Conference on Computer Vision and Pattern Recognition, pp. 3642–3649 (2012)
22. Yang, Y., Luo, H., Xu, H., Wu, F.: Towards real-time traffic sign detection and classification. Trans. Intell. Transp. Syst. **17**(7) (2016)
23. Zhang, Q., Zhang, M., Chen, T., Sun, Z., Ma, Y., Yu, B.: Recent Advances in Convolutional Neural Network Acceleration (2018)
24. Mittal, S., Sangwan, O.P.: Big Data analytics using Machine Learning techniques. In: 9th International Conference on Cloud Computing, Data Science & Engineering, pp. 203–207 (2019)
25. Deng, L., Yu, D., Platt, J.: Scalable stacking and learning for building deep architectures. In: International Conference on Acoustics, Speech and Signal Processing (ICASSP) (2012)
26. Heigold, G., McDermott, E., Vanhoucke, V., Senior, A., Bacchiani, M.: Asynchronous stochastic optimization for sequence training of Deep Neural Networks. In: International Conference on Acoustics, Speech and Signal Processing (ICASSP), pp. 5587–5591 (2014)
27. Raina, R., Madhavan, A., Ng, A.: Large-scale deep unsupervised learning using graphics processors. In: Proceedings of International Conference on Machine Learning (2009)
28. Gao, J., Li, P., Chen, Z., Zhang, J.: A survey on Deep Learning for multimodal data fusion. Neural Comput. **32**(5), 829–864 (2020)
29. Sarwar, S.S., Ankit, A., Roy, K.: Incremental learning in deep convolutional neural networks using partial network sharing. Access **8**, 4615–4628 (2020)
30. Dai, W., Yoshigoe, K., Parsley, W.: Improving data quality through Deep Learning and statistical models. Springer, Cham (2018)
31. Kumar, P., Bodade, A., Kumbhare, H., Ashtankar, R., Arsh, S., Gosar, V.: Parallel and distributed computing for processing big image and video data. In: Multimodal Analytics for Next-Generation Big Data Technologies and Applications, pp. 337–360. Springer, Cham (2019)
32. Xu, H., Srivastava, G.: Automatic recognition algorithm of traffic signs based on Convolution Neural Network. Multimed. Tools Appl. **79**, 11551–11565 (2020)
33. Islam, M.T.: Traffic sign detection and recognition based on Convolutional Neural Networks. In: International Conference on Advances in Computing, Communication and Control (ICAC3), pp. 1–6. IEEE (2019)
34. Vennelakanti, A., Shreya, S., Rajendran, R., Sarkar, D., Muddegowda, D., Hanagal, P.: Traffic sign detection and recognition using a CNN ensemble. In: International Conference on Consumer Electronics (ICCE), pp. 1–4, IEEE (2019)
35. Nguyen, H.: Fast traffic sign detection approach based on lightweight network and Multilayer Proposal Network. J. Sensors **2020**, 1–13 (2020)
36. William, M., et al.: Traffic signs detection and recognition system using Deep Learning. In: Ninth International Conference on Intelligent Computing and Information Systems, pp. 160–166, IEEE (2019)
37. Gregor, K., Lecun, Y.: Emergence of complex-like cells in a Temporal Product Network with local receptive fields (2010)
38. Chen, Li, Fei, H., Xiao, Y. He, J., Li, H.: Why Batch Normalization works? A buckling perspective. In: International Conference on Information and Automation (ICIA) (2017)
39. Shen, J., Shafiq, O.: Deep Learning Convolutional Neural Networks with Dropout – A parallel approach. In: International Conference on Machine Learning and Applications (ICMLA), (2018)
40. Shrestha, A., Mahmood, A.: Review of Deep Learning algorithms and architectures. IEEE Access **7**, 53040–53065 (2019)

41. Paoletti, M.E., Haut, J.M., Plaza, J.: An investigation on Self-Normalized Deep Neural Networks for hyperspectral image classification. In: International Geoscience and Remote Sensing Symposium (2018)
42. Stallkamp, J., Schlipsing, M., Salmen, J., Igel, C.: The German traffic sign recognition benchmark: a multi-class classification competition. In: The International Joint Conference on Neural Networks (IJCNN) (2011)
43. Mathias, M., Timofte, R., Benenson, R., Gool, L.V.: Traffic sign recognition – how far are we from the solution? In: The International Joint Conference Neural Networks (IJCNN) (2013)
44. United Nations, E.C.: Convention on road signs and signals: done at Vienna on 8 November 1968, amendment 1 (Incorporating the amendments to the Convention which entered into force on 30 November 1995) (1995)
45. Bousarhane, B., Bensiali, S., Bouzidi, D.: Road signs recognition: state-of-the-art and perspectives. Int. J. of Data Anal. Tech. Strat. Special Issue: Advances and Applications in Optimization and Learning, **13**(1–2), 128–150 (2021)
46. Bousarhane, B., Bouzidi, D.: Convolutional Neural Networks for traffic signs recognition. In: ACOSIS 2019, CCIS 1264 proceedings, Springer Nature, 1264, pp. 73–91 (2020)
47. Bousarhane, B., Bouzidi, D.: Map-CNNs: thin Deep Learning models for accelerating traffic signs recognition. Adv. Dyn. Syst. Appl. **16**(2), 1777–1798 (2021)
48. Bangquan, X., Xiong, W.: Real-time embedded traffic sign recognition using efficient Convolutional Neural Network (2019)

Implicit JSON Schema Versioning Triggered by Temporal Updates to JSON-Based Big Data in the τJSchema Framework

Zouhaier Brahmia[1](✉), Safa Brahmia[1], Fabio Grandi[2], and Rafik Bouaziz[1]

[1] University of Sfax, Sfax, Tunisia
zouhaier.brahmia@fsegs.rnu.tn, rafik.bouaziz@usf.tn
[2] University of Bologna, Bologna, Italy
fabio.grandi@unibo.it

Abstract. Schema versioning of JSON-based Big Data is driven either explicitly by schema changes or implicitly by updates. In the τJSchema framework, we have previously investigated implicit JSON Schema versioning, by dealing with implicit schema changes driven by updates of JSON-based conventional Big Data. Since τJSchema supports not only conventional but also temporal JSON-based Big Data, in this paper, we complete our investigation by focusing on the temporal side of implicit schema versioning in τJSchema. To this end, we propose an approach for handling implicit schema changes triggered by temporal updates of JSON-based Big Data. More precisely, when a user specifies a temporal JSON update operation that modifies a snapshot JSON component assigning a valid-time timestamp to its new value, the execution of such an operation requires the JSON component to become temporal, which is for all intents a schema change. Thus, a new version of the τJSchema temporal characteristics document is generated, with the addition of a new valid-time characteristic. New versions of the temporal JSON schema and of the temporal JSON document are also accordingly created.

Keywords: τJSchema · Temporal characteristics document · Temporal JSON document · Temporal JSON update operation · Schema versioning · Implicit schema change

1 Introduction

Big Data [1, 2] are being extensively exploited by contemporary applications (e.g., social networks, Internet of Things, cloud systems, Geographical Information Systems, big science projects). NoSQL databases [3, 4] have been conceived and built to manage mainly Big Data [5, 6]. Furthermore, the JSON format [7] is being widely used to store and represent NoSQL data, particularly in document-oriented NoSQL databases, and the JSON Schema language [8, 9] has been proposed to allow NoSQL Data Base Administrators (NSDBAs) and applications developers to define the structure of their JSON documents.

M. Lazaar et al. (Eds.): BDIoT 2021, LNNS 489, pp. 32–47, 2022.
https://doi.org/10.1007/978-3-031-07969-6_3

By their nature, Big Data are changing very rapidly. In fact, not only the values/instances of these data are evolving but also their structures (or schemata), in order to reflect changes in the real world or in the users' requirements. Besides, several of the above applications (like e-education, e-health, e-tourism, e-government, and e-business applications) require keeping a fully fledged history of changes involving both Big Data instances and Big Data schemata [10], in order to answer temporal and multi-schema-versions queries, either analytical [11, 12] or transactional [13], to track Big Data instance/schema changes over time, and to recover any Big Data instance/schema version. To satisfy such a requirement, we have proposed a framework [14, 15] named τJSchema (Temporal JSON Schema), which supports, in an integrated and consistent manner, both temporal aspects and schema versioning [16–18] of JSON-based Big Data. τJSchema has been defined by applying, in the NoSQL world, a subset of the temporal database concepts [19–21] (e.g., valid time) and the well-known technique of database schema versioning that allows keeping all database schema versions along with their corresponding data avoiding data loss and obsolescence of existing applications. Our framework allows the construction and validation of time-varying JSON documents conforming to a temporal JSON schema, starting from conventional JSON documents (i.e., standard JSON documents). The temporal JSON schema is created from a conventional JSON schema (i.e., a JSON Schema file) and a set of temporal characteristics, which are stored in a temporal characteristics document associated to the conventional JSON schema. Temporal characteristics specify which components of a JSON document can vary over time. A temporal JSON document [14, 15, 22] is a sequence of versions of the same conventional JSON document, each one with a distinct timestamp.

Moreover, τJSchema supports temporal schema versioning [23–26], with distinct timestamps associated to each version of the same τJSchema schema. Similarly to its originator in other database settings (e.g., relational, object-oriented, XML), schema versioning in the τJSchema NoSQL setting is the result of the execution of schema change operations. A change to a τJSchema schema could involve either the conventional JSON schema (e.g., adding a new object or a new array, adding or removing an object member or changing the type of an object member) or the temporal characteristics document (e.g., adding a new temporal characteristic or changing an existing temporal characteristic). Changing the conventional JSON schema or the temporal characteristics document produces a new version of it and, consequently, requires an update of the corresponding temporal JSON schema. At instance level, the change of a conventional JSON schema, with respect to which some conventional JSON instance documents are valid, also produces a new version for each of these, which must be valid with respect to the new conventional JSON schema version. Hence, the temporal JSON documents associated to such conventional JSON documents must also be accordingly updated. Notice that, according to the τJSchema philosophy of schema versioning [23], the update of the temporal JSON schema (document, respectively) involves the creation of a new temporal version of such a schema (document, respectively), even if there is no creation of a new JSON file for each new version of such a schema (document, respectively).

In τJSchema, schema changes are either explicitly specified by NSDBAs or implicitly triggered by non conservative data updates that are executed by end users. Hence, from this point of view, in τJSchema we have two types of schema versioning: explicit and implicit. Notice that non conservative updates are those that do not respect the structure of the updated JSON document (e.g., renaming an instance of an object member, deleting an instance of a required object member, or assigning a value to an instance of an object member that is non-compatible with the type of the object member as declared in the corresponding JSON schema).

In our previous work [23–26], we have studied explicit JSON schema versioning within the τJSchema framework. A lot of research work has also dealt with explicit schema versioning in semi-structured (XML) databases (e.g., in [27–29]). As far as implicit JSON Schema versioning is concerned, it had not been considered in the literature before our previous work [30], where we have partially investigated it in the τJSchema framework, by dealing with implicit schema changes generated by updates to JSON-based conventional Big Data (which are stored in conventional JSON documents). Besides, since the τJSchema framework supports not only conventional Big Data (and their conventional schema) but also JSON-based temporal Big Data (and their temporal schema), in this paper we complete our previous work on implicit schema versioning in τJSchema, by focusing on the temporal aspects. In fact, we propose an approach for handling implicit schema changes triggered by temporal updates to JSON-based Big Data (which are stored in temporal JSON documents). More precisely, we consider implicit schema changes occurring when a user executes a temporal JSON update operation that modifies the value of a snapshot (i.e., non-temporal) JSON component (i.e., a JSON component for which there is no temporal characteristic in the current version of the temporal characteristics document; in this paper, "JSON component" is used to designate a JSON object, a JSON object member, a JSON array, or a JSON array element) and specifies a valid-time timestamp to be assigned to the new value. Such an operation will automatically and transparently generate: a new version of the modified conventional JSON document (which will contain the modified JSON component); a new version of the temporal characteristics document, with a new temporal (valid-time) characteristic associated to the involved JSON component; a consequent update of the temporal JSON schema and of the corresponding temporal JSON document. In this way, the temporal NoSQL Data Base Management System (DBMS) fulfills the requirement of the end user (and avoids his/her frustration) by forcing the association of a valid-time period to the new value of a snapshot JSON component and handing all necessary schema versioning operations and consequent instance update operations without the intervention of the NSDBA and transparently to the end user. To the best of our knowledge, we are the first to deal with implicit schema versioning triggered by temporal data updates, in a JSON-based NoSQL context.

The rest of the paper is organized as follows. Section 2 describes the background of our work. Section 3 proposes our approach for managing implicit schema changes triggered by temporal updates to JSON-based Big Data, in τJSchema. Section 4 illustrates the use of our approach, through an example. Section 5 summarizes the paper and gives some remarks about our future work.

2 Background

In this section, we briefly recall the τJSchema framework, and present the τJUpdate [31] language that we propose to update temporal JSON data in τJSchema.

2.1 The τJSchema Framework

τJSchema [14, 15] is a temporal NoSQL framework (featuring a data model, a language and a set of tools), which allows the definition and validation of time-varying JSON documents with respect to a temporal JSON schema. This latter is based on a conventional JSON schema (a JSON Schema [8, 9] file) and a set of temporal characteristics that are stored in a temporal characteristics document (a standard JSON file). To each conventional JSON schema corresponds a set of conventional JSON (instance) documents.

The NSDBA starts by creating the conventional JSON schema. Then, he/she annotates this schema with temporal characteristics that allow him/her to specify all temporal aspects concerning the components of the conventional JSON schema: whether a conventional JSON schema component varies over valid time [32] or not; whether its lifetime is described as a continuous state or a single event; whether the component may appear at certain times and not at others and whether its content can change.

After that, when the NSDBA annotates the conventional JSON schema and asks the system to commit his/her work, the system generates the temporal JSON schema that associates the conventional schema to its temporal characteristics document.

Besides, after creating the temporal JSON schema, the system generates a temporal JSON document that links each conventional JSON document, which is valid with respect to a conventional JSON schema, to its temporal JSON schema and, thus, to its set of temporal characteristics. The temporal JSON document contains all the versions of the conventional JSON document with their corresponding timestamps and specifies the temporal JSON schema associated to these versions.

Finally, the squashed JSON document, which is automatically obtained from the temporal JSON document, stores the temporal JSON instances. This means that temporal instances can be obtained from the conventional instances, which are stored in conventional JSON documents, and the temporal JSON schema, by applying temporal characteristics on conventional instances.

Running Example. In the following, we provide an example that illustrates the functioning of τJSchema. Let us consider a JSON NoSQL database used by a scholarly publisher for managing scientific journals' details. Assume that on February 01, 2020, the NSDBA defined the first version of a conventional JSON schema, named "journals_ConventionalJSONSchema_V1.json" (as shown in Fig. 1), and the first version of a conventional JSON document, named "journals_ConventionalJSONDocument_V1.json" (as shown in Fig. 2), conformant to this schema. Hence, when the NSDBA asked the temporal and multi-version NoSQL DBMS to commit his/her work, the DBMS performed the following tasks, within the same creation transaction:

- First, it created the temporal JSON schema, as shown in Fig. 3, based on "journals_ConventionalJSONSchema_V1.json"; this schema is saved in a standard JSON file, named "journals_TemporalJSONSchema.json".
- Second, it used the temporal JSON schema (see Fig. 3) and the conventional JSON document version (as shown in Fig. 2) to generate the temporal JSON document,

named "journals_TemporalJSONDocument.json" (as shown in Fig. 4), which lists the first version of the conventional JSON document with its associated timestamp. The squashed version of the temporal JSON document is not generated since it coincides with "journals_ConventionalJSONDocument_V1.json".

```
{ "$schema": "http://json-schema.org/draft-04/schema#",
  "id": "http://jsonschema.net",
  "type": "object",
  "properties":
    { "journals":
        { "id": "http://jsonschema.net/journals",
          "type": "array",
          "items":
            { "type": "object",
              "properties":
                { "journal":
                    { "type": "object",
                      "properties":
                        { "title": { "type": "string" },
                          "SJR": { "type": "number" } },
                      "required": ["title", "SJR"] } },
              "required": ["journal"] } } },
  "required": ["journals"] }
```

Fig. 1. The first version of the conventional JSON Schema ("Journals_ConventionalJSONSchema_V1.json") on February 01, 2020

```
{ "journals":[
    { "journal":{
        "title": "NoSQL and NewSQL Databases",
        "SJR": 0.21 } } ] }
```

Fig. 2. The first version of the conventional JSON document ("Journals_ConventionalJSONDocument_V1.json") on February 01, 2020

```
{ "temporalJSONSchema":{
    "conventionalJSONSchema":{
      "sliceSequence":[
        { "slice":{
            "location":"journals_ConventionalJSONSchema_V1.json",
            "begin":"2020-02-01" } } ] },
    "temporalCharacteristicSet":{} } }
```

Fig. 3. The temporal JSON schema ("journals_TemporalJSONSchema.json") on February 01, 2020

```
{ "temporalRoot":{
    "temporalJSONSchema":{
      "location":"journals_TemporalJSONSchema.json"},
    "sliceSequence": [
      { "slice":{
          "location":"journals_ConventionalJSONDocument_V1.json",
          "begin":"2020-02-01" } } ] } }
```

Fig. 4. The temporal JSON document ("journals_TemporalJSONDocument.json") on February 01, 2020

2.2 Temporal JSON Data Updates

In [33], we have proposed JUpdate, a SQL-like language for updating (conventional) JSON documents. It provides a set of fourteen user-friendly high-level operations that allow creating, deleting or changing such documents; changing a JSON document consists in inserting, modifying, deleting, replacing, copying, moving etc., portions of data in such a document. The proposed operations are as follows: CreateDocument, DropDocument, InsertValue, DeleteValue, UpdateValue, CopyValue, MoveValue, InsertMember, DeleteMember, RenameMember, ReplaceMember, CopyMember, MoveMember, and UpdateObject.

Moreover, since there is no standard or consensual language for updating temporal JSON documents [14, 15, 22], we have decided to fill this gap and to propose a temporal extension of the JUpdate language for the τJSchema framework, named τJUpdate [31]. The syntax and the semantics of the JUpdate operations are extended to support temporal aspects. Each JUpdate statement is enriched with a new valid clause to specify the valid-time period of the updated (i.e., inserted, deleted or modified) JSON component, as shown in Fig. 5. Notice that `temporalValue` is used to express time literals (e.g., corresponding to the DATE or DATETIME data type of SQL). Notice also that the manner used to specify temporal JSON data updates with our τJUpdate language is similar to the manner used to specify temporal data updates with existing temporal languages like τXUF [34], in the temporal XML setting, and TSQL2 [13] and SQL:2011 [35], in the temporal relational setting. In fact, we have only augmented the JUpdate statements, which are used for updating non-temporal JSON data, with a valid clause to declare the applicability period of the operation, or more precisely the valid-time timestamp of the new value of the involved JSON component.

```
τJUpdateStatement   ::=  JUpdateStatement "valid" validPeriod
validPeriod         ::=  "in [" validBegin "," validEnd "]"
                       | "from" validBegin  |  "to" validEnd
validBegin          ::=  "Beginning" | "Now" | temporalValue
validEnd            ::=  "Forever" | "Now" | temporalValue
```

Fig. 5. The syntax of the τJUpdate statements

Running Example Reprise. We resume now our example presented as "Running Example" in Sect. 2.1 in order to illustrate the use of the τJUpdate language. We assume that, effective from August 01, 2020, the NSDBA updated the SJR from 0.21 to 0.54 (based on a message received from the "SCImago Journal & Country Rank" platform), by specifying and executing the following τJUpdate statement:

```
UPDATE journals_TemporalJSONDocument.json
OBJECT $.journals[?(@.title=='NoSQL and NewSQL Databases')]
SET SJR=0.54
VALID FROM "2020-08-01";
```

The temporal and multi-version NoSQL DBMS, in response, modified the first version of the conventional JSON document and produced a new version thereof, named "journals_ConventionalJSONDocument_V2.json" (as shown in Fig. 6). Furthermore, it updated the temporal JSON document (as shown in Fig. 7), in order to include the new slice associated to the new conventional JSON document version. Changes are presented in red bold type. This example will be continued in Sect. 4, to show how to manage implicit schema changes that are triggered by temporal JSON data updates.

```
{ "journals":[
    { "journal":{
        "title": "NoSQL and NewSQL Databases",
        "SJR": 0.54 } } ] }
```

Fig. 6. The second version of the conventional JSON document ("Journals_ConventionalJSONDocument_V2.json") on August 01, 2020

```
{ "temporalRoot":{
    "temporalJSONSchema":{
      "location":"journals_TemporalJSONSchema.json"},
    "sliceSequence": [
      { "slice":{
          "location":"journals_ConventionalJSONDocument_V1.json",
          "begin":"2020-02-01" } },
      { "slice":{
          "location":"journals_ConventionalJSONDocument_V2.json",
          "begin":"2020-08-01" } } ] } }
```

Fig. 7. The temporal JSON document ("journals_TemporalJSONDocument.json") on August 01, 2020

3 The Proposed Approach

In this section, we present our approach that allows an end user to execute temporal updates to JSON-based Big Data initially specified as snapshot in τJSchema, which gives rise to an implicit schema change involving the corresponding temporal characteristics document and temporal JSON schema. The implicit creation of a new schema version enables the execution of the explicit temporal update requested by users via the addition of new versions to both the temporal JSON document and the squashed JSON document.

Let us assume that an end user specifies, using the τJUpdate language, a temporal JSON instance update operation (TJIUO) that modifies the value of a JSON component for which there is no temporal characteristic in the current version of the temporal characteristics document (i.e., the value of a snapshot/non-temporal JSON component like the "title" and "SJR" members of the "journal" object in the example in Fig. 1 and 3). If the user also assigns a valid-time timestamp to the new value, then the temporal and multi-version NoSQL DBMS processes such an operation by calling the procedure **ApplyTemporalUpdatesWithImplicitSchemaVersioning**, whose algorithm is shown in Fig. 8.

This procedure uses some variables and calls a function and some other procedures, which are all presented in the following. Besides, more details on the τJSchema components, like the conventional JSON document, the temporal JSON schema, the temporal JSON document, and the squashed JSON document, can be found in [14, 15].

- Variables:
  - ConvJDoc_Vn (input): the current conventional JSON instance document version number n.
  - ConvJDoc_Vn+1 (output): the new conventional JSON instance document version number n+1.
  - TJIUO (input): a temporal JSON instance update operation/statement specified by the end user using the τJUpdate language.
  - TempJD (input): the temporal JSON document that ties the conventional JSON document versions and the temporal JSON schema together.
  - TCD_Vm (input): the current temporal characteristics document version number m, associated to the current conventional JSON schema version with respect to which ConvJDoc_Vn is valid.
  - TCD_Vm+1 (output): the new temporal characteristics document version number m+1, associated to the conventional JSON schema version with respect to which ConvJDoc_Vn+1 is valid.
  - TempJS (input): the temporal JSON schema that ties the conventional JSON schema versions and the temporal characteristics document versions together.
  - SquashJD_Vk (input): the squashed JSON document number k, associated to the temporal JSON document TempJD.
  - SquashJD_Vk+1 (output): the new squashed JSON document number k+1, associated to the updated temporal JSON document TempJD.
  - ConvJSch_Vh (input): the current conventional JSON version number h.

```
PROCEDURE ApplyTemporalUpdatesWithImplicitSchemaVersioning
Inputs: ConvJDoc_Vn, TJIUO, TempJD, TCD_Vm, TempJS,
        SquashJD_Vk, ConvJSch_Vh
Outputs: ConvJDoc_Vn+1, TCD_Vm+1 or TCD_V1,
         SquashJD_Vk+1 or SquashJD_V1, ConvJSch_Vh+1
BEGIN
01: If (TJIUO = JIUO VALID FROM VT) /* TJIUO is a temporal update,
        JIUO is the non temporal part, and VT is the valid time */
02: Then If (TCD_Vm exists) /* JSON document already temporal */
03:      Then ApplyTemporalUpdateOp(ConvJDOC_Vn, TJIUO, TempJD,
                          TempJS, SquashJD_Vk, ConvJDOC_Vn+1,
                          SquashJD_Vk+1);
04:      Else ApplyUpdatesWithImplicitSchemaVersioning(
                          ConvJDOC_Vn, ConvJSch_Vh, JIUO,
                          ConvJDOC_Vn+1, ConvJSch_Vh+1);
05:           If (CompareJInstanceDocs(ConvJDoc_Vn+1,
                                ConvJDoc_Vn))
06:           Then DeleteJInstanceDocument(ConvJDoc_Vn+1);
07:                Display("No updates actually applied to"+
                          ConvJDoc_Vn);
08:           Else UpdateTemporalJDocument(TempJD,
                                  ConvJDoc_Vn+1);
09:                CreateTemporalCharacteristicDocument(TCD_V1);
10:                AddLogicalItem(TCD_V1, targetItem(JIUO));
11:                AddValidTimeToLogicalItem(TCD_V1,
                          targetItem(JIUO), state, varying,
                          constant, );
12:                UpdateTemporalJSchema(TempJS, TCD_V1);
13:                CreateSquashedJDocument(SquashJD_V1, TempJS,
                                  ConvJDoc_Vn+1);
14:           End If;
15:      End If;
16: End If;
END
```

Fig. 8. The procedure for applying temporal updates to conventional JSON-based Big Data with implicit schema versioning

– ConvJSch_Vh+1 (output): the new conventional JSON schema version number h+1.

- Procedures:

  – ApplyTemporalUpdateOp(cjd_Vn, tjiuo, tjd, tjs, sjd_Vk, cjd_Vn+1, sjd_Vk+1): as defined in [31], it applies the temporal JSON instance document update operation (tjiuo), i.e., the τJUpdate operation/statement, on the conventional JSON instance document version number n passed as argument (cjd_Vn), while taking into account the corresponding temporal JSON document (tjd), temporal JSON schema (tjs), and the squashed JSON document version number k associated to tjd (sjd_Vk), to produce a new conventional JSON document version number n+1 (cjd_Vn+1) and a new squashed JSON document version number k+1 (sjd_Vk+s1).

- ApplyUpdatesWithImplicitSchemaVersioning(cjd_Vn, cjs_Vh, jiuo, cjd_Vn + 1, cjs_Vh + 1): as defined in [30], it applies a non temporal JSON instance update operation (jiuo) on the conventional JSON instance document version (cjd_Vn) that is valid to the conventional JSON schema version (cjs_Vh), both passed as arguments, and produces as outputs a new conventional JSON instance document version (cjd_Vn+1) that is valid to a new conventional JSON schema version (cjs_Vh +1).
- DeleteJInstanceDocument(cjd): it deletes the conventional JSON instance document whose name is passed as argument (cjd).
- Display(message): it displays the character string passed as argument (message).
- UpdateTemporalJDocument(tjd, cjd): it adds, to the temporal JSON document passed as argument (tjd), a new slice associated to a new conventional JSON instance document version (cjd).
- CreateTemporalCharacteristicDocument(tcd_V1): as defined in [24], it creates the first version of a temporal characteristics document (tcd_V1), as an empty JSON file.
- AddLogicalItem(tcd_Vo, targetItem): as defined in [24], it adds, to the temporal characteristics document version passed as argument (tcd_Vo), a new temporal (logical) characteristic for a JSON component whose path, which is a JSONPath [36] expression, is provided (targetItem).
- AddValidTimeToLogicalItem(tcd_Vo, targetItem, validTimeKind, validTimeContent, validTimeExistence, validTimeFrequency): as defined in [24], it adds, to the temporal characteristics document version passed as argument (tcd_Vo) and more precisely to the temporal (logical) characteristic of the JSON component located at targetItem, a "validTime" property with specified "validTimeKind", "validTimeContent", "validTimeExistence", and "validTimeFrequency".
- UpdateTemporalJSchema(tjs, tcd): it adds, to the temporal JSON schema passed as argument (tjs), a new slice associated to a new temporal characteristics document version (tcd).
- CreateSquashedJDocument(sjd, tjs, cjd): it creates a new squashed JSON document (sjd) based on a new conventional JSON instance document version (cjd) and its temporal JSON schema (tjs).

• Function:

  - CompareJInstanceDocs(jid1, jid2): it compares two JSON instance documents (jid1 and jid2) and returns true or false according to the fact that these documents are identical or not.

As shown in Fig. 8, the temporal JSON data update algorithm is based mainly on (i) the current conventional JSON document version, (ii) the current temporal characteristics document version, and (iii) the temporal JSON instance document update operation (i.e., the τJUpdate operation/statement, specified by the end user). These inputs allow the temporal and multi-version NoSQL DBMS to determine whether or not the τJUpdate operation requires a change to the current temporal characteristics document version or, more specifically, whether it requires or not the addition of a temporal (valid-time) characteristic to this document version. In this paper, among the τJUpdate operations that could trigger a change to the temporal characteristics document, we only focus on the temporal modification of the value of an object member that is performed via the τJUpdate statement τUpdateValue [31] whose SQL-like syntax is as follows:

```
UPDATE doc
PATH path VALUE value
VALID FROM ValidBegin
```

and its procedural form is as follows:

```
τUpdateValue(doc, path, value, ValidBegin)
```

This operation is used to replace, in the conventional JSON instance document “doc”, the value denoted by “path” [36] with a new (possibly complex) value “value”, effective from “ValidBegin”. Notice that, for simplicity, we only consider the case of update validity in the form “VALID FROM ValidBegin”. An extension to the most general update validity expression is straightforward following the same lines used for the definition of the ApplyTemporalUpdateOp procedure in [31].

In case the modified value is for a snapshot object member, the DBMS automatically executes the change operation DefineTimeVaryingLogicalItem [25] whose complete definition is as follows:

```
DefineTimeVaryingLogicalItem(TCD.json, target, "state",
    "varying", "constant", , , , , , , , , , , , , , , )
```

such that “TCD.json” is the name of the JSON file that represents the temporal characteristics document version on which the operation will be executed, “target” is the JSONPath path in the corresponding conventional JSON Schema version of the concerned object member; “state”, “varying”, and “constant” are the values of the three properties “kind”, “content”, and “existence”, respectively, of the “validTime” object representing the property of the temporal characteristic that indicates that the involved object member varies over valid time.

4 Illustrative Example

In order to illustrate the functioning of our approach and show how temporal JSON data updates could transparently trigger changes to the temporal characteristics of data, which are specified at schema level, we resume our running example at the point where we left off. In addition to the operations performed in the “Running Example Reprise” presented in Sect. 2.2, the DBMS detected the following situation that requires some changes at schema level:

- from one hand, the object member "SJR" is not declared as a time-varying member, since it has no temporal characteristic; from a temporal database point of view, it is considered a snapshot object member;
- from the other hand, the NSDBA specified, for the new value of this object member, a valid time period representing the "applicability period" of the update operation (i.e., the time period in which the update has to be in effect, e.g., from 2020–08-01 on) or, more precisely, the valid-time timestamp of such a new value.

Thus, in order to force the execution of that τJUpdate statement although the involved object member is not temporal, an implicit schema change is triggered. In particular, the temporal and multi-version NoSQL DBMS must, first of all, create the first version of the temporal characteristics document (as shown in Fig. 9) associated to the current version of the conventional JSON schema of journals, in order to define the above object member as a time-varying one, and more precisely as a valid-time object member. To this end, a temporal characteristic must be defined for this object member: the "target" of the characteristic is the path of such an object member (in the conventional JSON schema version) and its "validTime" property (i.e., the valid-time timestamp) is defined with the following specifications: kind = "state" (i.e., its lifetime will be a continuous valid-time interval), content = "varying" (i.e., its content will change over time), and existence = "constant" (i.e., the JSON component will exist over time in a constant manner).

Once the first version of the temporal characteristics document (i.e., "journals_TemporalCharacteristics_V1.json") had been created (as shown in Fig. 9), the temporal JSON schema was accordingly updated (as shown in Fig. 10).

Moreover, since the DBMS has also updated the temporal JSON document (as shown in Fig. 7), in order to take into account the new conventional JSON document version (see Fig. 6) with a valid-time timestamp, the squashed JSON document, corresponding to this temporal JSON document, has been generated in order to store the information concerning the new temporal object member (as shown in Fig. 11). Changes and contents of new documents are presented in red bold type.

```
{ "temporalCharacteristicSet":{
    "logical":{
      "logicalItems":[
        { "target":"$.properties.journals..journal.properties.SJR",
          "validTime":{
            "kind":"state",
            "content":"varying",
            "existence":"constant" } } ] } } }
```

Fig. 9. The first version of the temporal characteristics document ("journals_TemporalCharacteristics_V1.json") on August 01, 2020

```
{ "temporalJSONSchema":{
    "conventionalJSONSchema":{
      "sliceSequence":[
        { "slice":{
            "location":"journals_ConventionalJSONSchema_V1.json",
            "begin":"2020-02-01" } } ] },
    "temporalCharacteristicSet":{
      "sliceSequence":[
        { "slice":{
            "location":"journals_TemporalCharacteristics_V1.json",
            "begin":"2020-08-01" } } ] } } }
```

Fig. 10. The temporal JSON schema ("journals_TemporalJSONSchema.json") on August 01, 2020

```
{ "journals":[
    { "journal":{
        "title":"NoSQL and NewSQL Databases",
        "SJR":[
          { "validTimestamp":{
              "begin":"2020-02-01",
              "end":"2020-07-31" },
            "value":0.21 },
          { "validTimestamp":{
              "begin":"2020-08-01",
              "end":"Now" },
            "value":0.54 } ] } } ] }
```

Fig. 11. The squashed JSON document ("journals_SquashedJSONDocument.json") on August 01, 2020

The necessary sequence of (high-level) operations that have been implicitly generated by the temporal and multi-version NoSQL DBMS and, transparently to the end user, executed on August 01, 2020 in order to produce the first temporal characteristics document version (as shown in Fig. 9) and to update the temporal JSON schema (as shown in Fig. 10), is as follows:

```
(i) UpdateTemporalJSONSchema("journals_TemporalJSONSchema.json", , ,
        empty, "journals_TemporalCharacteristics_V1.json")
(ii) DefineTimeVaryingLogicalItem(
        "journals_TemporalCharacteristics_V1.json",
        "$.properties.journals..journal.properties.SJR", "state",
        "varying", "constant", , , , , , , , , , , , , , )
```

5 Conclusion and Future Work

In this paper, we have proposed an approach for implicit temporal JSON schema versioning driven by temporal updates to JSON-based Big Data instances, in the τJSchema framework. This approach completes our previous work [30] dealing with implicit temporal JSON schema versioning driven by non-temporal updates to JSON-based Big Data instances. Our proposal is threefold: (i) it allows migrating a JSON component from the snapshot format to the temporal one, without the intervention of the NSDBA; (ii) it provides more flexibility, to end users, in JSON data management; (iii) it deals with the required schema versioning details in an implicit way and transparently to the end users.

In order to show the feasibility of our approach, we are planning to develop in the near future a tool that supports our proposal, to be integrated with the tool that is being developed to support non-temporal implicit schema changes [30]. Moreover, since in our present work we have focused only on the effects of the τUpdateValue operation, we also intend to extend this work to cover more τJUpdate operations.

References

1. Information Resources Management Association (IRMA): Big data: Concepts, Methodologies, Tools, and Applications. IGI Global, Hershey, PA, USA (2016)
2. Davoudian, A., Liu, M.: Big Data systems: a software engineering perspective. ACM Comput. Surv. **53**(5), 1–39 (2020)
3. Davoudian, A., Chen, L., Liu, M.: A Survey on NoSQL Stores. ACM Comput. Surv. **51**(2), Article 40 (2018)
4. NoSQL Databases List by Hosting Data – Updated 2020. https://hostingdata.co.uk/nosql-database/. Accessed 18 Jan 2021
5. Sharma, S., Tim, U.S., Gadia, S.K., Wong, J., Shandilya, R., Peddoju, S.K.: Classification and comparison of NoSQL big data models. Int. J. Big Data Intell. **2**(3), 201–221 (2015)
6. Corbellini, A., Mateos, C., Zunino, A., Godoy, D., Schiaffino, S.N.: Persisting big-data: the NoSQL landscape. Inf. Syst. **63**, 1–23 (2017)
7. Internet Engineering Task Force (IETF): The JavaScript Object Notation (JSON) Data Interchange Format, Internet Standards Track document (December 2017). https://tools.ietf.org/html/rfc8259. Accessed 18 Jan 2021
8. IETF: JSON Schema: A Media Type for Describing JSON Documents. Internet-Draft, 19 Mar 2018. https://json-schema.org/latest/json-schema-core.html. Accessed 18 Jan 2021
9. Pezoa, F., Reutter, J.L., Suarez, F., Ugarte, M., Vrgoč, D.: Foundations of JSON schema. In: Proceedings of the 25th International World Wide Web Conference (WWW 2016), Montreal, Canada, 11–15 Apr 2016, pp. 263–273 (2016)
10. Cuzzocrea, A.: Temporal aspects of big data management: state-of-the-art analysis and future research directions. In: Proceedings of the 22nd International Symposium on Temporal Representation and Reasoning (TIME 2015), Kassel, Germany, 23–25 Sep 2015, pp. 180–185 (2015)
11. Franciscus, N., Ren, X., Stantic, B.: Answering temporal analytic queries over big data based on precomputing architecture. In: Nguyen, N.T., Tojo, S., Nguyen, L.M., Trawiński, B. (eds.) ACIIDS 2017. LNCS (LNAI), vol. 10191, pp. 281–290. Springer, Cham (2017). https://doi.org/10.1007/978-3-319-54472-4_27
12. Zheng, X., Liu, H.K., Wei, L.N., Wu, X.G., Zhang, Z.: Timo: in-memory temporal query processing for big temporal data. In: Proceedings of the 7th International Conference on Advanced Cloud and Big Data (CBD 2019), Suzhou, China, 21–22 Sep 2019, pp. 121–126 (2019)
13. Snodgrass, R.T. (ed.), et al.: The TSQL2 Temporal Query Language. Kluwer Academic Publishers, Norwell, MA, USA (1995)

14. Brahmia, S., Brahmia, Z., Grandi, F., Bouaziz, R.: τJSchema: a framework for managing temporal JSON-based NoSQL databases. In: Hartmann, S., Ma, H. (eds.) DEXA 2016. LNCS, vol. 9828, pp. 167–181. Springer, Cham (2016). https://doi.org/10.1007/978-3-319-44406-2_13
15. Brahmia, S., Brahmia, Z., Grandi, F., Bouaziz, R.: A disciplined approach to temporal evolution and versioning support in JSON data stores. In: Ma, Z., Yan, L. (eds.) Emerging Technologies and Applications in Data Processing and Management, pp. 114–133. IGI Global, Hershey, PA, USA (2019)
16. Brahmia, Z., Grandi, F., Oliboni, B., Bouaziz, R.: Schema versioning. In: Khosrow-Pour, M. (ed.), Encyclopedia of Information Science and Technology, 3rd edn, pp. 7651–7661. IGI Global, Hershey, PA, USA (2015)
17. Brahmia, Z., Grandi, F., Oliboni, B., Bouaziz, R.: Schema Versioning in conventional and emerging databases. In: Khosrow-Pour, M. (ed.), Encyclopedia of Information Science and Technology, 4th edn, pp. 2054–2063. IGI Global, Hershey, PA, USA (2018)
18. Roddick, J.F.: Schema versioning. In: Liu, L., Özsu, M.T. (eds.), Encyclopedia of Database Systems, 2nd edn. Springer, New York, NY, USA (2018)
19. Jensen, C.S., et al.: The consensus glossary of temporal database concepts – February 1998 version. In: Etzion, D., Jajodia, S., Sripada, S. (eds.), Temporal Databases – Research and Practice, LNCS 1399, pp. 367–405. Springer, Berlin, Germany (1998)
20. Grandi, F.: Temporal databases. In: Khosrow-Pour, M. (ed.), Encyclopedia of Information Science and Technology, 3rd edn, pp. 1914–1922. IGI Global, Hershey, PA, USA (2015)
21. Jensen, C.S., Snodgrass, R.T.: Temporal database. In: Liu, L., Özsu, M.T. (eds.), Encyclopedia of Database Systems, 2nd edn. Springer, New York, NY, USA (2018)
22. Goyal, A., Dyreson, C.: Temporal JSON. In: Proceedings of the 5th IEEE International Conference on Collaboration and Internet Computing (CIC 2019), Los Angeles, CA, USA, 12–14 Dec 2019, pp. 135–144
23. Brahmia, S., Brahmia, Z., Grandi, F., Bouaziz, R.: Temporal JSON schema versioning in the τJSchema framework. J. Digital Inf. Manage. **15**(4), 179–202 (2017)
24. Brahmia, S., Brahmia, Z., Grandi, F., Bouaziz, R.: Managing temporal and versioning aspects of JSON-based Big Data via the τJSchema framework. In: Proceedings of the International Conference on Big Data and Smart Digital Environment (ICBDSDE'2018), Casablanca, Morocco, 29–30 Nov 2018, Studies in Big Data, vol. 53, pp. 27–39. Springer Nature Switzerland AG (2019)
25. Brahmia, Z., Brahmia, S., Grandi, F., Bouaziz, R.: Versioning schemas of JSON-based conventional and temporal big data through high-level operations in the τJSchema framework. Int. J. Cloud Comput. **10**(5/6), 442–479 (2021)
26. Brahmia, S., Brahmia, Z., Grandi, F., Bouaziz, R.: Versioning temporal characteristics of JSON-based big data via the τJSchema framework. Int. J. Cloud Comput. **10**(5/6), 406–441 (2021)
27. Snodgrass, R.T., Dyreson, C.E., Currim, F., Currim, S., Joshi, S.: Validating quicksand: schema versioning in τXSchema. Data Knowl. Eng. **65**(2), 223–242 (2008)
28. Currim, F., et al.: τXSchema: Support for Data- and Schema-Versioned XML Documents. Technical Report TR-91, TimeCenter, 8 Sep 2009. http://timecenter.cs.aau.dk/TimeCenterPublications/TR-91.pdf. Accessed 18 Jan 2021
29. Brahmia, Z., Grandi, F., Oliboni, B., Bouaziz, R.: Schema change operations for full support of schema versioning in the τXSchema framework. Int. J. Inf. Technol. Web. Eng. **9**(2), 20–46 (2014)
30. Brahmia, Z., Brahmia, S., Grandi, F., Bouaziz, R.: Implicit JSON schema versioning driven by big data evolution in the τJSchema framework. In: Proceedings of the 3rd International Conference on Big Data and Networks Technologies (BDNT'2019), Leuven, Belgium, 29 Apr – 2 May 2019, LNNS, vol. 81, pp. 23–35. Springer Nature Switzerland AG (2020)

31. Brahmia, Z., Grandi, F., Brahmia, S., Bouaziz, R.: τJUpdate: A Temporal Update Language for JSON Data. Manuscript in preparation (2022)
32. Jensen, C.S., Snodgrass, R.T.: Valid time. In: Liu, L., Özsu, M.T. (eds.), Encyclopedia of Database Systems, 2nd edn. Springer-Verlag, New York, USA (2018)
33. Brahmia, Z., Brahmia, S., Grandi, F., Bouaziz, R.: JUpdate: a JSON update language. Electron. **11**(4), 508 (2022). https://doi.org/10.3390/electronics11040508
34. Brahmia, Z., Grandi, F., Bouaziz, R.: tauXUF: A temporal extension of the XQuery update facility language for the tauXSchema framework. In: Proceedings of 23rd International Symposium on Temporal Representation and Reasoning (TIME 2016), Lyngby, Denmark, 17–19 Oct 2016, pp. 140–148 (2016)
35. Kulkarni, K.G., Michels, J.-E.: Temporal features in SQL:2011. ACM SIGMOD Rec. **41**(3), 34–43 (2012)
36. Gössner, S.: JSONPath – Xpath for JSON, 21 Feb 2007. http://goessner.net/articles/JsonPath/. Accessed 18 Jan 2021

A Fuzzy Meta Model for Adjusting Ant Colony System Parameters

Safae Bouzbita(✉) and Abdellatif El Afia

Smart Systems Laboratory, ENSIAS, Mohammed V University, Rabat, Morocco
safae.bouzbita@gmail.com, a.elafia@um5s.net.ma

Abstract. Metaheuristic algorithms have become an important choice for solving complex optimization problems which are difficult to solve by conventional methods. But, like many other metaheuristic algorithms, ant colony system (ACS) has the problem of parameters setting. In the last few years, different approaches have been proposed to deal whit this problem. Recently the use of fuzzy logic in dynamic parameters adaptation of metaheuristic algorithms is gaining a considerable interest from the researchers. In this paper, a meta model for modifying the parameters of ACS during runtime based on fuzzy logic concept is presented. The main idea is to study the effect of modifying all the parameters of the ACS on the same time on its performance. To compare the efficiency of the proposed approaches, they were applied to a set of traveling salesman problem instances. Also, a comparison with the standard ACS and some literature results are discussed.

Keywords: Parameter adaptation · Fuzzy logic controller · Ant colony system · Travelling salesman problem

1 Introduction

The performance of ant colony system (ACS) as a well-known metaheuristic optimization algorithm depends on the values assigned to the parameters. In ACS algorithm, artificial ants construct solutions probabilistically, biasing the search by pheromone and heuristic information. Pheromones are usually presented by small values that are modified while solving the problem or iteratively to reflect the search experience, while the heuristic information is derived from the problem to represent the desirability of using a particular component of solution.

Traditionally, the parameters of a metaheuristic algorithm are remained fixed during the runtime, which can worsen its efficiency. To overcome this problem, many adaptive methods have been proposed over the last few years, which a large part relies on the machine learning algorithms.

Among these adaptive methods, there is the hidden markov model (HMM) algorithm that has become a powerful tool to predict and adapt parameters values [1–13].

Fuzzy logic also is one of the artificial intelligence techniques that gained wide attention in the field of controlling parameters as well as in other domains like prediction, decision making, and classification.

M. Lazaar et al. (Eds.): BDIoT 2021, LNNS 489, pp. 48–58, 2022.
https://doi.org/10.1007/978-3-031-07969-6_4

Many researchers have been suggested parameters adaptation techniques based on fuzzy logic concept. For ACO algorithms, Neyoy et al. [14] and Castillo et al. [15] proposed a dynamic parameter tuning for ACO based on fuzzy logic and its application in TSP problems and autonomous mobile robot respectively.

El Afia et al. [16], have developed an approach based on fuzzy logic controller to study the effect of varying the local pheromone parameter on the ACS performance. Bouzbita et al. [17] have proposed an online fuzzy controller method for adjusting the population size of ACS algorithm.

AS for Olivas et al. [18, 19], they proposed in the first work a fuzzy logic controller to adjust the evaporation parameter dynamically, while in their second work a 2-type fuzzy controller was developed to dynamically update two parameters of ACO algorithm.

Besides ACO algorithms, Lalaoui et al. proposed a neighborhood adaptation for simulated annealing algorithm based on fuzzy logic controller in [20] and a 2-type fuzzy controller for mixed-model assembly line balancing problem in [21].

For their side Mezouar et al. [22–25] have proposed new concepts and techniques of intelligence to predict and optimize electric supply chain management.

As for khaldi et al. [26–30] they suggested the artificial neural network algorithm for prediction and forecasting of blood demand, time series, and supplier performance.

In the same context sarhani et al. [31–33] proposed hybrid machine learning approach and support vector regression for electric load and supply chain demand forecasting.

Kabbaj et al. afia [34] proposed a learning integral strategy of branch and bound.

In this paper, a fuzzy meta model for updating ACS parameters is developed. The aim of this approach is comparing between the performance of the ACS algorithm when updating the parameters separately and its performance when updating all the parameters at the same time. The test bed travelling salesman algorithm was chosen as the testing application for the proposed approach.

The remainder of this article is organized as follows: Sect. 2 recalls the notion of ant colony system and fuzzy logic controller. Section 3 explains the proposed approach. Section 4 discusses the experimental results. Section 5 gives the statistical test. Finally, Sect. 6 outlines the coming out of the proposed approach.

2 Background

2.1 Ant Colony System

The Ant Colony System (ACS) metaheuristic is a cooperative learning algorithm, in which a set of artificial ants cooperate through an indirect communication form mediated by the pheromone they put on the ground (graph edges in case of TSP) while building a solution.

Informally, each ant builds a solution by repeatedly applying the state transition rule.

$$s = argmax_{u \in J_k(r)}[\tau(r,u)][\eta(r,u)]^{\beta}, \quad if\ q \leq q_0 \tag{1}$$

That is, the edge is chosen accumulated knowledge about the problem with a probability q_0. Otherwise, with probability $(1 - q_0)$, an edge is selected using biased exploration according to the following equation:

$$P^k_{rs} = \begin{cases} \frac{[\tau(r,s)].[\eta(r,s)]^\beta}{\sum_{u \in J_k(r)} [\tau(r,u)][\eta(r,u)]^\beta} & ifs \in j_k(r) \\ 0 & otherwise \end{cases} \quad (2)$$

The state transition rule obtained from (1) and (2) is called pseudo-random-proportional rule, and it balances directly between exploration of new solutions and exploitation of a priori knowledge about the problem.

To provide a better use of pheromone information, a local pheromone updating rule is applied by the ants while building their solutions to change dynamically the desirability of edges.

$$\tau(r_k, s_k) = (1 - \xi)\tau(r_k, s_k) + \xi\tau_0 \quad (3)$$

After all ants have completed the construction of their solutions, the pheromone is updated one more time by the globally best ant using the global updating rule:

$$\tau(r_k, s_k) = (1 - \rho)\tau(r_k, s_k) + \frac{\rho}{L_{best}} \quad (4)$$

2.2 Fuzzy Logic Controller

The Fuzzy concept was introduced by Lotfi Zadeh in 1965 (Zadeh L.A., 1965) to deal with an environment of uncertainty, vagueness, imprecision and imperfect information. Fuzzy Logic Controllers (FLCs) exploit the expert knowledge base represented by a linguistic form, to perform a robust control of an imprecise and uncertain system. Over the past few years, FLCs have known a widespread use in the field of controlling parameters.

The process of FLCs can be recapitulated in three main stages: i) Fuzzification, ii) Fuzzy inference engine, and iii) Defuzzification (Fig. 1).

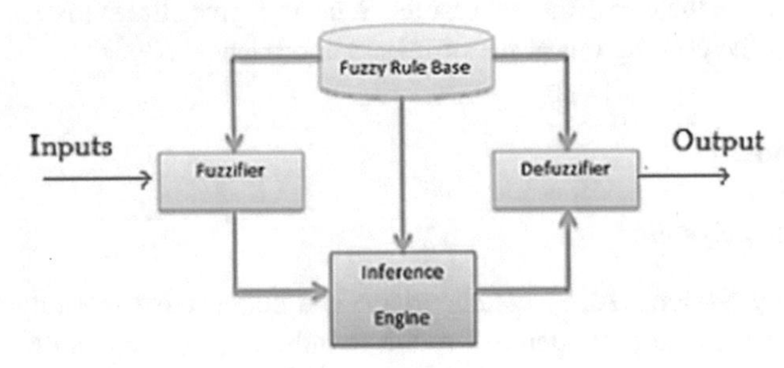

Fig. 1. Fuzzy Logic meta model for ACS parameters adaptation.

The inputs of a fuzzy system are often crisp values that are converted to fuzzy variables using membership functions; this process is called fuzzification, then the fuzzy variables are passed through the rule base which is a set of IF-THEN rules, and each rule has a fuzzy output. The outputs of all the rules are combined and aggregated using an inference engine to represent one fuzzy output. This fuzzy output needs to be converted into a crisp value by using a defuzzification method.

3 Proposed Method

To perform the all in the same time dynamic adaptation for ACS parameters, a fuzzy system with two inputs is used. As inputs, we used the percentage of passed iterations expressed by Eq. (5), and the measure of the ant's diversity expressed by Eq. (6).

$$\mathbf{Iteration} = \frac{\mathbf{Current\ Iteration}}{\mathbf{Total\ of\ Iterations}} \tag{5}$$

$$\mathbf{Diversity} = \frac{1}{m}\sum_{i=1}^{m}\sqrt{\sum_{j=1}^{n}\left(x_{ij}(t) - \bar{x}_j(t)\right)^2} \tag{6}$$

where, Current iteration is the number of past iterations, and total of iterations is the entire number of iterations needed for testing the algorithm, $\boldsymbol{m}$ is the number of ants, $\boldsymbol{i}$ is the index of the ant, $\boldsymbol{n}$ is the total number of dimensions, $\boldsymbol{j}$ is the index of the dimension, x_{ij} is the $\boldsymbol{j}$ dimension of the ant $\boldsymbol{i}$, $\bar{x}_j$ is the jth dimension of the current best ant of the colony.

Each input is mapped to values from 0 to 1 using three membership functions which are: {Low, Medium, and High}, these values are called degrees of membership.

Each output (ACS parameter) is mapped to values from 0 to 1 using five membership functions which are: {Low, Medium Low, Medium, Medium High, and High} (Figs. 2 and 3).

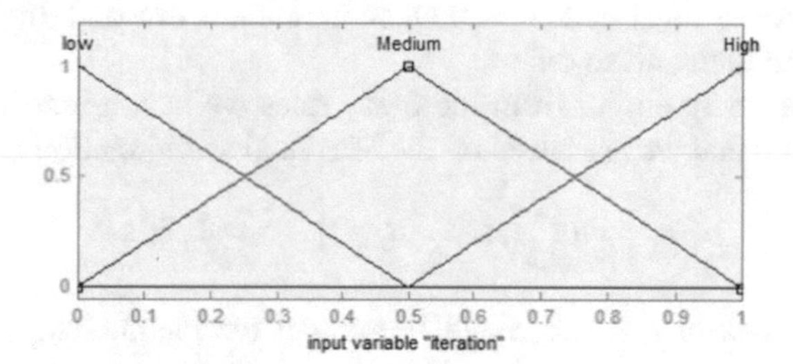

Fig. 2. Iteration membership functions.

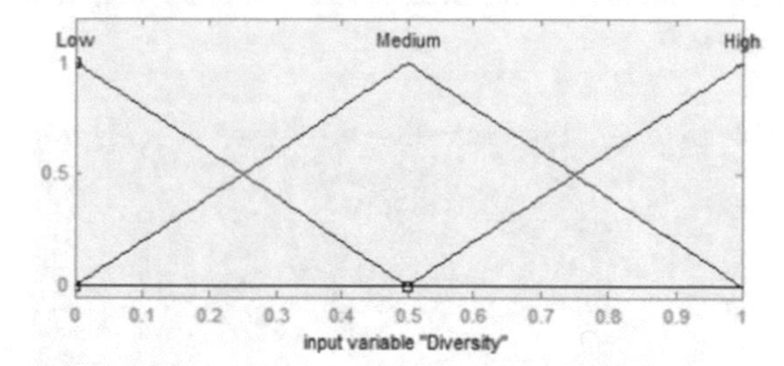

Fig. 3. Diversity membership functions.

For the inference engine, we built a rule matrix for each parameter according to the extracted knowledge from the literature, and then we used an aggregation method to extract the fuzzy output (Table 1).

Table 1. The fuzzy rule base of the proposed fuzzy system.

	ξ	ρ	α	β	q_0	M
LL	L	H	L	H	L	H
LM	ML	MH	ML	MH	ML	MH
LH	M	M	M	M	M	M
ML	ML	MH	ML	MH	ML	MH
MM	M	M	M	M	M	M
MH	MH	ML	MH	ML	MH	ML
HL	M	M	M	M	M	M
HM	MH	ML	MH	ML	MH	ML
HH	H	L	H	L	H	L

As mentioned previously, the inputs can have three different degrees of membership which are: low, medium, and high. By combining those using IF-THEN statements we obtain nine rules for each parameter. The symbols in the table are the concatenation of the first letter of the membership of each input. For example LL means iteration is low and diversity is low. So the table can be interpreted as follows: when iteration is low and diversity is low then ξ must be low, and when iteration is low and diversity is medium ρ must be medium high and so on.

To extract the fuzzy output from the fuzzy rules we have used the Min fuzzy set operation, knowing that we are based on the Mamdani's conjunction operator (AND).

$$\mu_c = \min_{i,j=1,2,3} \{\mu_i(x), \mu_j(y)\} \quad \mathrm{c} = 1, 2, .., 9 \tag{7}$$

where, c is the index of the rule, i and j represent the membership functions {Low, Medium, and High}, x and y are crisp values.

Using the rule base, the membership functions of the output, the fuzzy inference, and a defuzzification method we can finally obtain a crisp output value which is the controller value (Fig. 4).

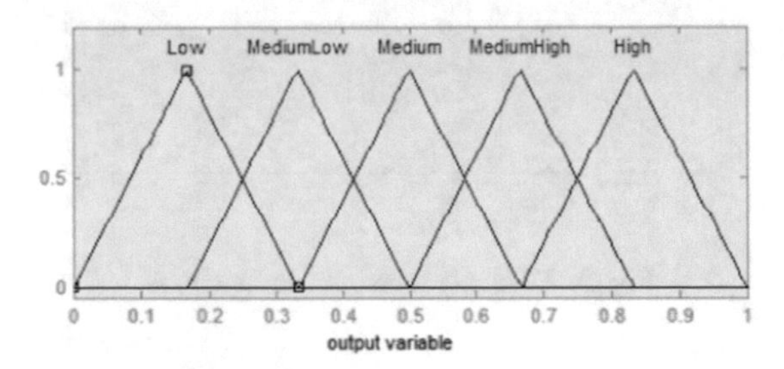

Fig. 4. Output Membership functions.

For the defuzzification method we have chosen the Centroid of Area (COA) Method. The defuzzified value z* is defined as follows:

$$z^* = \frac{\sum_{i=1}^{k} A_i * \overline{z}_i}{\sum_{i=1}^{k} A_i} \quad (8)$$

where, A_i represents the fuzzy surface area of ith rules, k is the total number of rules, and $\overline{z}_i$ is the center of the firing area.

4 Experimental Results

The proposed algorithm was compared to the standard ACS algorithm to test its efficiency. The ACS parameters were initialized by the best known values from the literature [35] which are $\beta = 2$, $\rho = 0.1$, $\mathbf{q}_0 = 0.9$, and $m = 10$. The ants are positioned randomly on all experiments. The TSP benchmark instances tested in this work were chosen from the TSPLIB [36] according to the most common used instances in the literature. Each instance was run for 1000 iterations. The stop condition is: defining a maximum number of iterations or reaching the optimal solution. The proposed algorithm was developed on C.

The following table summarizes the results from testing the proposed method and the standards ACS on some TSP benchmark instances.

Table 2. Summary of results of the proposed method and the standards ACS on TSP benchmarks.

Problem	Fuzzy ACS		Standard ACS		Best known solution
	Solution	CPU time	Solution	CPU time	
Eil51	426 [10]	1.00	426 [28]	1.07	426
St70	675 [14]	1.38	675 [28]	1.14	675
Eil76	538 [11]	1.19	538 [33]	1.07	538
Rat99	1211 [17]	1.15	1211 [47]	1.29	1211
kroA100	21282 [16]	1.06	21282 [29]	1.06	21282
Lin105	14379 [7]	1.04	14379 [5]	1.00	14379
Rat195	2323 [214]	1.28	323 [215]	1.06	2323
Pr264	49135 [15]	1.05	49135 [20]	1.11	49135
Lin318	42029 [430]	0.44	42029 [484]	1.09	42029
Pr439	107217 [180]	6.72	107217 [325]	0.37	107217
U1060	224094 [1127]	144	224094 [1066]	114	224094
U2152	64253 [1489]	199	64253 [1406]	196	64253

The numbers between brackets represents the numbers of iterations in which the algorithms found the best solution, and the CPU time is the amount of time that was spend to find this best solution.

From Table 2 we can see that most instances have achieved the best known solution, except the last two instances that have approached to the best known solution without reaching it, but with giving it more time they could achieve it.

The results from testing the meta model method on the TSP benchmark instances show that the proposed method gives a better performance in term of the number of iterations except few ones, so that the modification of the parameters according to the reflected information of the search space by the number of iterations and the diversity between solutions can reduce the number of iterations compared to the standards ACS.

In term of the elapsed time for both algorithms, we can see that the standards ACS algorithm keeps its advantage comparing to the proposed method, which is acceptable because in the proposed method we hybridize two algorithms: ACS and Fuzzy logic controller.

The figures below show the result of running both the standard and the modified ACS on two TSP instances which are: pr439.tsp and lin318.tsp.

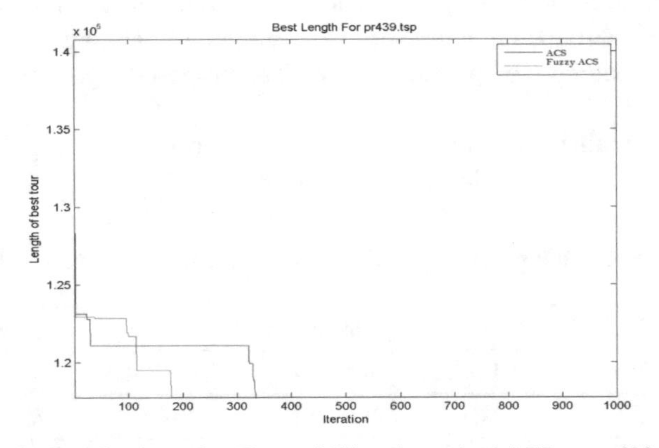

Fig. 5. Result of running Fuzzy ACS and standard ACS on pr439.tsp.

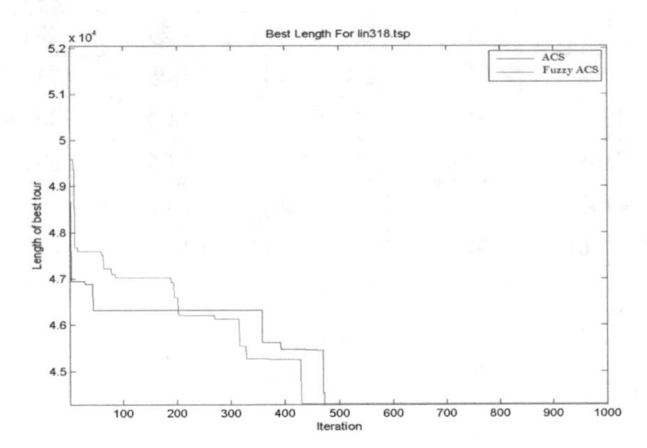

Fig. 6. Result of running Fuzzy ACS and standard ACS on lin318.tsp.

Figures 5 and 6 confirm the obtained results cited in Table 2. Thus, in both figures we can observe that although the two algorithms achieve the best solution, but the proposed method achieves it in a less number of iterations compared to the standards one.

5 Statistical Test

To compare the efficiency between the updated ACS and the standard one, A Wilcoxon Rank Test in a pair-wise comparison procedure was used with a significance level $\alpha = 0.05$.

The Wilcoxon Rank Test is a recommended test procedure for comparing dependent examples and for considering the significant differences [37, 38].

The following table represents the calculated p-value of the test.

Table 3. Statistical validation for the TSP benchmark instances with Fuzzy ACS as control algorithm.

TSP	Eil51	St70	Eil76	Rat99	kroA100	Lin105
Standard ACS	2.17E−02	3.56E−02	4.02E−02	1.09E−02	3.35E−02	4.12E−02
TSP	Rat195	Pr264	Lin318	Pr439	U1060	U2152
Standard ACS	2.33E−02	1.56E−02	1.56E−02	2.47E−03	2.47E−03	3.55E−02

From Table 3 we can notice that the modified ACS approach can reach better solutions with level of significance of 5% compared to the standard ACS.

Thus, the calculated p-value is under the significance level in most benchmark samples.

6 Conclusion and Future Work

This paper proposed a meta model for adjusting all the ACS parameters that have a crucial importance on the performance of the ACS algorithm using the well-known Fuzzy Logic Controller algorithm. The simulation results on TSP and the statistical tests have shown that the proposed method outperforms the standards ACS in term of the number of iterations, but regarding the elapsed time, we can say that the standards ACS gives a better performance which is explained by the processing time taken by the fuzzy system. In our future work we will try to reduce the processing time taken by the proposed method.

References

1. El Afia, A., Lalaoui, M., Chiheb, R.: A self controlled simulated annealing algorithm using hidden Markov model state classification. Procedia Comput. Sci. **148**, 512–521 (2019)

2. Lalaoui, M., El Afia, A., Chiheb, R.: A self-tuned simulated annealing algorithm using hidden markov model. Int. J. Electr. Comput. Eng. **8**(1), 291 (2018)
3. Lalaoui, M., El Afia, A., Chiheb, R.: A self-adaptive very fast simulated annealing based on Hidden Markov model. In: 3rd International Conference of Cloud Computing Technologies and Applications (CloudTech), pp. 1–8. IEEE (2017). https://doi.org/10.1109/CloudTech.2017.8284698
4. Lalaoui, M., El Afia, A., Chiheb, R.: Hidden Markov Model for a self-learning of Simulated Annealing cooling law. In: 5th international conference on multimedia computing and systems (ICMCS), pp. 558–563. IEEE (2016). https://doi.org/10.1109/ICMCS.2016.7905557
5. Bouzbita, S., El Afia, A., Faizi, R.: A novel based Hidden Markov Model approach for controlling the ACS-TSP evaporation parameter. In: 5th international conference on multimedia computing and systems (ICMCS), pp. 633–638. IEEE (2016). https://doi.org/10.1109/ICMCS.2016.7905557
6. Bouzbita, S., El Afia, A., Faizi, R., Zbakh, M. (2016, May). Dynamic adaptation of the ACS-TSP local pheromone decay parameter based on the Hidden Markov Model. In: 2nd international conference on cloud computing technologies and applications (CloudTech), pp. 344–349. IEEE (2016). https://doi.org/10.1109/CloudTech.2016.7847719
7. Bouzbita, S., El Afia, A., Faizi, R.: Hidden markov model classifier for the adaptive ACS-TSP pheromone parameters. In: Talbi, E.-G., Nakib, A. (eds.) Bioinspired Heuristics for Optimization. SCI, vol. 774, pp. 153–169. Springer, Cham (2019). https://doi.org/10.1007/978-3-319-95104-1_10
8. Bouzbita, S., El Afia, A., Faizi, R.: Parameter adaptation for ant colony system algorithm using hidden markov model for TSP problems. In: Proceedings of the International Conference on Learning and Optimization Algorithms: Theory and Applications, pp. 1–6. ACM (2018). https://doi.org/10.1145/3230905.3230962
9. El Afia, A., Aoun, O., Garcia, S.: Adaptive cooperation of multi-swarm particle swarm optimizer-based hidden Markov model. Prog. Artif. Intell. **8**(4), 441–452 (2019)
10. Aoun, O., Sarhani, M., Afia, A.E.: Hidden markov model classifier for the adaptive particle swarm optimization. In: Amodeo, L., Talbi, E.-G., Yalaoui, F. (eds.) Recent Developments in Metaheuristics. ORSIS, vol. 62, pp. 1–15. Springer, Cham (2018). https://doi.org/10.1007/978-3-319-58253-5_1
11. Aoun, O., Sarhani, M., Afia, A.E.: Particle swarm optimisation with population size and acceleration coefficients adaptation using hidden Markov model state classification. Int. J. Metaheuristics **7**(1), 1–29 (2018)
12. El Afia, A., Sarhani, M., Aoun, O.: Hidden markov model control of inertia weight adaptation for Particle swarm optimization. IFAC-PapersOnLine **50**(1), 9997–10002 (2017)
13. Aoun, O., Sarhani, M., El Afia, A.: Investigation of hidden markov model for the tuning of metaheuristics in airline scheduling problems. IFAC-PapersOnLine **49**(3), 347–352 (2016)
14. Neyoy, H., Castillo, O., Soria, J.: Dynamic fuzzy logic parameter tuning for ACO and its application in TSP problems. In: Castillo, O., Melin, P., Kacprzyk, J. (eds.) Recent Advances on Hybrid Intelligent Systems, pp. 259–271. Springer Berlin Heidelberg, Berlin, Heidelberg (2013). https://doi.org/10.1007/978-3-642-33021-6_21
15. Castillo, O., Neyoy, H., Soria, J., García, M., Valdez, F.: Dynamic fuzzy logic parameter tuning for ACO and its application in the fuzzy logic control of an autonomous mobile robot. Int. J. Adv. Rob. Syst. **10**(1), 51 (2013)
16. El Afia, A., Bouzbita, S., Faizi, R.: The effect of updating the local pheromone on acs performance using fuzzy logic. Int. J. Electr. Comput. Eng. **7**(4), 2161 (2017)
17. Bouzbita, S., El Afia, A., Faizi, R.: Adjusting population size of ant colony system using fuzzy logic controller. In: Nguyen, N.T., Chbeir, R., Exposito, E., Aniorté, P., Trawiński, B. (eds.) ICCCI 2019. LNCS (LNAI), vol. 11684, pp. 309–320. Springer, Cham (2019). https://doi.org/10.1007/978-3-030-28374-2_27

18. Olivas, F., Valdez, F., Castillo, O.: Ant colony optimization with parameter adaptation using fuzzy logic for TSP problems. In: Melin, P., Castillo, O., Kacprzyk, J. (eds.) Design of Intelligent Systems Based on Fuzzy Logic, Neural Networks and Nature-Inspired Optimization. SCI, vol. 601, pp. 593–603. Springer, Cham (2015). https://doi.org/10.1007/978-3-319-17747-2_45
19. Olivas, F., Valdez, F., Castillo, O., Gonzalez, C.I., Martinez, G., Melin, P.: Ant colony optimization with dynamic parameter adaptation based on interval type-2 fuzzy logic systems. Appl. Soft Comput. **53**, 74–87 (2017)
20. Lalaoui, M., El Afia, A., Chiheb, R.: Simulated annealing with adaptive neighborhood using fuzzy logic controller. In: Proceedings of the International Conference on Learning and Optimization Algorithms: Theory and Applications, pp. 1–6. ACM (2018). https://doi.org/10.1145/3230905.3230963
21. Lalaoui, M., El Afia, A.: A versatile generalized simulated annealing using type-2 fuzzy controller for the mixed-model assembly line balancing problem. IFAC-PapersOnLine **52**(13), 2804–2809 (2019)
22. Mezouar, H., El Afia, A., Chiheb, R., Ouzayd, F.: Toward a process model of Moroccan electric supply chain. In: International Conference on Electrical and Information Technologies (ICEIT), pp. 184–191. IEEE (2015). https://doi.org/10.1109/EITech.2015.7162990
23. Mezouar, H., El Afia, A.: A process simulation model for a proposed Moroccan supply chain of electricity. In: International Renewable and Sustainable Energy Conference (IRSEC), pp. 647–654. IEEE (2016). https://doi.org/10.1109/IRSEC.2016.7983999
24. Mezouar, H., El Afia, A., Chiheb, R.: A new concept of intelligence in the electric power management. In: International Conference on Electrical and Information Technologies (ICEIT), pp. 28–35. IEEE (2016). https://doi.org/10.1109/EITech.2016.7519596
25. Mezouar, H., El Afia, A.: Proposal for an approach to evaluate continuity in service supply chains: case of the Moroccan electricity supply chain. Int. J. Electr. Comput. Eng. **9**(6), 2088–8708 (2019)
26. Khaldi, R., El Afia, A., Chiheb, R.: Forecasting of weekly patient visits to emergency department: real case study. Procedia Comput. Sci. **148**, 532–541 (2019)
27. Khaldi, R., Chiheb, R., El Afia, A.: Feedforward and recurrent neural networks for time series forecasting: comparative study. In: Proceedings of the International Conference on Learning and Optimization Algorithms: Theory and Applications, pp. 1–6. ACM (2018). https://doi.org/10.1145/3230905.3230946
28. Khaldi, R., El Afia, A., Chiheb, R., Faizi, R.: Forecasting of Bitcoin daily returns with EEMD-ELMAN based model. In: Proceedings of the International Conference on Learning and Optimization Algorithms: Theory and Applications, pp. 1–6. ACM (2018). https://doi.org/10.1145/3230905.3230948
29. Khaldi, R., El Afia, A., Chiheb, R., Faizi, R.: Artificial neural network based approach for blood demand forecasting: fez transfusion blood center case study. In: Proceedings of the 2nd international Conference on Big Data, Cloud and Applications, pp. 1–6. ACM (2017). https://doi.org/10.1145/3090354.3090415
30. Khaldi, R., Chiheb, R., El Afia, A., Akaaboune, A., Faizi, R.: P rediction of supplier performance: a novel DEA-ANFIS based approach. In: Proceedings of the 2nd international Conference on Big Data, Cloud and Applications, pp. 1–6. ACM (2017)
31. Sarhani, M., El Afia, A.: Electric load forecasting using hybrid machine learning approach incorporating feature selection. In: BDCA, pp. 1–7 (2015)
32. Sarhani, M., El Afia, A.: Intelligent system based support vector regression for supply chain demand forecasting. In: 2014 Second World Conference on Complex Systems (WCCS), pp. 79–83. IEEE (2014). https://doi.org/10.1109/ICoCS.2014.7060941

33. Sarhani, M., El Afia, A.: Feature selection and parameter optimization of support vector regression for electric load forecasting. In: 2016 International Conference on Electrical and Information Technologies (ICEIT), pp. 288–293. IEEE (2016). https://doi.org/10.1109/EITech.2016.7519608
34. Kabbaj, M.M., El Afia, A.: Towards learning integral strategy of branch and bound. In: 2016 5th International Conference on Multimedia Computing and Systems (ICMCS), pp. 621–626. IEEE (2016). https://doi.org/10.1109/ICMCS.2016.7905626
35. Stützle, T., et al.: Parameter adaptation in ant colony optimization. In: Hamadi, Y., Monfroy, E., Saubion, F. (eds.) Autonomous Search, pp. 191–215. Springer, Heidelberg (2011). https://doi.org/10.1007/978-3-642-21434-9_8
36. Reinelt, G.: TSPLIB – a traveling salesman problem library. ORSA J. Comput. **3**(4), 376–384 (1991)
37. LaTorre, A., Muelas, S., Peña, J.M.: A comprehensive comparison of large scale global optimizers. Inf. Sci. **316**, 517–549 (2015)
38. Veček, N., Črepinšek, M., Mernik, M.: On the influence of the number of algorithms, problems, and independent runs in the comparison of evolutionary algorithms. Appl. Soft Comput. **54**, 23–45 (2017)

Ontology Engineering Methodologies: State of the Art

Jalil ElHassouni[1](✉) and Abderrahim El Qadi[2]

[1] LRIT-CNRST (URAC'29), Faculty of Sciences, Rabat IT Center, Mohammed V University, Rabat, Morocco
jalil.elhassouni@gmail.com

[2] High School of Technology in Sale, Mohammed V University in Rabat, Rabat, Morocco
abderrahim.elqadi@um5.ac.ma

Abstract. Recently, the use of ontologies has become more popular in both academia and industry fields. However, many ontology projects have failed due to, at least in part, a lack of discipline in the development process; that is, the poorly specified, underspecified, or lack of requirements and evaluation criteria. Therefore, it is reasonable to ask: what is the most prominent methodology to develop an ontology? To answer this question in this paper, we reviewed the most prominent ontology engineering methodologies, and analyzed the most mature and suitable approaches for the development of ontology based on a set of criteria. We also provide a better understanding of ontology engineering methodology, most used ontology engineering methodologies, current prominent methods, and future research scope for standard ontology engineering methodology. This study shows that no methodology enjoys consensus among the community, and none of these methodologies are mature enough and without limitations. Finally, we concluded that Krisnadhi & Hitzler methodology remains very practical and much more detailed in a step-by-step manner. The fact that it is based on ontology design patterns ensures that there is a trade-off between interoperability, on the one hand, and over-commitment and conflicting requirements on the other hand.

Keywords: Developing ontology · Knowledge engineering · Ontology engineering methodology

1 Introduction

Ontologies enable and facilitate knowledge sharing. While academic and industry fields generate knowledge, they may not often effectively reuse and share that knowledge. Researchers and experts working in the same fields sometimes reinvent the wheels to recreate the same or similar type of knowledge and intellectual work. A software engineering methodology based on formal specifications of shared resources, reusable components, and standard services is needed. Ontology is a system of vocabulary for describing the problem-solving structure of all the existing tasks domains independently. It is obtained by analyzing problem solving processes of domain experts and task structures of real-world problems. Ontologies aim at capturing domain knowledge in a generic

M. Lazaar et al. (Eds.): BDIoT 2021, LNNS 489, pp. 59–72, 2022.
https://doi.org/10.1007/978-3-031-07969-6_5

way and provide a commonly agreed understanding of a domain, which may be reused and shared with others who have similar needs for knowledge representation in that domain [1], thereby, saving time and costs, minimize risks, and increase effectiveness.

An ontology is defined by [2] as "a formal, explicit specification of a shared conceptualization in terms of concepts (i.e., classes), properties and relations". The term "formal" entails that ontologies should be expressed in a machine-readable format such as semantic language (RDF, OWL, PLIB…) allowing automatic reasoning, consistency-checking capability, to be shared by a particular community, and able to be referenced from any environment [3]. In the last decades, the use of ontologies has become more popular in both academia and industry fields ranging from e-learning to financial, medicine, military and engineering systems. However, many ontology projects have failed due to, at least in part, a lack of discipline in the development process; that is, the poorly specified, under specified, or lack of requirements and evaluation criteria [4]. Developing a successful ontology depends on how far its methodology supports collaborative construction and allows for reusability. The methodology should also take into consideration the degree of application dependency, life cycle recommendation, strategy for identifying concepts, methodology details, and interoperability support. Therefore, ontologies shall be well developed to fit purpose and to be re-used.

In this paper we aim to focus on the stat of art of ontology engineering methodologies. The remainder of this paper is organized as follows: in Sect. 2, we introduce the related works to Ontology development methods. In Sect. 3, we discuss the summary analysis of methodologies, and in the last section we conclude with discussing future work.

2 Ontology Development Methods Review

Ontologies have been used in various contexts and domains, which explains the difficulty to identify and enumerate the different roles of ontologies in knowledge-based systems. Nonetheless, several authors have focused on one or more roles of ontologies in a specific context including knowledge representation and reasoning (KR&R), knowledge sharing and reusing (KS&R), knowledge integration (KI), and information retrieval (IR). The development of ontology is an important step in the development of knowledge-based systems.

However, the development of an ontology is a tedious and time-consuming task. It requires significant and qualified skills as an art and creativity rather than technology or understood engineering process. Although a variety of ontology building methodologies have been proposed, they are still at the level of general guidelines and none of them are mature enough to be standardized or receive wide consensus in the ontology community [5, 6]. Generally speaking, most of these methodologies are described as an iterative ontology engineering process; i.e., the deliverable of each step of the process can be modified, revised, evaluated, and refined at any stage of the process. This process may include, albeit with different labeling, the following steps: identifying the ontology scope and the ontology capture, choosing the ontology encoding language, integrating the ontology with the existing one, evaluating the developed ontology, and documenting the ontology [7]. They also follow the same scenarios depending on the knowledge resources available: without reusing existing resources (i.e., the ontology developed from scratch), non-ontological resources or ontological resources, e. g., ontology design patterns reuse as is the case in this work (see Fig. 1).

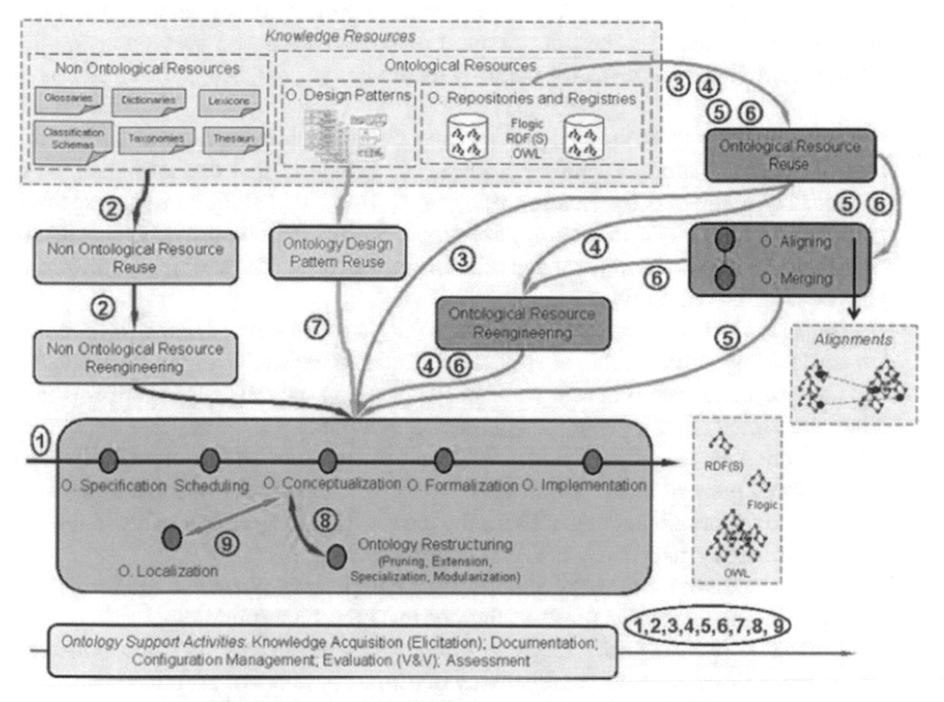

Fig. 1. Scenarios for building ontology networks [8].

2.1 Methodologies for Building Ontologies

There are many methodologies inspired by software development methodologies proposed for the development and maintenance of ontologies. Some of these methodologies are designed for building ontologies from scratch, reusing non-ontological resources, or ontological resources. In this section, based on the same evaluation criteria used by Iqbal et al. [9], we provide a review of the most popular methodologies used by developers in order to come up with the most suitable approach for our work (see Table 1 below). The choice of adopting these criteria is owed to the fact that the study, apart from being the most cited one in Google Scholar, is as conclusive as possible, which allows for greater flexibility in the evaluation and choosing of the best methodology. It is also noteworthy that we have added five more methodologies (most of them came out after the study) to the two derived from the study (the most prominent ones, namely METHONTOLOGY and Gruninger and Fox) in order to compare them before finally selecting our approach.

A. **Gruninger and Fox methodology:**

Gruninger and Fox methodology [10] is developed based on experiences in the development of TOVE (Toronto Virtual Enterprise) project. It is a semi-independent application [11] inspired by the development of knowledge-based systems using first order logic. It proposes six steps for building an ontology:

(1) Identify motivating scenarios: the starting point for ontology development is a set of scenarios which are story problems or examples that cannot be solved or adequately addressed by existing ontologies. A set of solutions to the scenario problems may include motivating scenarios; these solutions provide an informal intended semantics for objects and relationships between those objects that will be included in the ontology.
(2) Formulate informal competency questions (CQs): given a set of scenarios obtained in the first step, an ensemble of CQs are formulated to provide an initial evaluation of the new or extended ontology. These CQs are in natural language forms until they are expressed in a given formal language of ontology.
(3) Specification in First-Order Logic – Terminology: once the informal CQs have been posed, a set of terms is extracted to serve as a basis for specification of the ontology in a formal language. Then, the terminology is specified using first-order logic (formal language such as OWL or similar).
(4) Formal competency questions: the informal competency questions are formalized using the same formal language as the one used for the terminology.
(5) Specification in First-Order Logic – Axioms: define the axioms using first-order logic guided by the formal competency questions in iterative process (i.e., update the set of axioms added to the ontology until it is sufficient to express the formal competency questions). These axioms will be used to determine the semantics or meaning of ontology's terms.
(6) Completeness theorems: the last step is the evaluation stage; once the competency questions have been formally stated, the conditions under which the solutions to the questions are completed must be defined.

B. **METHONTOLOGY methodology:**

Based on the experience acquired in the development of ontology in the chemistry domain, the Ontology Engineering Group at Universidad Politécnica in Madrid developed the METHONTOLOGY framework [12] which is used to build ontologies from knowledge level. It proposes two levels of ontology construction: (i) an ontology development process and (ii) an ontology life cycle. The ontology development process identifies a set of activities to be carried out to develop ontologies: planning tasks, specifying purpose and scope, acquiring knowledge using Knowledge-based Systems (KBSs), conceptualizing this knowledge in a conceptual model, formalizing this model in order to turn it into a formal or semi-compatible model, integrating existing ontologies, implementing the ontology in a formal language, evaluating it, documenting it, and carefully maintaining it. The ontology life cycle allows the ontologist to revise and modify the deliverable of any development process steps: specification, conceptualization, formalization, integration, implementation, and maintenance, regardless of the current stage of development process. However, regardless of its main limitations, the methodology enables the construction of ontologies at knowledge level and does not take in consideration the scenarios of the reuse and reengineering of non-ontological resources and the ontological resources.

C. **NeOn methodology**

Contrary to previous methods that help to develop ontologies from scratch, the NeOn methodology [8] has initiated a new paradigm whose emphasis is the reuse and possible subsequent reengineering of knowledge aware resources. It proposes nine flexible ontology-building scenarios-these scenarios can be combined in different ways-(see Fig. 1). To build ontologies and ontology networks, there are three activities and processes to be carried out during the whole ontology and ontology network development: (i) knowledge acquisition and elicitation, (ii) configuration management, and (iii) evaluation and assessment.

The methodology supports the argumentation process (the collaboration is mentioned but not treated in detail) which permits the collaborators to reach a consensus or explain a decision about the requirements of the ontology networks and how they should be implemented during different scenarios of building ontologies. Each discussion or decision during the entire ontology building process can be documented or even recorded. The methodology proposes a set of guidelines and tasks to carry out the ontology specification activity:

(1) Identifying purpose, scope, and level of formality: obtaining the main goal, scope, and granularity of ontology.
(2) Identifying intended users: determining the main users of the ontology.
(3) Identifying intended uses: obtaining a list of intended uses via a set of scenarios.
(4) Identifying requirements: identifying a set of requirements or needs that an ontology should fulfill once formally implemented using CQs, tools as mind map tools, excel, and collaborative tools techniques.
(5) Group requirements: a set of groups including different CQs that are relevant to specific features of the ontology.
(6) Validating the set of requirements: identifying possible disagreements between CQs, missing CQs, and contradictions in CQs.
(7) Prioritizing requirements: determining a set of priorities related to each group of CQs and for each CQ in the group.
(8) Extracting terminology and its frequency: extracting the terms that will be formally represented in the ontology.

D. **Basic Formal Ontology methodology**

The Basic Formal Ontology (BFO) methodology [13] is a top-down approach to the problem of automatically administrating scientific information located in heterogeneous data sources. To ensure widespread accessibility and usability, it proposes to start from defining the general scientific concepts then encoding detailed terminological content of a specific science. To achieve this goal the methodology proposes a set of steps. In order to design a useful ontology, the methodology proposes four general principles:

(1) Realism: the idea behind the development of ontology is to represent reality, not people's concepts or mental representations or use of languages. Science can help to determine general features of reality in the form of terms and relation between them.
(2) Perspectivalism: it comes from the postulate that reality is too complex and variegated. It needs two or more scientific theories to be covered in its totality. The ontology developers should respect a modular approach rather than seek to represent all portions and features of reality in one single ontology. The ontology developers can develop multi-ontologies to accurate descriptions of reality in which each one should be maintained by experts in the corresponding scientific discipline.
(3) Fallibilism: it flows from the fact that scientific theories may be subject to correction. Therefore, the implication of fallibilism on ontology design is that every ontology:

    - Must have strategies of versioning.
    - Needs to have tracking service for its users.

(4) Adequatism: given the fact that each scientific discipline provides a representation of what exists in reality. Hence, the ontology in a given domain should be designed to represent the entities, not kinds of entities.

The methodology proposes to take into consideration other four concrete guidelines concerning the design process itself:

(1) The principle of reuse: the first step in ontology development should be searching, examining, and evaluating existing ontological resources in the domain and around for possible reutilization or at least for recommendation.
(2) The ontology design process should balance utility and realism: the ontology should be designed to meet specific local purposes provided that the ontology design process should not be on the detriment of adequacy to the reality which the ontology is being developed to represent. The ontology should be developed to be re-used in neighboring domains to which an ontology is constructed for.
(3) The ontology design process is Open-Ended: scientific ontologies should be designed in such a way as to be expandable and adjust to neighboring ontologies through time; i.e., they should be in continuous maintaining, evaluating, updating, correcting, and adjusting.
(4) The principle of Low-Hanging fruit: the ontology designer should start by identifying the general terms most commonly used that are easier to understand and define, then move step by step to more complex and controversial terms.

The BFO methodology is a top-down approach. In the first stage, to design a good ontology it proposes, as previously outlined, to follow general principles in order to determine general features of the ontology. In the second stage, it proposes a set of steps to follow for the development of a domain ontology:

(1) Demarcate the subject matter of the ontology: determine the intended scope of the ontology.
(2) Gather information: select the most common general terms from relevant ontologies, such as Basic Formal Ontology (BFO), and from standard textbooks.
(3) Order these terms hierarchically from the more to the less general ones.
(4) Regiment the results: proceed to an iterative process to understand well the domain, starting from the deliverable of phase 3 in order to guarantee:

- The hierarchy coherence which will serve as a core to the ontology being developed.
- The human understandable definitions for the selected terms.
- The coherence with neighboring ontologies.

(5) Formalize the ontology: encode the ontology using the formal language via an iterative task. Once the terms have been extracted, hierarchized, and defined in natural language, they are specified using first-order logic (formal language such as OWL or similar).

E. **Agile methodology for ontology development (AMOD)**

AMOD [14] is one of agile methodologies which is used in ontology engineering. It aims to adapt the agile principles and practices from software engineering to the development of ontologies. The AMOD framework consists of the following parts:

(1) The ontology development:

- Pre-game phase includes (i) identification of ontology goal and scope, (ii) specification of the tools and techniques used for the formalization and the implementation of the ontology, (iii) identification of ontology requirements using competency questions, and (iv) selection of available sources allowing the extraction of domain knowledge.
- Development phase is organized in multiple and iterative cycles named sprints, each sprint includes the following stages: (i) identification of items and how they are implemented during the sprint planning; (ii) capturing the terms and relations between them using some techniques such as interviewing and brainstorming (knowledge acquisition); (iii) organizing the terms in hierarchy as the core of ontology being developed; (iv) transforming the ontology conceptual model into a formal model using a formal language such as OWL or similar; (v) once the ontology is developed in this sprint, it must integrate those developed in the previous sprints; (vi) and finally reviewing the sprint by the ontology engineer and ontology owner.
- Post-game phase includes the following activities: (i) evaluation of ontology with regard to ontology consistency, answering CQs, and ontology content; and (ii) maintenance of ontology.

(2) Support activities: Supporting activities are achieved in parallel with the development of ontology; they include the following activities:

- Ontology documentation includes three main aspects: creating a human-readable representation of the ontology content, creating machine-readable annotations of documentation metadata, and making the documentation files available as a web resource.
- Configuration management includes four activities: configuration identification, configuration control, and configuration control and configuration audits.

The work [15] presents a simplified agile methodology for ontology development in the form of quick, small, and iterative steps to produce ontologies ready to be used and easily understandable. It contains the same framework as AMOD on miniature scale which aims at developing the final model through a series of small steps. Thus, it can be used to solve and model a small and simple enterprise project in which the ontology being developed is composed by a limited amount of ontological entities.

However, the work [16] affirms that using agile in software and development of safety critical applications requires more efforts than just applying it. Therefore, the use of agile may impact negatively the product innovations.

F. **Krisnadhi and Hitzler methodology**

The methodology proposed in [17–20] is an approach based on a worked example for developing the so-called modular ontologies based on Ontology Design Patterns (ODPs). Developing ODPs in a particular domain typically requires domain expert team. The methodology proposes to adopt the following workflow (see Fig. 2 below):

(1) Use case(s) or scope of use case(s): Define the use case(s) or a set of potential use case(s) for which the ODP is intended and that shall drive the modeling of ontology. The modeling should be as general as possible which will permit to create an underlying schema which is robust, expandable, and easy to maintain and update.
(2) Competency Questions and data sources: Formulate in natural language a set of CQs that shall be answered once the ontology is populated. Inventory the initial list of available data sources that can be used for the intended purpose. This list can be updated with time.
(3) Key Notions to model: Build an ontology for the data we already have, leading to an ontology that is really only useful for a specific use-case, so as soon as data or data formats change, the ontology will quickly lose its value leaving no way to maintain it and to expand it or reuse it by another community or enterprise. To attempt to counteract these issues and to construct a successful ontology, the structure of the latter will not be informed only by the data or data formats but by general ontology design patters. Based on data sources and CQs, identification of the pattern which will be used is better borrowed from best practices in the library [21] in order to realize the very common ontology design pattern for each purpose. The starting point for this step is identifying the pattern which will be the core of modeling,

adapting/changing it as needed, and adding axioms to each module informed by the pattern axioms, obtaining as result a set of modules for the final ontology.

(4) Putting things together: Assemble the modules developed in step 3 together and check module axioms for consistency.

(5) Create OWL files.

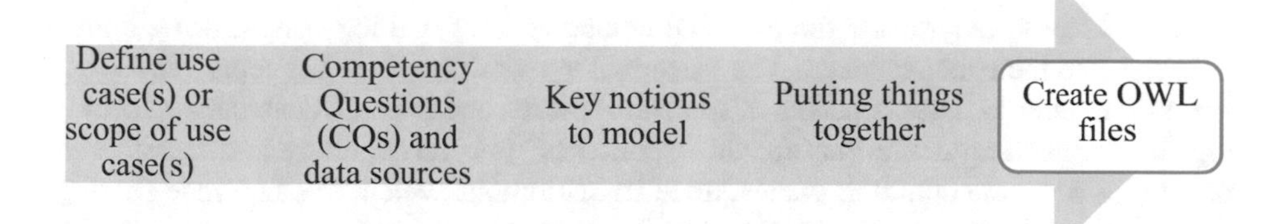

Fig. 2. Krisnadhi and Hitzler methodology life cycle.

G. **Kendall and McGuinness methodology**

Based on their most cited ontology engineering method [22], according to Google Scholar, and on their experience in academic and industry fields, McGuinness and Kendall wrote the book Ontology Engineering (2018) [4] in order to demystify the holistic ontology development process. In chapter 3 of the book, they deal with the first step of ontology engineering methodology; they propose a set of steps to determine requirements and use cases:

(1) Use cases [18] and [19] are used to capture what is needed in terms of domain knowledge reference materials, ontology-specific requirements, basic functional requirement and success criteria, and to determine scoping boundaries of ontology development. The deliverable of this step is iterative and up to date, so it can be modified, revised, evaluated, and refined at any stage of the ontology development process. They are documented to reflect preliminary requirements, requirements learned through development, testing, deployment, maintenance, and evolution.

(2) Gathering resources and potentially reusable ontologies.

(3) Extracting domain terminology.

(4) Creating usage scenarios as they help in describing the use case. They provide the frame for the set of steps that together form the normal flow of use case operations.

(5) Determining the flow of events: the normal or basic flow of events (also known as the primary scenario) defines the process whereby a system that is implemented is executed from start to end. Alternate flows, however, remain complementary to the basic flow. They are often invoked by valid, but not typical, situations or rules.

(6) Creating Competency Questions (CQs): an ensemble of questions that a knowledge base or application must be able to address correctly [6, 20]. CQs are derived from user stories, usage scenarios, normal and alternative process steps, as well as from reference materials.

(7) Describing additional resources: additional resources are composed of miscellaneous notes (information about non-human actors) collected at the very end of the use case and then moved around once we understand where they fit in the overall requirements.
(8) Integrating an ontology into a larger business requirements document (BRD) to provide a basic vocabulary, related definitions, and the seed for elaboration not only of the ontology itself but for other project components.

In chapter 4, they tackle the essential components of ontology engineering terminologies and their management. The "terminology work involves concepts, terms that are used to refer to those concepts in specific contexts, and formal vocabularies, including the establishing definitions for those concepts" [4]. Terminology is imperative in providing adequate common vocabularies in institutions which will facilitate sharing data within and across institutions, and enable them to measure their performance. For instance, the main problem of the 2008 financial crisis is the lack of ground-truth in many banking data assertions or the decisions that depend on the data, creating a confusion in terms used to manage risks. Though the terminology is sine qua non for the development of ontology, this methodology, however, does not expand on this part as to include proper documentation of the terminology identification.

In the final chapter of the book, they present the primary steps involved in the conceptual modeling aspects of ontology development, starting from the deliverable of the previous stages namely: one or more use case(s), CQs, list of curated terms, and a business architecture, and other business models if available. Given these inputs, the ontologist can formalize the ontology in formal language following a set of activities:

- Reaching existing ontological resources to determine reusability.
- Identifying relationships between the concepts from the term list and CQs.
- Testing, evaluation and validating of the ontology

2.2 Summary Analysis of Methodologies

In this section we present a review of some methodologies for the development and maintenance of ontologies; most of them came out after the study [9].

The Gruninger and Fox methodology follows stage-based model. A stage-based model approach is best used when the requirements and scope of an ontology are well identified. The methodology supports reusability, allowing ontology engineers to make use of existing knowledge resources and reducing the ontology development efforts and time. For the degree of application dependency, the methodology opts for a semi-dependent approach. However, the methodology does not offer any support for collaborative construction and interoperability. Therefore, using this methodology to develop an ontology will restrict the collaborative construction and geographical location, plus it will complicate the communication and sharing knowledge in neighboring domains. Also, the methodology does not provide life cycle recommendations. As for identifying concepts, the methodology uses the middle-out strategy, but it provides only some details for its employed techniques and activities.

Table 1. Comparison of methodologies based on the established criteria

Methodologies	Type ofDevel-opment	Collaborative construction	Reusability support	Degree of application dependency	Life cycle recommendation	Strategies for identify-ing con-cepts	Methodology details	Interoperability support
Gruninger and Fox (1995)	Stage based model	No	Yes	Application Semi-dependent	No	Middle-out approach	some details	No
METHONTOLOGY (1997)	Evolving prototype	No	Yes	Application independent	Yes	Middle-out approach	sufficient details	No
NeoN (2008)	Stage based model	Yes	Yes	Application dependent	Yes	Middle-out approach	some details	Yes
BFO (2015)	Evolving prototype	No	Yes	Application independent	Yes	Top-down approach	Some details	Yes
AMOD (2016)	Stage based model	No	No	Application dependent	Yes	Middle-out approach	Some details	No
Krisnadhi, & Hitzler (2016)	Evolving prototype	Yes	Yes	Application semi-dependent	Yes	Middle-out approach	Some details	Yes
Kendall and McGuinness (2019)	Evolving prototype	Yes	Yes	Application independent	No	Developer's consent	Some details	No

Unlike the *Gruninger and Fox*, the *METHONTOLOGY* methodology provides sufficient details regarding the deployed techniques, but it does not allow for collaborative construction and interoperability.

Similar to these methods, the *BFO* method does not allow for collaborative construction as well, though it supports interoperability. The problem; however, with the *BFO* is that it adopts the top-down approach. While the top-down approach is the ideal approach for ontology development, it runs the risk of remaining highly abstract and difficult to be adapted to and reused in specific domains.

In addition to its lack of support for collaborative construction and interoperability, the *AMOD* method also lacks the reusability support. The *AMOD* methodology resembles in many ways *Gruninger and Fox*except for life cycle recommendation and the degree of application dependency.

Kendall and McGuinness methodology stands out from the other methodologies by adopting the developer's consent strategy for identifying concepts. Regardless of being tested, on the whole, it remains vague as to how to define the terminology, collaborative construction, and the tools deployed in ontology development process.

The *NeOn* methodology proposes nine flexible ontology-building scenarios based on the reuse and reengineering of ontological, non-ontological, and ODPs resources. Theoretically, it exceeds greatly the methodology employed in this work, but the main weaknesses of this methodology are: (i) it does not explicitly and sufficiently detail the extraction of terminologies, (ii) and its application time consuming [25].

Krisnadhi & Hitzler methodology shares with the previous two methodologies (*Kendall & McGuinness and NeOn*) the fact that they are all based on worked and tested examples; however, this methodology is very practical and much more detailed in a step-by-step manner-which is lacking in the NeOn methodology-for those who wish to develop an ontology based on ODPs. In fact, the main strength of this methodology is that it relies on ODPs for modeling ontologies, which permits the ontology to be reused and expanded, and prevents it from being too specific.

3 Conclusion

The main factor for a well-developed ontology rests on the choice of the ontology engineering methodology adopted. The successful, however, of any ontology is measured by its adoption in a given community/domain. Several methodologies have been proposed by both academics and professionals since their emergence in the early nineties with Lenat & Guha, 1990. However, none of the methodologies developed so far enjoy consensus in the community. This is primarily due to the lack of some evaluation criteria or/and insufficient details regarding certain criteria. These shortcomings remain, up to this date, the main reason behind the failure of many ontology projects.

Despite all the efforts done in methodology development, ontology engineering methodologies are still a difficult process, and many challenges remain to be solved. According to our review of the most prominent ontology engineering methodologies, the most mature and suitable approach for the development of ontology, especially in a critical domain such as credit risk management, is Krisnadhi & Hitzler methodology; the latter is perceived as being more practical and helpful than others such as NeOn and METHONTOLOGY. Krisnadhi & Hitzler methodology, based on ontology design patterns, ensures that there is a trade-off between interoperability, on the one hand, and over-commitment and conflicting requirements on the other hand. This is not to say though that this methodology is perfect and without limitations. The main weaknesses of this methodology can be summed up thus: (i) it proposes the development of ontologies based only on OPDs, rather than on ontological and non-ontological resources, and (ii) as most other methodologies, it does not document well the extraction of terminologies. Therefore, in our future work we will work with rigor to propose a new ontology engineering methodology.

References

1. Chandrasekaran, B., Josephson, J., Benjamins, V.R.: What are ontologies, and why do we need them? Intell. Syst. Their Appl. IEEE **14**, 20–26 (1999)
2. Studer, R., Benjamins, V.R., Fensel, D.: Knowledge engineering: principles and methods. Data Knowl. Eng. **25**(1), 161–197 (1998)
3. Jean, S., Pierra, G., Ait-Ameur, Y.: Domain ontologies: a database-oriented analysis. In: Filipe, J., Cordeiro, J., Pedrosa, V. (eds.) Web Information Systems and Technologies. LNBIP, vol. 1, pp. 238–254. Springer, Heidelberg (2007). https://doi.org/10.1007/978-3-540-74063-6_19
4. Tamma, V., Dragoni, M., Gonçalves, R., Ławrynowicz, A. (eds.): OWLED 2015. LNCS, vol. 9557. Springer, Cham (2016). https://doi.org/10.1007/978-3-319-33245-1
5. Keet, M.: An Introduction to Ontology Engineering, vol. 1. Maria Keet (2018)
6. Aminu, E.F., Oyefolahan, I.O., Abdullahi, M.B., Salaudeen, M.T.: A review on ontology development methodologies for developing ontological knowledge representation systems for various domains. Int. J. Inf. Eng. Electron. Bus. **12**(2), 28–39 (2020)
7. Subhashini, R., Akilandeswari, J.: A survey on ontology construction methodologies. Int. J. Enterp. Comput. Bus. Syst. **1**(1), 60–72 (2011)
8. Suarez-Figueroa, M.C., et al.: NeOn methodology for building contextualized ontology networks. NeOn Deliv. D **5**, 150 (2008)
9. Iqbal, R., Murad, M.A.A., Mustapha, A., Sharef, N.M.: An analysis of ontology engineering methodologies: a literature review. Res. J. Appl. Sci. Eng. Technol. **6**(16), 2993–3000 (2013)
10. Grüninger, M., Fox, M.S.: Methodology for the design and evaluation of ontologies. Workshop on Basic Ontological Issues in Knowledge Sharing, IJCAI-95, Montreal (1995)
11. Corcho, O., Fernández-López, M., Gómez-Pérez, A.: Methodologies, tools and languages for building ontologies. Where is their meeting point? Data Knowl. Eng. **46**(1), 41–64 (2003)
12. Fernández-López, M., Gómez-Pérez, A., Juristo, N.: Methontology: from ontological art towards ontological engineering (1997)
13. Arp, R., Smith, B., Spear, A.D.: Building Ontologies with Basic Formal Ontology. The MIT Press (2015). https://doi.org/10.7551/mitpress/9780262527811.001.0001
14. Abdelghany, A.S., Darwish, N.R., Hefni, H.A.: An agile methodology for ontology development. Int. J. Intell. Eng. Syst. **12**(2), 170–181 (2019)
15. Peroni, S.: A simplified agile methodology for ontology development. In: Dragoni, M., Poveda-Villalón, M., Jimenez-Ruiz, E. (eds.) OWL: Experiences and Directions – Reasoner Evaluation, pp. 55–69. Springer International Publishing, Cham (2017). https://doi.org/10.1007/978-3-319-54627-8_5
16. Axelsson, J., Papatheocharous, E., Nyfjord, J., Törngren, M.: Notes on agile and safety-critical development. ACM SIGSOFT Softw. Eng. Notes **41**(2), 23–26 (2016)
17. Hitzler, P.: Modeling with ontology design patterns: chess games as a worked example. Ontol Eng. Ontol. Des. Patterns Found. Appl. **25**, 3 (2016)

18. Hitzler, P., Krisnadhi, A.: A Tutorial on Modular Ontology Modeling with Ontology Design Patterns: The cooking recipes ontology. ArXiv Prepr. ArXiv180808433 (2018)
19. Elhassouni, J., El Qadi, A., Bazzi, M., El Haziti, M.: Modeling with ontologies design patterns: credit scorecard as a case study. Indones. J. Electr. Eng. Comput. Sci. **17**(1), 429 (2020)
20. Elhassouni, J., El Qadi, A., El madani El alami, Y., El Haziti, M.: The implementation of credit risk scorecard using ontology design patterns and BCBS 239. Cybern. Inf. Technol. **20**(2), 93–104 (2020)
21. Shimizu, C., Hirt, Q., Hitzler, P.: Modl: A modular Ontology Design Library. ArXiv Prepr. ArXiv190405405 (2019)
22. Noy, N.F., McGuinness, D.L.: Ontology development 101: A guide to creating your first ontology. Stanford knowledge systems laboratory technical report KSL-01-05 and … (2001)
23. Jacobson, I., Christerson, M.J., Jonsson, P., Vergaard, G.: Object-Oriented Software Engineering, A Use Case Driven Approach. Addison-Wesley, Wokingham (1992)
24. Jacobson, I., Spence, I., Bittner, K.: USE-CASE 2.0. The Guide to Succeeding with Use Cases, Ivar Jacobson International sa. (2011)
25. Suárez-Figueroa, M.C., Gómez-Pérez, A., Fernández-López, M.: The NeOn methodology framework: a scenario-based methodology for ontology development. Appl. Ontol. **10**(2), 107–145 (2015)

Conception of an Automatic Decision Support Platform Based on Cross-Sorting Methods and the Fuzzy Logic for General Use

Manal Tamir[1](✉), Raddouane Chiheb[1], Fatima Ouzayd[2], and Kawtar Retmi[3]

[1] ADMIR, ENSIAS, Mohamed V University Rabat, Rabat, Morocco
{manal.tamir,r.chiheb}@um5s.net.ma

[2] Smart Systems Laboratory, ENSIAS, Mohamed V University Rabat, Rabat, Morocco
fatima.ouzayd@um5.ac.ma

[3] CIAM, UM6P Benguerir, Morocco
Kawtar.RETMI@um6p.ma

Abstract. Problems related to the classification of a set of data (methods, tools...), by using the opinions of a committee members, challenge the need for an effective treatment based on the theoretical methods developed through the literature. Indeed, this study was conducted to answer the need identified during the realization of a prioritization study of a graphic modeling methods set in order to develop it into a computing platform. This platform is based on Multi-Criteria Decision Analysis and is destined for practical and general use. The developed platform uses fuzzy pairwise comparisons and cross-sorting methods. It consists of the calculation of two main results: the overall weight of each data to be classified and the consistency index. Regarding the use of fuzzy logic for calculating weights, we opt for the triangular function in the extension of the principle of least-squares logarithmic regression for taking into account the inaccuracy. Finally, the calculation of the normalized weight values can generate an irrational ordering of fuzzy number elements. In our previous work, we have tried to find the conditions on pairwise comparisons values to get rational outcomes, but the calculation was very heavy with the Matlab software. In addition, the modifications to make on the comparisons are multiple, not automatic and depend on the user's appreciation. The present work gives a classification solution of results in order to keep only input data (comparisons) that respect the rational order of the weights values. Then, an optimization algorithm is programmed to maximize the consistency indexes of this part of results, by allowing a limited margin of modification without greatly altering the initial data.

Keywords: Classification · Prioritization study · Computing platform · Multi-criteria decision analysis · General use · Fuzzy pairwise comparisons · Consistency index · Optimization algorithm

1 Introduction

Decision-making is, in most situations, a delicate matter for decision-makers, especially when they have to produce a collaborative decision as part of a group. For example, it

M. Lazaar et al. (Eds.): BDIoT 2021, LNNS 489, pp. 73–81, 2022.
https://doi.org/10.1007/978-3-031-07969-6_6

involves prioritizing different scenarios, choosing between candidates, deciding within the value analysis methods, practicing the budget allocations, ... This study completes and improves our work [1]. In fact, it is a classification study based on the cross sorting methods and the fuzzy pairwise comparisons that has been developed as part of a computer platform. At first, six criteria (Implementation, Relevance, Exploitation, Granularity, Description & Organization (Structure) and Validation) were developed and it has been requested from three experts to fill the comparison matrices concerning criteria and thereafter the six modeling methods by considering the hospital supply chain context. The principle goal of the contribution was to facilitate the selection of the best modeling methods for the hospital supply chain and the best alternative in a general context. After several iterations in [1], it has been concluded that BPMN and Petri Networks methods had the best notations. Otherwise, the developed platform had limitations regarding the normalized weights values that can present an irrational ordering of the fuzzy number's elements [2, 3]. The adopted solution consisted to find the conditions on pairwise comparison values in order to get rational outcomes. However, this technique has shown its limitation because the calculation was very heavy and the modifications to make on the comparisons are multiple, not accurate, not automatic, does not guarantee consistency and depend on the user's appreciation. In order to overcome this, we propose, on the one hand, to improve the platform by automating the calculation as much as possible and optimizing the developed algorithms on Matlab. On the other hand, we propose to calculate the consistency index in order to get results in which experts were on one side consistent with each other and on the other side consistent with themselves. In this context, authors of [4] urged to improve the consistency of the decision group insofar as it does not oppose its own consistency. In fact, the final formula of calculation is inspired from this study [4]. Finally, a classification study is also developed by generating a number of random perturbations which are between -5% and 5% and carried out on the entered data. Thereafter, we create an iterations table that shows whether the order of the fuzzy weights is respected or not: the number of the iteration is affected to column 0 if the order is not respected and to column 1 if it is resected.

Otherwise, the consistency problem of input data interests the majority of researchers dealing with decision support methods. The authors of [4] propose to define a coherent matrix from that supplied by the decision makers and to compare the values of similar cells of the two matrices by calculating their ratio and seeing if it is less than or greater than 1. Finally, depending on this result, the facilitator suggests either increasing or decreasing the values of the comparison cells. Other works suggest methods of supporting decision-makers during the filling of the matrices in order to guarantee a certain consistency between the members of the group. In our study, we use the Simulated Annealing method in order to maximize the consistency index of data that gives a correct order of the fuzzy weights.

The remainder of this paper is structured as follows: Sect. 2 is dedicated to a reminder of the formulas for calculating weights in fuzzy numbers. In Sect. 3, a literature review of the consistency index will be given as well as the calculation method. Thereafter, in Sect. 4, we give an overview about the adopted optimization algorithm and its adaptation to our study case. Finally, Sect. 5 is devoted to the presentation of the algorithms simulation results on Matlab.

2 The Calculation Algorithm

To determine the final scores of the studied modeling methods considering the six performance criteria of decision, it is opted for the fuzzy multi criteria method proposed by [5] and modified by [4]. The choice of this method was not made arbitrarily but was based on the originality of his theory in terms of taking into account the inaccuracy in spite of the extensions which have been proposed later and which merely adopt other logics which also have their limitations or sometimes violate the assumptions of validity of the initial approach (for example the adoption of FWA algorithm in deterministic methods [4]).

In order to attack the allocation of triangular fuzzy pairwise comparisons, a decision committee of 3 members is defined A1, A2, A3.

The method will be applied in three main phases: Firstly, fuzzy weights $\tilde{\alpha}_{\mathrm{i}} = (\alpha_{\mathrm{il}}, \alpha_{\mathrm{im}}, \alpha_{\mathrm{iu}})$ i = 1,…,m will be assigned to the performance criteria of decision based on fuzzy pairwise comparisons $\widetilde{\mathrm{r_{ijk}}} = \left(\mathrm{r_{ijkl}}, \mathrm{r_{ijkm}}, \mathrm{r_{ijku}}\right)$ (i, j = 1,…,m) given by the committee members k = 1,2,3. Secondly, fuzzy weights $\widetilde{\beta_{\mathrm{ij}}} = \left(\beta_{\mathrm{ijl}}, \beta_{\mathrm{ijm}}, \beta_{\mathrm{iju}}\right)$ (i, j = 1, …, n) will be estimated for modeling methods under each one of the criteria separately. Lastly, the final scores of methods ω_{j} j = 1,…,n are calculated by the aggregation of the calculated weights according to the formula below:

$$\omega_j = \sum\nolimits_{i=1}^{m} \alpha_i \beta_{ij},\ j = 1, \ldots \ldots, n \tag{1}$$

In fact, the weights will be estimated by minimizing a logarithmic regression function as shown in the formula (1) below and the fuzzy weights are deduced by the following developed formulas (2, 3, 4, 5):

$$\begin{aligned} &\sum\nolimits_{i=1}^{m} \sum\nolimits_{j=i+1}^{m} \sum\nolimits_{AD_{ij}} \left\{ln\left(r_{ijkl}\right) - ln(\alpha_{il}) + ln\left(\alpha_{ju}\right)\right\}^2 \\ &+\left\{ln\left(r_{ijkm}\right) - ln(\alpha_{im}) + ln\left(\alpha_{jm}\right)\right\}^2 + \left\{ln\left(r_{ijku}\right) - ln(\alpha_{iu}) + ln\left(\alpha_{jl}\right)\right\}^2, \end{aligned} \tag{2}$$

$$ln(\alpha_i) \sum_{j \neq i}^{m} \delta_{ij} - \sum_{j \neq i}^{m} \delta_{ij} ln\left(\alpha_j\right) = \sum_{j \neq i}^{m} \sum_{k \in D_{ij}} ln\left(r_{ijk}\right)\ i = 1, \ldots \ldots, m \tag{3}$$

$$ln(\alpha_{il}) \sum\nolimits_{j \neq i}^{m} \delta_{ij} - \sum\nolimits_{j \neq i}^{m} \delta_{ij} ln\left(\alpha_{ju}\right) = \sum\nolimits_{j \neq i}^{m} \sum\nolimits_{k \in D_{ij}}^{m} ln\left(r_{ijkl}\right)\ i = 1, \ldots \ldots, m \tag{4}$$

$$ln(\alpha_{im}) \sum\nolimits_{j \neq i}^{m} \delta_{ij} - \sum\nolimits_{j \neq i}^{m} \delta_{ij} ln\left(\alpha_{jm}\right) = \sum\nolimits_{j \neq i}^{m} \sum\nolimits_{k \in D_{ij}}^{m} ln\left(r_{ijkm}\right)\ i = 1, \ldots \ldots, m \tag{5}$$

$$ln(\alpha_{iu}) \sum\nolimits_{j \neq i}^{m} \delta_{ij} - \sum\nolimits_{j \neq i}^{m} \delta_{ij} ln\left(\alpha_{jl}\right) = \sum\nolimits_{j \neq i}^{m} \sum\nolimits_{k \in D_{ij}}^{m} ln\left(r_{ijku}\right)\ i = 1, \ldots \ldots, m \tag{6}$$

$$\begin{aligned} &\tilde{\alpha}_i = (aexp(x_{il}), bexp(x_{im}), aexp(x_{iu}))i = 1, \ldots, m, \\ &x_{il} = ln(\alpha_{il})\, x_{im} = ln(\alpha_{im})\, x_{iu} = ln(\alpha_{iu}) \end{aligned} \tag{7}$$

$$a = \frac{1}{\left(\sum_{i=1}^{m} exp(x_{il}) \sum_{i=1}^{m} exp(x_{iu})\right)^{\frac{1}{2}}},\ b = \frac{1}{\sum_{i=1}^{m} exp(x_{im})} \tag{8}$$

$$\tilde{\alpha}_i = \begin{pmatrix} \alpha_{il} \\ \alpha_{im} \\ \alpha_{iu} \end{pmatrix} = \begin{bmatrix} \frac{exp(x_{il})}{(\sum_{j=1}^{m} exp(x_{jl}) \sum_{j=1}^{m} exp(x_{ju}))^{\frac{1}{2}}} \\ \frac{exp(x_{im})}{\sum_{j=1}^{m} exp(x_{jm})} \\ \frac{exp(x_{iu})}{(\sum_{j=1}^{m} exp(x_{jl}) \sum_{j=1}^{m} exp(x_{ju}))^{\frac{1}{2}}} \end{bmatrix} \quad i = 1, .., m \tag{9}$$

3 The Consistency Index

In order to assess the consistency of the input data, two indicators will be calculated at two levels [4]: the indicator of consistency of data for each member separately (indicator of consistency of the member) and the indicator of consistency of the group data (indicator group consistency). Moreover, high values of these indicators ensure that the results obtained are based on consistent data. To do this, an optimization algorithm (simulated annealing), whose objective function is to maximize the consistency indicator, will be developed thereafter.

The formula for calculating the consistency indicator to be used is as follows:

$$I.C._{Genericcorrected}^{a,b} = \frac{1}{n^2}\left(\left(n^2 - n\right)I.C._{Regressioncorrected}^{a} + \sum_{i=1}^{n} I.C._{Diagonalcorrected,i}^{b}\right) \tag{10}$$

with: a ϵ {0, 1, 2, 3, 4} and b ϵ {0, 1}.

We will opt for the following correction case: $I.C._{Genericcorrected}^{0,1}$ because we consider that any inconsistency penalizes our consistency indicator, in particular the following two types of errors:

- The calibration errors of an opinion r_{ijk} with its symmetric r_{jik} are not tolerated because we have opted for reciprocal comparison matrices.
- Pure errors relating to multiple opinions associated with the same r_{ij} comparison are also not tolerated.

Whence:

$$I.C._{Regressioncorrected}^{a} = I.C._{Regressioncorrected}^{0} = I.C._{Regressioncorrected}$$

For $v_{\in} = v_{\in 0}$ and: $I.C._{Diagonalcorrected,i}^{1} = 1$.
With:

$$I.C._{Regressioncorrected} = \frac{\sum_{i=1}^{n-1} ln(\alpha_i)(\sum_{j\neq i}^{n} \sum_{k=1}^{d}(\alpha'_{ijk} ln(r_{ijk}) - \alpha'_{ijk} ln(r_{jik})))}{\sum_{i=1}^{n-1} \sum_{j>i}^{n} \sum_{k=1}^{d}(\left(\alpha'_{ijk} ln(r_{ijk})\right)^2 + \left(\alpha'_{ijk} ln(r_{jik})\right)^2) - v_{\in}} \tag{11}$$

i = 1, 2,..., n, $ln(\alpha_n) = 0$.
And: $v_{\in} = v_{\in 0} = 0$.

Using fuzzy logic, we get:

$$I.C._{Regressioncorrected} = (I.C._l, I.C._m, I.C._u)$$

$$I.C._l = \frac{\sum_{i=1}^{n-1} \ln(\alpha_{il})(\sum_{j\neq i}^{n} \sum_{k=1}^{d} (\alpha'_{ijk}\ln(r_{ijkl}) - \alpha'_{ijk}\ln(r_{jikl})))}{\sum_{i=1}^{n-1} \sum_{j>i}^{n} \sum_{k=1}^{d}((\alpha'_{ijk}\ln(r_{ijkl}))^2 + (\alpha'_{ijk}\ln(r_{jikl}))^2)} \quad (12)$$

$$I.C._m = \frac{\sum_{i=1}^{n-1} \ln(\alpha_{im})(\sum_{j\neq i}^{n} \sum_{k=1}^{d} (\alpha'_{ijk}\ln(r_{ijkm}) - \alpha'_{ijk}\ln(r_{jikm})))}{\sum_{i=1}^{n-1} \sum_{j>i}^{n} \sum_{k=1}^{d}((\alpha'_{ijk}\ln(r_{ijkm}))^2 + (\alpha'_{ijk}\ln(r_{jikm}))^2)} \quad (13)$$

$$I.C._u = \frac{\sum_{i=1}^{n-1} \ln(\alpha_{iu})(\sum_{j\neq i}^{n} \sum_{k=1}^{d} (\alpha'_{ijk}\ln(r_{ijku}) - \alpha'_{ijk}\ln(r_{jiku})))}{\sum_{i=1}^{n-1} \sum_{j>i}^{n} \sum_{k=1}^{d}((\alpha'_{ijk}\ln(r_{ijku}))^2 + (\alpha'_{ijk}\ln(r_{jiku}))^2)} \quad (14)$$

with:

$$r_{jikl} = \frac{1}{r_{ijku}};\ r_{jikm} = \frac{1}{r_{ijkm}};\ r_{jiku} = \frac{1}{r_{ijkl}}$$

n: Number of modeling methods;
d: Number of committee members;

According to the formula, the consistency index $I.C._i$ depends on the input values of r_{ijk}, with i = 1,…,n; j = 1, …, n; k = 1, …,d.

4 Optimization Algorithm

4.1 Literature Review

Optimization issues have been addressed for a long time and several techniques are used to do this. In fact, researchers in different fields have formulated their own optimization models according to the nature of their needs. In general, it is about solving analytically or numerically a model based on one or a set of variables to be minimized or maximized. Also, the quality of the results depends on the choice of these variables, the relevance of the model as well as the technical or computer support intended for the resolution. According to [6], there are different types of optimization problems namely: continuous optimization versus discrete optimization, unconstrained optimization versus constrained optimization, deterministic optimization versus stochastic optimization and single or multi-objective optimization. Furthermore, the author of [7] discussed the functionality of optimization methods in the sense that they treat optimization as part of the global or local search. Moreover, the optimization method has a global component (also called explorer), which aims to explore the search space, and a local component (called exploiter), which allows to exploit the resulting data (relationship between x and f (x)). In other words, the performance of the optimization method increases by integrating local search after global search and vice versa; through the probabilized re-initialization (depending on the objective considered) of the local search by considering the starting points and the convergences performed at each operation. Therefore, the optimization method that we are going to use in this work, and which aims to improve the consistency of the group through the maximization of the consistency indicator, will focus on the search for the global maximum without however falling in a local maximum, not being able to look for other solutions and thus limiting our optimization process.

4.2 Simulated Annealing: Algorithm Description

4.2.1 General Algorithm

The simulated annealing method is an optimization algorithm inspired mainly by thermodynamics. It reflects the slow cooling process of a metal heated previously and maintained at a high temperature in order to change its characteristics. This process is repeated as many times as required for the goal set. On the other hand, the main feature of this method is that it is developed in such a way that it avoids local minimums and searches for the global minimum. The basic algorithm of the simulated annealing method is based on the following steps:

Step 1: Definition of the problem: It is about defining the initial field of work.
Step 2: Selection of the parameters of the algorithm: This step consists in fixing the number of input data (starting points) for which it is necessary to calculate the energy (objective function), the initial temperature, the minimum temperature or the maximum number of iterations and the degree of temperature decrease by ensuring its slowness.
Step 3: System energy measurement: As its name suggests, this phase is devoted to the calculation of the total energy from the initiated parameters.
Step 4: Creation of fluctuations in the system: This function allows you to create disturbances in relation to the starting points.
Step 5: Implementation of the algorithm: This is the step of calculating the energies of the disturbances and their comparison with the previous results to find the optimal solution.
Step 6: The law of cooling: In this calculation step, it is imperative to run the temperature variation equation.
Step 7: Regeneration of the compute loop.

4.2.2 The Adapted Algorithm

The integration of the simulated annealing algorithm in the initial program (see [1]), which permit to calculate the weights and the consistency indicator, is done in order to optimize the group consistency indicator. Moreover, this implementation is characterized by the creation of the following new functions:

- **Recuit_simule_mod/ Recuit_simule_mod/ Recuit_simule_mod:** Those give results as fuzzy values of the optimized cosistency indicator.
- **Perturbation_recuit_criteres et perturbation_recuit_candidats:** These two intermediate functions have for objective the realization of the disturbances compared to the initial data. (step 4)

5 Simulation and Results

As already mentioned in the introduction, a problem linked to the order of the lower, medium and upper elements of the results in fuzzy numbers was observed in the study [1]. In order to overcome this problem, we generate 1000 iterations by making disturbances on the starting comparisons (between −5% and 5%). Afterwards, we create a two column table that shows whether the order of the fuzzy weights is respected or not: the number of the iteration is affected to column 0 if the order is not respected and to column 1 if it is resected (see Figs. 1 and 2). Disruptions are made first on the entries for the criteria. If the order is respected in the criteria weights for the iteration, the weights are calculated for the candidates and the iteration number is recorded in column 1 if both orders are respected. Otherwise, it is recorded in column 0. Finally, the consistency indicator is calculated for each iteration that respects the order (see Fig. 3) and the maximum value of this indicator will be displayed and used as initial data for our simulated annealing program.(see Figs. 4 and 5).

```
les poids finaux du candidat 1 par rapport aux critères est :
le poids final lower du candidat 1 par rapport aux critères est : 0.022295
le poids final modal du candidat 1 par rapport aux critères est : 0.247179
le poids final upper du candidat 1 par rapport aux critères est : 0.149796
-----------
les poids finaux du candidat 2 par rapport aux critères est :
le poids final lower du candidat 2 par rapport aux critères est : 0.093577
le poids final modal du candidat 2 par rapport aux critères est : 0.740130
le poids final upper du candidat 2 par rapport aux critères est : 0.660910
-----------
les poids finaux du candidat 3 par rapport aux critères est :
le poids final lower du candidat 3 par rapport aux critères est : 1.504261
le poids final modal du candidat 3 par rapport aux critères est : 1.937097
le poids final upper du candidat 3 par rapport aux critères est : 2.971984
*******************************************
BETA final du candidat 1 :
    0.0131    0.1233    0.0426

BETA final du candidat 2 :
    0.0565    0.3786    0.1949

BETA final du candidat 3 :
     1     1     1

-----------
les poids finaux du candidat 1 par rapport aux critères est :
le poids final lower du candidat 1 par rapport aux critères est : 0.019735
le poids final modal du candidat 1 par rapport aux critères est : 0.238936
le poids final upper du candidat 1 par rapport aux critères est : 0.126692
-----------
les poids finaux du candidat 2 par rapport aux critères est :
le poids final lower du candidat 2 par rapport aux critères est : 0.084989
le poids final modal du candidat 2 par rapport aux critères est : 0.733398
le poids final upper du candidat 2 par rapport aux critères est : 0.579309
```

Fig. 1. Impact of disturbances on weights

```
     0        391
     0        392
     0        393
     0        394
     0        395
     0        396
   397          0
   398          0
   399          0
   400          0
   401          0
```

Fig. 2. Extract of the iteration table

```
Les indicateurs de coherence de iteration  140 respectant la logique des nombres flous est
Notre indicateur de coherence lower est : 0.160722
Notre indicateur de coherence modal est : 0.428775
Notre indicateur de coherence upper est : 0.905606
Les indicateurs de coherence de iteration  141 respectant la logique des nombres flous est
Notre indicateur de coherence lower est : 0.172765
Notre indicateur de coherence modal est : 0.438231
Notre indicateur de coherence upper est : 0.940638
Les indicateurs de coherence de iteration  142 respectant la logique des nombres flous est
Notre indicateur de coherence lower est : 0.174152
Notre indicateur de coherence modal est : 0.442868
Notre indicateur de coherence upper est : 0.933563
Les indicateurs de coherence de iteration  143 respectant la logique des nombres flous est
Notre indicateur de coherence lower est : 0.161544
Notre indicateur de coherence modal est : 0.441770
Notre indicateur de coherence upper est : 0.944210
Les indicateurs de coherence de iteration  144 respectant la logique des nombres flous est
Notre indicateur de coherence lower est : 0.170182
Notre indicateur de coherence modal est : 0.448359
Notre indicateur de coherence upper est : 0.963776
Les indicateurs de coherence de iteration  145 respectant la logique des nombres flous est
Notre indicateur de coherence lower est : 0.183234
Notre indicateur de coherence modal est : 0.453624
Notre indicateur de coherence upper est : 0.946274
```

Fig. 3. Extract of the calculated consistency index of some iteration

```
Iteration ayant indicateur de coherence maximal est :303

Son indicateur de coherence lower est : 0.356607
Son indicateur de coherence modal est : 0.590968
Son indicateur de coherence upper est : 0.597267
```

Fig. 4. The maximum consistency index

```
Son indicateur de coherence maximise est :
Son indicateur de coherence lower maximise est : 0.359453
Son indicateur de coherence modal maximise est : 0.590968
Son indicateur de coherence upper maximise est : 0.648288
```

Fig. 5. The optimized consistency index

The results above show that the problem of the irrational order of the results fuzzy numbers was resolved through the applied disturbance on the 397 iteration without really altering the starting comparisons. By comparing the present realization with what was carried out in the work [1], we can confirm that the heaviness of the calculation and the difficulty of precisely determining the comparison to modify with the most adequate value have been exceeded and corrected through automatic disturbances which are faithful to the input data. Finally, the application of the Simulated Annealing Algorithm, an approach that has not been studied before, showed a good performance in relation to the maximization of the group consistency indicator.

6 Conclusion

The proposed approach brings major complements to our decision support platform. Indeed, the problem of the irrational order of the elements lower, medium and upper of the results in fuzzy numbers has been corrected through very small automatic corrections of the comparisons entered initially by the members of the committee. This contribution saved the effort of calculating the conditions on the initial comparisons. Then, the effectiveness of the results was improved through the calculation of the group consistency indicator and its optimization to take into account the results calculated from the inputs for which the consistency indicator is maximum. This second contribution is a considerable advance for our decision support tool intended for general use. Moreover, the implementation of this platform was carried out on Matlab and the number of iterations is chosen so as not to burden the calculation process considering the capacity of our machine. Towards the end and in order to seek better performance, an artificial intelligence model is planned in order to exploit the input data that gives good results in relation to the order and the consistency index (the one which is greater than a value that we fixe).

References

1. Tamir, M., Chiheb, R., Ouzayd, F.: A decision support platform based on cross-sorting methods for the selection of modeling methods. Int. J. Adv. Comp. Sci. Appli. (IJACSA) **9**(10) (2018)
2. Gogus, O., Boucher, T.O.: A consistency test for rational weights in multi-criterion decision analysis with fuzzy pairwise comparisons. Fuzzy Sets and Systems **86**, 129–138 (1997)
3. Boender, C.G.E., de Graan, J.G., Lootsma, F.A.: Multi-criteria decision analysis with fuzzy pairwise comparisons. Elsevier Seience Publishers B.V., North-Holland. 0165–0114/89/$3.50. 1989
4. Limayem, F. : Modèles de pondération par les méthodes de tri croisé pour l'aide à la décision collaborative en projet. Sciences de l'ingénieur [physics]. Ecole Centrale Paris, Français (2001). <tel-00011948>
5. Van Laarhoven, P.J.M., Pedrycz, W.: A fuzzy extension of Saaty's priority theory. Fuzzy Sets Syst. **11**, 229–241 (1983)
6. Neos Guide: https://neos-guide.org/
7. Luersen, M.A.: GBNM: un algorithme d'optimisation par recherche directe. Application à la conception de monopalmes de nage. Modélisation et simulation. INSA de Rouen, Français (1993). ffNNT : 2004ISAM0014ff. fftel-00850658f

Cyber Security

Classification of URLs Using N-gram Machine Learning Approach

Abdelali Elkouay[1(✉)], Najem Moussa[2], and Abdellah Madani[1]

[1] LAROSERI, Department of Computer Science, University of Chouaib Doukkali, Faculty of Science EL Jadida, El Jadida, Morocco
elkouayabdelali@gmail.com, madani.a@ucd.ac.ma

[2] Faculty of Sciences, Mohammed V University in Rabat, Rabat, Morocco

Abstract. Nowadays, the internet is growing so rapidly in a lightning way that changes our daily behaviors, from online shopping, online learning to online banking and more activities that make our lives easier. However, using such of ways imposed sharing personal informations such as email, password, credit card information etc. Cybercriminals try to find their victims in the cyberspace by tricking the user using the anonymous structure of the internet. Cybercriminals set out new techniques such as phishing, to deceive victims with the use of false websites, in order to collect their sensitive informations. Understanding whether a web page is legitimate or phishing is a very challenging problem that requires our attention. In this work, we propose a new model that classify whether a web page is legitimate or phishing, based on URLs natural language processing and by applying the n-gram model. We analyze the model with different machine learning algorithms and our system achieves an accuracy of 96.41% with 97% precision.

Keywords: Cybersecurity · Phishing attack · Machine learning · Text classification · N-gram

1 Introduction

The cyberspace has become a huge container of a lot of our daily activities, in a way that requires an efficient security control; otherwise it will be a huge platform for cyber attacks. URL phishing attack is a kind of cyberattack that is used recently by many attackers and therefore has become a field of attention from many researchers in recent years. According to Anti-Phishing Working Group, the number of detected phishing attacks in the first half of the year 2018 was 496,578, compared to 371,519 in the second part of 2017, resulting in over 33% growth in half a year.

Phishing is one of the efficient ways that even an experienced user can be victim of. It is an attractive technique for phishers who try, by the way of generating a fraudulent URL, to capture some sensitive and personal information of the victim like username, password, credit card information. In a study carried

M. Lazaar et al. (Eds.): BDIoT 2021, LNNS 489, pp. 85–99, 2022.
https://doi.org/10.1007/978-3-031-07969-6_7

out on the experiences of phishing attacks, Volkamer *et al.* [4] formulated five main reasons for Internet users to fall into phishing:

- Users do not have enough knowledge about URL's structure.
- Users are not aware which web pages can be trusted.
- Users don't see the whole address of the web page, due to the redirection or hidden URLs.
- Often, users are not interested in viewing the content of the URL or they do not have enough time to observe the URL pattern.
- User cannot distinguish phishing web pages from the legitimate ones.

In this paper, we are focused on the classification of phishing web page based on the structure of URLs, by applying the N-gram technique with different machine learning algorithms.

The aim of our contribution can be stated as follows:

- we propose a new model that classify whether a web page is legitimate or phishing, based on URLs natural language processing and by applying the n-gram model. We analyze the model with different machine learning algorithms. Our system is trained and tested on a new dataset published recently using the n-gram technique (Unigram, Bigram, combination of Unigram and Bigram).
- Analysing which combination of features (unigram, bigram, unigram+ bigram) will distinguish between classes.
- We present also an analysis that compare the effectiveness of CountVectorizer and TF-IDF which doesn't exist in the state of art.

The structure of the paper is defined as follows: Sect. 3 presents our methodology while Sect. 4 is devoted to give the implementation and results. Finally, Sect. 5 concludes the paper and presents the scope for future work.

2 Related Work

Many researchers based their works on URLs vectorization as features extraction technique in their proposed system.

Choon Lin *et al.* [9] proposed a phishing webpage detection technique called PhishWHO, where the differences between the target and actual identities of a webpage are exploited for classification. Identity keywords are extracted from the textual contents of the webpage, using the proposed novel weighted URL tokens system based on the N-gram model. These keywords are then searched using a search engine, and the returned domain name are compared against the suspicious domain name using the proposed 3-tier identity matching system, including full string matching, ccTLD matching and IP alias matching. An accuracy of 96.10% is achieved. Case-based reasoning phishing detection system(CBR-PDS) is a system proposed by Abutair H et al. [3] that relies on precedent cases in order to detect phishing attacks. CBR-PDS count tow stage hybrid procedure using Information gain and Genetic algorithms. CBR-PDS has been tested on various balanced datasets and different scenarios. The system achieved a suitable performance carrying the proposed hybrid feature selection procedure and the accuracy rates that surpass 95% however,it is time-consuming that require a reduction.

URL classification techniques aim to extract features from URLs to track the phishing URLs. This uses blacklisted words, count and binary based features to check if a suspicious URL is legitimate or not. Wang and Shirley(2015) and Feng et al.(2018) [19,20] techniques use count-based features such as number of characters, '_', numbers, digits. Similarly Moghimi and Varjani (2016) and Abutair et al. (2018) [3,14] use binary features such as presence of IP address, special characters (*, –, /, ?), blacklisted words, brand names and http/https protocol. Other methods use third party-based features to detect suspicious domains more efficiently. These approaches, presented by Feng et al. (2018) [20], use third party services like WHOIS, page rank, search results, and DNS record etc.

Content-based techniques extract features from source code of the suspicious URLs. Features like hyperlinks-based features, text-based features, tag-based features, image based features are used by various approaches. The hyperlink-based features, used by Shirazi et al. (2018), Rao and Pais 2018, Marchal et al. 2017, Jain and Gupta 2018 [11,13,15,21], include ratio of foreign links, broken links, common URLs and null links.

Marchal et al. (2016) [12] used Text-based features consist of extracting prominent keywords from the web page and TF-IDF is one such method to extract important keywords. Tag-based features include extraction of DOM trees from suspicious websites and compared with existing DOM tree databases. The third party services comprehend the use of search engine, page ranking, WHOIS, etc. The techniques such as Varshney et al. (2016), Tan et al. (2016), Chiew et al. (2015) and Jain and Gupta (2017) [8,10,16,17] use search engine results,Rao and Pais (2018) [21] use WHOIS for the detection of phishing sites.

3 Methodology

The methodology of our work consists of three main steps. The first step consists to collect data and processing while the next step is to construct the N-gram approach and to scale the data. The last step is to present the classifier evaluation based on evaluation metrics. We depicted these steps in Fig. 1.

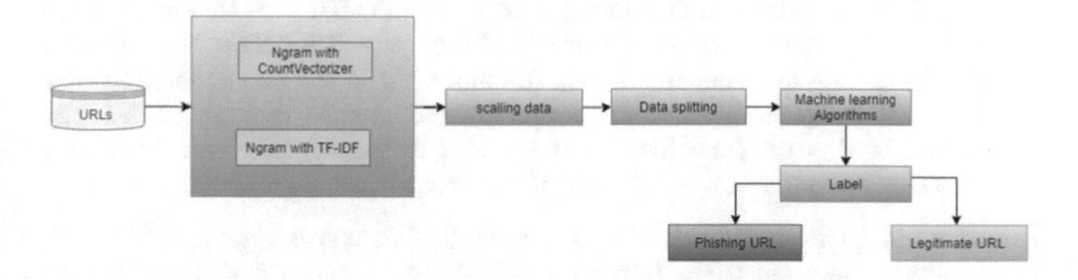

Fig. 1. Approach used in our work

3.1 Step1: Data Collection and Processing

Data Collection: The dataset used has been retrieved from (Ebbu2017 Phishing Dataset, 2017), which contains 73,575 URLs. This dataset totally contains 36.400 legitimate URLs and 37,175 phishing URLs. This dataset is created and published by Sahingoz *et al.* in their phishing URL detection work [2].

Development Environment: The table below shows the environment used to build and run the proposed model

Operating system	Windows 10
Memory	12.0 GB
Processor	Intel core i7
Data set storage location	OS file system
Development tool	Python 3.7
Used libraries	Sklearn, NumPy, pandas, matplotlib

Data Processing: Our dataset first requires significant processing in order to establish a good model. Due to the high computation of data, 6000 URLs been selected, 3000 URLs as legitime and 3000 as phishing. After that, we need to remove all special characters that a URL can contains, such as ("?", "/", ".", "=", "&"). Once all this is done, we can easily transform the textual data into matrix of number.

3.2 Step2: N-gram Approach and Scaling

URLs classification can be taken as binary classification, where each URL $\{d_i\}$ in D = $\{d_1, d_2 \ldots .. d_n\}$ is classified as a label C where C = (legitimate, Phishing). Mainly these URLs are text format, but for classification using machine learning, numerical matrices are required. Thus, the task of conversion of URLs into numerical matrices are made using the following methods:

- CountVectorizer: This method converts the given URL into a matrix of integers. In other words, it helps to generate a sparse matrix of counts.
- Term Frequency-Inverse Document Frequency (TF-IDF): in this method we focus on the importance of a word in the corpus. TF-IDF value increases with increasing in frequency of a particular word in the document. In order to control the generality of more common words, term frequency (TF) is the number of times a particular term appears in the text. Inverse document frequency measures the occurrence of any word in all documents.

In our work, we analyze both of these methods, CountVectorizer and TF-IDF, to transform the given URLs into a numerical vector, which is then considered as input to a supervised machine learning algorithm. WordVector technique has been analyzed with Sahingoz et al. (2019) however, it showed poor classification results with the same dataset .

N-gram model: from a given sequence of text this model check n continuous words which helps to predict the next item in a sequence. Applying such model in our case helps to analyse the sequence component of a given URL. Note that when n=1 we are referring to Unigram and when n = 2 it's referred to Bigram model. As an example, we apply those models on the URL depicted in Fig. 2.

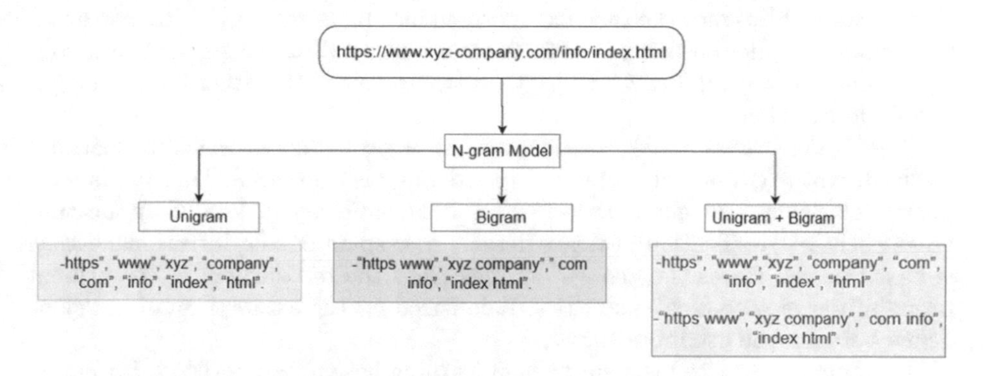

Fig. 2. N-gram model applied on URL

The Standard Scaler: After the processing phase depicted in step 1, now we apply the n-gram model with different variations, unigram when n = 1, bigram when n = 2 and a combination of unigram and bigram. Then we construct the numeric matrix vector using the countVectorizer and TF-IDF. In most cases, dataset contain features with highly varying in magnitudes, units and range. But since, Eucledian distance is often used in machine learning's computation one needs to bring all features to the same level of magnitudes. This can be acheived by feature scaling. In our approach, we used the standardization method which replace the values by their Z-score. This task is made by using "StandardScaler" from "sklearn" which is a module in python.

$$standarization : z = \frac{x - \mu}{\sigma}$$

$$mean : \mu = \frac{1}{N}\sum_{i=1}^{N} x_i, standardDeviation : \sigma = \sqrt{\frac{1}{N}\sum_{i=1}^{N}(x_i - \mu)^2}$$

Step 3: Machine Learning Algorithms: Many algorithms can take the classification task. In our work, we used the same algorithms that showed a good performance in the previous works such as Logistic regression (LR), Support vector Machine (SVM), Naïve Bayes (NB), Random forest (RF), Decision Tree (DT), KNN (k = 3), Adaboost Support vector machine (SVM): in this algorithm, each data item is plot as a point in n-dimensional space, n represent the number of features, the value of each feature being the value of a particular coordinate. In order to classify the data, we try to find the hyperplane that differentiates the two classes very well. The optimal hyperplane, for SVM, is the one that maximizes the margin from both tags. Which means, the hyperplane whose distance to the nearest element of each tag is the largest.

Logistic regression (LG) measures the relationship between the dependent variable, labels or what want to predict, and one or more independent variable (features). By estimating probabilities using its underlying logistic function. These probabilities must be then transformed into binary values in order to actually make a prediction using logistic function, also called the sigmoid function. This values between 0 and 1, will be then transformed into either 0 or 1 using a threshold classifier.

The Naïve Bayes (NB) classification is a probabilistic machine learning method, which is not only straightforward but also powerful. Due to its simplicity, efficiency and good performance, it is preferred in lots of application areas such as classification of texts, detection of spam emails/intrusions, etc. It is based on the Bayes theorem, which describes the relationship of conditional probabilities of statistical quantities. It is based on the assumption of independence between the attribute values.

Random Forest (RF) is a supervised learning ensemble algorithm. Ensemble algorithms are those which combine more than one algorithms of the same or different kind for classifying objects. random forest Classifier constructs a set of decision trees, mostly trained with the "bagging" method which is that a combination of learning models that increases the global result. From a random selected subset of the training set the classifier constructs a set of decision trees and then aggregates the votes from different decision trees in order to decide the class below to the tested object.

Adaboost or adaptive boosting is an iterative ensemble method that combines multiple weak classifiers to increase the accuracy of classifiers. Adaboost assign weights for every observation and train the weak classifier, most often a decision tree. After that ,for every weak classifier and for each observation the weight increase or decrease if the observation is correctly or incorrectly predicted.

The k nearest neighbors (k-NN) algorithm is supervised learning and non-parametric algorithm used for classification and regression. In both cases, it is a matter of classifying the entry in the category to which the k nearest neighbors belong in the space of the characteristics identified by learning. In k-NN classification, the result is a membership class. An input object is classified according to the majority result of the membership class statistics of its k nearest neighbors, (k is a generally small positive integer). If $k = 1$, then the object is assigned to the membership class of its close neighbor.

The Decision Tree classification is one among the favored supervised learning processes that's used not just for classification however additionally for regression tasks. This classifier repetitively divides the training dataset into subparts to spot the separation lines in an exceedingly arboresque structure. Then these lines area unit wont to find the acceptable category for the target item. Every call node splits the info into 2 or a lot of classes consistent with one attribute worth. Every leaf node is appointed to a category (especially by calculative a probability) within the classification rule.

3.3 Performance Evaluation

Our model's performance was evaluated using relevance measures such as accuracy, f-measure, recall, and precision. First of all, we need to construct contingency matrix for the confusion matrix where we depict its components as follows (Table 1):

Table 1. Confusion matrix

	Correct label	
	Positive	Negative
Positive	TP	FP
Negative	FN	TN

$$accuracy: \frac{TP+TN}{TP+TN+FP+FN}, Precision = \frac{TP}{TP+FP}$$

$$Recall: \frac{TP}{TP+FN}, F-score = 2 * \frac{Precsion * Recall}{Precsion + Recall}$$

TP and TN stand for True positive and True negative respectively, which represent the correct classified URLs. However FP and FN stand for false positive and false negative and they refer to the misclassified URLs. By completing the matrix above we can calculate the following performance measures: Accuracy: show the performance of the system, the closer is to 1 the highest the performance is. Precision: in classification problem, closer the precision value to 1 implies that the predicted labels are closer to truth. Recall: refers to the ratio of total relevant results correctly classified by the model. F-score: It is the harmonic mean of precision and recall. It is required to optimize the system towards either precision or recall, which have more influence on final result.

Another metric used to visualize the quality of a classifier which is the Receiver Operating Characteristic (ROC) and known as the ROC Curve. ROC curve is an excellent method of measuring the performance of a classification method. The true positive rate is plotted against false positive rate for different probabilities of a classifier predictions. The larger the area under the curve, the better the model is for classification.

4 Implementation and Results

As mentioned before, we choose many machine learning algorithms that showed a good performance on previous proposed systems. The tables from 1 to 7 summarize the entire results of the 7 machine learning algorithms with countVectorizer and TF-IDF configuration. By training the model on a dataset, we can determine the best parameters for each predictor variable. This can be seen clearly in a linear regression, where the coefficients are determined for each variable used in the model. The coefficients in this case are the parameters they are found via the training process. Hyperparameters on the other hand are parameters that are independent of the training process. Using cross-validation, one will be able to obtain a model that performs the best hyperparameters. We used the GridSearch which is the most basic hyperparameter tuning method. Using this technique, we can build a model for each possible combination of all the hyperparameter values provided. Then, each model is evaluated in order to select the architecture that will produce the best results.

The "unigram" and "unigram bigram" configuration with countVectorizer and TF-IDF separate very well the URLs into phishing and legitimate, with a reasonable F-measure and good precision. The best model is unigram with TF-IDF using the logistic regression with an accuracy of 96.25% and 97% precision. We can see that KNN with n = 3 got the highest precision with 98.81%. However, it is less accurate than the other algorithms used in the analysis. Using SVM, we achieved roughly the same precision (95%) for both "Unigram with CountVectorizer" and "Unigram Bigram" models. With TF-IDF, the difference is very big in terms of precision and F measurement, where the model, "Unigram" with CountVectorizer, is more precise than "Unigram Bigram", with a difference of 3.83% and 0.49% in precision and F-measure respectively. In order to further increase the precision of our "Unigram with CountVectorizer", a GridSearch is applied with 5 cross validations. Figures 4 and 5 summarize the results before and after the GridSearch.

In Fig. 6, we can see a remarkable increase in "bigram" and "unigram bigram" with CountVectorizer 2.58% and 2.41% respectively and a decrease with 0.08% in the "unigram-countVectorizer". Another part of the analysis shows the results of the "unigram bigram with TF-IDF" and shows an increase in precision after the GridSearch as in CountVectorizer. However, it affects the bigram with a decrease of 1.75%. In term of precision we can see an increase specially in "bigram countVectorizer" and "unigram bigram countVectorizer" model with 13% and 6% respectively.

In Fig. 7, we can see that the only increase is achieved with the "unigram bigram TF-IDF". Based on these results, we can conclude that our model achieved a good result in terms of precision and accuracy. We can more emphasize those results with Roc curve as depicted in Fig. 3.

To prove more the performance of our model, we used the Roc curve discussed before. First of all, we calculate the probabilities of classification and True positive and False positive rates. The packages used are sklearn.metrics.roc_curve and sklearn.metrics.auc. As we can see from Fig. 3 that the curve is closer to left-top corner which means that our test is more accurate. Also, the training and testing is made using 5-cross validation, where each fold achieved a 100% of detection as presented in legend in Fig. 3.

Table 1 : Results of RF with CountVectorizer

Method		Confusion matrix		Evaluation Parameter			Accuracy	Training Time (s)	Testing time (s)
Unigram		Correct labels		Precision	Recall	F-mesure	90.91%	31.88011908531189	0.24943113327026367
		Positive	Negative	93%	88%	91%			
	Positive	521	40	89%	93%	91%			
	Negative	69	570						
Bigram		Correct labels		Precision	Recall	F-mesure	86.58%	101.51801753044128	0.5296964645385742
		Positive	Negative	90%	82%	86%			
	Positive	476	54	84%	91%	87%			
	Negative	107	563						
Unigram+Bigram		Correct labels		Precision	Recall	F-mesure	90.33%	83.62547993659973	0.355792760848999
		Positive	Negative	88%	92%	90%			
	Positive	537	70	92%	89%	90%			
	Negative	46	547						

Table 2 : Results of SVM(C=100)with CountVectorizer

Method		Confusion matrix		Evaluation Parameter			Accuracy	Training Time (s)	Testing time (s)
Unigram		Correct labels		Precision	Recall	F-mesure	95%	196.72702431678772	44.613308906555176
		Positive	Negative	94.91%	95.72%	95.31%			
	Positive	560	25	95.90%	95.12%	95.51%			
	Negative	30	585						
Bigram		Correct labels		Precision	Recall	F-mesure	88%	395.6583368778229	95.8100643157959
		Positive	Negative	68.61%	99%	81.05%			
	Positive	446	4	99.35%	77.01%	86.76%			
	Negative	137	613						
Unigram+Bigram		Correct labels		Precision	Recall	F-mesure	95%	592.3106887340546	144.31831336021423
		Positive	Negative	91.08%	98.88%	94.82%			
	Positive	531	6	99.02%	92.15%	95.46%			
	Negative	52	611						

Table 3 : Results of LR(C=100) with CountVectorizer

Method		Confusion matrix		Evaluation Parameter			Accuracy	Training Time (s)	Testing time (s)
Unigram		Correct labels		Precision	Recall	F-mesure	96.00%	1.3122344017028809	0.017995834350585938
		Positive	Negative	96%	96%	96%			
	Positive	567	25	96%	96%	96%			
	Negative	23	585						
Bigram		Correct labels		Precision	Recall	F-mesure	89.08%	1.535097360610962	0.03598523139953613
		Positive	Negative	82%	100%	90%			
	Positive	581	129	100%	79%	88%			
	Negative	2	488						
Unigram+Bigram		Correct labels		Precision	Recall	F-mesure	94.41%	2.5456299781799316	0.0700995922088623
		Positive	Negative	90%	99%	95%			
	Positive	579	63	99%	90%	95%			
	Negative	4	554						

Table 4 : Results of NB with CountVectorizer

Method		Confusion matrix		Evaluation Parameter			Accuracy	Training Time (s)	Testing time (s)
Unigram		Correct labels		Precision	Recall	F-mesure	92.25%	0.8786230087280273	1.1424717903137207
		Positive	Negative	87.62%	96.27%	91.74%			
	Positive	517	20	96.72%	88.98%	92.69%			
	Negative	73	590						
Bigram		Correct labels		Precision	Recall	F-mesure	86.92%	1.6810367107391357	2.1817500591278076
		Positive	Negative	73.75%	99.07%	84.56%			
	Positive	430	4	99.35%	80.02%	88.64%			
	Negative	153	613						
Unigram+Bigram		Correct labels		Precision	Recall	F-mesure	93.33%	2.3806612491607666	3.1162543296813965
		Positive	Negative	89.19%	96.83%	92.85%			
	Positive	520	17	97.24%	90.49%	93.75%			
	Negative	63	600						

Table 5 : Results of KNN(K=3)with CountVectorizer

Method		Confusion matrix		Evaluation Parameter			Accuracy	Training Time (s)	Testing time (s)
Unigram		Correct labels		Precision	Recall	F-mesure	79.50%	13.835811614990234	89.53878879547119
		Positive	Negative	98.81%	71.18%	82.75%			
	Positive	583	236	61.31%	98.16%	75.47%			
	Negative	7	374						
Bigram		Correct labels		Precision	Recall	F-mesure	61.75%	28.037535905838013	192.30340385437012
		Positive	Negative	100%	55.95%	71.75%			
	Positive	583	459	25.60%	100%	40.77%			
	Negative	0	158						
Unigram+Bigram		Correct labels		Precision	Recall	F-mesure		43.34791111946106	288.5707416534424
		Positive	Negative	100%	56.60%	72.28%	62.75%		
	Positive	583	447	27.55%	100%	43.20%			
	Negative	0	170						

Table 6 : Results of Decision Tree with CountVectorizer

Method		Confusion matrix		Evaluation Parameter			Accuracy	Training Time (s)	Testing time (s)
Unigram		Correct labels		Precision	Recall	F-mesure	90.25%	14.501118183135986	0.03697848320007324
		Positive	Negative	87.28%	92.45%	89.79%			
	Positive	515	42	93.11%	88.33%	90.66%			
	Negative	75	568						
Bigram		Correct labels		Precision	Recall	F-mesure	86.91%	79.67367911338806	0.07095909118652344
		Positive	Negative	81.98%	90.18%	85.89%			
	Positive	478	52	91.57%	84.32%	87.80%			
	Negative	105	565						
Unigram+Bigram		Correct labels		Precision	Recall	F-mesure	91.50%	40.37585663795471	0.1110849380493164
		Positive	Negative	92.10%	90.55%	91.32%			
	Positive	537	56	90.92%	92.42%	91.66%			
	Negative	46	561						

Table 7 : Results of Adaboost with CountVectorizer

Method		Confusion matrix		Evaluation Parameter			Accuracy	Training Time (s)	Testing time (s)
Unigram		Correct labels		Precision	Recall	F-mesure	85.75%	50.15170383453369	1.7989730834960938
		Positive	Negative	81.69%	88.44%	84.93%			
	Positive	482	63	89.67%	83.51%	86.48%			
	Negative	108	547						
Bigram		Correct labels		Precision	Recall	F-mesure	79.00%	133.9463279247284	3.404043436050415
		Positive	Negative	69.98%	84.12%	76.40%			
	Positive	408	77	87.52%	75.52%	81.08%			
	Negative	175	540						
Unigram+Bigram		Correct labels		Precision	Recall	F-mesure	86.50%	249.2019031047821	5.544311046600342
		Positive	Negative	85.42%	86.60%	86.01%			
	Positive	489	77	87.52%	86.40%	86.95%			
	Negative	85	540						

Table 1 : Results of RF with TF-IDF

Method		Confusion matrix		Evaluation Parameter			Accuracy	Training Time (s)	Testing time (s)
Unigram		Correct labels		Precision	Recall	F-mesure	90.75%	29.198968410491943	0.20491790771484375
		Positive	Negative	91%	90%	91%			
	Positive	529	50	90%	92%	91%			
	Negative	61	560						
Bigram		Correct labels		Precision	Recall	F-mesure	85.16%	107.49537539482117	0.5567116737365723
		Positive	Negative	89%	79%	84%			
	Positive	462	57	82%	91%	86%			
	Negative	121	560						
Unigram+Bigram		Correct labels		Precision	Recall	F-mesure	90%	79.36006426811218	0.3428030014038086
		Positive	Negative	88%	92%	90%			
	Positive	534	71	92%	88%	90%			
	Negative	49	546						

Table 2 : Results of SVM(C=100)with TF-IDF

Method		Confusion matrix		Evaluation Parameter			Accuracy	Training Time (s)	Testing time (s)
Unigram		Correct labels		Precision	Recall	F-mesure	93%	197.12814283370972	44.670459270477295
		Positive	Negative	88.64%	97.21%	92.73%			
	Positive	523	15	97.54%	89.87%	93.55%			
	Negative	67	595						
Bigram		Correct labels		Precision	Recall	F-mesure	84%	395.75115609169006	108.77010869979858
		Positive	Negative	68.61%	99%	81.05%			
	Positive	400	4	99.35%	77.01%	86?76%			
	Negative	183	613						
Unigram+Bigram		Correct labels		Precision	Recall	F-mesure	92%	634.3117344379425	143.63499760627747
		Positive	Negative	83.19%	99.79%	90.73%			
	Positive	485	1	99.83%	86.27%	92.56%			
	Negative	98	616						

Table 3 : Results of LG(C=100) with TF-IDF

Method		Confusion matrix		Evaluation Parameter			Accuracy	Training Time (s)	Testing time (s)
Unigram		Correct labels		Precision	Recall	F-mesure	96.25%	1.3334345817565918	0.018861770629882812
		Positive	Negative	97%	96%	96%			
	Positive	565	20	96%	97%	96%			
	Negative	25	590						
Bigram		Correct labels		Precision	Recall	F-mesure	93.41%	1.6547000408172607	0.03499150276184082
		Positive	Negative	96%	91%	93%			
	Positive	528	24	92%	96%	94%			
	Negative	55	593						
Unigram+Bigram		Correct labels		Precision	Recall	F-mesure	96.16%	2.4446465969085693	0.0540766716003418
		Positive	Negative	96%	97%	96%			
	Positive	563	26	97%	96%	96%			
	Negative	20	591						

Table 4 : Results of NB with TF-IDF

Method		Confusion matrix		Evaluation Parameter			Accuracy	Training Time (s)	Testing time (s)
Unigram		Correct labels		Precision	Recall	F-mesure	91.42%	0.7545666694641113	1.0204179286956787
		Positive	Negative	87.28%	94.84%	90.90%			
	Positive	515	28	95.40%	88.58%	91.87%			
	Negative	75	582						
Bigram		Correct labels		Precision	Recall	F-mesure	87.00%	1.7413771152496338	2.3287765979766846
		Positive	Negative	73.92%	99.08%	84.67%			
	Positive	431	4	99.35%	80.13%	88.71%			
	Negative	152	613						
Unigram+Bigram		Correct labels		Precision	Recall	F-mesure	92.83%	2.606539011001587	3.410043954849243
		Positive	Negative	89.19%	95.46%	92.36%			
	Positive	520	23	96.27%	90.41%	93.24%			
	Negative	63	594						

Table 5 : Results of KNN(K=3) with TF-IDF

Method		Confusion matrix		Evaluation Parameter			Accuracy	Training Time (s)	Testing time (s)
Unigram		Correct labels		Precision	Recall	F-mesure	66.91%	14.800068378448486	94.62715935707092
		Positive	Negative	97.42%	59.78%	74.10%			
	Positive	568	382	38.08%	94%	54.20%			
	Negative	15	235						
Bigram		Correct labels		Precision	Recall	F-mesure	66.91%	28.267566680908203	206.47737336158752
		Positive	Negative	100%	56.05%	71.84%			
	Positive	583	457	25.93%	100%	41.18%			
	Negative	0	160						
Unigram+Bigram		Correct labels		Precision	Recall	F-mesure	62.75%	39.93971920013428	268.1403543949127
		Positive	Negative	100%	56.60%	72.28%			
	Positive	583	447	27.55%	100%	43.20%			
	Negative	0	170						

Table 6 : Results of Decision Tree with TF-IDF

Method		Confusion matrix		Evaluation Parameter			Accuracy	Training Time (s)	Testing time (s)
Unigram		Correct labels		Precision	Recall	F-mesure	88.58%	7.833627939224243	0.031981706619262695
		Positive	Negative	90.90%	86.31%	88.55%			
	Positive	530	84	86.38%	90.95%	88.61%			
	Negative	53	533						
Bigram		Correct labels		Precision	Recall	F-mesure	86.08%	71.31711292266846	0.06196475028991699
		Positive	Negative	80.78%	89.54%	84.94%			
	Positive	470	59	91.08%	83.38%	87.06%			
	Negative	113	558						
Unigram+Bigram		Correct labels		Precision	Recall	F-mesure	88.83%	20.267635345458984	0.10693717002868652
		Positive	Negative	92.79%	85.46%	88.98%			
	Positive	541	92	85.08%	92.59%	88.86%			
	Negative	42	525						

Table 7 : Results of Adaboost with TF-IDF

Method		Confusion matrix		Evaluation Parameter			Accuracy	Testing Time (s)	Testing time (s)
Unigram		Correct labels		Precision	Recall	F-mesure	86.16%	49.378706216812134	1.685487985610962
		Positive	Negative	85.76%	85.76%	85.76%			
	Positive	500	83	86.54%	86.54%	86.54%			
	Negative	83	534						
Bigram		Correct labels		Precision	Recall	F-mesure	78.75%	138.8884792327881	3.3054513931274414
		Positive	Negative	69.63%	83.88%	76.10%			
	Positive	406	78	87.35%	75.25%	80.87%			
	Negative	177	539						
Unigram+Bigram		Correct labels		Precision	Recall	F-mesure	86.66%	247.05723404884338	5.249084711074829
		Positive	Negative	86.44%	86.15%	86.30%			
	Positive	504	81	86.87%	87.15%	87.01%			
	Negative	79	536						

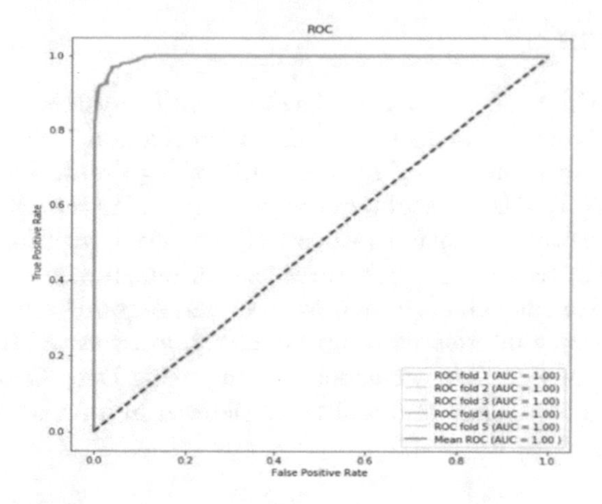

Fig. 3. Roc curve of Logistic regression (LR)

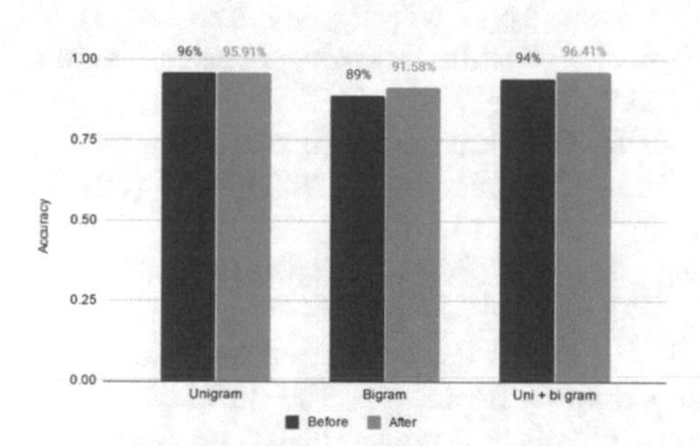

Fig. 4. CountVectorizer Grid Search accuracy of LR

Fig. 5. TF-IDF Grid Search accuracy of LR

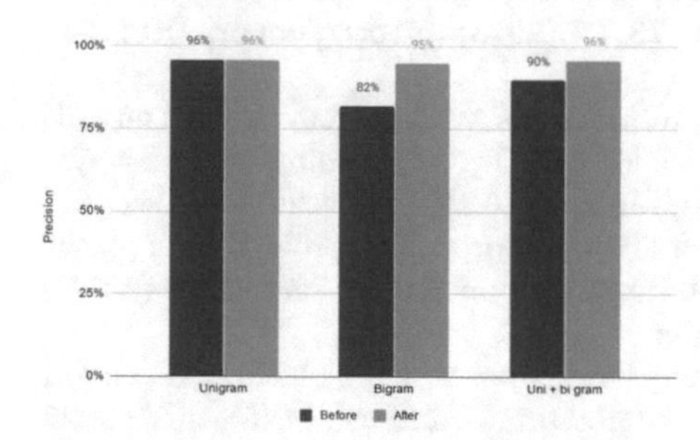

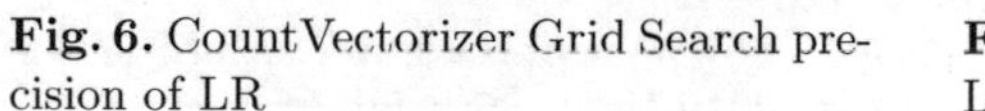

Fig. 6. CountVectorizer Grid Search precision of LR

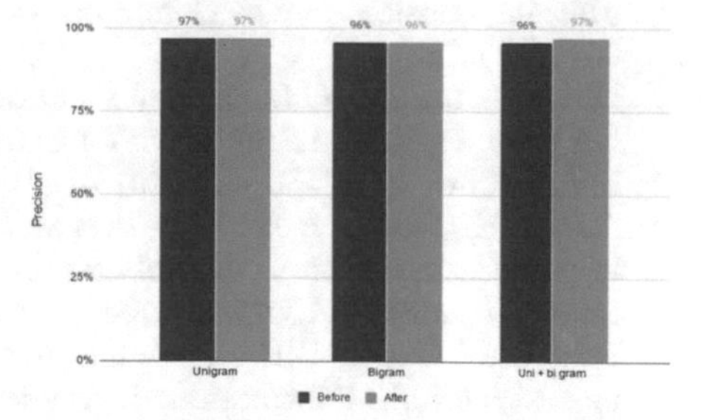

Fig. 7. TF-IDF Grid Search precision of LR

5 Conclusion

In this paper, we have presented a URLs classification system based on the URL with the use of coutvectorizer and TF-IDF vectorization technique. Our model has been tested on a dataset of 6000 records of legitimate and phishing. The "unigram bigram TF-IDF" results are very encouraging, specially with 96.41% and 97% accuracy and precision respectively achieved with logistic regression. Both vectorization techniques as features extraction, showed a good result with our approach. Also such combination of "unigram bigram" has been never used before. It will be very interesting to apply the proposed model in a robust architecture such as CNN or Reinforcement learning using Deep Q-Learning (DQL). These ideas are under study and will be considered in a future work.

References

1. Tripathy, A., Agrawal, A., Rath, S.K.: Classification of sentiment reviews using n-gram machine learning approach. Expert Syst. Appl. **57**(15), 117–126 (2016)
2. Sahingoz, O.K., Buber, E., Demir, O., Diri, B.: Machine learning based phishing detection from URLs. Expert Syst. Appl. **117**, 345–357 (2019)
3. Abutair, H., Belghith, A., AlAhmadi, S.: CBR-PDS: a case-based reasoning phishing detection system. J. Ambient Intell. Hum. Comput. **10**(7), 2593–2606 (2018). https://doi.org/10.1007/s12652-018-0736-0
4. Ebbu2017 Phishing Dataset. Accessed 1 Apr 2020, https://github.com/ebubekirbbr/pdd/tree/master/input
5. Volkamer, M., Renaud, K., Reinheimer, B., Kunz, A.: User experiences of torpedo: tooltip-powered phishing email detection. Comput. Secur. **71**, 100–113 (2017)
6. Peng, T., Harris, I., Sawa, Y.: Detecting phishing attacks using natural language processing and machine learning. In: IEEE 12th International Conference on Semantic Computing (ICSC), pp. 300–301 (2018)
7. Tan, C.L., et al.: PhishWHO: phishing webpage detection via identity keywords extraction and target domain name finder. Decis. Supp. Syst. **88**, 18–27 (2016)
8. Chiew, K.L., Choo, J.S.F., Sze, S.N., Yong, K.S.: Leverage website favicon to detect phishing websites. Secur. Commun. Netw. **78**, 95 (2018). https://doi.org/10.1155/2018/7251750
9. Chiew, K.L., Tan, C.L., Wong, K., Yong, K.S., Tiong, W.K.: A new hybrid ensemble feature selection framework for machine learning based phishing detection system. Inf. Sci. **484**, 153–166 (2019). https://doi.org/10.1016/j.ins.2019.01.064
10. Jain, A.K., Gupta, B.B.: Two-level authentication approach to protect from phishing attacks in real time. J. Ambient Intell. Hum. Computi. **9**(6), 1783–1796 (2017). https://doi.org/10.1007/s12652-017-0616-z
11. Jain, A.K., Gupta, B.B.: A machine learning based approach for phishing detection using hyperlinks information. J. Ambient Intell. Hum. Computi. **10**(5), 2015–2028 (2018). https://doi.org/10.1007/s12652-018-0798-z
12. Marchal, S., Saari, K., Singh, N., Asokan, N.: Know your phish: novel techniques for detecting phishing sites and their targets. In: 2016 IEEE 36th International Conference on Distributed Computing Systems (ICDCS), pp. 323–333. IEEE (2016)
13. Marchal, S., Armano, G., Gröndahl, T., Saari, K., Singh, N., Asokan, N.: Off-the-Hook: an efficient and usable client-side phishing prevention application. IEEE Trans. Comput. **66**(10), 1717–1733 (2017)

14. Moghimi, M., Varjani, A.Y.: New rule-based phishing detection method. Expert Syst. Appl. **53**, 231–242 (2016). https://doi.org/10.1016/j.eswa.2016.01.028
15. Shirazi, H., Bezawada, B., Ray, I.: Know thy domaln name: unbiased phishing detection using domain name based features. In: Proceedings of the 23nd ACM on Symposium on Access Control Models and Technologies, pp. 69–75. ACM (2018)
16. Tan, C.L., Chiew, K.L., Wong, K., Sze, S.N.: Phishwho: phishing webpage detection via identity keywords extraction and target domain name finder. Decis. Support Syst. **88**, 18–27 (2016). https://doi.org/10.1016/j.dss.2016.05.005
17. Varshney, G., Misra, M., Atrey, P.K.: A phish detector using lightweight search features. Comput. Secur. **62**, 213–228 (2016). https://doi.org/10.1016/j.cose.2016.08.003
18. Garera, S., Provos, N., Chew, M., Rubin, A.D.: A framework for detection and measurement of phishing attacks. In: Proceedings of the ACM Workshop on Rapid Malcode (WORM), Alexandria, VA (2007)
19. Wang, W., Shirley, K.: Breaking bad: detecting malicious domains using word segmentation. arXiv preprint arXiv:1506.04111 (2015)
20. Feng, F., Zhou, Q., Shen, Z., Yang, X., Han, L., Wang, J.Q.: The application of a novel neural network in the detection of phishing websites. J. Ambient Intell. Hum. Comput. (5), 1–15 (2018). https://doi.org/10.1007/s12652-018-0786-3
21. Rao, R.S., Pais, A.R.: Detection of phishing websites using an efficient feature-based machine learning framework. Neural Comput. Appl. **31**(8), 3851–3873 (2018). https://doi.org/10.1007/s00521-017-3305-0

Denial of Service Attack Detection in Wireless Sensor Networks and Software Defined Wireless Sensor Networks: A Brief Review

Hamza Belkhiri[1(✉)], Abderraouf Messai[1], André-Luc Beylot[2], and Farhi Haider[1]

[1] Department of Electronics, University Freres Mentouri Constantine 1, Constantine, Algeria
hamza.Belkhiri@umc.edu.dz
[2] IRIT/ENSEEIHT, INP Toulouse, Toulouse, France
andre-luc.beylot@enseeiht.fr

Abstract. Wireless Sensor Network (WSN) is a communication technology that aims at connecting remote sensors to aggregation devices, and is an important building block for the Internet of Things (IoT) infrastructure. However, WSN is faced by many challenges such as network management, energy consumption, and security. A potential solution was introduced to mitigate these challenges, which is the integration of software-defined networks (SDN) concept with WSN to form what is known as a software-defined wireless sensor network (SDWSN). The SDWSN model allowed a simplification of network management and configuration with more control and programmability. However, and due to the use open interfaces and standard protocols, securing SDWSN communications against various attacks is still a real challenge. In this paper, we present a comprehensive review of one of the most devastating, and yet very challenging attack on the WSN/SDWSN, namely Denial of Service (DoS) attack, as well as the recent solutions for detecting this attacks, and which we noticed that it have differed in terms of its complexity, accuracy and also its applied. Moreover, we provide an evaluation of the proposed solutions, as well as our recommendations to improve the protection level in WSN/SDWSN with regards to future trends.

Keywords: WSN · SDWSN · WSN/SDWSN security · DoS · DDoS

1 Introduction

The Internet environment has shifted from being the Internet of communications to the Internet of Things (IoT). This technological advance made our lives much easier, by providing assistance in various fields through a large number of interconnected devices that exchange data without any human intervention [1]. Experts at NIC [2], predicted that by 2025, Internet will reach our everyday's life objects, and this mainly due to advances of techniques that allow the integration of these objects to the internet network, such as the Wireless Sensor Network (WSN) technology. WSNs are considered as one of the best emerging technologies of the last years, in which wireless sensors communicate with each other using radio signals in order to perform a variety of tasks without having

M. Lazaar et al. (Eds.): BDIoT 2021, LNNS 489, pp. 100–115, 2022.
https://doi.org/10.1007/978-3-031-07969-6_8

any infrastructure [3]. With the growth of the IoT technology, the scope of demand on WSNs has increased considerably, and which made it the backbone of IoT. However, and looking at the core of the network, we see that it still suffers from many challenges, most notably; when deployed, this network becomes unreliable and vulnerable to harmful attacks and cybercriminal activities. Indeed, and since it is connected to the Internet, this will increase the risk of compromising the nodes, and posing a serious threat regarding data integrity and confidentiality especially if it is deployed on sensitive infrastructures (i.e. military, health services, etc.). In addition, the network has limited resources such as processing capacities, memory, and energy of sensor nodes [4]. These restrictions also become increasingly significant due to the lack of suitable network management and the inability to achieve the heterogeneous node network. All these constraints reduce the efficiency of WSNs, and make them, so far, unable to meet, completely, the requirements of the IoT [5]. Software Defined Networks (SDN) [6], which are a new paradigm shift in network architectures, emerged as a solution to face these challenges. The whole idea behind SDNs, is to separate the control and data plane [7]. The combination of SDN and WSN can bring many advantages such as programmability of the network, management flexibility, etc. Software Defined Wireless Sensor Network (SDWSN) was introduced to play a critical role in the traditional network architecture and also to face the challenges and inflexibility issues, and this by moving many tasks performed by the sensor node to the console and thus contributing in the energy saving which increases the lifetime of the network. However, the security of the network remains an issue especially since SDWN has inherited attacks from both SDN and WSN, and the most important of them is the DoS attack. For this reason, ensuring an efficient protection against this attack in SDWSN is indispensable and plays a critical role in the development of the IoT infrastructure [8].

In this work, we provide a brief comprehensive survey of DoS attacks detection in WSN and SDWSN, by providing an overview of the most notable and recent works published in the literature in the last three years. We focused on this specific attack for two main reasons: 1) it is so far the most common attacks that target the IoT infrastructure including the sensors networks (WSN/SDWSN). 2) It is also, one of the most devastating attacks and one of the hardest to detect and/or prevent from.

Existing surveys [9–11] have provided detailed studies of the various routing attacks (including DoS attack) that target the standard WSN, with a focus on those targeting only the network layer of the WSN architecture. However, in our paper, we conducted a comprehensive examination of the different methods and scenarios that the DoS attacks comes, through reviewing the literature containing existing solutions for DoS attacks targeting the different layers of the WSN architecture. Moreover, we conducted the same study on the SDWSN, which we believe to be the first review paper that mainly focuses on DoS attack detection in SDWSN. Indeed, the only review paper that discussed security aspect in SDWSN, is the one published by M. Bongeni et al. [12], in which they provided a general insight regarding SDWSN issues including security ones and do not focus on attacks and their countermeasures, contrarily to our work, which is meant to be very specific. Finally, in this review we tried to highlight the advantages and limitations of the proposed solutions for DoS attack detection in both WSN and SDWSN, and tried to provide some recommendations that we think can help providing better security solutions.

The rest of the paper is organized as follows: The structure of the WSN/SDWSN is studied in the second section. In the third section, we study attacks targeting WSN/SDWSN and we will discuss the existing solutions. In Sect. 4, we offer some recommendations and advice with the determination of our future direction. Final remarks from previous studies are given in Sect. 5.

2 Background

2.1 Wireless Sensor Networks (WSN)

Wireless Sensor Networks can be defined as a set of devices sensors used in certain tasks. It has also the ability to collect and transmit data over wireless links using radio signals until it reaches the data processing node [13] and the process of routing data from the sensors to the base station can take four forms. The sensors can communicate directly with the base station or via a multi-hop mode. In flat architectures, as for the hierarchical ones, the cluster-head node transmits the data directly to the base station or via a multi-hop mode between Cluster Heads [14]. A sensor also includes a sensing unit, radio transceivers, power and processing units [15], as shown in Fig. 1 [12].

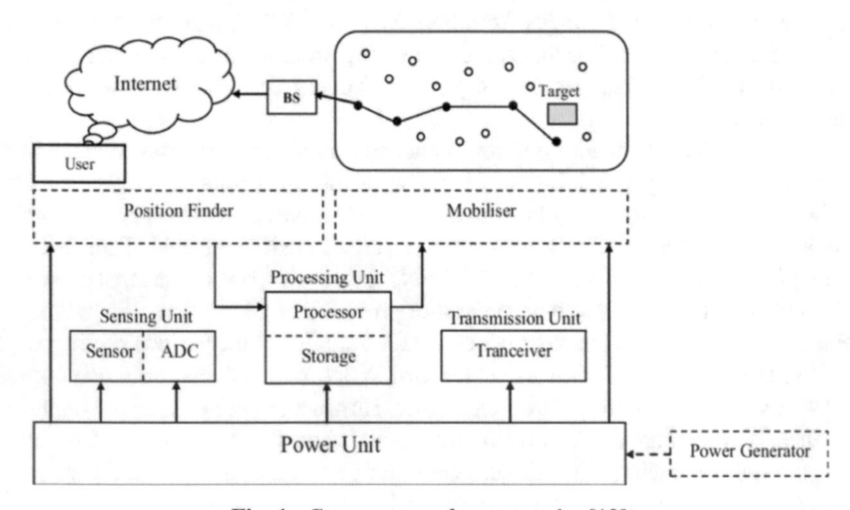

Fig. 1. Components of sensor nodes [12].

The layered WSN architecture is composed of a physical layer, data link layer, network layer and application layer.

2.2 Software Defined Wireless Sensor Network (SDWSN)

The term Software-Defined Networking (SDN) [16] refers to an emerging architecture developed to liberalize traditional network management, through optimize flow management and configuration and support service-user requirements of scalability and flexibility and to foster interoperability with other networks [17]. Given the restrictions imposed on WSN regarding limited resources and energy, lack of efficient management, and other challenges like topology discovery mechanisms, routing protocols to steer the network traffic, etc. [18], the SDWSN model consists of a platform that works to face this challenges, i.e. Due to the control level separation from the data plane thus moving the control logic from the forwarding node to control plane (controller) that is the data plane becomes programmable [19] and therefore the nodes do not need to send broadcast message to identify the neighbors periodically for topology discovery. Moreover, it does not need each node to store network routing tables within its limited memory, nor computing the path for other nodes [20]. This makes the model play a critical role in speeding up basic network management (plane data) and its flexibility through communication and the deployment of the network [21].

Opinions differ on the implementation of the SDWSN architectures, but they all conform to the fundamentals of SDN decoupling [19]. Figure 2 shows the basic SDWSN architecture [22], and which consists of the following logical planes: data plane, control plane, and application plane.

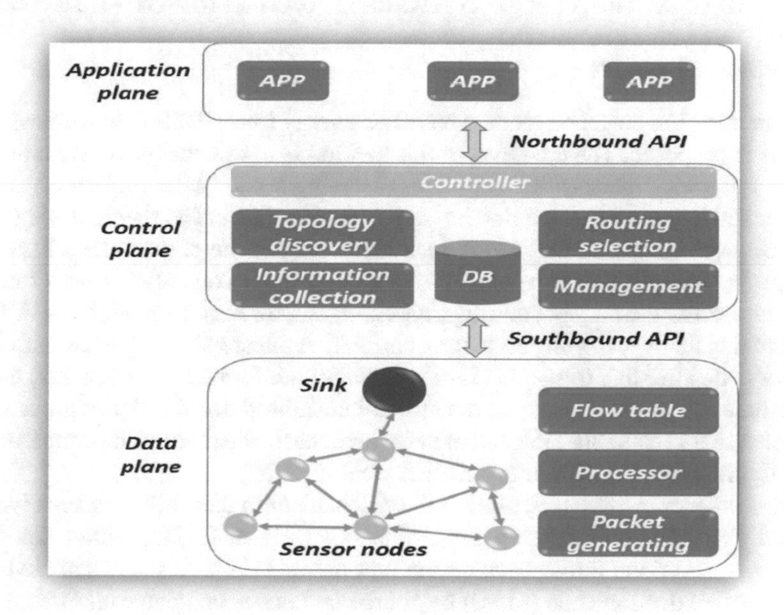

Fig. 2. Basic SDWSN architecture [22].

- **Data plane:** The data plane consists of sensors nodes that work on forwarding the data that have been sensed with each other in the network [23].
- **Control plane:** the control plane consists of one controller or more and it is the intelligent part which is performing the whole network control and its management

and made the data plane programmable by using the communication protocol Sensor OpenFlow [20, 24].

- **Application plane:** consists of a set of network applications that are input to the controllers to install appropriate rules on the data paths. Examples of network applications are routing network monitoring, network configuration and management, network address translation, network policies and security, etc. [20].

2.3 Denial of Service Attack

Denial of Service attack (DoS) is one of the most common attacks in WSNs, where in this attack, a huge amount invalid packets are sent to the target system (sinks) [25], and the main aim is to waste the network resources and disrupt the normal operations of WSNs, as such attacker can be malicious node outside the network or it can be one of the legitimate nodes in network that has been compromised, and harms the network by attracting or dropping packets [26]. In the case the network was attacked by multiple attackers (sources), we call it Distributed Denial of Service (DDoS) attack [27].

3 Existing Solutions for DoS Attack Detection in WSN and SDWSN

3.1 Solutions for WSN

In [28] an intrusion detection method based on energy trust (IDSET) in wireless sensor networks is proposed. The core idea of this method is to use energy consumption trust algorithm to verify the security state after receiving sensor data in cluster head node (CH) and that by applying the following steps: identity check; CH checks the legitimacy of the message senders in the case it does not belong to the cluster, CH will drop the message. In case the sender belongs to the cluster the second step is performed and which transforms the energy consumption check to power consumption check by CH after checking them, and with designation correlation threshold to judge the abnormality of sensor node after that trust value is calculated with information aggregation and sent them to base station. The results of the evaluation showed that the algorithm has a high detection rate for malicious node but in case But in case of targeting and tampering with energy information, the algorithm becomes weak.

C.K. Marigowda et al. [29] proposed a mechanism to detect DoS attacks based on Modified Constrained Function based Authentication (MCFA) algorithm. The verification process is done through packet encrypt using AES with the calculation of the Message authentication code (MAC) by source node after computes h(M) hash value, and is calculated verification number and calculate the subsequent verification difference after that by the intermediate node, and by computing the polynomial while passing the packet to the destination, it is possible to know whether the attack occurred or not. The proposed method provides less packet loss, thus increasing the network lifetime. However, the metrics used in the proposed work did not indicate the performance of the proposed algorithm.

A. Amouriet et al. [30] proposed an approach for DoS attack detection using Machine Learning The core idea in this work can be summarized in two steps. The first is by using dedicated sniffers (DSs) the data is collected from MAC and network layer after that by a random forest classifier is generates correctly classified instances (CCIs). When receiving it from the super node (SN) it begins second stage; in it Malignant and normal nodes are separated. And after basing the detection threshold in it, after that it performs linear regression process by super node (SN) on a parameter accumulated measure of fluctuation (AMoF) which was calculated during this process. The experimental results show that the proposed IDS is effective, however it suffers from a deterioration of detection accuracy at the early stages of the fitted slope calculation.

In the work of A.I. Al-issa et al. [31], Machine learning was used to detect denial of service attacks in wireless sensor networks, based on two techniques which are decision trees and Support Vector Machines. The authors concentrate on using each technique (decision trees, Support Vector Machines) with Full Dataset and with a Dataset of Selected Attacks (flooding and Grayhole attacks) and they note that the latter is better than the use of the full dataset from where the number of correctly and incorrectly classified Instances, time taken to build the model.

The obtained results show that the proposed method is effective. However, it has been evaluate don two attack types only.

H. Chen et al. [32] proposed a new method to deal with a Low-rate Denial of Service (LDoS) attack, which is characterized by the difficulty in distinguishing between LDoS attack traffic and normal traffic in WSN. The authors used Combining Hilbert-Huang transformation and Trust Evaluation to detect the Ldos attack on two stages, The first is WSN routing traffic capture unit, and which depends the ZigBee sniffer node the second part is the security analysis server Which works on analyzes the sniffed data which in turn depends on HHT time-frequency approach and which work on analyzing the network traffic time-serial dataset to intrinsic mode function (IMF) components, and by trust evaluation approach IMF components are evaluated after based Correlation coefficient and Kolmogorov_Smirnov (KS) to get on detection precision high on LDOS attack. The approach has been highly effective in detecting LDoS attack after seeing the experimental results However, The proposed approach was not subject to the parameter computational cost consumption with that wireless sensor networks is limited resources.

F. Afiantiet al. [33] employed a new approach to deal with DoS attacks by applying a multi-user dynamic cipher puzzle (M-DCP) for message authentication in WSNs The authors concentrated in this work on proposes M-DCP Who works on the safety and security of the sent messages by dependent on RC5 encryption, with TinySet and there the sender's information is inserted and this information consists from the public key and the user ID the core idea in this paper; when the registered user starts to communicate with the sensor nodes It creates signature constructing by use elliptic curve digital signature algorithm (ECDSA). Then the puzzle is built by session key generation the process is continued by encryption some parameters and puzzle construction. As for the receiver, tags verification begins and checks the hash result of id and publishes key if it exists in the TinySet with session key verification is used to verify the puzzle. To verify the signature in the latter and experimental results show that the mechanism is able to respond to DOS attack with a short time to verification for authentication and the storage overhead decreases, but no parameter was provided on of the mechanism computational cost consumption Although it is important parameter in WSN.

C. Lyu et al. [34] proposed approach for DoS attacks detection in wireless sensor networks based on reliable data delivery with detecting and dropping invalid packets And that by relying on elliptic curve digital signature algorithm (ECDSA) used for the packets signature (private key) and a public key for verification in this paper the authors focused on use this protocol (SelGOR) in which is contains three major components: trust-based geographic opportunistic routing, selective authentication algorithm and cooperative verification scheme. The latter works to share the results of checking between the candidate nodes on that the packets are invalid to avoid over verification after do the selective authentication algorithm verification which in turn works to block bogus signatures while the SSI-based trust model works on adjusting the trust of wireless link between nodes. The results of the evaluation showed that the protocol (SelGOR) works it on quick isolate the attackers with low computational cost. but this proposed mechanism the end-to-end delay could become quite long when a high node verification probability is decided.

P. Nayak et al. [35] proposed a mechanism to detect DoS attack in Wireless Sensor Networks (WSN) using IDS Which is installed in each sensor node, that consists of three units; monitoring level, level detection and it is based on an algorithms for analysis the data to find out the behavior of each sensor node and response level in this paper the authors focused on methods to identify infiltration through Anomaly based IDS and Signature etc., which through the threshold value or signature illegitimate Can find out the presence of intrusion the proposed work is capable of dealing with attacks dos but, installing IDS into each sensor node can be computationally costly.

In their work, Y. Cao et al. in [36] used a mechanism working to isolate Low-rate TCP DoS attacks in WSN by using AccFlow security protocol, this protocol drops packets during the high rate of loss and congestion caused by excessive transmission of packets. AccFlow is based on Software defined networking (SDN) centralized controller which includes two unit; Aggressive Detection and Early Drop in Aggressive detection offensive flows are detected and prevented using the Uniform Loss Rate (ULR), and he "the usage rate of one flow is the ratio of the number of its transmitted packets in one detection period over the total number of arriving packets from all flows in this detection period". While early drop is used to address periodic attacks which its traffic comes at a different rate and it is based on an algorithm that drops the packet according to the rate of loss of each flow's. The results of the evaluation showed that the proposed approach is effective in defend against DoS attacks, but an attacker can participate in sending legitimate packages and then works on the injection of illegal packages.

A.E. Guerrero-Sanchez et al. [37] Propose that the blockchain paradigm provides increased security for networks (IoT) in terms of threat detection, remediation, after relying on the use of encryption tools in that the authors agree it can be implemented in most infrastructure of the Internet of Things. The accuracy of the proposed methodology was measured on a temperature and humidity sensing IoT-based Wireless Sensor Network (WSN). The experimental results showed the ability of the proposed solution can be implemented in various IoT contexts with low resource consumption however when looking at the security in blockchain itself it can be hack (Tables 1 and 2).

Table 1. Summary of the recent solutions for DoS attack detection in WSN.

Work	Year	Description	Advantage	Disadvantage	Suggestions
A.E. Guerrero-Sanchez et al. [37]	2020	The use of blockchain mechanism in a (WSN) provides increased security through based on the use of symmetric encryption	Fulfils the fundamental requirement s of security with a low-resource consumption	When looking at the security in blockchain itself it can be hacked	It should include security in itself, for respond and counter many attacks
A. Amouri et al. [30]	2020	The proposed IDS provide use dedicated sniffers (DSs), it collects data from the MAC and the network layer after that by a random forest classifier (Machine Learning) is generates correctly classified instances (CCIs), Then, based on the detection threshold a malicious and normal nodes are sorted out	The proposed scheme (IDS) effective and perfect	It suffers from a deterioration of detection at the early stages of the fitted slope calculation	Use classification algorithms,it contributes to dealing with the problem more
C. Lyu et al. [34]	2019	The SelGOR protocol was used to ensure reliable delivery of data with preventing and isolating DOS attacks with speed, at low computational cost through; three major components: trust-based geographic opportunistic routing, selective authentication algo-rithm and cooperative verification scheme	The protocol (SelGOR) works on quick isolate the attackers with low computa-tional cost	The end-to-end delay could become quite long when a high node verification probability is decided	Using the linear function to contribution in reduce the probability of verification and thus reducing the delay

(continued)

Table 1. (*continued*)

Work	Year	Description	Advantage	Disadvantage	Suggestions
F.Afianti et al. [33]	2019	Uses a multi user dynamic cipher puzzle (M-DCP) equipped with Tiny-Set, (M-DCP) depends on RC5 en-cryption and the elliptic curve digital signature algorithm (ECDSA) for prevention, block DoS attacks and provides guaranteed confidentiality in the multiuser WSN authentication	The mechan- ism is able to respond to DOS attack with a short time	No parameter was provided on of the mechanism computational cost consump-tion Although it is important parameter in wsn	Use cooperative verification scheme, it is works on share the verification information of invalid signa-tures between nodes, for fast isolation of attackers
H. Chen et al. [32]	2019	When capturing WSN traffic by the ZigBee sniffer node and based on the security analysis server Which works on analyzes the sniffed data which in turn depends on HHT time-frequency approach to it works on analyze the network traffic time-serial dataset to intrinsic mode function (IMF) com-ponents, than by trust evaluation, can get on detection precision high on LDOS attack	The approach has a high detection accuracy for an LDOS attack	The proposed approach was not subject to the parameter computational cost consump-tion with knowing that wirelesssensor networks is limited resources	Establish the behavior model of LDoS attackers We can control the attack more
A.I. Al-issa et al. [31]	2019	Based on two techniques which is decision trees and Support Vector Machines and by use with Full Data-set and with a Data set of Selected Attacks (flooding and grayhole attacks), the data are handled by analyzing and processing it, selecting feature and regression	The proposed technique effective	This technique has been used in two types onely of attack, but DOS comes in various forms attacks	The use of other classifiers and data mining approaches and processing, contribute to tuning on other types of DoS attacks scenrios

(*continued*)

Table 1. (*continued*)

Work	Year	Description	Advantage	Disadvantage	Suggestions
Y. Cao et al. [36]	2018	Uses AccFlow security protocol to isolate Low-rate TCP DoS attacks in WSN and that is through dropping packets during the high rate of loss and congestion caused by excessive transmission of packets	The effectiveness of the proposed approach in defense against DoS attacks	Attacker can participate in sending legitimate packages and then works on the injection of illegal packages	Through it is deploying proposed approach (AccFlow) in a real networks This attack can be prevented and isolated
C.K. Marigowda et al. [29]	2018	Using a algorithm Modified Con-strained Function based Authentication (MFA)and through; encrypt the source node of the packet using AES after computes h(M) hash value, and also it is calculated verification number and calculate the sub-sequent verification difference after that by the intermediate node, can know if an attack DOS occurred or not	It provides less packet loss, thus increasing the network lifetime	The metrics used in the proposed work did not indicate the performance of the proposed algorithm	Applying the method proposed in base station to avoid flooding her by the attacks
T. Yang et al. [28]	2018	Using the energy consumption algorithm for intrusion detection (hybrid DoS Attacks) to check the security status after receiving sensor data in the cluster head node (CH) through; CH checks the legitimacy of the message senders and dropping the message in the case not be node a member of the cluster	A high detection rate for malicious node	The algorithm is weak if energy information is forged	Imposing on each node a public and a private key to verify and sign data packets for isolate illegitimate packets
P. Nayak et al. [35]	2017	Using both anomaly detection and signature-based detection on WSN network behavior data, and through the threshold value or signature illegitimate Can find out the presence of intrusion	Accurately collect and analyze data	Installing IDS into each sensor node can be Coputationally costly	Installing IDS in sink (base station) can avoid receiving much invalid data from DoS attackers with a reduction in the computational cost

3.2 Solutions for SDWSN

Miranda et al., suggested in [38] a Collaborative Security Framework for Software-Defined Wireless Sensor Networks to address DoS attacks, it combines three layers; in the data plane An effective and lightweight authentication system has been proposed allowing only correct information to be inserted by the authenticated nodes, based on the Intrusion Prevention System (IPS). An IDS-enabled energy prediction model, It was use to provide additional protection in the layer connecting the data plane and the control plane, then is applied Smart Monitoring System (SMS) with the aid of a machine learning algorithm in the control plane after correlating the decisions taken by an IDS for anomaly detection and isolation malicious nodes.

The results of the evaluation showed that the proposed approach provides high security, low computational complexity, but, relying solely on the Support Vector Machine algorithm (SVM) at the control plane, not all anomaly are detected, especially as the algorithm is weak in dealing with large data sets.

G.A. Nunez Segura et al. in [39] provided a solution to detection DDoS attack in SDWSN and that through examining whether a change has occurred in the mean value of the time series of the metrics; the mean data packet delivery rate and the mean control packets overhead, based on this change point (CP) detection algorithm. The authors focused on the algorithm to arrive at a real-time estimate of the existence, the number, the magnitude and the direction of changes in a time series through "off-lineand on-line CP schemes; an improved measurements window segmentation heuristic for the detection of multiple CPs; and a variation of the moving average convergence divergence (MACD) indicator to detect the direction of changes".

The experimental results indicate that the proposed algorithm has a high detection rate however; the algorithm was implemented to only two metrics.

In the work of M. Huang and B. Yu in [40] a mechanism has been designed to deal with DOS attacks in SDWSN, called Fuzzy Guard. and that by designing Components such as control network construction, fuzzy attack detection and probabilistic flow suppression Control Network Construction contribute to protect the forwarding of legitimate data flows and It is created by select part of the sensor nodes from the data plane as the control nodes, Once a control network is formed becomes responsible for control flows routing independently. While the fuzzy attack detection component adopts on fuzzy inference method to detect the attack and to know security state of the network. After reading the statistical data related to network security status after obtaining the attack probability by fuzzy inference the probabilistic flow suppression component are adopted for adjusts the Packet-in flow suppression probability through different suppression modes. After looking at the resultants we note the effectiveness of detection of the attack as well prevent rapid saturation (Resulting from the attack) of the control plane. However, too much Packet-in flows causes the bandwidth of the communication to be consumed thus it causes normal control flows to delay.

V.M. Vishnu and P. Manjunath. In [41] introduced a mechanism to guarantee security at all levels of the SDWSN architecture. Where the sensor nodes are handled by secure hash tree based clustering (SHTC) algorithm and with adaptive spider monkey optimization (ASMO) algorithm, the optimal path is chosen. For ensuring data security and transmitted to optimal switch and then to the SDN controllers after based on PAES-CBC algorithm.

The results of the evaluation showed that the proposed approach is effective and ensures high-level security but, it consumes a lot of energy, we believe that the use of opportunistic routing for forwarding the data at the first tier works to save energy and thus ensure security and quality of service.

Table 2. Summary of the recent solutions for DoS attack detection in SDWSN.

Work	Year	Description	Advantage	Disadvantage	Suggestions
C. Miranda et al. [38]	2020	Uses a collaborative framework which combines IPS, IDS and an IDS-enabled energy prediction model in addition to applied Smart Monitoring System (SMS), and with the aid of a machine learning algorithm in the control plane anomaly are detected and isolation malicious nodes	Provides high security, low compu-tational complexity	Only SVM algorithm is relied on at the control plane, so it is difficult to detect all anomaly	Use of other classifications, to accurately identify unknown anomalies in SDWSN environments
G.A. Nunez Segura et al. [39]	2020	Through examining whether a change has occurred in the mean value of the time series of the metrics; the mean data packet delivery rate and the mean control packets overhea, and based on this change point (CP) detection algorithm, a DDoS attack is detected in SDWSN	The algorithm has a high detection rate	The algorithm is implemented on only two metrics	The change or increase in the number of metrics can further improve detection performance

(continued)

Table 2. (*continued*)

Work	Year	Description	Advantage	Disadvantage	Suggestions
M. Huang and B. Yu [40]	2019	The FuzzyGuard mechanism was used to deal with DOS attacks in SDWSN, based on idea of control flow, and a control network which is constructed at data plane. then through relying on fuzzy inference, the attack is revealed, to be dealt with by the probabilistic flow suppression component are adopted for adjusts the Packet-in flow suppression probability	The effectiveness of detection of the attack as well prevent rapid saturation (Resulting from the attack)	The Communications bandwidth consumption	Deploying multiple controllers, the Packet-in flows can be mitigated, there by reducing bandwidth consumption for communication
V.M. Vishnu and P. Manjunath [41]	2019	Security and QoS are improved in the SeC-SDWSN environment through ensures high-level security in each of its three tier, after based on several algorithms as PAES-CBC	The proposed approach is effective and ensures high-level security	It consumes a lot of energy	Use of Opportunistic routing for forwarding the dataat the first tier

4 Discussion

In this paper, and after reviewing the literature in the previous sections, it became clear to us that WSN technology provided efficient solutions through its wide use in many applications. However, it faces many challenges such as network management and configuring heterogeneous node networks, resources constraints and limitations, energy, security...etc. It also became clear to us that the current solutions discussed in advance and which related to the detection of attacks DoS in WSN/SDWSN it is despite the diversity of its strategy in detecting attacks DoS in all levels of architecture; as well as its distinctiveness with high and accuracy detection rates. But it is still very expensive in terms of high consumption of computational resources and memory, and also the end-to-end delay in number of messages exchanged on the network could become quite long when a high nodes verification probability is decided. Thus with standard WSN technology; it is difficult to implement efficient and complex security algorithms due to the existence of constraint resources. For instance, and in case of insufficient residual power from the nodes, or failure of the wireless links, it may take some time to inform the controller by multi-hops communications. However, with SDWSN, the intelligence of the network or the console is stable in SDN and thus reduces the burden on the sensor knot, and thus maintains the energy.

All these features have made SDWSN a great platform for such communications. However, they still suffer from many limitations. If we consider the security aspect as the main evaluation criterion the SDWSN model is still not receiving that attention especially since the solutions available for both SDN and WSN separately cannot be applied or adapted to SDWSN. In addition, the model SDWSN lacks of major security components like middle boxes and transport layer security (TLS) [42]. For this reason, it is therefore necessary to provide security in the SDN and WSN individually and to reconsider its integration into the SDWSN model, and also there must be multiple and effective security mechanisms in every plane of the paradigm with maximizing usability and necessity of its work in real-time, by developing lightweight effective encryption algorithms to preserve network resources, and also the use of intrusion detection systems (IDS), and anti-malware systems [43], as a second line of defense. and this is a critical factor in the success and development of the model and to meet the requirements of any network and also make them gain wide applicability, and as a future work we hope establish the behavior model an DoS/DDoS attackers and exploiting deep learning techniques to accurately classify and identify unknown anomalies in SDWSN environments and deploy also distributed SDN controllers at the edges the network to protects end-users when a malicious attack is detected by exploiting flow rules which based on OpenFlow standard messages.

5 Conclusion

In view of its wide applications, SDWSN become a scientific revolution in the field of wireless communications and embedded systems, especially as it plays a crucial role in the development of Internet of Things technology, And Given the many advantages that the integration of SDN into WSN model brought in from programming, network management and control it but there are still many challenges in this model Especially in terms of security. The security of the SDWSN model is very difficult, because it consists of two models in which security has not yet been fully achieved so it is necessary to review the security of the SDWSN model itself or integrate the individual security aspects of the SDN and WSN that can be adapted to the SDWSN model.

When reviewing the literature, it has become clear to us that the SDWSN network is still suffering from many problems notably from the security side, although it has made great strides so far. This Paper provided an overview of more destructive attacks DoS in WSN/SDWSN and current solutions as well as our recommendations to improve these solutions. As a perspective, due to the importance of SDWSN, we are working on providing our own solution for Network security; this solution will focus on the detection of DoS and DDoS attacks.

References

1. Cheng, N., et al.: Big data driven vehicular networks. IEEE Network **32**(6), 160–167 (2018)
2. Atzori, L., Iera, A., Morabito, G.: The internet of things: a survey. Comput. Netw. **54**(15), 2787–2805 (2010)

3. Alsace, H., Domaine, F., Math, L.M.D., Fili, I., Sp, I.: T hèse de D octorat 3 ème Cycle M r K amel SADDIKI D enial of services attack in wireless networks. 2018–2019 (2019)
4. Amish, P., Vaghela, V.B.: Detection and prevention of wormhole attack in wireless sensor network using AOMDV protocol. Procedia Comput. Sci. **79**, 700–707 (2016)
5. Kobo, H.I., Abu-Mahfouz, A.M., Hancke, G.P.: A survey on software-defined wireless sensor networks: challenges and design requirements. IEEE Access **5**, 1872–1899 (2017)
6. Ndiaye, M., Hancke, G.P., Abu-Mahfouz, A.M.: Software defined working for improved wireless sensor network management: a survey. Sensors. **17**(5:1031), 1–32 (2017)
7. Saqib, M., Khan, F.Z., Ahmed, M., Mehmood, R.M.: A critical review on security approaches to software-defined wireless sensor networking. Int. J. Distrib. Sens. Netw. **15**(12), 155014771988990 (2019)
8. Pritchard, S.W., et al.: Security in software-defined wireless sensor networks: threats, challenges and potential solutions. In: Proceedings – 2017 IEEE 15th International Conference on Industrial Informatics, vol. 2017, pp. 168–73 (2017)
9. Venkatraman, K., Daniel, J.V., Murugaboopathi, G.: Various attacks in wireless sensor network: survey. Int. J. Soft Comput. Eng. **31**, 2231–2307 (2013)
10. Patel, M.M., Aggarwal, A.: Security attacks in wireless sensor networks: a survey. In: 2013 International Conference Intelligent System Signal Processing ISSP 2013, pp. 329–333 (2013)
11. Dubey, R., Jain, V., Thakur, R.S., Choubey, S.D.: Attacks in wireless sensor networks. Comput. Sci. **3**(3), 2–5 (2012)
12. Manuel, M., Isong, B., Esiefarienrhe, M., Abu-Mahfouz, A.M.: Analysis of notable security issues in SDWSN. In: IECON 2018-44th Annual Conference of the IEEE Industrial Electronics Society, pp. 4706–4711. IEEE (2018)
13. Derhab, A., Bouras, A., Belaoued, M., Maglaras, L., Khan, F.A.: Two-hop monitoring mechanism based on relaxed flow conservation constraints against selective routing attacks in wireless sensor networks. Sensors **20**(21), 6106 (2020)
14. Lehsaini, M.: Thèse de Doctorat Diffusion et couverture basées sur le clustering dans les réseaux de capteurs : application à la domotique. Université A.B Tlemcen (2009)
15. Matin, M.A., Islam, M.M.: Overview of wireless sensor network. In: Matin, M. (ed.) Wireless Sensor Networks – Technology and Protocols. InTech (2012). https://doi.org/10.5772/49376
16. Bull, P., Austin, R., Popov, E., Sharma, M., Watson, R.: Flow based security for IoT devices using an SDN gateway. In: Proceedings of the 2016 IEEE 4th International Conference Future Internet Things Cloud, pp157–163. FiCloud (2016)
17. Fraser, B., et al.: Introduction: What Is Software Defined Networking? pp. 36–43 (2013)
18. Ndiaye, M., Hancke, G., Abu-Mahfouz, A.: Software defined networking for improved wireless sensor network management: a survey. Sensors **17**(5), 1031 (2017)
19. Kobo, H.I., Abu-Mahfouz, A.M., Hancke, G.P.: A survey on software-defined wireless sensor networks: challenges and design requirements. IEEE Access **5**(12), 1872–1899 (2017)
20. Mostafaei, H., Menth, M.: Software-defined wireless sensor networks: a survey. J. Netw. Comput. Appl. **119**, 42–56 (2018)
21. Hassan, M.A., Vien, Q.-T., Aiash, M.: Software defined networkingfor wireless sensor networks: a survey. Adv. Wireless Commun. Netw. **3**(2), 10–22 (2017)
22. Egidius, P.M., Abu-Mahfouz, A.M., Hancke, G.P.: Programmable node in software-defined wireless sensor networks: a review. In: Proceedings of the IECON 2018 – 44th Annual Conference IEEE Industrial Electronics Society, pp. 4672–4677 (2018)
23. Kipongo, J., Olwal, T.O., Abu-Mahfouz, A.M.: Topology discovery protocol for software defined wireless sensor network: solutions and open issues. In: IEEE International Symposium on Industrial Electronics, vol. 2018, pp. 1282–1287 (2018)
24. Kobo, H.I., Hancke, G.P., Abu-Mahfouz, A.M.: Towards a distributed control system for software defined wireless sensor networks. In: IECON 2017 – 43rd Annual Conference IEEE Industrial Electronic Society, vol. 2, pp. 6125–6130 (2017)

25. Belkhiri, H., et al.: Security in the internet of things : recent challenges and solutions. In: Fourth International Conference on Electrical Engineering And Control Applications ICEECA. Constantine (2019)
26. Gavric, Z., Simic, D.: Overview of DOS attacks on wireless sensor networks and experimental results for simulation of interference attacks Visión general de los ataques de DOS en redes de sensores de ataques de interfrencia. Ing. Invest. **38**(1), 130–138 (2018)
27. Maylyn, B.: What Is the Difference Between DoS and DDoS Attacks?, Bisend. https://www.bisend.com/blog/difference-between-dos-and-ddos-attack (2019)
28. Yang, T., et al.: Intrusion detection system for hybrid attack using energy trust in wireless sensor networks. Procedio Comput. Sci. **131**, 1188–1195 (2018)
29. Marigowda, C.K., Thriveni, J., et al.: An efficient secure algorithms to mitigate DoS, replay andjamming attacks in wireless sensor network. Lect. Notes Eng. Comput. Sci. **22**, 166–171 (2018)
30. Amouri, A., Alaparthy, V.T., Morgera, S.D.: A machine learning based intrusion detection system for mobile internet of things. Sensors **20**(2), 461 (2020)
31. Al-issa, A.I., et al.: Using machine learning to detect DoS attacks in wireless sensor networks. In: 2019 IEEE Jordan International Joint Conference on Electrical Engineering and Information Technology, pp. 107–112 (2019)
32. Chen, H., Meng, C., Shan, Z., Fu, Z., Bhargava, B.K.: A novel low-rate denial of service attack detection approach in ZigBee wireless sensor network by combining hilbert-huang transformation and trust evaluation. IEEE Access **7**, 32853–32866 (2019)
33. Afianti, F., Wirawan, Suryani, T.: Lightweight and DoS resistant multiuser authentication in wireless sensor networks for smart grid environments. IEEE Access **7**, 67107–67122 (2019)
34. Lyu, C., Zhang, X., Liu, Z., Chi, C.H.: Selective authentication based geographic opportunistic routing in wireless sensor networks for internet of things against DoS attacks. IEEE Access **7**, 31068–31082 (2019)
35. Nayak, P.: A Review on DoS attack for WSN: defense and detection mechanisms. In: 2017 International Conference on Energy, Communication Data Analytics and Soft Computing, pp. 453–461 (2018)
36. Cao, Y., Han, L., Zhao, X., Pan, X.: AccFlow : defending against the low-rate TCP DoSAttack in wireless sensor networks. Comput. Sci. **X**(XX), 1–14 (2018)
37. Guerrero-Sanchez, A.E., Rivas-Araiza, E.A., Gonzalez-Cordoba, J.L., Toledano-Ayala, M., Takacs, A.: Blockchain mechanism and symmetric encryption in a wireless sensor network. Sensors (Switzerland) (2020). https://doi.org/10.3390/s2010279
38. Miranda, C., Kaddoum, G., Bou-Harb, E., Garg, S., Kaur, K.: A collaborative security framework for software-defined wireless sensor networks. IEEE Trans. Inf. Forensics Secur. **15**, 2602–2615 (2020)
39. Nunez Segura, G.A., et al.: Denial of Service Attacks Detection in Software-Defined Wireless Sensor Networks. http://www.larc.usp.br/users/cbmargi/w/it-sdn/
40. Huang, M., Yu, B.: Fuzzyguard: a ddos attack prevention exte-nsion in software-defined wireless sensor networks. KSII Trans. Internet Inf. Syst. **13**(7), 3671–3689 (2019)
41. Vishnu, V.M., Manjunath, P.: SeC-SDWSN: secure cluster-based SDWSN environment for QoS guaranteed routing in three-tier architecture. Int. J. Commun. Syst. **32**(14), 1–22 (2019)
42. Rawat, D., Reddy, S.: Recent advances on software defined wireless networking. In: IEEE SoutheastCon, pp. 1–8 (2016)
43. Belaoued, M., Derhab, A., Mazouzi, S., Khan, F.A.: MACoMal: a multi-agent based collaborative mechanism for anti-malware assistance. IEEE Access **8**, 14329–14343 (2020)

An RGB Image Encryption Algorithm Based on Clifford Attractors with a Bilinear Transformation

Fouzia El Azzaby[1(✉)], Nabil El Akkad[2], Khalid Sabour[1], and Samir Kabbaj[1]

[1] Department of Mathematics, Faculty of Sciences, Ibn Tofail University, 14000 Kenitra, Morocco
fouzia_099@hotmail.com

[2] Laboratory of Engineering, Systems and Applications, ENSA of Fez., Sidi Mohamed Ben Abdellah University of Fez-Morocco, Fes, Morocco
nabil.elakkad@usmba.ac.ma

Abstract. This work represents a new contribution in the field of encryption of RGB images, using Clifford attractors, which are the basic constituents of chaos theory, considering the dynamic behavior and random appearance, we generated three substitution boxes $K^{(1)}, K^{(2)}, K^{(3)}$ are enerated, each of them helps us to successively encrypt the channels of the original RGB image (red, green, blue). This random permutation is done by digram (sequence of two digits), which means that each pixel of the image to be encrypted with the red color will be replaced by the cell of position (i, j). In fact, the value of i is extracted from the channel $K^{(1)}$ with an increasing step and j is obtained from the channel $K^{(2)}$ of decreasing step. While the modification of the green matrix is done by the choice of the channels $K^{(2)}$ and $K^{(3)}$, on the other side, the blue color uses the channels $K^{(3)}$ and $K^{(1)}$ while following the same principle. On the purpose of making the present approach more efficient in the unreadable and fuzzy image while increasing the decryption time, we resort to the application of an XOR mask between each substituted matrix and each examined matrix with a bilinear transformation of different steps. Based on several criteria such as histogram, differential attacks, entropy, and correlation analysis, this approach has yielded experimental results that justify its performance and reliability in protecting data against any malicious attack. In addition, it marks its competence in front of other existing methods in the literature.

Keywords: Chaos · Clifford attractors · Security · S-Box · Image encryption · XOR

1 Introduction

Security and the threats that come with it are the biggest obstacle to gaining people's trust and participation in the advancement of internet. The question of

M. Lazaar et al. (Eds.): BDIoT 2021, LNNS 489, pp. 116–127, 2022.
https://doi.org/10.1007/978-3-031-07969-6_9

maintaining internet security by adopting simple and economical means remains of the main problems which currently poses a great challenge.

In order to face this security challenge and in order to satisfy Internet users, several researchers have looked into the choice and development of techniques in recent years, among which the chaotic cryptography.

And since chaos is sensitive to initial conditions and also to control parameters, pseudo-randomness and ergodicity [1,2], it proves its basis for masking or rearranging information in a secure transmission. First in 1980, Chaos was used by Matthews to encrypt information, and he was able to design a key flow based on a logistics map [3].

Habustu et al. (1991) proposed the first chaotic block cipher method [4].

Baptisa (1998) wrote an article on chaotic encryption algorithm [5]. A sum of 1D chaotic maps has been exploited in data encryption and to encrypt the image, and chaotic maps of 2D dimension or greater have been adopted.

Friedrich, (1998) proposed that an image encryption scheme should be iterated in two phases: diffusion and confusion [6,7]. And even chaos-based cipher schemes consist of two stages: one is the chaotic confusion of pixel positions by the process of substitutions, and the other is the distribution of the gray values of the pixels by a diffusion process. The pixels of the image rearrange with a chaotic 2D map. The so-called map changes the values of each of the pixels one by one so that a change, even a small one, can affect almost all the pixels of the entire image.

In (2004), during the substitution phase of image encryption, the 3D chaotic map of Chat and that of Baker were adopted by Chen et al. and Mao et al. [8,9]. Guam et al. (2005) took a 2D cat map and Chen's chaotic system for rearranging the position of pixels and scrambling their value [1].

Giesl (2009) [10] proposed a new image encryption technique that uses a strange attractor. Hua, Z.Y., Zhou, Y.C., Pun, C.M. and Chen, C.L. (2015) [16] created a two-dimensional sinusoidal logistic modulation map (2D-SLMM) and they proposed a chaotic magical transformation (CMT) to encrypt the images The latter is done by permutation of the positions of pixels of the original image using the new proposed 2D-SLMM map.

HEGUI, Z. YIRAN, Z. and YUJIA S. (2019) [18] suggested an image encryption method based on the LSMCL modulated two-dimensional logistic chaotic map-sinusoidal coupling with two cycles of permutation and diffusion operation.

Currently, in consideration of the properties of chaotic systems that has been mentioned previously, chaos is becoming widespread in cryptography [19–26] and it has proven its feasibility and its coding speed power [12,13]. The rest of the article is organized as follows:

Section 2: presents the operation of Clifford attractors.

Section 3: describes the main steps of our new image encryption scheme.

Section 4: shows the experimental results and statistical analyses of this new approach and its effectiveness against other techniques existing in the literature.

Section 5: presents the conclusion of our work.

2 Clifford Attractor

In mathematics, a dynamic system is a set of functions that describes the time dependence of a point in a geometric space.

An attractor is a set of numerical values towards a dynamic system which tend to evolve in discrete time whatever their starting conditions. Simple attractors can be sets of points or limit cycles.

The notion of attractor can be demonstrated from the iterations of a mathematical function, i.e. the behavior of the sequence $x = U_k, f(x) = U_{k+1}, f(f(x)) = U_{k+2}$, where f is a particular function. The initial value of x is called the seed.

The most interesting attractors are the strange attractors called chaotic attractors, that cover a range of possible states in which a dynamic system can move without repeating itself. Strange attractors are graphs of relatively simple formulas. They are created by repeating (or iterating) a formula over and over and using the results on each iteration to plot a point. The result of each iteration is fed back into the equation. After millions of points have been drawn, fractal structures appear. The repeated points are part of a basin of attraction (they are attracted by the points that make up these shapes). All points, regardless of their starting point, tend to end by following the same path. Another interesting property of strange attractors is that the points that are very close at one time can differ considerably in the next step.

In our work, Clifford attractors are used and defined by these equations:

$$\begin{cases} U_{k+1} = \sin(aV_k) + c\cos(bU_k) \\ V_{k+1} = \sin(bU_k) + d\cos(aV_k) \\ W_{k+1} = \sin(bU_k) + \mathrm{f}\cos(aV_k) \end{cases} \tag{1}$$

Given six values (one for each parameter) and a starting point (U_0, V_0, W_0), the previous equation defines the exact location of the point in step n, which is defined simply by its location in $n - 1$. An attractor can be thought of as the trajectory described by a particle. Moreover, there is an infinity of attractors, since the real numbers a, b, c, d, e and f are parameters. These attractors were discovered by Clifford Pickover, an American science discloser.

3 Proposed Approach

This part shows step by step the process followed by our new RGB image encryption scheme, which is based on Clifford attractors. Due to the random behavior of the latter, it has been possible to amalgamate any resemblance between the source image and the recipient image.

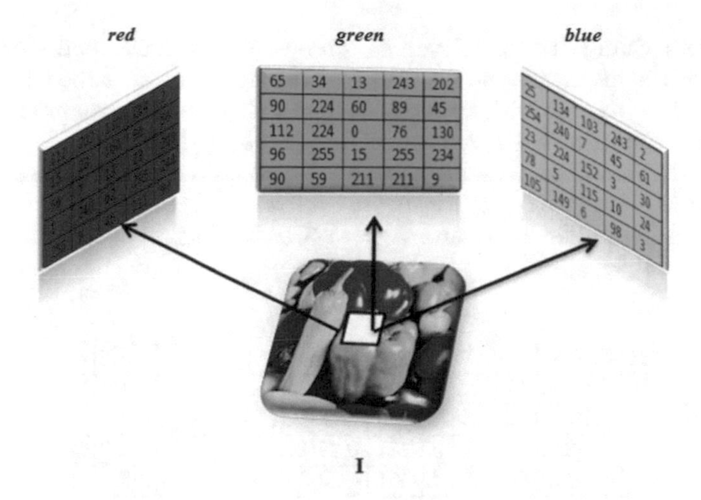

Fig. 1. Fig. Example of splitting a source image into three planes R, G and B.

Step1: Load an original RGB I image of size $N \times M$ (Fig. 1) and cut it into three planes R, G and B.

Step2: generate the key $= a, b, c, d, e, f, \sigma_1, \sigma_2, \sigma_3$ where a, b, c, d, e and f represent the initial conditions of the Clifford attractors and the integers $\sigma_1, \sigma_2, \sigma_3$ are the control parameters of these bilinear applications.

Step3: Generate respectively three matrices U_k, V_k and W_k, of size $N \times M$ of the Clifford attractors using Eq. (1) With $k = 0, 1, ..., (N \times M - 1)$.

Step4: Create three matrices $K^{(1)}, K^{(2)}, K^{(3)}$ used for the iterative rearrangement respectively of the R, G, B channels.

Each of them contains the order of each pixel according to the values of each column respectively for each matrix U_k, V_k and W_k as shown in Fig. 2.

Step5: Mix each color channel by two boxes of substitutions by the P2SB transformation according to the following formula:

$$T = P2SB(I)/ \quad I(i,j) = k \text{ with } k \in [0, 255]$$

P2SB is the permutation of the pixels by the 2 substitution boxes, where Each pixel value to coordinate (i, j) in the red color matrix is permuted by $(i^{'}, j^{'})$ existing in the same channel of this image whose the point of the coordinates of abscissa $(i^{'})$ corresponds to the pixels of the first matrix $K^{(1)}$ and of ordinate $(j^{'})$ corresponds to the pixels of the second matrix $K^{(2)}$, with the choice of pixels which is done in an increasing way in the substitution box $K^{(1)}$ and decreasing in the box $K^{(2)}$. For the green color, we take the channels $K^{(2)}, K^{(3)}$, on the other hand the blue color uses the channels $K^{(3)}, K^{(1)}$ see (Fig. 3).

Step6: change the pixel values, and apply an XOR operator for each channel with a mathematical sequence of different steps defined by the relation $S_{k\sigma}(i,j) = \sigma i + j$, with i represents the i th row and j represents the j th column of the matrix, $k \in 1,2,3$ and $\sigma \in \sigma_1, \sigma_2, \sigma_3$ (Fig. 4).

Example

The parameters we used are: $a = 1.5, b = -1.2, c = 0.4, d = 1.7, e = -1.3, f = -1.4$

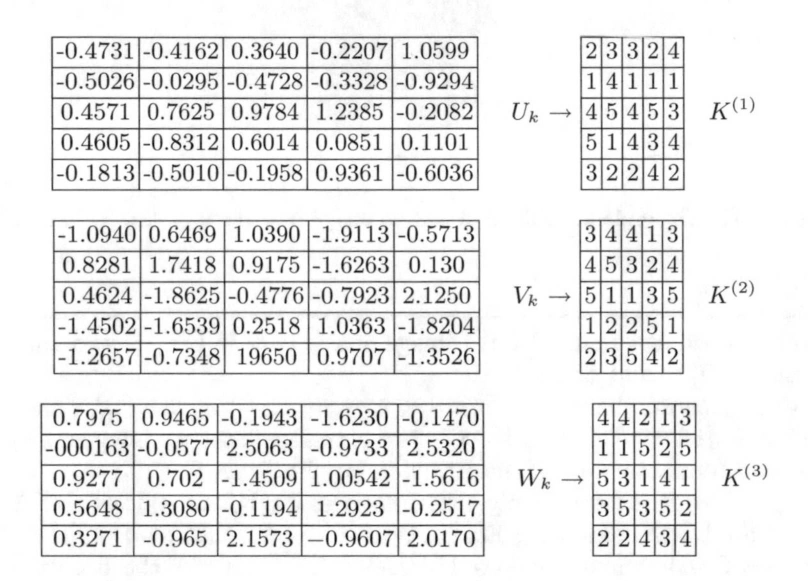

-0.4731	-0.4162	0.3640	-0.2207	1.0599
-0.5026	-0.0295	-0.4728	-0.3328	-0.9294
0.4571	0.7625	0.9784	1.2385	-0.2082
0.4605	-0.8312	0.6014	0.0851	0.1101
-0.1813	-0.5010	-0.1958	0.9361	-0.6036

$U_k \rightarrow$ $K^{(1)}$

2	3	3	2	4
1	4	1	1	1
4	5	4	5	3
5	1	4	3	4
3	2	2	4	2

-1.0940	0.6469	1.0390	-1.9113	-0.5713
0.8281	1.7418	0.9175	-1.6263	0.130
0.4624	-1.8625	-0.4776	-0.7923	2.1250
-1.4502	-1.6539	0.2518	1.0363	-1.8204
-1.2657	-0.7348	19650	0.9707	-1.3526

$V_k \rightarrow$ $K^{(2)}$

3	4	4	1	3
4	5	3	2	4
5	1	1	3	5
1	2	2	5	1
2	3	5	4	2

0.7975	0.9465	-0.1943	-1.6230	-0.1470
-000163	-0.0577	2.5063	-0.9733	2.5320
0.9277	0.702	-1.4509	1.00542	-1.5616
0.5648	1.3080	-0.1194	1.2923	-0.2517
0.3271	-0.965	2.1573	−0.9607	2.0170

$W_k \rightarrow$ $K^{(3)}$

4	4	2	1	3
1	1	5	2	5
5	3	1	4	1
3	5	3	5	2
2	2	4	3	4

Fig. 2. An example of the construction of substitution boxes.

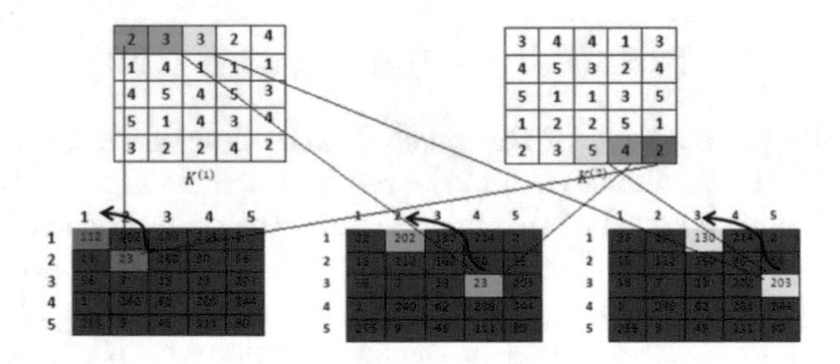

Fig. 3. Example of the transformation of the image red color to be encrypted by the two substitution boxes $K^{(1)}, K^{(2)}$

(112) is associated with the cording point $(1, 1)$ and it is replaced by (23) which is in position $(2, 2)$.

(2) is taken from the matrix $K^{(1)}$ which is generated from the Clifford attractors and (2) is taken from the end of the matrix $K^{(2)}$. Notice that during this substitution the choice of pixels is done in an increasing way in the channel $K^{(1)}$ and decreasing in the channel $K^{(2)}$.

So as you see the first row of the matrix red $(112, 202, 130, 214, 2)$ is replaced by the points of lines $((2, 2), (3, 4), (3, 5), (2, 3), (4, 2))$ which gives as result $(23, 23, 203, 160, 240)$.

For the second line $(15, 23, 160, 90, 56)$ will be substituted by $((1, 1), (4, 5), (1, 2), (1, 2), (1, 1))$ which result in $(112, 244, 202, 202, 112)$, and the same treatment is repeated for the following lines. and so on for the other colors.

Flowchart of the proposed method

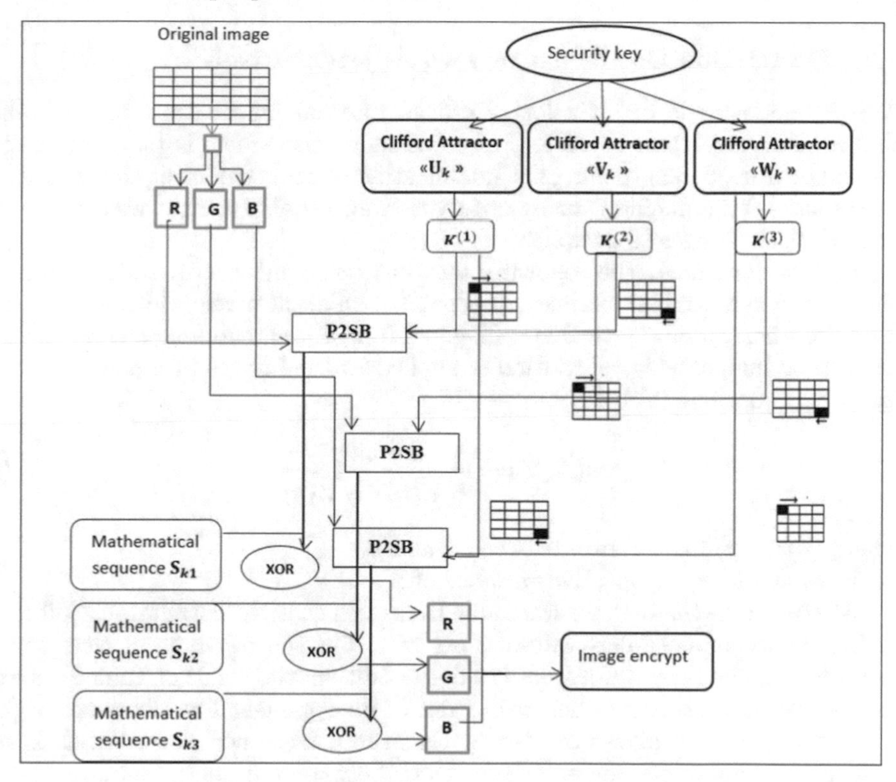

Fig. 4. The flowchart of the proposed method

4 Statistical Analyzes

In order to be able to verify how reliable a cryptographic system is and also to assess its resistance, it has been confronted with different attacks,

namely: analytical cryptography, differential attack and statistical analysis.

4.1 Histogram

To be able to make a comparison between the graphic representation of the distribution of the intensities of the encrypted images and the original ones, various criteria can be used and among which histogram is mentioned. Therefore, comparison is made between the histogram of 4 original images with completely different content and the histogram of their images encrypted by the new adopted scheme.

From Fig. 4, we can easily deduce that the distribution of pixels in the histograms of the encrypted images is almost uniform, while the distribution of the original images has both very high and very short values. The difference between the two histograms is very apparent and can only lead us to say that this new approach demonstrates a great performance with regard to the security of the images against any statistical attack.

4.2 Correlation Coefficient of Two Adjacent Pixels

Any image is made up of a lot of crucial internal information, among other things "the correlation coefficient of adjacent pixels" which is targeted by statistical attacks. Strengthening the qualification of said coefficient thus proves to be essential, by reducing it or by clearly getting rid of it so that we can get the most difficult images to decipher.

In this document, the following test has been put into practice: we have chosen and in a random manner, 2000 pairs of adjacent pixels which follow three directions horizontal, vertical and diagonal from the source image and from the 'encrypted image. This calculation is applied to the images in Fig. 5, using the following equations (Table 1):

$$\rho(X;Y) = \frac{\mathrm{Cov}(x;y)}{\sqrt{V(x)} \cdot \sqrt{V(y)}} \tag{2}$$

where $Cov(x;y)$: covariance between x and y.

$V(x)$ and $V(y)$: respective variance of x and y

At the end of the test, we were able to deduce that the correlation coefficient of the source image takes values close to 1. On the other hand that of the encrypted image takes values close to 0. In addition, the result of Table 2 showed us that our new scheme is better in front of its opponent like Behrouz FV [17] and Tong [14]. This allows one to say that the present new contribution is the most competent to destroy any dependency between adjacent pixels.

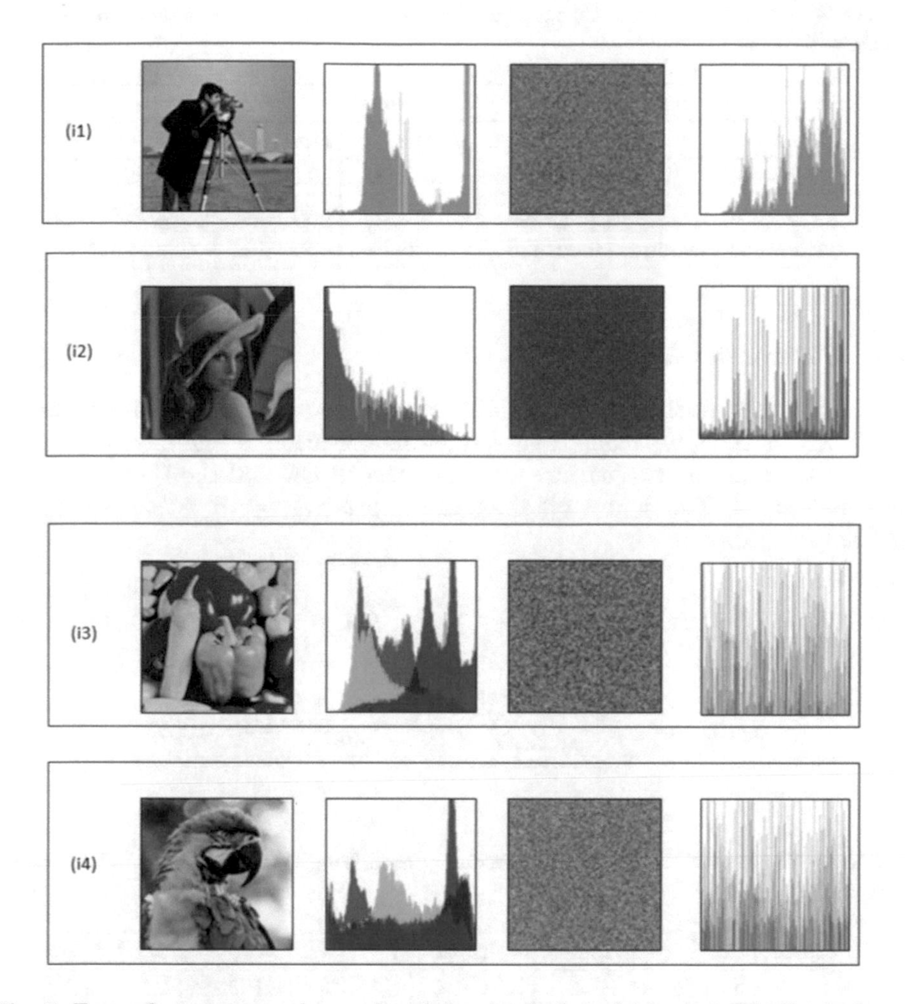

Fig. 5. Example 1 presents the results of the simulation of 4 images. The first column denotes the original images, the second column shows their histograms, and the third and fourth columns denote the encrypted images and their histograms. ($i1$) cameraman image, ($i2$) Lena image, ($i3$) Image peppers, ($i4$) parrot image.

Table 1. Correlation coefficient of the original images and the corresponding numbered images

Image	Original image			Encrypted image		
	Horizontal	Vertical	Diagonal	Horizontal	Vertical	Diagonal
Peppers	0.957970	0.9387402	0.948944	9.9790E–4	1.48022 E–4	0.0034479
Lena	0.9760746	0.9337169	0.9386612	0.001875	1.23753E–4	3.82480E–4

Table 2. Comparison of the Correlation Coefficient of two adjacent pixels of our approach with other algorithms.

Encrypted image	Directions			Average
Lena	Horizontal	Vertical	Diagonal	
Behrouz F-V [17]	-0.009264	-0.000945	-0.000284	-0.0034976
Tong [14]	0.003800	0.005800	0.013300	0.007633
Proposed technique	0.001875	1.23753E-4	3.82480E-4	0.0007937

4.3 Differential Attacks

In order to ensure that this method is robust against differential attacks, on the image of Lena, we made a very small modification at the level of a single pixel of the said image, then, we measured the NPCG and the UACI as shown by Eqs. 3 and 4. The figures obtained are respectively greater than 99.6% and 33.8% (Table 3).

$$NPCR = \frac{1}{M \times N} \sum_{i=1}^{M} \sum_{j=1}^{N} D(i,j) \times 100\% \tag{3}$$

$$UACI = \frac{1}{M \times N} \sum_{i=1}^{M} \sum_{j=1}^{N} \frac{|C_1(i,j) - C_2(i,j)|}{255} \times 100\% \tag{4}$$

Table 3. The sensitivity of our approach by modifying a single pixel of the original image.

Image	Proposed method		Behrouz F-V [17]		Norouzi [15]	
	NPCR	UACI	NPCR%	UACI%	NPCR%	UACI%
Lena	99.71633	33.64781	99.626668	33.567139	99.66890	33.55610

4.4 Information Entropy Analysis

The measure of the disorganization degree of the image gray values can be obtained by the information image entropy. Whitch is defined as:

$$H(s) = \sum_{i=0}^{2^n-1} P_i \log_2[P_i] \tag{5}$$

P_i denotes the pixel occurrence probability.

Table 4. Entropy of the encrypted image and the original image.

Images	Original	Encrypted
	Entropy	Entropy
Lena	7.610197	7.99779
Peppers	7.737281	7.993955

Table 4 shows the obvious approach of the entropy of the encrypted images towards the value 8 which is the ideal value, which leads us to justify the efficiency of our approach.

5 Conclusion

This work evokes a very important contribution in the field of RGB images encryption which is based on one of the fundamental constitutive elements of the chaos theory (the attractors of Cliford). With the dynamic attitude and the random aspect of this last, three substitution channels were created $K^{(1)}, K^{(2)}$ and $K^{(3)}$, each of them allowed us to encrypt the matrices in the original RGB image one after the other (red, green, blue). This substitution means that each pixel of the source image for each color will be replaced by the cell which has as position (i, j), the value of i is taken from the channel $K^{(1)}$ and the value of j is extracted from the end of the channel $K^{(2)}$ knowing that the encryption of the th position of a matrix to be encrypted needs to take the values of increasing steps for $K^{(1)}$ and a decreasing step for $K^{(2)}$. On the other hand, changing the green matrix is done by choosing the channels $K^{(2)}$ and $K^{(3)}$, whereas that of the blue matrix is done using the channels $K^{(3)}$ and $K^{(1)}$ and always following the same principle. In order to strengthen the performance of our method, by making the image confidential, we implemented the application of an XOR mask between each of the rearranged matrices and a matrix generated with a bilinear application of different pitches.

References

1. Arrowsmith, D., Place, C.: Dynamical Systems: Differential Equations, Maps, and Chaotic Behavior. Chapman and Hall, New York (1992)
2. Kocarev, L., Lian, S.: Chaos-Based Cryptography: Theory, Algorithms and Applications. Springer, Heidelberg (2011). https://doi.org/10.1007/978-3-642-20542-2
3. Matthews, R.: On the derivation of a chaotic encryption algorithm. Cryptology **8**, 29–41 (1989)
4. Habutsu, T., Nishio, Y., Sasase, I., Mori, S.: A secret key cryptosystem by iterating a chaotic map. In: Davies, D.W. (ed.) EUROCRYPT 1991. LNCS, vol. 547, pp. 127–140. Springer, Heidelberg (1991). https://doi.org/10.1007/3-540-46416-6_11
5. Baptista, M.S.: Cryptography with chaos. Phys. Lett. A **240**, 50–54 (1998)
6. Fridrich, J.: Image encryption based on chaotic maps. In: Proceedings of IEEE Conference on Systems, Man, and Cybernetics, pp. 1105–1110 (1997)

7. Fridrich, J.: Symmetric ciphers based on two-dimensional chaotic maps. Int. J. Bifurcat. Chaos **8**(6), 1259–1284 (1998)
8. Chen, G., Mao, Y.B., Chui, C.K.: A symmetric image encryption scheme based on 3D chaotic cat maps. Int. J. Chaos Solitons Fractals **21**, 749–761 (2004)
9. Mao, Y.B., Chen, G., Lian, S.G.: A novel fast image encryption scheme based on the 3D chaotic baker map. Int J Bifurcat. Chaos **14**(10), 3613–3624 (2004)
10. Giesl, J., Vlcek, K.: Image encryption based on strange attractors. ICGST-GVIP J. **9**(2), 19–26 (2009)
11. Lian, S., Sun, J., Wang, Z.: A block cipher based on a suitable use of the chaotic standard map. Chaos Solitons Fractals **26**(1), 117–129 (2005)
12. Elazzaby, F., El Akkad, N., Kabbaj, S.: A new encryption approach based on four-square and zigzag encryption (C4CZ). In: Bhateja, V., Satapathy, S.C., Satori, H. (eds.) Embedded Systems and Artificial Intelligence. AISC, vol. 1076, pp. 589–597. Springer, Singapore (2020). https://doi.org/10.1007/978-981-15-0947-6_56
13. Elazzaby, F., EL akkad, N., Kabbaj, S.: Advanced encryption of image based on S-box and chaos 2D (LSMCL). In: 1st International Conference on Innovative Research in Applied Science, Engineering and Technology (IRASET) (2020)
14. Tong, X.J., Wang, Z., Zhang, M., Liu, Y., Xu, H., Ma, J.: An image encryption algorithm based on the perturbed high-dimensional chaotic map. Nonlinear Dyn. **80**(3), 1493–1508 (2015). https://doi.org/10.1007/s11071-015-1957-9
15. Norouzi, B., Seyedzadeh, S.M., Mirzakuchaki, S., Mosavi, M.R.: A novel image encryption based on hash function with only two-round diffusion process. Multimedia Syst. **20**(1), 45–64 (2013). https://doi.org/10.1007/s00530-013-0314-4
16. Hua, Z.Y., Zhou, Y.C., Pun, C.M., Chen, C.L.: 2D Sine logistic modulation map for image encryption. Inf. Sci. **297**, 80–94 (2015)
17. Fathi-Vajargah, B., Kanafchian, M., Alexandrov, V.: Image encryption based on permutation and substitution using Clifford Chaotic System and logistic map. J. Comput. **13**(3), 309–326 (2018). https://doi.org/10.17706/jcp.13.3.3
18. Zhu, H., Zhao, Y., Song, Y.: 2D logistic-modulated-sine-coupling-logistic chaotic map for image encryption. IEEE Access **7**, 14081–14098 (2019). https://doi.org/10.1109/ACCESS.2019.2893538
19. El Akkad, N., Merras, M., Saaidi, A., Satori, K.: Camera self-calibration with varying parameters from two views. WSEAS Trans. Inf. Sci. Appl. **10**(11), 356–367 (2013)
20. El Akkad, N., Merras, M., Saaidi, A., Satori, K.: Robust method for self-calibration of cameras having the varying intrinsic parameters. J. Theor. Appl. Inf. Technol. **50**(1), 57–67 (2013)
21. El Akkad, N., El Hazzat, S., Saaidi, A., Satori, K.: Reconstruction of 3D scenes by camera self-calibration and using genetic algorithms. 3D Res. **7**(1), 1–17 (2016)
22. Es-Sabry, M., El Akkad, N., Merras, M., Saaidi, A., Satori, K.: A novel text encryption algorithm based on the two-square cipher and caesar cipher. In: Tabii, Y., Lazaar, M., Al Achhab, M., Enneya, N. (eds.) BDCA 2018. CCIS, vol. 872, pp. 78–88. Springer, Cham (2018). https://doi.org/10.1007/978-3-319-96292-4_7
23. Es-sabry, M., El Akkad, N., Merras, M., Saaidi, A., Satori, K.: A new color image encryption algorithm using random number generation and linear functions. In: Bhateja, V., Satapathy, S.C., Satori, H. (eds.) Embedded Systems and Artificial Intelligence. AISC, vol. 1076, pp. 581–588. Springer, Singapore (2020). https://doi.org/10.1007/978-981-15-0947-6_55
24. Es-Sabry, M., El Akkad, N., Merras, M., Saaidi, A., Satori, K.: Grayscale image encryption using shift bits operations. In: International Conference on Intelligent Systems and Computer Vision, pp. 1–7 (2018)

25. Es-Sabry, M., El Akkad, N., Merras, M., Saaidi, A., Satori, K.: A new image encryption algorithm using random numbers generation of two matrices and bit-shift operators. Soft Comput. **24**(5), 3829–3848 (2019). https://doi.org/10.1007/s00500-019-04151-8
26. Essaid, M., Akharraz, I., Saaidi, A., Mouhib, A., Mohamed, E., Ismail, A., Abderrahim, S., Ali, M.: A new color image encryption algorithm based on iterative mixing of color channels and chaos. Adv. Sci. Technol. Eng. Syst. J. **2**, 94–99 (2017)
27. Touil, H., El Akkad, N., Satori, K.: Text encryption: hybrid cryptographic method using vigenere and hill ciphers. In: 2020 International Conference on Intelligent Systems and Computer Vision (ISCV), Fez, Morocco, pp. 1–6 (2020)
28. Touil, H., El Akkad, N., Satori, K.: H-rotation: secure storage and retrieval of passphrases on the authentication process. Int. J. Saf. Secur. Eng. **10**(6), 785–796 (2020)
29. El Akkad, N., Merras, M., Baataoui, A., Saaidi, A., Satori, K.: Camera self-calibration having the varying parameters and based on homography of the plane at infinity. Multimedia Tools Appl. **77**(11), 14055–14075 (2017). https://doi.org/10.1007/s11042-017-5012-3
30. El Akkad, N., Saaidi, A., Satori, K.: Self-calibration based on a circle of the cameras having the varying intrinsic parameters. In: Proceedings of 2012 International Conference on Multimedia Computing and Systems, ICMCS, pp. 161–166 (2012)

An Overview of Security in Vehicular Ad Hoc Networks

Nada Mouchfiq(✉), Ahmed Habbani, Chaimae Benjbara, and Halim Berradi

SSLab, ENSIAS, Mohammed V University in Rabat, Rabat, Morocco
nada_mouchfiq@um5.ac.ma

Abstract. In recent decades, the importance of security has increased more and more especially in the field of research dealing with the Internet of Things (IoT) and implicitly ad hoc networks since they are part of it, considering they undergo several changes and that following these changes this area is exposed to several risks. Many researchers have shed light on different techniques to ensure the security of this intelligent universe. In this study, we identify the risks to which Vehicular ad hoc networks (VANETs) are exposed according to several criteria and metrics. For this purpose, we do bench-marking by carrying out research on the types of attacks that harm VANETs, after that we study the solutions proposed to ensure a high level of security in this type of network and finally we discuss these solutions to find out one that is most adapted to our teamwork and which is based on a relevant and recent technology.

Keywords: Attacks · VANETs · Security · Blockchain · Ad hoc · IoT

1 Introduction

The new era of communication that invades our world increases the trend of the existence of the Internet of Things (IoT) [1]. Because it is based on the intelligence of communication between different devices with the physical world and the human being, in order to carry out tasks in a more advanced way through system integration, and this necessitates, of course, that the transmitted information containing sensitive data must protect the privacy of individuals and organizations.

Especially since the IoT helps people lead an intelligent life considering these technologies offer devices that automate homes, hospitals, businesses .... It also provides a real-time summary of the data and performance of these devices [2], via increasing efficiency and accountability, this enhances performance [3].

The Internet of Things refers to a type of network that can connect anything to the Internet. Figure 1 illustrates a set of domains influenced by this type of IoT technology, including the automotive domain, which is part of ad hoc networks where nodes are mobile vehicles constituting a highly predictable architecture as they move, called Vehicular Ad Hoc Networks (VANETs).

M. Lazaar et al. (Eds.): BDIoT 2021, LNNS 489, pp. 128–140, 2022.
https://doi.org/10.1007/978-3-031-07969-6_10

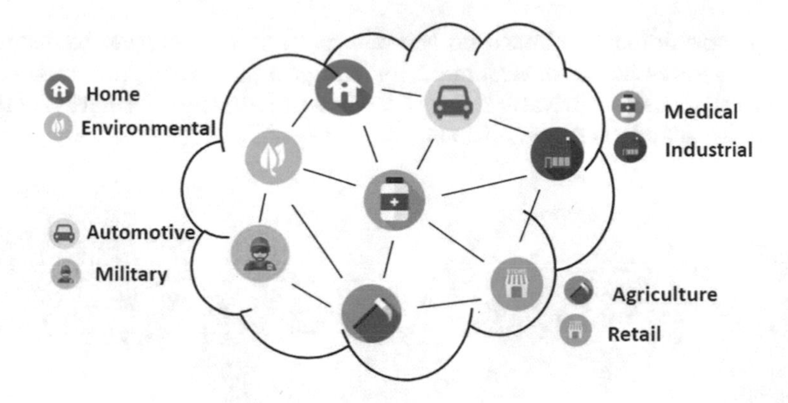

Fig. 1. IoT application domains

Like every network, the ad hoc network has its own challenges, including routing and multicasting, quality of service (QoS), energy, security and mobility. So VANETs will face the same challenges. In our article we will focus on security within this type of network.

This paper is organized as follows. Section 2 gives an overview about Vehicular ad hoc networks (VANETs), Sect. 3 discusses security inside VANETs, which lead us to the discussion in Sect. 4, and finally, our research is concluded in Sect. 5.

2 Vehicular Ad Hoc Networks "VANETs"

The intelligent transportation system is progressing and adapting to the pace of modern life. In this context, ad hoc vehicle networks are a fundamental element, as they are gaining in popularity due to the unique features they contain and the services they efficiently provide.

VANET is a network that is characterized by a heterogeneous architecture (Fig. 2) grouping three domains distributed according to the source and destination of the information exchanged [4]:

- The mobile domain is essentially composed of two types of components: Vehicle component (buses, cars, ...) and Mobile component including smartphones with GPS ...
- The infrastructure domain includes two main units that exchange information necessary to ensure the safety of vehicles and the proper functioning of the entire network: Roadside Units (RSUs), represented by sensors installed on the roads to collect information about the vehicles in their network. These data will be collected by the central infrastructure unit, which uses them for traffic lights to analyze data useful for traffic forecasting.

– The generic domain is linked on the one hand to the internet to ensure the storage of data in the environment for analysis purposes. And on the other hand to the local (or private) network to disseminate the information collected on accidents and traffic conditions.

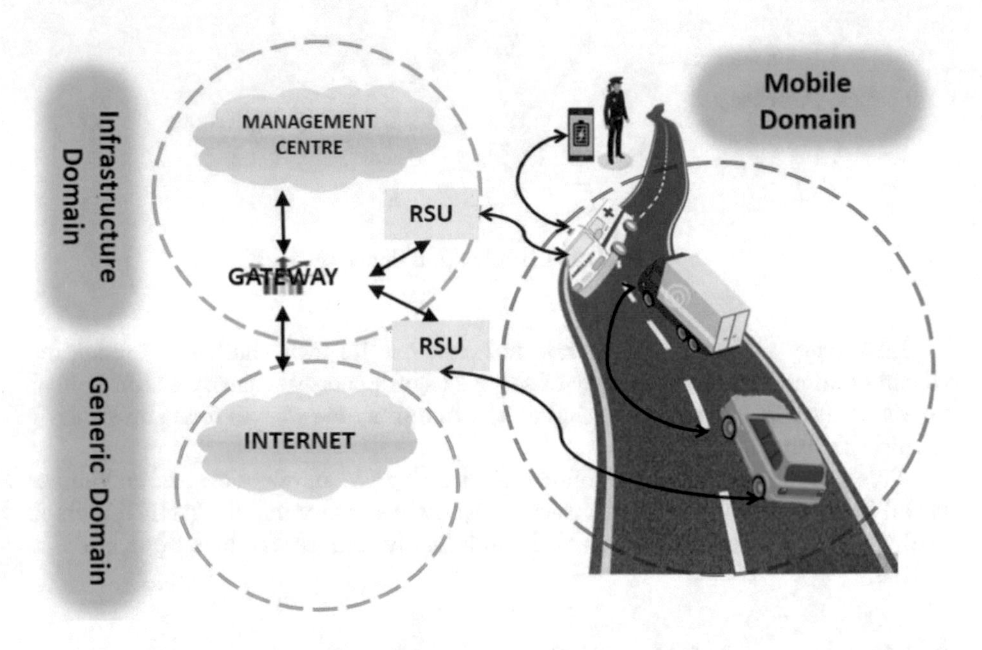

Fig. 2. VANETs architecture

This special architecture dedicated to vehicular networks offers them their own characteristics [5] like:

– High Mobility
– Frequent exchange of information
– Rapidly changing network topology
– Wireless Communication
– Time Critical
– Better Physical Protection
– Unbounded network size
– Sufficient Energy

And also allows them to have a diversity of communication within the same network linked to the elements that exchange information. So, we find [6]:

– Vehicle-to-vehicle (V2V) communication, whose main objective is to prevent possible accidents by allowing vehicles on the road to transfer data on their position and speed within the network. One of the biggest challenges of V2V technology is to coordinate network security, automation of alarm and braking systems in case of emergency.

- Vehicle-to-Infrastructure (V2I) which represents a communication model that ensures the movement of vehicles to relay information to other components of the network in question, including: signals, panels, cameras, RFID (Radio Frequency IDentification), sensors. For example, in the case where RSUs regularly send messages to avoid speed limits if there are vehicles or alarms nearby, the driver must react and reduce the speed of his car.

Table 1. VANETs applications

Layer	Applications
Traffic control	Route navigation
	Red lights
	Traffic signs
	Ambulances on the road
	Civilian surveillance
	Vehicle tracking
	Weather information
	Information on parking
	Road Map update
	Dangers on the road
	Collisions on the road
	Road condition
Environment control	Home control
	Internet access
	File sharing
	Gas stations availability and prices
	Restaurant locations
	Service shops with updated deals
	Tourist information
	Signs information
Assistance information	Emergency breaking
	Crashes notification
	Work zones warnings
	Speed limits
	Road slip warnings
	Lanes changing
	Traffic updates

The decentralized nature of VANETs and the fact that they allow users to connect and exchange data regardless of their geographical location or adjacency to the infrastructure, makes them more versatile and robust and extends their application domains as shown in (Table 1) [7].

The architecture of the VANETs network, its characteristics and also its variety of communication make it demanding on several aspects [8]: high latency, need of high computational ability, irregular network density, constraint related to variation of link and device capabilities, security, privacy...

But, they also make it vulnerable to multiple attacks, so many threats can damage VANET networks. In the table below (Table 2), we can find different attacks in each of the 5 VANET layers [9].

Table 2. VANETs attacks

Layer	Attacks
Physical Layer	Overhearing
	Interference
	Denial of Service (DoS)
	Jamming
	Intercepting
MAC	Adding or removing bits in ongoing transmission
	Sparse of bandwidth for transmission
	Transmit unwanted information
Network Layer	Routing cache poisoning attack
	Rushing attack
	Wormhole attack
	Black hole attack
	Resource consumption attack
Transport Layer	SYN flooding attack
	Session Hijacking
	TCP ACK Storm DoS attacks
Application Layer	Malicious code attack
	Repudiation attack
	DoS attack

To remedy this security problem within this network, several solutions have been proposed to certain security problems based on transmission techniques [10] represented in the below table (Table 3):

Table 3. VANETs security solutions

Transmission technique	Security solutions
Routing based	Dynamic source routing
	Advanced Encryption Standard
	Cuckoo search algorithm
	Stochastic routing algorithm
Trust based	Blow fish
	Trace back
	Meta heuristic
	Secure Hash algorithm
Cluster based	Black hole
	Fuzzy
	Gravitational search algorithm
	Genetic algorithm
	Ant Colony algorithm
Location based	Dynamic privacy protection
	Triangle based security algorithm
	Source location privacy
	Confused arc Service Location Protocol

According to the literature, full insurance covering all components [11] should be included in the security solution. Arguably, safety becomes the most critical challenge in VANETs. So, to meet this need we must find a solution that can be able to organize all aspects of security:

- Confidentiality (C): Ensure that data packets and configuration parameter information are not accessed or appropriated to an attacker or disclosed to unknown entities.
- Integrity (I): ensure that the packets exchanged (or stored) have not been modified in an unauthorized manner (ensure that the data is not modified by unauthorized agents).
- Availability (AV): Ensure that the packets exchanged are always available and that devices and agents are not prevented from having access to the information.
- Authentication (Au): Authentication makes it possible to decide if a person is really what he claims to be in terms of information security by following an identity authentication procedure by checking the identification information presented compared to that stored.
- Non-Repudiation (NR): Ensure that any person or any other entity engaged in communication by computer, cannot deny having received or sent a message.
- Access control (AC): refers to the selective restriction of access to a place or other resource or network, while access management describes the process.

The act of access can manifest itself as consumption, access or use; the operation of accessing a resource is called authorisation and in this process, two analogous access control mechanisms are locks and login identifiers.

3 Security of Vehicular Ad Hoc Networks

Several approaches and propositions have been put forward to deal with the security challenge in vehicular ad hoc networks in various ways. For example, the authors of [12] proposed using overhearing and autonomous agents in wireless Ad-Hoc networks for the detection of node-misbehavior, for this purpose, they opted for two basic techniques in order to detect a malicious node. Firstly they were based on the heating theory to which they gave the name Neighbor Misconduct Misbehavior (OMD), this approach is dedicated to node listening and packet transmission rate (PFR) measurement for each node by overhearing the transmissions of its neighbors and calculating packet forwarding ratio of its own as well as its neighbors. Source node uses the calculated information to identify a misbehaving node, and secondly, they concentrated on the detection of behavioral anomalies in the detection of improper behavior based on an autonomous agent that they baptized (AAMD), an agent residing at a node is activated using the activation key generated by a trusted third party to check the misbehavior of the node. The authors of [13], for their part, have chosen to enhance the Efficiency of Routing Protocols in VANETS by defending Dos Attacks using BF-IPCM (Bloom-Filter based IP Choke Mechanism). The Bloom filter-based DoS detection scheme is a class using a combination of reactive and proactive approaches. The proactive approach takes place in order to maintain the new nodes and the reactive approach is used to determine all connected vehicles (nodes). This type of detection scheme promotes an increase in the system's ability to function properly with higher workloads through a near area by the nearest vehicles and this leads to the reduction of the bandwidth, collisions and calculated values. This proposition allows localisating, identifying and blocking the malicious vehicle form the routing process in order to defend the attacked vehicle from it.

For ensuring authentication and preserving privacy in VANETs, the authors of [14] proposed a ticket-based scheme, their approach is based on temporary ticket to guarantee the privacy of the vehicles and for the authentication of nodes in VANETs, they used identity-based signature mechanism, so that each message is signed by the nodes (vehicles) and each one has its own set of tickets but only one of these tickets can be used for each time slot so malicious nodes causing DoS attacks are avoided. The operation of creating ticket is divided in two phases. The first is the offline process, in which the vehicle gets the identification information from the appropriate trusted authority (TA). The second step is an online part, in which the vehicle requests to sign its certificate and form a ticket based on it. When a vehicle behaves improperly, the TA recovers his true identity, decides on his sanction depending on the seriousness of the offence of his misconduct and includes his title(s) in cancellation list(s). At the beginning of every time period, the respective cancellation list is sent to the RSUs. These exclude misconduct of vehicles by not signing credentials from the revocation lists they received. The authors of [15] propose a privacy-preserving authentication scheme with full aggregation in VANET, to achieve this, they opted for using a signature method without creating an overhead and thus authentication is ensured. This method guarantees that the duration of the aggregates is constant, reduces storage and communication while the reduction of the calculations performed is ensured by the pre-calculation mechanism. Once the real identity RIDi of a vehicle i has been captured, it uniquely identifies the vehicle (e.g., the licence plate number), TAs generate the pseudo identity of the vehicle. By entering a security parameter, the TAs respectively generate the secret/public key pair and publish the public system parameters. At the input, an aggregated signature without certificate on n messages, n pseudo identities and public keys, outputs from the application server decide then whether the signature without certificate is valid or not.

The method proposed by the authors of [16] follow the track of detection and prevention of distributed denial of service attacks in VANETs, this advanced process makes it possible to differentiate between a normal node and a malicious node by taking as a benchmark the comparison between their periods of communication during a precise delay by controlling the number of packets injected into the network in such a way that the streams of these packets is minimized and that this practically does not cause overloads on the network resources. To achieve this, they took advantage of the fact that in an ordinary scenario, two nodes communicate for a short period of time, whereas in the case of a malicious and victimised node, their communication will either be extended for an unusually long time or for an infinite period of time. And it is in this direction that the authors of this work have thought of differentiating between an attacker and a normal node on the basis of a comparison of their communication period with a previously determined threshold time.

Recently with the increasing preponderance of integrated intelligent systems and the apparition of the notion of IOT, and the crucial importance of the some applications (such as surveillance, eHealth, and network control, traffic control, emergency ...) require reliable security as vulnerabilities are becoming more and more intense, hence the need to increase the level of network security to protect against possible attacks. And it is in this context that blockchain technology has emerged in recent years in order to benefit from the high level of security it offers to the network in which it is integrated, and as a part of the IOT environment, ad hoc networks have not been excluded from this evolution and have gained from the advantages provided by the blockchain.

The blockchain technology was conceived by Satoshi Nakamoto [17] in 2008, the first time use of this technology was in the banking and finance sector to support digital transactions and, provide access to the distributed ledger in a secure and trusted way. It was implemented as the main component of Bitcoin, where it serves as the public register for all transactions on the network.

A blockchain is a chain of blocks which contain specific information (database). In the case of the assigned network of blockchain, every contributor in the network manages, approves, and restores new accesses. The figure above summarizes the principle of blockchain (Fig. 3). All connected networks have a blockchain copy in stock. Due to the obvious high degree of security and accuracy, blockchain has been spread in various scenarios of functions and is recognized as one of the main approaches to evolving the world's evolution.

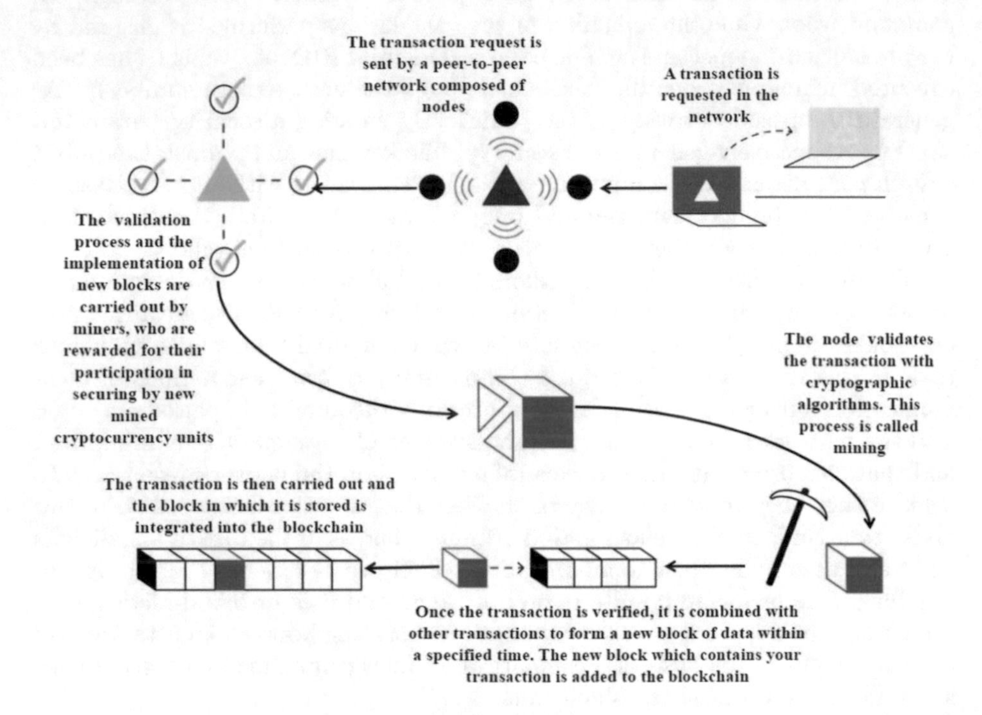

Fig. 3. Blockchain principle

Since the blockchain technology provides transparency, trustless, and secure transactions in the decentralized network, which helps in attaining robust and auditable records of all transactions, it is recently applied to different scenarios for Internet of Things (IoT) and especially Vehicular Adhoc Networks (VANETs).

The integration of this technology into VANETs has been the subject of interest for several researchers. For example, in [18] the authors propose a secure identity management framework for ad hoc vehicle networks using the Hyperledger fabric, which is an authorised blockchain developed in the framework of the Hyperledger project. Hyperledger Fabric implements a decentralised and distributed framework. According to the proposed framework, the participants in the block chain network are the authentication parties and the RSUs. With the proposed method, they intend to provide a decentralised and distributed framework for VANET, to provide a lightweight authentication system using pseudo-IDs, public/private keys and digital signatures, and to generate transactions for registering, validating and revoking vehicles. In addition, as the register is distributed, the registration of the vehicle in the network is visible to all the participants (including other authorities and RSUs) and not only to the authority registering the vehicle. This enhances the scalability of the network, allowing new vehicles to join the network easily and without the need to re-register with another authority in case they travel to a region or domain managed by another authentication party. From their own perspective, the authors of [19] opted for the proposal of a blockchain-based solution in order to make communication between smart vehicles secure, they used the protocol based on public keys in order to verify the mechanism proposed by the cryptography which combines both the use of side channels and the blockchain public key. They have used various forms of communication to improve the security of the vehicle network even more. propose a new secure inter-vehicle communication system focalising on lateral channels (visual light and ultrasonic sound) that is highly resistant to interference and attacks; the proposed system checks the location and the identity of the vehicle being communicated with, and then integrates this vehicle identifier into the exchanges of the cryptographic configuration. An original handshake protocol for establishing the key is provided as well. The system operates both physical side channels and uses a block-chain public key infrastructure for interoperability between untrusted vehicles and manufacturers. The side channels provide increased physical (directional) security for transmissions and are directly useful for transmitting messages between vehicles and, for example, for maintaining the distance between vehicles in the traffic. In their own interest, and in order to guarantee a transparent, self-managed and decentralized system, the authors of [20] combined VANETs and application concepts based on the Ethereum blockchain. In order to achieve their goal, they used Ethereum's intelligent contract system for the execution of all types of applications on an Ethereum Blockchain to guarantee the secure distribution of messages in vehicular ad hoc networks.

4 Discussion

It has been clearly noted from the research we have done and the literature review we have conducted in the previous section that there are new security approaches applied to other networks, such as the blockchain concept in the VANETs that we listed in our article (Table 4). After reviewing these studies, we found that the proposed solutions based on the concept of blockchain ensure the two security criteria: integrity and confidentiality, which are the most important security criteria. And availability can be easily guaranteed by the integration of the IDS (Intrusion Detection System); on the other hand, only one of the protection criteria is supported by the majority of the remaining proposals we have dealt with.

The blockchain integration in the networks is the most appropriate proposal to protect and secure our VANETs network in the most adequate way; since it can offer many advantages to increase and boost security such as: elimination of trust theaters, reduction of costs through, automation and simplification, dematerialisation of a number of processes, decentralization and dematerialisation of transactions and assets...

Table 4. Summary of related works

Article	Proposed solution	Security criteria
[12]	Detection of node-misbehavior using overhearing and autonomous agents in wireless Ad-Hoc networks	(C)
[13]	Enhancing the Efficiency of Routing Protocols in VANETS by defending Dos Attacks using BF-IPCM	(AV)
[14]	A ticket-based authentication scheme for vanets preserving privacy	(AU) and (I)
[15]	Privacy-preserving authentication scheme with full aggregation in VANET	(AU)
[16]	Detection and prevention of distributed denial of service attacks in VANETs	(AU)
[18]	Secure Identity Management Frame-work for Vehicular Ad-hoc Network using Blockchain	(C) and (I)
[19]	Securing vehicle to vehicle communications using blockchain through visible light and acoustic side-channels	(AV) and (I)
[20]	Self-managed and blockchain-based vehicular ad-hoc networks	(C) and (I)

On the basis of the research we have conducted in this paper, it is remarkable that it would be more efficient to lean on these networks to implement blockchain technology, making them safe and secure and robust. As far as we are concerned, we join our viewpoint to that of the authors who opted for the integration of blockchain technology in VANETs. Not only does it enable this, but it also ensures the immutability and stability of data even after connectivity losses, which means that it is the most suitable solution for in a network where the components are of a particular level of mobility such as VANETs. We propose to integrate the principle of blockchain in the RSU in the case where mobility is not high at the instant (t) because they are the most stable elements in VANETs network, and if the security impact of this on the whole network is high and does not affect the performance of the network in terms of sending time, energy or loss of information, we adopt it, otherwise we will redirect to another approach to get the maximum benefits offered by the blockchain. And in the case of high mobility, we propose that the integration of the blockchain will not be restricted to the RSU but all the components of our VANETs network, whether they are RSUs or others, will be able to accomplish this mission, it is only necessary that they ensure some conditions so that this does not have a detrimental effect on the network and that the blockchain keeps its characteristics. And we will also look at the impact of our approach on the performance and efficiency of the network to decide whether to adopt it or improve it for better results.

5 Conclusion

In this article, we provided an overview of the VANETs networks and their architecture, infrastructure and components. Then we addressed the security of these networks by citing the security requirements and talking about the offensive intrusions that can damage these systems. A literature review of the work already done in the security of VANETs was also presented, which led us to conduct a comparative study of different solutions proposed by several authors for ensuring security within the VANETs network, and it was found that the proposals integrating and implementing the blockchain are the most efficient in the network. And on the basis of this study, we found that the implementation of bolckchain technology since it makes the network more secure and more suited to the needs of our Research team. And then as a planned task, and in order to ensure the continuity of team work and in particular, to ensure network security using the enhanced OLSR protocols built in our team, we will suggest and develop a security solution that is more suited to our needs based on the Blockchain concept.

References

1. Khan, M.A., Salah, K.: IoT security: review, blockchain solutions, and open challenges. Future Gener. Comput. Syst. **82**, 395–411 (2018)
2. Khan, U.A., Lee, S.S.: Multi-layer problems and solutions in VANETs: a review. Electronics **8**, 204 (2019)
3. Hussain, R., Lee, J., Zeadally, S.: Trust in VANET: a survey of current solutions and future research opportunities. IEEE Trans. Intell. Transp. Syst. (2020). https://doi.org/10.1109/TITS.2020.2973715
4. Balu, M., Kumar, G., Lim, S.: A review on security techniques in VANETs. Int. J. Control Autom. **12**(4), 1–14 (2019). https://doi.org/10.33832/ijca.2019.12.4.01
5. Bashir, N., Boudjit, S.: An energy-efficient collaborative scheme for UAVs and VANETs for dissemination of real-time surveillance data on highways. In: 2020 IEEE 17th Annual Consumer Communications & Networking Conference (CCNC), Las Vegas, NV, USA, pp. 1–6 (2020). https://doi.org/10.1109/CCNC46108.2020.9045425
6. Triwinarko, A., et al.: A PHY/MAC cross-layer design with transmit antenna selection and power adaptation for receiver blocking problem in dense VANETs. Veh. Commun. **24**, 100233 (2020)
7. Yeferny, T., Sofian, H.: Vehicular ad-hoc networks: architecture, applications and challenges, arXiv preprint arXiv:2101.04539 (2021). https://doi.org/10.48550/arXiv.2101.04539
8. Xie, L., Ding, Y., Yang, H., Wang, X.: Blockchain-based secure and trustworthy internet of things in SDN-enabled 5G-VANETs. IEEE Access **7**, 56656–56666 (2019). https://doi.org/10.1109/ACCESS.2019.2913682

9. Arif, M., et al.: A survey on security attacks in VANETs: communication, applications and challenges. Veh. Commun. **19**, 100179 (2019)
10. Mahapatra, S.N., Singh, B.K., Kumar, V.: A survey on secure transmission in internet of things: taxonomy, recent techniques, research requirements, and challenges. Arab. J. Sci. Eng. **45**(8), 6211–6240 (2020). https://doi.org/10.1007/s13369-020-04461-2
11. Obaidat, M., Khodjaeva, M., Holst, J., Ben Zid, M.: Security and privacy challenges in vehicular ad hoc networks. In: Mahmood, Z. (ed.) Connected Vehicles in the Internet of Things, pp. 223–251. Springer, Cham (2020). https://doi.org/10.1007/978-3-030-36167-9_9
12. Agarwal, D., Rout, R.R., Ravichandra, S.: Detection of node-misbehavior using over-hearing and autonomous agents in wireless Ad-Hoc networks. In: Applications and Innovations in Mobile Computing (AIMoC), 12 February, pp 152–157. IEEE (2015)
13. Pagadala, P.K., Saravana Kumar, N.M.: Enhancing the efficiency of routing protocols in VANETS by defending dos attacks using BF-IPCM. Int. J. Adv. Network. Appl. **9**(4), 3510–3514 (2018)
14. Chikhaoui, O., Chehida, A.B., Abassi, R., Fatmi, S.G.E.: A ticket-based authentication scheme for VANETs preserving privacy. In: Puliafito, A., Bruneo, D., Distefano, S., Longo, F. (eds.) ADHOC-NOW 2017. LNCS, vol. 10517, pp. 77–91. Springer, Cham (2017). https://doi.org/10.1007/978-3-319-67910-5_7
15. Zhong, H., et al.: Privacy-preserving authentication scheme with full aggregation in VANET. Inf. Sci. **476**, 211–221 (2019)
16. Shabbir, M., et al.: Detection and prevention of distributed denial of service attacks in VANETs. In: 2016 International Conference on Computational Science and Computational Intelligence (CSCI). IEEE (2016)
17. Nakamoto, S.: Bitcoin. A Peer-to-Peer Electronic Cash System (2008). http://bitcoin.org/
18. George, S.A., Jaekel, A., Saini, I.: Secure identity management framework for vehicular ad-hoc network using blockchain. In: 2020 IEEE Symposium on Computers and Communications (ISCC), Rennes, France, pp. 1–6 (2020). https://doi.org/10.1109/ISCC50000.2020.9219736
19. Rowan, S., et al.: Securing vehicle to vehicle communications using blockchain through visible light and acoustic side-channels, arXiv preprint arXiv:1704.02553 (2017)
20. Leiding, B., Parisa, M., Dieter, H.: Self-managed and blockchain-based vehicular ad-hoc networks. In: Proceedings of the 2016 ACM International Joint Conference on Pervasive and Ubiquitous Computing: Adjunct (2016)

Adaptive Approach of Credit Card Fraud Detection Using Machine Learning Algorithms

EL Khyati Bouchra(✉), Abdellatif Ezzouhairi, and Haddouch Khalid

Laboratory: Engineering, Systems and Applications, National School of Applied Sciences, Sidi Mohamed Ben Abdellah University, Fez, Morocco
{bouchra.elkhyati,abdellatif.ezzouhairi, khalid.haddouch}@usmba.ac.ma

Abstract. With the evolution of the electronic payment systems, fraudsters always find an illegal way to steal people's money. Credit card fraud has become a critical issue for businesses and individuals as all online transactions can be easily done by just entering credit card information's. For this reason, fraud detection is of utmost importance for all financial institutions. Different techniques and approaches are used to secure online transactions, as well as rule-based fraud detection methods, EMV technology, 3D-secure protocol. However, fraud rates still increasing in card not present transactions. Researchers started using different machine learning methods to detect and prevent frauds in online transactions as well as Logistic Regression (LR), Naïve bayes (NB), Random Forest (RF) and Multilayer perceptron (MLP) algorithms. In this paper, the aim objective is to elaborate a comparative study of credit card fraud detection methods. For this objective, we used Google-Colab as an experimentation platform and studied the performance of each machine learning techniques in term of accuracy, Forecast Error and time of prediction using the European cardholder's dataset that we have balanced by applying SMOTE method. Finally, basing in this comparison study, we proposed and discussed an adaptive approach for credit card detection system.

Keywords: Credit card fraud · EMV technology · 3D-secure · Logistic Regression (LR) · Naïve bayes (NB) · Random Forest (RF) and Multilayer perceptron (MLP) · SMOTE

1 Introduction

Credit card is a payment tool issued by a bank or financial services company. This card is in the form of a plastic card equipped with a magnetic strip and/or electronic chip, which allows cardholders to purchase goods and services and pay with an electronic payment terminal or with e-commerce website. It also allows cash withdrawals from ATM terminals. The Credit card is associated with one or more payment networks that can be deployed globally such as Visa, MasterCard, American Express, or locally, such than CB in France, CMI in Morocco, Interac in Canada.

Credit card fraud is an illegal use of a payment card by stealing cardholder's personal information. Thanks to the advancement of e-commerce websites, credit card fraudsters

M. Lazaar et al. (Eds.): BDIoT 2021, LNNS 489, pp. 141–152, 2022.
https://doi.org/10.1007/978-3-031-07969-6_11

now have an easier time than ever to theft cards details. According to the Federal Trade Commission (FTC), credit card fraud is one of the fastest growing forms of identity theft. Credit card fraud reports increased 104% between the first quarter of 2019 and the first quarter of 2020 and continue to grow. With a view to rapid growth, the number of declarations of card fraud between the first quarter of 2017 and the first quarter of 2019 only increased by 27%. Opening a new credit card account is the most reported type of credit card fraud in 2019, an 88% increase from 2018, with 246,000 reported cases [1] (Fig. 1).

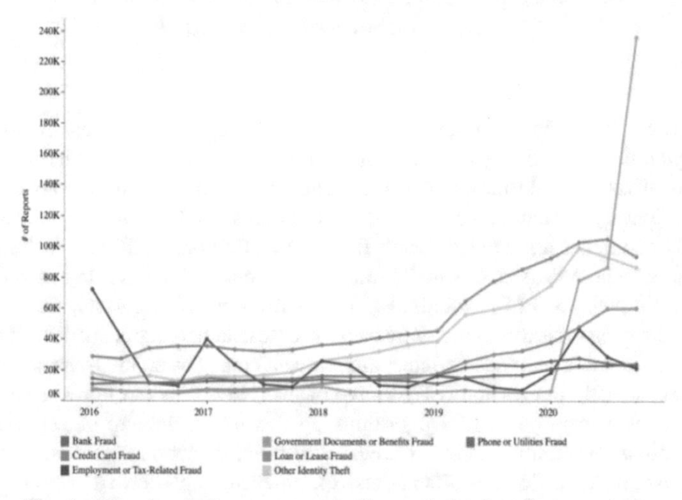

Fig. 1. Identity theft reports according to federal trade commission

To prevent fraudulent transactions, banks used a classical rule-based method of fraud detection. Where the decision rules are implemented manually as an historical analysis of past fraudulent behavior in the portfolio. But this kind of method is extremely time consuming and multiple verification methods are required, and it could only detect an obvious fraudulent transaction [9].

Thus, to improve its systems security, banks have opted for EMV card. Which is a smart card able to store card account data. Unlike magnetic cards, each time an EMV card is used for payment transaction, the chip creates a unique transaction code that cannot be reused in another transaction that helped to prevent card fraud and defend against skimming attacks [10].

However, fraud rates still increasing in manual and Card Not Present transactions. To reduce fraud attempts and make online transactions safer, Banks has implemented an additional security level called 3D-Secure or payer authentication, this protocol is initiated and created by Visa and MasterCard and it is branded as 'Verified by Visa' and 'MasterCard SecureCode' respectively. However, not all cards are currently enrolled 3D-Secure program, and this protocol does not restrict fraud to happen but just reduces the cost of fraudulent transactions.

The machine learning techniques are used in different fields such as Healthcare [11], Retail [12], Finance [13], Cyber Security [14], Smart City [15], Telecommunication

[16] and optimization fields [7, 8]. In recent years, these techniques have shown a very interesting performance compared to the classical methods of treatment. One of the areas that do not escape this observation is the area of credit card fraudulent activities. To overcome this issue, many studies that exploited the strength of the machine learning algorithms have been performed to prevent and detect the credit card fraudulent activities. Based on Logistic Regression (LR), Naïve bayes (NB), Random Forest (RF) and Neural Network Approach (NNA) algorithms, the obtained results were very important.

This paper is organized as follows: In Sect. 2, we present the important machine learning techniques used to handle the problem of credit card fraudulent activities. In the Sect. 3, the implementation details of the analytical and comparative study for the most known machine learning techniques are presented. Finally, in the Sect. 4, we propose and discuss an adaptive approach based on two stages. The first one consists of using the machine learning techniques to forecast the new transaction. However, the second one uses the obtained result and executes the validating scenario.

2 Related work

There are a lot of studies done on credit card fraud detection. Most of the credit card fraud detection systems are using machine learning algorithms such as Logistic Regression (LR) [2], Naïve bayes (NB) [3], Random Forest (RF) [4] and Multilayer perceptron (MLP) [6].

2.1 Logistic Regression

Logistic regression is one of the most used and useful classification algorithms in machine learning. It is a statistical method used to predict the outcome of a binomial or polynomial. When the target field is a defined field with two or more possible values, a polynomial logistic regression algorithm can generate a model. Thus, the binomial logistic regression algorithm is limited to models where the target field is a flag or a binary field [2].

Logistic regression uses the logistic sigmoid function; it returns a probability which takes values between 0 and 1. It is defined as:

$$\sigma(x) = \frac{1}{1 + e^{-x}} \tag{1}$$

Logistic regression is easy to implement, interpretable and very effective for training. It is easy to extend it to many categories (multinational regression) and a natural probability of class predictions. At the same time, regression cycles are limited in the assumption of linearity between a dependent variable and independent variables. It can only be used to predict different actions.

2.2 Naïve Bayes

Naive Bayes is a probabilistic classifier based on the application of Bayes' theorem, which strongly assumes independence between features. The naive Bayesian model is easy to build without complicated estimation of iteration of parameters. Despite its simplicity, Bayes' naive classifier generally shows surprisingly good performance and is widely used because it performs better than more complex classification methods [3].

Bayes theorem provides a way of calculating the posterior probability, the formula for Bayes' theorem is given as:

$$P(A\backslash B) = \frac{P(A \cap B)}{P(B)} \tag{2}$$

Naive Bayes is one of the fast and easy to predict ML algorithms for a class of datasets. It can be used for binary and multi-class classification. Compared to other algorithms, it can predict several categories at the same time. For text classification issues, this is the most popular choice. However, Naive Bayes assumes that all features are independent or unrelated, so it is impossible to understand the relationship between features.

2.3 Random Forest

Random forest is a classifier composed of a set of tree classifiers {h (x, k), k = 1,…}, where {k} is an independent random vector with the same distribution, and each tree has the most popular class [4].

Random forest can be used for classification and regression problems. It is made up of many decision trees. This algorithm works best when there are more trees in the forest and can prevent the model from over-adding itself. Every decision tree in the forest produces results. Combine these results to get a more accurate and stable forecast [5].

Due to the large number of decision trees involved in the process, Random Forest is considered a very accurate and sustainable method. It will not encounter the problem of too many devices. Random forests can also deal with missing values by using means instead of continuous variables or by calculating a weighted average of the proximity of missing values. However, since random forests have multiple decision trees, they will slowly produce predictions. At each prognosis, all trees must predict the same bet given and then vote on it. This whole process takes time. In addition, the model is difficult to interpret relative to the decision tree and can be easily determined by following the path of the decision tree.

2.4 Multilayer Perceptron

The multilayer perceptron is one of the most widely used neural network architectures, and it is a supervised neural network. The multilayer perceptron consists of a network of nodes arranged in layers. A typical MLP network consists of three or more layers of processing nodes: an input layer that receives external input, one or more hidden layers, and an output layer that produces classification results. Unlike other layers, the input layer does not involve any calculations.

The principle of the network is that when data is provided in the input layer, the network nodes will perform calculations in successive layers until the output value is obtained on each output node. The output signal must be able to indicate the appropriate category of the input data. In other words, it can be expected to have a higher output value on the correct class node and a lower output value in all other respects [6].

3 Experimental result

This section presents dataset used during experiment, technical tools and obtained results using different machine learning algorithms for credit card fraud detection.

3.1 Dataset Description

The dataset shows the credit card transactions made by European cardholders for two days in September 2013, where there are 492 fake transactions out of 284,807 transactions. The dataset is very unbalanced because the observation of the valid transaction is very higher than fraudulent transactions (positive class) which present 0.172% of all transactions [17] (Fig. 2).

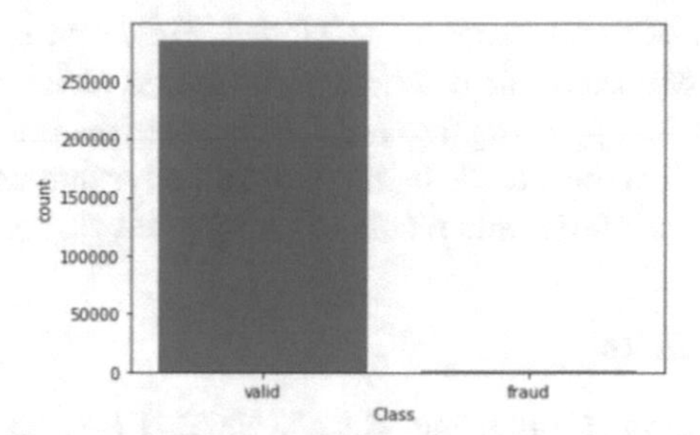

Fig. 2. Visualization of the class variables.

This dataset contains only numeric input variables that are the result of a PCA transformation. Unfortunately, for privacy reasons, original functionality and other general data information may not be provided. The functions V1, V2… V28 are the main components obtained with PCA, the only functions that are not transformed with PCA are 'Time' and 'Amount'.

The "Time" function contains the seconds that have elapsed between each transaction and the first transaction in the dataset. The "Amount" function is the amount of the transaction. This functionality can, for example, be used for cost-dependent learning. The "Class" function is the response variable and takes the value 1 in the case of fraud and 0 in the opposite case (Fig. 3).

3.2 Google Colaboratory

Google Colaboratory is a free environment that runs entirely in the cloud. it used without requiring any settings and the created notebook can be edited by team members at the same time. Colab supports many popular machine learning libraries, which can be easily loaded into notebooks. It provides many features to the data scientist such as writing and running python code; Recording codes that support mathematical equations; importing external data sets from Kaggle and integrating PyTorch, TensorFlow, Keras, OpenCV. The experimental analysis was performed using a computer HP EliteBook 840 G2; Intel (R) Core (TM) i7-5600U CPU @ 2.60 GHz; 2601 MHz; 2 cores; 4 logical processor

```
          Time            V1  ...        Amount          Class
count  284807.000000  2.848070e+05  ...  284807.000000  284807.000000
mean    94813.859575  3.919560e-15  ...      88.349619       0.001727
std     47488.145955  1.958696e+00  ...     250.120109       0.041527
min         0.000000 -5.640751e+01  ...       0.000000       0.000000
25%     54201.500000 -9.203734e-01  ...       5.600000       0.000000
50%     84692.000000  1.810880e-02  ...      22.000000       0.000000
75%    139320.500000  1.315642e+00  ...      77.165000       0.000000
max    172792.000000  2.454930e+00  ...   25691.160000       1.000000

[8 rows x 31 columns]
0.0017304750013189597
Fraud Cases: 492
Valid Transactions: 284315
```

Fig. 3. Dataset description

(s), and using a speed of the internet of 12 Mbits/s. However, since we do not have a high CPU, we have connected our machine with Google Colab which is a free cloud service hosted by Google supporting free GPU. Noted that, the implementation language used in this work is Python because it is the most attractive and powerful programming language for computer scientists and machine learning developers.

3.3 Performance Metrics

We have used the Accuracy, Precision, Recall, MCC, F1-Score, prediction time and forecast error formulas for the experimented models' evaluation.

Accuracy presents the number of the data points which is correctly predicted out of all the data points. More specifically, it is described the division of the true positives and true negatives.

$$Accuracy = \frac{TN + TP}{TP + FP + TN + FN} \tag{3}$$

Precision measures the number of the predicted data point that belong to the positive class.it is the fraction between the True Positives and all the Positives.

$$\Pr ecision = \frac{TP}{TP + FP} \tag{4}$$

Recall calculates the number of the positive predicted class made of all positive examples in the dataset.

$$\text{Recall} = \frac{TP}{TP + FN} \tag{5}$$

F-Score provides a single score that can balance precision and recall in a single number.

$$F1 - score = 2 \times \frac{\Pr ecision \times \text{Recall}}{\Pr ecision + \text{Recall}} \tag{6}$$

Matthews Correlation Coefficient (MCC) Used to evaluate the quality of the binary classification; it considers all cells of the Confusion Matrix in its formula.

$$MCC = \frac{TP \times TN - FP \times FN}{\sqrt{TP + FP \times (TP + FN) \times (TN + FP) \times (TN + FN)}} \quad (7)$$

Whereas,

TP = True Positive
TN = True Negative
FP = False Positive
FN = False Negative

Forecast Error measures the difference between the actual or real and the predicted data point.

Prediction time is the response time for the model to predict a given transaction.

3.4 Obtained Result

We have experimented previous machine learning algorithms on credit card transactions dataset. As well as Logistic Regression (LR), Naïve bayes (NB), Random Forest (RF) and Multilayer perceptron (MLP) methods. The results are tabulated, which shows great differences in accuracy, precision, Recall, F1-Score, forecast error and prediction time as well.

The following table presents the results on the imbalanced transactions dataset using the different machine learning methods (Figs. 4, 5 and 6 and Table 1).

Table 1. Performance result for different algorithms before SMOTE

Methods	Accuracy (%)	Precision (%)	Recall (%)	F1-Score (%)	MCC (%)	Forecast error	Prediction time (s)
LR	99.86	61.11	56.12	58.51	58.49	0.0013	0.0087
NB	99.30	14.62	63.26	23.75	30.19	0.0069	0.0315
RF	99.96	97.47	78.57	87.01	87.49	0.0004	0.6451
MLP	99.90	72.34	69.39	70.83	70.79	0.0018	1.157

Noted that our dataset is extremely imbalanced, which might significantly affect the performance metrics like accuracy of the model by giving high accuracy just by predicting the majority class, but not capturing the minority class. To overcome this classification issue, we used Synthetic Minority Oversampling Technique (SMOTE) which randomly sampling data from the minority class by generating data point on the road segment connecting a randomly selected point and one in all its K-nearest neighbors. [18]. The obtained result shows remarkable difference in the performance metrics (Fig. 7).

The table below describes the obtained results on the transactions dataset using the different machine learning methods after dataset balancing (Figs. 8, 9 and 10).

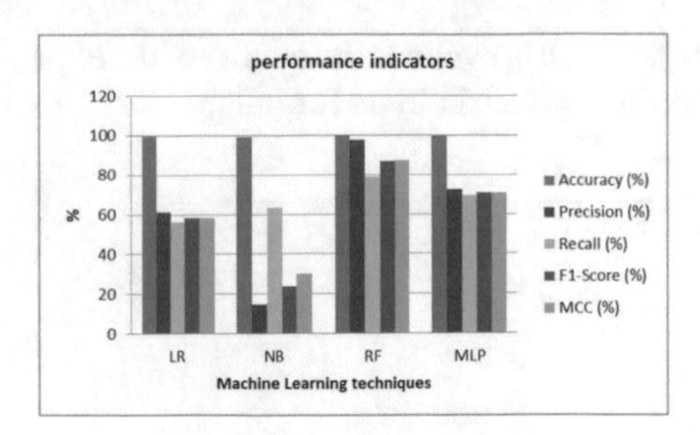

Fig. 4. Performance indicators

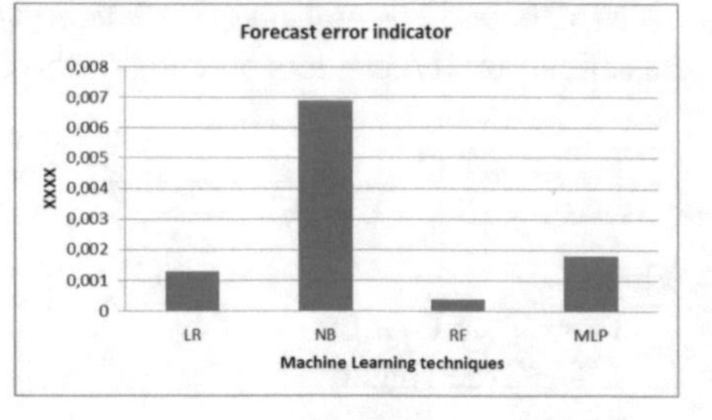

Fig. 5. Forecast error indicator

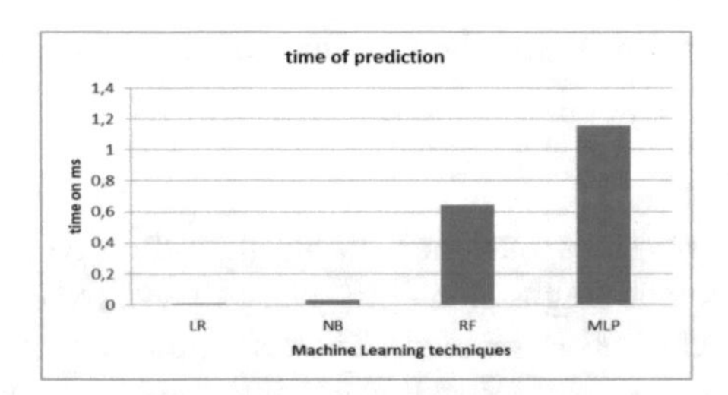

Fig. 6. Prediction Time

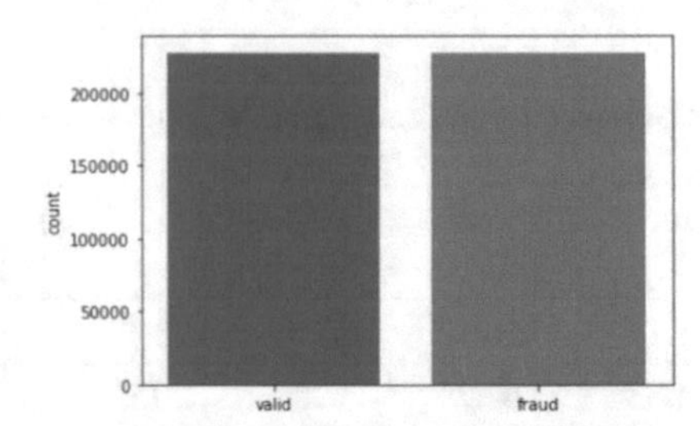

Fig. 7. Visualization of Class variable after SMOTE oversampling

Table 2. Performance result for different algorithms after SMOTE

Methods	Accuracy (%)	Precision (%)	Recall (%)	F1-Score (%)	MCC (%)	Forecast error	Prediction time (s)
LR	98.05	07.33	88.77	13.55	25.20	0.01948	0.00898
NB	99.22	14.31	70.40	23.79	31.52	0.0077	0.0323
RF	99.95	87.23	83.67	85.41	85.41	0.00049	0.6831
MLP	98.45	09.34	91.83	16.96	29.02	0.0	1.496

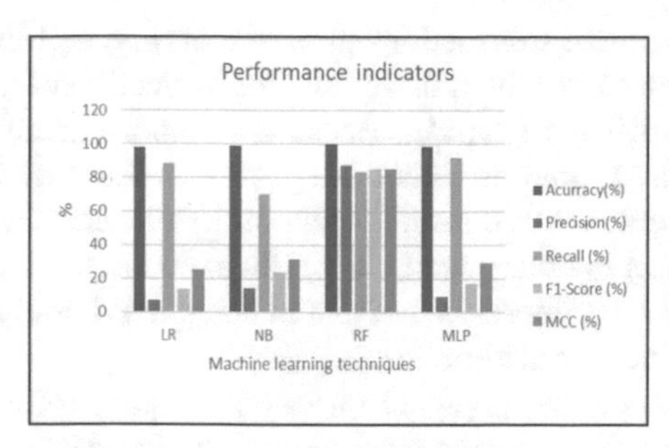

Fig. 8. Performance indicator after SMOTE

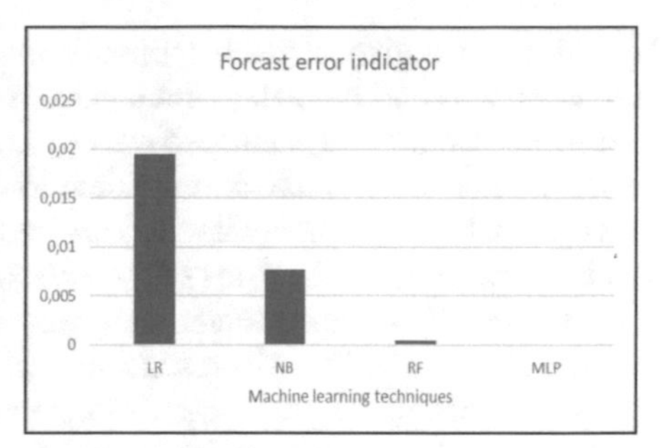

Fig. 9. Forecast error indicator after SMOTE

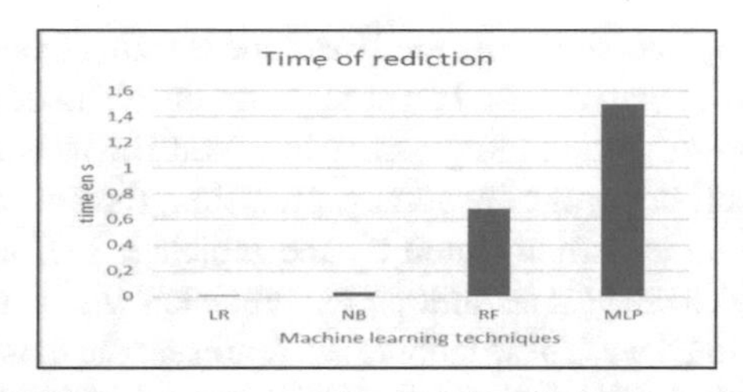

Fig. 10. Time of prediction after SMOTE

According to the obtained results, the model's performance metrics have considerably changed, and the accuracy has significantly decreased after applying the machine learning techniques on the dataset balanced using SMOTE oversampling method. The model that kept almost the same accuracy is Random Forest. Which additionally provides the best accuracy, precision, MCC and has a low forecast error. But it is more expensive in time of prediction compared to Logistic Regression that is the fastest model in prediction time opposed to which is the slowest model in prediction time.

The choice of the best fraud detection algorithm to implement depends generally on the approach and global strategy of the financial company. if the objective is to satisfy the customer in terms of the efficiency of investigation by minimizing the prediction and response time while taking the risk of having a false prediction and generating a non-negligible number of false positives, and therefore errors. The algorithm that can be used is a Logistic Regression. However, if the organization aims to block fraudulent transactions in real time and accept payments with confidence ensuring the certainty of results obtained and a height accuracy with a minimal error score and avoid the dissatisfaction of the customer in the case of an error. The best algorithm to implement is Random Forests.

4 Adaptive approach

To strengthen its lines of defense, secure the transactions of its customers and predict attempted fraud, the financial institutions have set up automated fraud detection systems which require the study, alerts and countermeasures associated. The challenge is to ensure a model which gives a fast response as the online transactions pass in real time and in a few seconds and if a delay of response is detected the transaction will be declined for time out reason. The system also must give sufficiently stringent results by reporting any suspicious behavior to block fraudulent transactions and limit false positives as much as possible, which are mainly due to non-fraudulent but unusual behavior. In all cases the decision-making mechanisms of the detection system operate in a critical context, with a direct impact on the customer which makes to choose the best fraud detection module more difficult. The Fig. 11 presents the online transaction flux.

In this section, we propose and discuss an adaptive approach based on two stages. The first one consists of using the performed machine learning techniques to be forecasting the new transaction. Then, the second one uses the obtained result and executes the validating scenario.

In the first stage, we select the performed machine learning techniques in term of performance metrics. In this context, based on comparing the obtained results, our approach adopts the Random forest, Logistic Regression and Naïf Bayes Networks algorithms to generate a suitable result in terms of accuracy, precision, forecast error prediction time. This choosing technique is demonstrated by the indicator performance values (show Table 2) while it provides the best accuracy, best precision with a minimal forecast error and an average prediction time. These techniques represent the most effective techniques and the best ratio between accuracy and prediction time. The integration of these three models with the payment system helps to secure transactions and hold the possibility of being fraudulent. In the second time, these techniques are used to classify the new transaction. In this context, you can define three adaptive strategies or scenarios to validate definitively the new transaction. They are the main objective for the second stage.

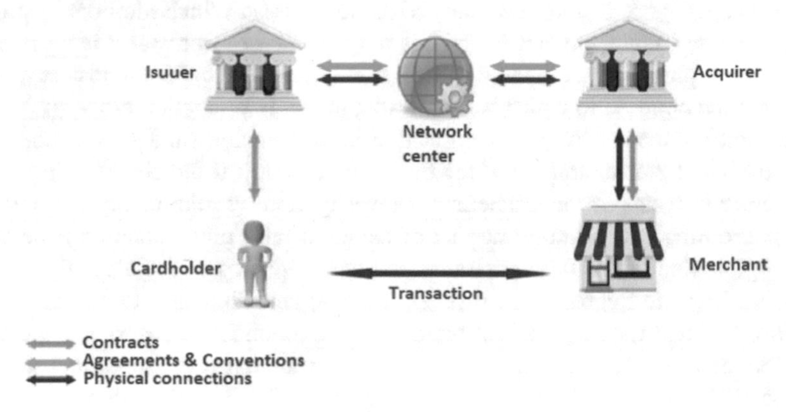

Fig. 11. Online transaction Flux

The first strategy can be named a distrustful strategy, it requires the validation of the new transaction by the customer if one of the three prediction techniques classifies the transaction as a fraudulent. The second strategy that can be named the optimistic strategy, it requests the validation of the new transaction from the client if two of the three prediction techniques classify the transaction as a fraudulent transaction. Finally,

For the third strategy, if the three models classify the new operation as fraudulent, the validation of this transaction is required by the costumer via the scenario of validation.

We can notice that this adaptive strategy can improve the performance of the credit card fraud detection. Moreover, we can enrich the data set of this problem by the validation phase assured by the costumer himself. This validation plays an important role to detect the new ways of the fraudulent transactions. Finally, this approach can be carried out in parallel to minimize the overall time of carrying out a purchase transaction via a card.

Finally, when the cardholder performs a purchase using an e-commerce website or a merchant's terminal, the process of the financial transaction starts, and the transaction data will be sent to the gateway. The fraud detection models will process the dataset and two scenarios are possible, if the system decides that the transaction is legitimate, the transaction will be authorized. However, if it identifies the transaction as suspicious or fraudulent. The transaction will be rejected, and an alert email or SMS will be sent the cardholder.

5 Conclusion

In this paper, we have elaborated a comparative study of credit card fraud detection using machine learning algorithms as well as Regression; Naïve bayes; Random Forest and Multilayer perceptron. Supervised learning algorithms are recently used in this type of anomaly detection problems. If these methods are applied to the bank credit card fraud detection system, the probability fraudulent transactions can be quickly predicted. And a series of fraud prevention modules can be implemented to secure online transactions and prevent banks from great losses. In this study we applied the different machine learning on transactions dataset that we balanced using SMOTE oversampling method and we used several criterions to evaluate model's performance as well as Accuracy, Precision, Recall, F1-Score, MCC, forecast error and the prediction. This all indicators can be used to select a good strategy to handle this problem in function of the global politic of financial company. In the future work, we aim to propose a collaborative learning and big data solutions to develop real time credit card fraud detection systems.

References

1. Identity Theft Reports according to Federal Trade Commission: Available at: 2020 Study: Credit Card Fraud Has Exploded in Recent Years (creditcardinsider.com)
2. Sahin, Y., Duman. E.: Detecting Credit Card Fraud by ANN and Logistic Regression. IEEE (2011)
3. Vembandasamy, R., Sasipriya, E.D.: Heart Diseases Detection Using Naive Bayes Algorithm. IJISET (Sept 2015)
4. Robert, E.: Schapire: Random Forests. Kluwer Academic Publishers, Manufactured in The Netherlands (2001)
5. Dejan, V., Mirjana, K., Srdjan, S., Marko, A., Andras, A.: Credit card fraud detection - machine learning methods. In : 18th International Symposium INFOTEH-JAHORINA, pp. 20–22 (March 2019)

6. Yana, H., Jiangb, Y., Zhenge, J., Pengc, C., Lid, Q.: A multilayer perceptron-based medical decision support system for heart disease diagnosis, Elsevier. Expert Syst. Appl. **30**, 272–281 (2006)
7. Haddouch, K., Elmoutaoukil, K.: New starting point of the continuous hopfield network. In: International Conference on Big Data, Cloud and Applications. Spinger, pp. 379–389 (2018)
8. Haddouch, K., Elmoutaoukil, K., Ettaouil, M.: Solving the weighted constraint satisfaction problems via the neural network approach. Int. J. Artifi. Intell. Intera. Multi. **4**(1), pp. 56–60 (2016)
9. Sushmito, G., Douglas, L.: Reilly: Credit Card Fraud Detection with a Neural-Network. IEEE (1994)
10. Ward, M.: EMV card payments – An update. Elsevier (2006)
11. Shailaja, K., Seetharamulu, B., Jabbar, M.A., Tech Scholar, M.: Machine Learning in Healthcare: A Review. IEEE (2018)
12. Jakob, H., Heiner, S.: Daily retail demand forecasting using machine learning with emphasis on calendric special days, Elsevier. International Institute of Forecasters (2020)
13. Thomas, R.: Sentiment analysis and machine learning in fnance: a comparison of methods and models on one million messages. Springer Nature Switzerland AG (2019)
14. Ozlem, Y., Murat, A.: A review on cyber security datasets for machine learning algorithms. In: IEEE International Conference on Big Data (2017)
15. Mehdi, M., Ala, A.-F.: Enabling Cognitive Smart Cities Using Big Data and Machine Learning: Approaches and Challenges. IEEE Communications Magazine (2018)
16. Saad, A.Q., Ammar, S.R., Ali, M.Q., Aatif, K.: Telecommunication Subscribers' Churn Prediction Model Using Machine Learning. IEEE (2013)
17. Credit card transactions dataset: Available at: Credit Card Fraud Detection | Kaggle
18. Elreedy, D., Atiya, A.F.: A comprehensive analysis of synthetic minority oversampling technique (SMOTE) for handling class imbalance. Elsevier Information Sciences **505**, 32–64 (2019)

Homomorphic Method Additive Using Pailler and Multiplicative Based on RSA in Integers Numbers

Hamza Touil[1(✉)], Nabil El Akkad[1,2], and Khalid Satori[1]

[1] LISAC, Faculty of Sciences, Dhar-Mahraz (FSDM), Sidi Mohamed Ben Abdellah University, Fez, Morocco
{Hamza.touil,nabil.elakkad}@usmba.ac.ma

[2] Laboratory of Engineering, Systems and Applications (LISA), National School of Applied Sciences (ENSA), Sidi Mohamed Ben Abdellah University, Fez, Morocco

Abstract. When we use conventional encryption, we are confronted with a situation we are used to and don't even consider it a problem. To work with encrypted data, we have to decipher it, and then it can become the property of the attackers. Homomorphic encryption implies that you can perform operations on encrypted text and get a perfect result without decrypting the text. For example, such a scheme can be used in e-elections (counting votes while preserving voters' anonymity), in cloud computing, insecure search (delivering the result without analyzing its real content), or feedback systems. This document aims to create a crypto-system that allows making calculations on encrypted data, more particularly on integers, and combining two encryption algorithms, additive homomorphic following paillier method, and multiplicative homomorphic applying RSA. All this to guarantee the security of calculations in cloud storage.

Keywords: Cryptography · Homomorphic · Computation · Storage · Cloud

1 Introduction

In the theoretical cryptography literature, homomorphic encryption refers to a cryptographic primitive that is an encryption function that satisfies the additional requirement of homomorphism concerning any algebraic operations on the plaintext. Let E(k, m) be an encryption function, where m is the plaintext, and k is encrypted. Note that forgiven fixed k and m, the cryptogram E(k, m) may generally be a random variable. In such cases, we speak of probabilistic encryption. E is homomorphic for operation on the plaintext if there is an efficient algorithm [1–5].

M, which receives any pair of cryptograms of the form E (k, m1), E(k, m2), yields a cryptogram c such that decrypting c will yield a plaintext text m1opm2 will be obtained.

Generally, the following crucial particular case of homomorphic encryption is considered. For a given encryption function E and operation op1 over the plaintexts, there is an operation op2 over the cryptograms. From the cryptogram E (k, m1) op2E (k, m2),

M. Lazaar et al. (Eds.): BDIoT 2021, LNNS 489, pp. 153–164, 2022.
https://doi.org/10.1007/978-3-031-07969-6_12

the public text m1opl m2 is extracted during decryption. The algorithm M is, generally speaking, probabilistic. There is a special modification of the primitive in question, called homomorphic encryption with rerandomization. In this case, given fixed E (k, m1) and E (k, m2), the cryptogram c is a random variable. It is required that given c, E (k, m1), E (k, m2), but with an unknown decryption key, it is impossible to effectively check that the cryptogram c is obtained from E (k, m1) and E (k, m2), i.e., it contains the plaintext m1 op m2 [1, 6–9].

Homomorphic encryption as a cryptographic primitive can find wide application in cryptography and, more generally, develop mathematical methods for protecting information. Here, one should here. First of all, it is necessary to allocate such, engaging from the application point of view, the task as calculations on the encrypted data. Confidential data are stored in encrypted form. To perform computations on them, the data can be decrypted, necessary operations can be performed, and the results are encrypted again. However, this requires secure hardware and at the very least secure hardware or at least arrangements for keeping secret keys. Calculations on encrypted data, if possible, helps to avoid all these problems. The task of computing on encrypted data may be thought of in different ways. For example, the data can be an array of natural numbers a1,..., and from range 1 to N. Suppose that the encrypted data calculation system allows the adversary to perform the following query. An arbitrary number a from the same range from 1 to N and index i ∈ {1, ..., n}. To the query (a, i), the adversary gets the answer 0 if a ≥ ai, and the answer is 1 if a < ai. No matter how highly resistant a cryptosystem, an adversary can always use the "split in half" method to determine ai's value. Thus, in this setting in this formulation, calculations over encrypted data are impossible. The most popular versions of a system of calculations over encrypted data considered in literature do not include comparison [4, 10]. In the rest of this paper, we will see the work in relations. Then we will dissect our method ending with the experiments.

2 Related Works

Hundreds, if not thousands, of researches, have been carried out in the field of homomorphic cryptography. On the other hand, in this paper, we will focus on the ones related to data computation. [11] Offers a Multi-Tasks technique in the form of a beneficial protocol in terms of speed of execution. It can build look-up tables that can evaluate multi-input functions to manage general functions. It adopts block instead of bit-by-bit coding with permutation techniques and data retrieval schemes through the cloud. On the other hand, [12] provides a new encryption system entirely homomorphic, non-bootable, and straightforward. It is based on matrices as symmetrical keys to perform the various possible calculations while keeping the encrypted data [13]. Proposes three improved axes of SK17 to improve efficiency, security, and flexibility. Then implements the first protocol to show its efficiency using the homomorphic encryption scheme while providing an analysis of security and an evaluation of the practical aspect in theory.

3 Homomorphic Encryption System

Homomorphic cryptography is a complex cryptographic approach permitting to carry out calculations on encrypted data. Formally, if c1 (respectively c2) is a cipher of m1 (respectively m2), there are two operations * and ○ such as [14, 15]:

Dec(c1*c2) = Dec(c1) ○ Dec(c2) = m1 ○ m2

Typically, it will be a modular addition or multiplication, but this is not always the case. A completely homomorphic encryption system is nothing more than a homomorphic encryption system to analyze any encrypted data function. Since any function can be expressed as a polynomial, and a polynomial consists of a series of additions and multiplications, a cryptosystem will be completely homomorphic when evaluating a random number of additions and multiplications on the encrypted data [16, 17].

3.1 Partially Homomorphic Cryptographic Scheme

This type of cryptography distinguishes between additive homomorphic (based on addition only), such as the Pailler and Goldwasser-Micalli cryptosystems, and multiplicative homomorphic ciphers (based on multiplication operations only), such as the RSA and El Gamal cryptosystems [18, 19].

3.1.1 Additive Homomorphic Paillier Encryption

This cryptosystem is the one with the largest bandwidth, also known as the expansion rate, which is the ratio between the clear and the cipher length. The Paillier cryptosystem concept is based on a calculation of n classes of residues is computationally intensive. The algorithm's operation mode allows homomorphic addition operations to respond once the data has been decrypted. The key generated by the Paillier cryptosystem is given in the following algorithm. To encrypt a message, the message is used as a superscript for g, a random value is carried to the other public key of value n, as shown in the algorithm. This produces a modulo n^2 encryption value. Decryption is again a simple equation and is given in the algorithm. The definition of L (x) is given with the key generation [20].

3.1.2 Homomorphic Multiplicative RSA Encryption

It has withstood years of intensive cryptanalysis and is still considered robust enough to protect banking exchanges and other critical data. This level of security resides in the difficulty of factoring in large numbers. Therefore, here we do not describe this cryptosystem but limit ourselves to notations.

Let N be a compound module (RSA modulus). Furthermore, e is an open exponent. Thus, the pair (N, e) is the public key of the cryptosystem. Further, let m ∈ ZN be the plaintext [21–23].

The encryption function E ((N, e), m) = me mod N of the cryptosystem RSA is homo-morphic for the multiplication of the plaintexts. Indeed, for any two public texts m1, m2, and any public key k, the cryptogram of the pro is equal to the product of the cryptograms of the multipliers: E (k, m1 − m2) = E (k, m1) − E (k, m2) [24–26].

4 Proposed Method

This method consists of securing the data transmitted (more precisely, integers) to remote systems. To do this, we will implement two homomorphic cryptosystems so that we can, towards the end, realize a crypto-system that allows us to make calculations on the encrypted data (Fig. 1) without decrypting them, using Paillier for the addition or RSA for the multiplication.

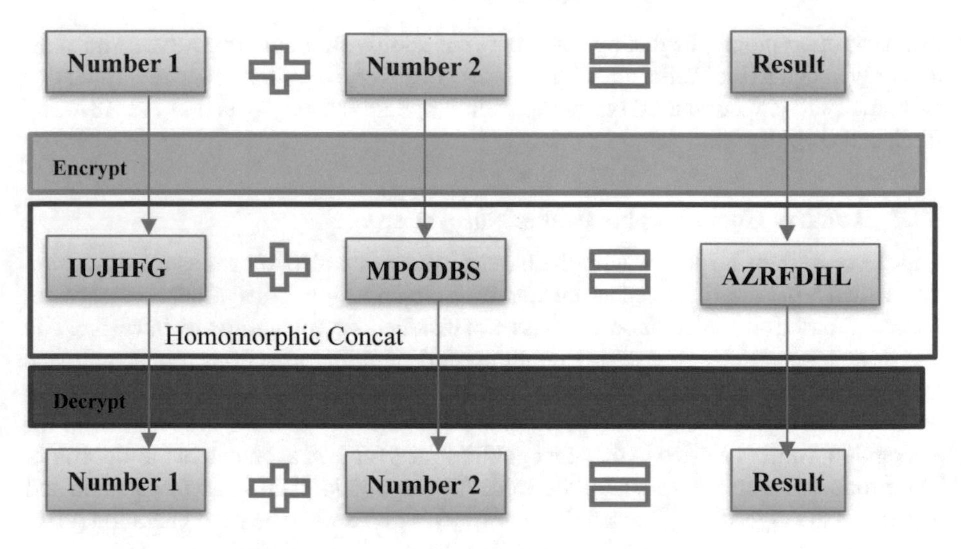

Fig. 1. Homomorphic calculation of two integers.

Implementing this solution is composed of a client who will enter the numbers on which we want to carry out the calculation and the chosen operation. A choice of the encryption algorithm applied to the numbers is made automatically; that is to say, if the client chooses the multiplication, the algorithm generated will be RSA; on the other hand, if he chooses the addition, the algorithm generated will be Paillier. The data will be sent to the server in an encrypted way, in a secure channel. Then the server receives the encrypted numbers (with RSA and Paillier) and stores them in the database. The next step is to perform the calculation desired by the client. The final step is to send the encrypted results to the client, who will decrypt them and display them, as shown below (Fig. 2).

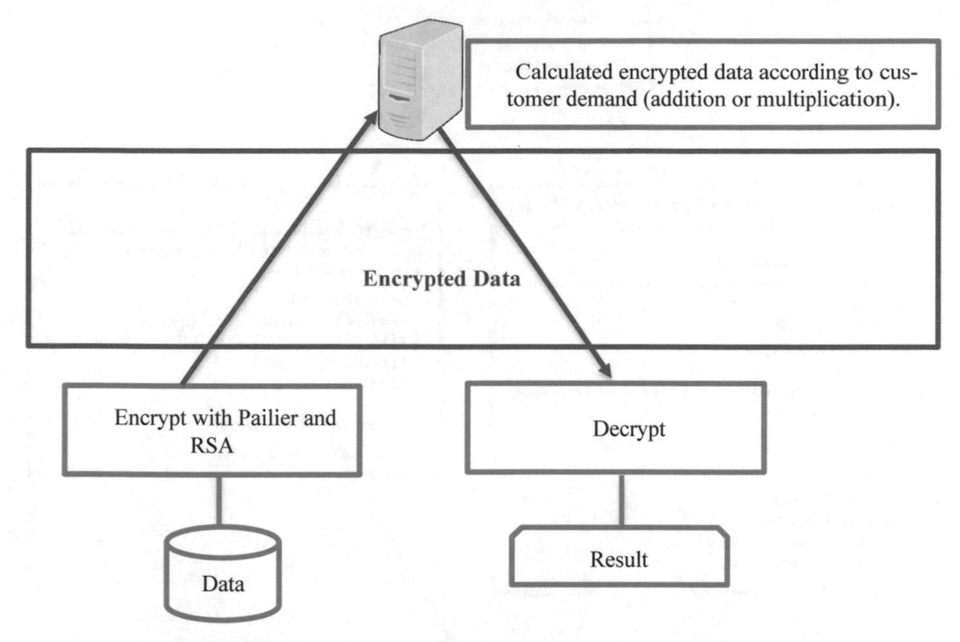

Fig. 2. Schema of homomorphic addition and multiplication cryptography.

4.1 Homomorphic Additive and Multiplicative Encryption/Decryption Algorithm

Here is an explanation of the different steps of our proposed method. It starts with choosing the user's operation, automatically choosing the appropriate encryption algorithm, and then sending them to the cloud. The calculation must be carried out in encrypted mode, and the final step is to send the encrypted results so that they can be displayed to the user (Fig. 3).

5 Experimentation

In this part, we will see the execution of the two encryption algorithms "RSA" and "paillier" using integers. We will see our method's implementation on an application knowing that we have used Java as a programming language. The functions of the two encrypted data calculations are predefined.

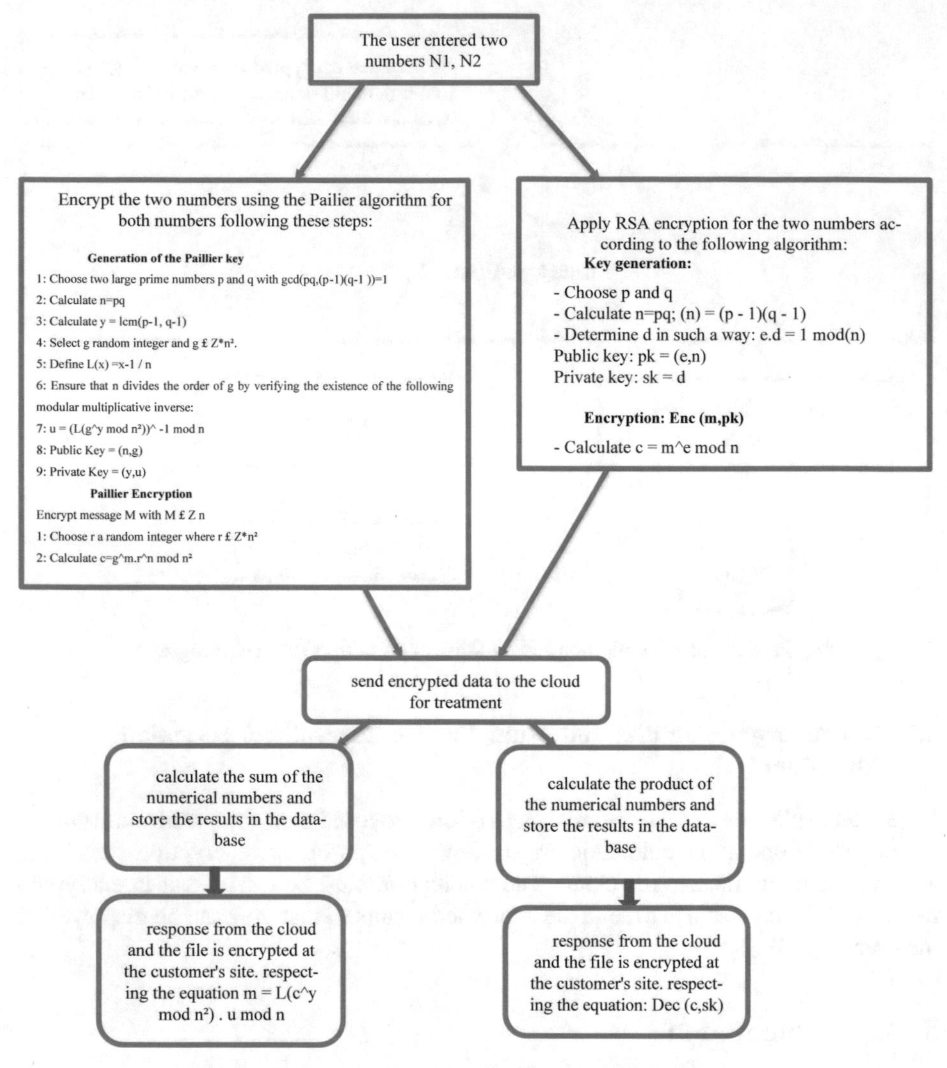

Fig. 3. The encryption and decryption steps of our method.

5.1 Paillier Encryption/Decryption

As explained if it is an addition operation, the system will directly choose the paillier encryption. We will apply this method on two digits m1 = 123 and m2 = 37.

Briefly, the steps are as follows:

- Generate a public-private key pair:

  1. Pick p = 13 and q = 17. (meets all requirements)
  2. Determine n = 221.
  3. Determine $\lambda = 48$.

4. Pick g = 4886.
5. Calculate $\mu = 159$. (It exists.)

- Encrypt number_1.

6. m1 = 123.
7. Pick r1 = 666.
8. Calculate c1 = 25889 mod 2212.

- Encrypt number_2.

9. m2 = 37.
10. Pick r2 = 999.
11. Calculate c2 = 30692mod2212.

- Calculate the sum.

12. Calculate Csum = 25889.30692 = 39800 mod 2212.

- Decrypt sum.

13. Calculate Msum = 160 = 123 + 37 = m1 + m2 mod 221.

5.2 Multiplication Based on RSA

Here is a numerical application of the homomorphic multiplicative RSA encryption:

Let p = 3, q = 5, e = 9 and d = 1 with a block size = 4;
m1 and m2 two clear messages and c1 and c2 their respective ciphers, obtained using RSA encryption.
m1 = 589625 c1 = 000500080009000600020005
m2 = 236491 c2 = 000200030006000400090001

After encrypting messages m1 and m2, Fig. 4 below shows the result of the binary conversion of the c1 and c2 ciphers of each block of 4, and Fig. 5 shows the process of multiplying c1 * c2 block by block, then converting the result of each multiplication into a decimal.

– Conversion c1 and c2 to binary

c1	c2
00 05 => 00 0101	00 02 => 00 0010
00 05 => 00 0101	00 02 => 00 0010
00 08 => 00 1000	00 03 => 00 0011
00 09 => 00 1001	00 06 => 00 0110
00 06 => 00 0110	00 04 => 00 0100
00 02 => 00 0010	00 09 => 00 1001
00 05 => 00 0101	00 01=> 00 0001

Fig. 4. Binary conversion of the c1 and c2 ciphers

00 0101×00 0010 = 00 1010	00 10
00 1000×00 0011 = 00 11000	00 24
00 1001×00 0110 = 00 110110	00 54
00 0110×00 0100 = 00 11000	00 24
00 0010×00 1001 = 00 10010	00 18
00 0101×00 0001 = 00 0101	00 05

Fig. 5. The multiplication process of c1 * c2

Multiplication of c1 * c2 block by block
Thus: c1c2 = 00100002000400050004000200040002000400010008005
By decrypting the c1 * c2 encryption with RSA's private key, we get:
m1m2 = 102454241805
Verification: m1 = 589625 and m2 = 236491
We multiply m1 * m2 block by block, we get
m1m2 = 10245424185
We notice that it is the same result as if the calculation was carried out on the plain text messages: m1 * m2.

5.3 Application

We used two extremities; a client that will encrypt the calculated integers then send them to the server. Moreover, the server that will perform calculations on the received metrics uses the encryption of Paillier on the case of addition and RSA on multiplication (Fig. 6).

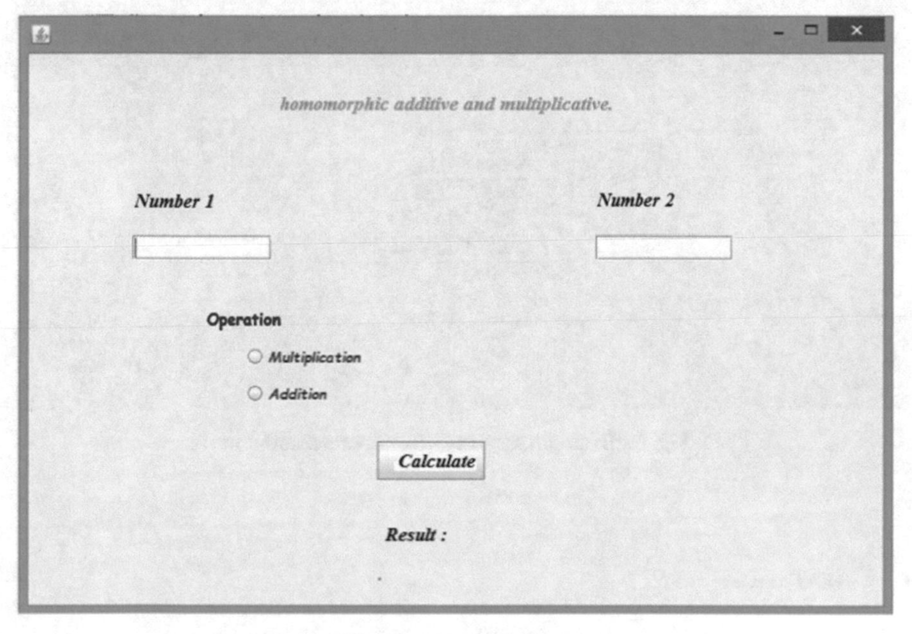

Fig. 6. User interface.

The customer can enter two numbers of his choice and the desired operation and then click on the calculation button.

At this step, a set of operations are carried out automatically.

- The customer encrypts the two numbers by the two methods RSA and Paillier.
- Take charge of the desired operation.
- Store the encrypted numbers in the database (Fig. 7).

- Send the parameters to the server.
- The server does the calculations.
- Store the results in a database (Fig. 8).

```
Veuillez saisir votre chiffre 1:
10
Veuillez saisir votre chiffre 2:
13
le chiffre 1 crypté paillier: 67708455868147730350400157740688382637070225757289076745239732153011506921127859177919882097522302757338
le chiffre 2 crypté: paillier :187948347063888077682186573328143205352687579598982147889830948248699018933935718801742198166246292556 2
le chiffre 1 décrypté paillier: 10
le chiffre 2 décrypté paillier: 13
le chiffre 1 crypté RSA: 26218aecc5694ddba69840cf99e6f255335360b707b39aad99316dbdf4ac4e539ad98ff3375471d28b228579e7c95030e9172f329a507
le chiffre 2 crypté RSA: 7cb30f5e4f4f6aa976c4a82293df1c2cdc6b095039ef09a8fa5de2bd87f11743fd48acd0b87706434381c645c9d2fa77f06a5352be8cc
le chiffre 1 decrypté RSA: 10
le chiffre 2 decrypté RSA: 13
```

Fig. 7. Paillier and RSA encryption/decryption

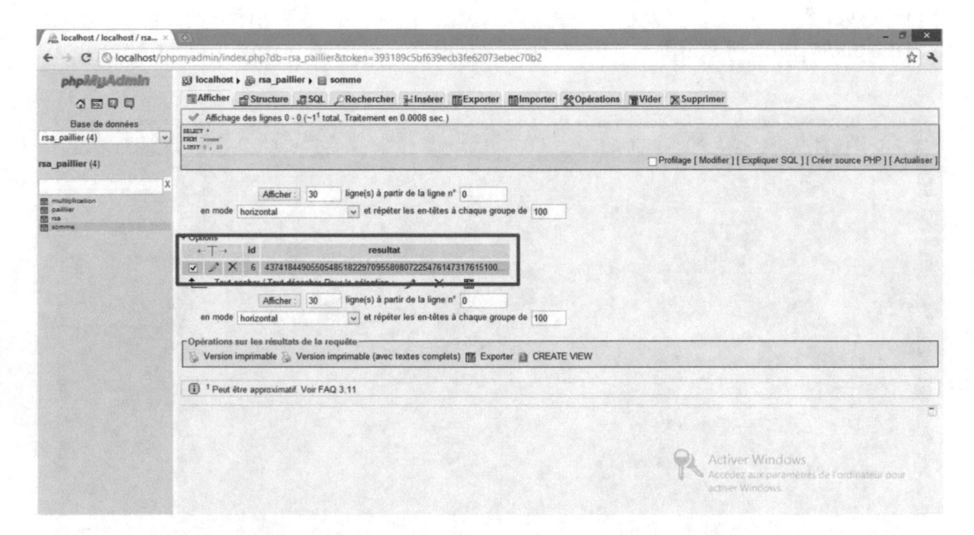

Fig. 8. Example result of the numerical addition

6 Conclusion

From the aspect of cryptographic applications, the encryption function's homomorphism property is not always appreciated unambiguously as a merit of a cryptosystem. The encryption function should not always be assessed explicitly as a merit of the cryptosystem. There are examples where this property should be considered a weakness, for example, the transformation inverse of a cryptosystem's encryption function. This document's objective was to create a solution that would allow a third party to calculate the numbers without deciphering them and that the result would be usable and exploitable. Before realizing this solution, we researched the different methods and algorithms, allowing the encryption while guaranteeing its data security.

References

1. Touil, H., El Akkad, N., Satori, K.: Secure and guarantee QoS in a video sequence: a new approach based on TLS protocol to secure data and RTP to ensure real-time exchanges. Int. J. Saf. Secur. Eng. **11**(1), 59–68 (2021)
2. Touil, H., El Akkad, N., Satori, K.: Text encryption: hybrid cryptographic method using Vigenere and Hill Ciphers. In: 2020 International Conference on Intelligent Systems and Computer Vision (ISCV), Fez, Morocco, pp. 1–6 (2020)
3. Touil, H., El Akkad, N., Satori, K.: H-Rotation: secure storage and retrieval of passphrases on the authentication process. Int. J. Saf. Secur. Eng. **10**(6), 785–796 (2020)
4. Touil, H., El Akkad, N., Satori, K.: Securing the storage of passwords based on the MD5 HASH transformation. In: International Conference on Digital Technologies and Applications (2021)
5. El Akkad, N.E., Merras, M., Saaidi, A., Satori, K.: Robust method for self-calibration of cameras having the varying intrinsic parameters. J. Theor. Appl. Inf. Technol. **50**(1), 57–67 (2013)
6. El Akkad, N.E., Merras, M., Saaidi, A., Satori, K.: Camera self-calibration with varying parameters from two views. WSEAS Trans. Inf. Sci. Appl. **10**(11), 356–367 (2013)
7. El Akkad, N., Saaidi, A., Satori, K.: Self-calibration based on a circle of the cameras having the varying intrinsic parameters. In: Proceedings of 2012 International Conference on Multimedia Computing and Systems, ICMCS, pp. 161–166 (2012)
8. Es-sabry, M., El Akkad, N., Merras, M., Saaidi, A., Satori, K.: Grayscale image encryption using shift bits operations. In: International Conference on Intelligent Systems and Computer Vision (ISCV), Fez, pp. 1–7 (2018). https://doi.org/10.1109/ISACV.2018.8354028
9. Es-sabry, M., El Akkad, N., Merras, M., Saaidi, A., Satori, K.: A novel text encryption algorithm based on the two-square cipher and Caesar cipher. Commun. Comput. Inf. Sci. **872**, 78–88 (2018)
10. Es-sabry, M., El Akkad, N., Merras, M., Saaidi, A., Satori, K.: A new color image encryption using random numbers generation and linear functions. Adv. Intell. Syst. Comput. **1076**, 581–588 (2020)
11. Li, R., Ishimaki, Y., Yamana, H.: Privacy preserving calculation in cloud using fully homomorphic encryption with table lookup. In: 2020 5th IEEE International Conference on Big Data Analytics, ICBDA 2020, May 2020, Article number 9101276, pp. 315–322 (2020)
12. Umadevi, C.N., Gopalan, N.P.: Privacy preserving outsourced calculations with symmetric fully homomorphic encryption. Int. J. Innov. Technol. Exploring Eng. **8**(10), 3012–3015 (2019)
13. Wang, L., Saha, T.K., Aono, Y., Koshiba, T., Moriai, S.: Enhanced secure comparison schemes using homomorphic encryption. In: Barolli, L., Li, K.F., Enokido, T., Takizawa, M. (eds.) NBiS 2020. AISC, vol. 1264, pp. 211–224. Springer, Cham (2021). https://doi.org/10.1007/978-3-030-57811-4_20
14. Smart, N.P., Vercauteren, F.: Fully homomorphic encryption with relatively small key and ciphertext sizes. Cryptology ePrint Archive, Report 2009/571 (2009)
15. Dijk, M., Gentry, C., Halevi, S., Vaikuntanathan, V.: Fully homomorphic encryption over the integers. Cryptology ePrint Archive, Report, 2009/616 (2009)
16. Canteaut, A., et al.: Stream ciphers: a practical solution for efficient homomorphic-ciphertext compression. Cryptology EPrint Archive, Report (2015)
17. Gentry, C., Halevi, S., Nigel, P.: Smart. Homomorphic evaluation of the AES circuit. Crypto (2012)
18. Mousa, A., Faragallah, O.S., El-Rabaie, S., Nigm, E.M.: Security analysis of reverse encryption algorithm for databases. Int. J. Comput. Appl. (0975–8887) **66** (14) (2013)

19. Mousa, A., Faragallah, O., Nigm, E., Rabaie, E.: Evaluating the performance of reverse encryption algorithm (REA) on the databases. Int. Arab J. Inf. Technol. **10**(6) (2013)
20. Paillier, P.: Public-key cryptosystems based on composite degree residuosity classes. In: Stern, J. (ed.) EUROCRYPT 1999. LNCS, vol. 1592, pp. 223–238. Springer, Heidelberg (1999). https://doi.org/10.1007/3-540-48910-X_16
21. Rao, G., Subba, V., Uma, G.: An efficient secure message transmission in mobile ad hoc networks using enhanced homomorphic encryption scheme. GJCST-E: Netw. Web Secur. **13**(9) (2013)
22. Potey, M.M., Dhote, C.A., Sharma, D.H.: Homomorphic encryption for security of cloud data (open access). Procedia Comput. Sci. **79**, 175–181 (2016)
23. Parmar, P.V., Padhar, S.B., Patel, S.N., Bhatt, N.I., Jhaveri, R.H.: Survey of various homomorphic encryption algorithms and schemes. Int. J. Comput. Appl. **91**, 26–32 (2014)
24. Hizkia, N.E.: Implementasi Algoritma KriptografiKunci Publik Okamoto-Uchiyama. Informatics Engineering, Bandung Institute of Technology (2013)
25. Kumar Arya, P., Singh Aswal, M., Kumar, V.: Comparative study of asymmetric key cryptographic algorithms. Int. J. Comput. Sci. Commun. Netw. **5**(1), 17–21 (2015)
26. Mohan, R., Dhruw, H.L., Raghvendra: An effective image encryption based on the combination of scan and Elgamal method. Int. J. Eng. Comput. Sci. **4**(5), 11793–11796 (2015)

Deep Learning

Hybrid Deep Learning Models for Diabetic Retinopathy Classification

Mounia Mikram[1,4](✉), Chouaib Moujahdi[2], Maryem Rhanoui[3,4], Majdouline Meddad[1], and Asmaa Khallout[1,2,3,4]

[1] LRIT Laboratory, Associated Unit to CNRST (URAC 29), Rabat IT Center, Faculty of Sciences, Mohammed V University in Rabat, Rabat, Morocco
mmikram@esi.ac.ma

[2] Scientific Institute of Rabat, Mohammed V University in Rabat, Rabat, Morocco

[3] IMS Team, ADMIR Laboratory, Rabat IT Center, ENSIAS, Mohammed V University in Rabat, Rabat, Morocco

[4] Meridian Team, LYRICA Laboratory, School of Information Sciences, Rabat, Morocco

Abstract. Diabetic retinopathy is a complication of diabetes in the eye. This disease is caused by the damage of the blood vessels of the back of eye (i.e., retina). Unfortunately, diabetic retinopathy can cause several symptoms, the most serious of which is complete vision loss. Indeed, the detection of diabetic retinopathy is a time-consuming manual process that requires a qualified clinician to examine and evaluate digital color photographs of the retina's fundus.

Currently, several researches are looking to employ artificial intelligence techniques, especially the Deep Learning, to deal with this issue. In this paper, we study some hybrid models for diabetic retinopathy severity classification in distributed and non-distributed environments. The studied models perform two main tasks: deep feature extraction and then classification of diabetic retinopathy according to its severity. The models were trained and validated on a publicly available dataset of 80,000 images and they achieved an accuracy of 80.7%.

Keywords: Diabetic retinopathy · Deep learning · CNN · Inception v3

1 Introduction

Diabetes is a serious disease that affects more than 415 million people and retinal pathologies are responsible for millions of cases of blindness worldwide [10]. The main causes of blindness are age-related glaucoma (4.5 million cases), macular degeneration (3.5 million cases) and diabetic retinopathy (2 million cases).

Diabetic retinopathy (DR) is one of the leading causes of blindness in the world today [13]. Increasing life expectancy, sedentary lifestyles and other factors will continue to increase in the numbers of people with diabetes. Regular monitoring of DR in diabetic patients has proven to be a cost-effective and important

M. Lazaar et al. (Eds.): BDIoT 2021, LNNS 489, pp. 167–178, 2022.
https://doi.org/10.1007/978-3-031-07969-6_13

aspect for their care. The accuracy and timing of this care is of great importance to both the cost and effectiveness of treatment. If detected early enough, effective DR treatment is available, which making it an essential process.

Classifying dead zones involves weighting many features and the location of these features, which is time consuming for already overwhelmed clinicians. Artificial intelligence models are able to achieve much faster classifications once trained, helping clinicians with real-time classification. The effectiveness of automated classification for DR has been actively researched in the field of computer imaging and the results are encouraging.

Deep Learning is rapidly becoming the leading-edge technology for medical image analysis [15]. Given the importance of medical archives, where each image is associated with a diagnosis, it is possible to train effective pathology detectors or classifiers with virtually no specialized knowledge of the targeted pathologies. Convolutional Neural Networks (CNNs), a branch of deep learning, have a long and proven track record in image analysis and interpretation, particularly in medical imaging.

Deep neural networks were proven to have high potential for identifying and classifying diabetic retinopathy in retinal fundus images with high sensitivity and high specificity [16].

The fundus image analysis system described in this work is designed to help ophthalmologists establish a diagnosis by providing a second opinion, along with being a tool for mass screening and detection for diabetic retinopathy.

The remainder of this paper is organized as follow: in Sect. 2 we present the related works. Section 3 presents an overview of the diabetic retinopathy and the motivation of this work. Section 4 explains the proposed models and the experimental results. Finally, Sect. 5 concludes the paper.

2 Motivation: Diabetic Retinopathy

Diabetic retinopathy, characterized by damage to the retina of the eye, is a serious complication of diabetes that affects 50% of type 2 diabetic patients (at different stages). Diabetic retinopathy is the leading cause of blindness in the active age group. Among 23 million Americans, 59 million Europeans and up to 50 million Indians with diabetes, the prevalence of the disease is estimated between 18% and 28%.

Several factors contribute to the onset of diabetic retinopathy and accelerate its progression: the duration of diabetes, blood sugar levels, unstable diabetes, high blood pressure, smoking, etc. This pathology can accelerate the onset of other eye diseases such as glaucoma or cataracts, and can even lead to blindness in the absence of appropriate treatment.

2.1 Diabetic Retinopathy Characteristics

Receiving light waves and transmitting them to the brain via the optic nerve, the retina is a thin membrane of the eye crossed by a multitude of small vessels, the capillaries. Excess sugar in the blood - as in the case of diabetes - weakens the walls of these vessels, causing them to become leaky. The vessels then rupture and burst ("micro-aneurysms"). Over time, large areas of the retina are no longer oxygenated. In reaction, the retina produces new, even more fragile vessels.

The phenomenon amplifies and extends to the macula (area in the middle of the retina), where the center of vision is located. The macula thickens and a macular edema occurs, responsible for a decrease in visual acuity that can be very significant and only partially reversible. In addition, the vessels newly produced by the retina can bleed into the vitreous (area in front of the retina), leading in some cases to a risk of tearing and therefore detachment of the retina, responsible for a permanent loss of vision.

A regular eye examination is necessary to diagnose DR at an early stage, when it can be treated with the best prognosis. Currently, the detection of DR is a time-consuming, manual process that requires a qualified clinician to examine and evaluate digital color photographs of the retina's fundus.

The clinical classification process involves detecting subtle features such as microaneurysms, exudates, intraretinal hemorrhages, and sometimes their position relative to each other on the eye's images [20] (Fig. 1).

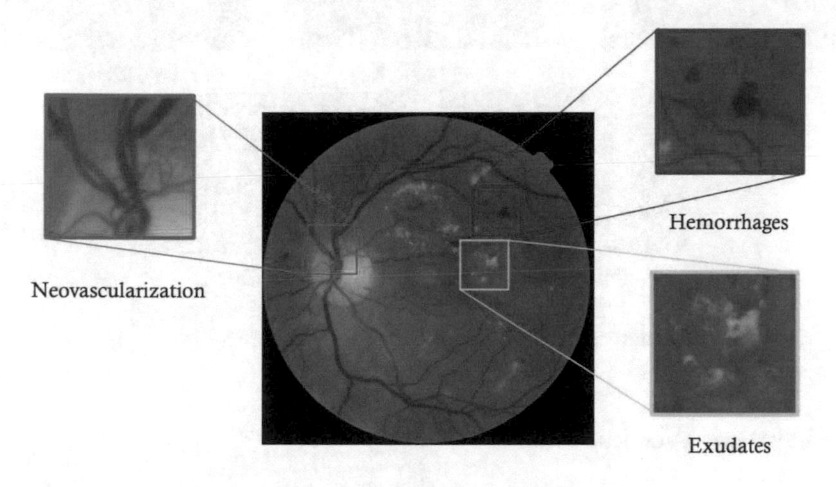

Fig. 1. Diabetic retinopathy [6]

2.2 Diabetic Retinopathy Classification

The classification of DR is based on the ocular risk associated with the severity of retinal ischemia. It begins with a stage of minimal and then moderate nonproliferative diabetic retinopathy and progresses to severe (or pre-proliferative) nonproliferative diabetic retinopathy characterized by extensive retinal ischemia. It then progresses to proliferative diabetic retinopathy characterized by the proliferation of neo-vessels on the surface of the retina and/or on the papilla.

Macular edema can be associated with all stages of DR. However, its incidence increases with the severity of retinopathy. In general, there are 5 classes of diabetic retinopathy [35].

- No diabetic retinopathy.
- Mild non-proliferative DR: presence of microaneurysms and punctiform hemorrhages at the posterior pole: on examination of the fundus, microaneurysms appear as small punctiform, red, lesions at the limit of visibility.
- Moderate non proliferative DR: hemorrhage in few spots.
- Severe non-proliferative DR: numerous spotted haemorrhages all around the periphery.
- Proliferative DR (intravitreal hemorrhage, traction retinal detachment, neovascular glaucoma) (Fig. 2).

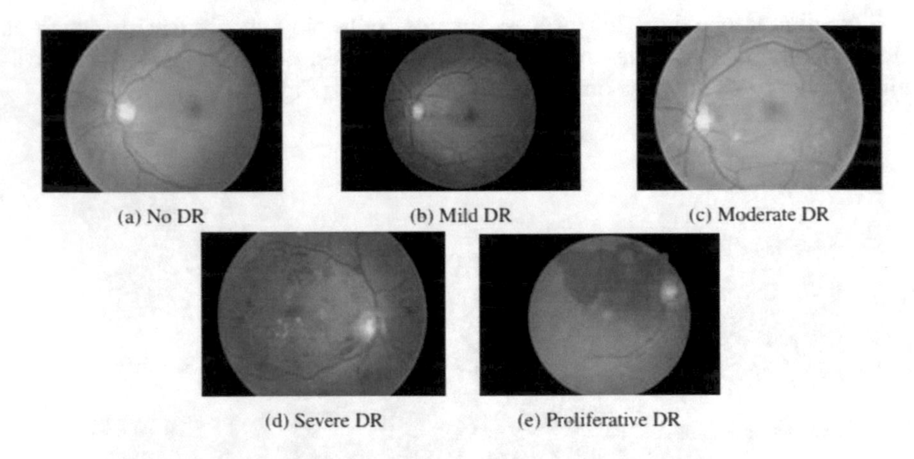

Fig. 2. Increasing severity of diabetic retinopathy levels [4]

3 Related Works

Convolutional Neural Network (CNN) is a powerful model of artificial intelligence successfully used in various applications such as image and video recognition, recommendation systems, image classification, natural language processing and medical image analysis.

In the literature, according to Asiri et al. [5], the proposed solutions for Diabetic Retinopathy Recognition can be divided into two main categories: Software Solutions and Hardware Solutions. Each category can be further divided into several sub-categories.

- **Software solutions** are techniques for detecting whether a patient is suffering from diabetic retinopathy.

- **Hardware solutions** aim to create a hardware system that can visualize the problem of diabetic retinopathy.

In this paper, we are interested in the first category. The software solutions can be further divided into two sub-categories: traditional vision methods and artificial intelligence methods.

- **Traditional vision methods** aims to formulate a way of representing the image by coding the existence of various features using image processing techniques. These features can be corners, color schemes, image texture, etc.
- **Artificial intelligence methods** automate the detection/recognition process by obtaining relevant features/separation classes by learning models.

3.1 Traditional Vision Methods

For traditional vision methods, Chaudhuri et al. [8] and Al-Rawi et al. [3] used 2 D matched filters to detect retinal vasculature. [32] developed an exudate detection technique for poor quality images of the eyes using morphological operators optimally adjusted without pupil dilation. Soares et al. [31] proposed a vessel detection method using a Gaussian Mixture Model (GMM) and a 2-D Gabor wavelet.

Sinthanayothin et al. [30] used the adaptive local contrast enhancement to enhance the image quality then applied a Recursive Region-Growing Technique (RRGT) on a 10×10 window using selected threshold values in grayscale images to localizes other anatomical regions such as OD, blood vessels, and fovea. Fleming et al. [12] proposed a multi-scale morphology method for locating exudates in fundus images of the retina and uses Gaussian and median filtering of green channel images to correct contrast and shading variations.

The main disadvantage of these approaches is that they are time-consuming and focuses on a single issue, which is the detection of a specific feature.

3.2 Artificial Intelligence Methods

Artificial Intelligence methods can be divided into three sub-categories: Machine Learning approaches, Deep Learning approaches, and Hybrid approaches.

- **Machine Learning approaches** automate the classification based on the well-extracted features before using a machine learning classification model.
- **Deep Learning approaches** automatically extract the pertinent features and then perform a classification task.
- **Hybrid approaches** extract features using deep learning models and then classify them using machine learning models.

Machine Learning Approaches. Rahim et al. [25] used fuzzy image processing techniques, namely fuzzy histogram equalization and fuzzy edge detection for the image processing phase. Then, DR classification and detection was performed using the SVM and KNN machine learning algorithms. Acharya et al. [2] proposed to use morphological image processing techniques to extract the characteristics of exudates, blood vessels, microaneurysms and hemorrhages. An SVM model was then trained on a data set of 331 images.

Carrera et al. [7] proposed efficient algorithms for the detection of blood vessels, microaneurysms, the optic disc, and hard exudates then using an SVM classification to classify the grade of non-proliferative diabetic retinopathy at any retinal image. Jayashree et al. [18] analyzed the diabetic retinopathy using Particle swarm optimization Feature Selection algorithm on three different Classifiers SVM, Random Forest, decision Tree.

The main disadvantage of these approaches is that if feature extraction is not correctly done, the machine learning model cannot properly classify the images.

Deep Learning Approaches. Several deep learning models have been proposed for the classification of Diabetic Retinopathy images.

Alban et al. [4] proposed an automated detection system of diabetic retinopathy by using fluorescein angiography photographs.

Lim et al. [19] proposed a nine-layer CNN model to segment the Optical Cup (OC) and Optical Disk (OD) evaluated on MESSIDOR and SEED-DB in four phases. They first locate the region around the optical disk (OD) and then enhance this region using a CNN model, then they obtain the pixel-level classification of the enhanced region to produce a probability map, which is finally segmented to predict the cup and disk boundaries. Guo et al. [17] segmented the Optic Cup (OC) in the classification of each pixel patch and post-processing by using a large pixel patch based on the CNN model which was assessed on the DRISHTI-GS dataset.

Nayak et al. [22] used image processing techniques to extract the area of blood vessels, the area of exudate and texture features which are then input into a small neural network that reports 90% accuracy and 90% sensitivity on the binary classification task on a data set of 140 images. Sevastopolsky [29] proposed to segment both OC and OD by using modified U-net CNN presented in [27].

Tan et al. [34] proposed to first normalize the colored images, then to formulate a segmentation of the Optical Disc (OD), blood vessels and fovea as classification problems by assuming 4 classes: blood vessels, OD, background and fovea, using a 7-layer CNN model. Wu et al. [36] proposed to extract discriminant features from the CNN model and then estimate the distribution of the local structure using PCA-based nearest neighbor search to segment blood vessels.

Hybrid Approaches. Qomariah et al. [24] proposed to use the high-level features of the last fully connected layer based on transfer learning from Convolutional Neural Network (CNN) as the input features for classification using the support vector machine (SVM) algorithm.

These approaches leverage the combined power of Deep Learning methods for feature extraction and the efficiency of Machine Learning methods for classification.

Thus, in this paper, we evaluate hybrid models for the classification of diabetic retinopathy, and we also study the impact of processing distribution in a Big Data environment on performance.

4 Diabetic Retinopathy Recognition in Big Data Environment

In this section, we present and evaluate hybrid models based on pretrained Convolutional neural network with KNN or SVM Algorithms for Diabetic Retinopathy classification.

4.1 Database Description

The Diabetic retinopathy database [11] includes 35126 images in the training set and 53576 images in the test set (Table 1).

Table 1. Dataset distribution

	Training	Test
No Dr	25810	39533
Mild	2443	3762
Moderate	5292	7861
Severe	873	1214
Proliferative DR	708	1206
Total	35126	53576

The data is divided into five categories [23]:

- **No DR** (Normal: No abnormalities)
- **Mild** (Microaneurysms only)
- **Moderate** (Microaneurysms and one or more of exudates)
- **Severe** (Microaneurysms and exudates in all four quadrants as fovea is a center of four quadrants/Intra retinal micro vascular abnormalities in one or more quadrants/Venous beading in at least two quadrants)
- **Proliferative DR** (New vessels elsewhere/New vessels on the optic disc/Neovascularisation).

4.2 Hybrid CNN-KNN/SVM Model

The hybrid model is composed of Inception-v3 Deep Learning model for feature extraction and a KNN or SVM Machine Learning model for classification.

- **Feature Extraction:** CNN is a powerful feature-extracting model and gives better results when it is integrated and combined into larger network [14]. Therefore, hybrid systems combining CNN with other more suitable models for classification have been proposed for various purposes [9,26,28].
- **Classification**: Once the features have been extracted in the previous step, a machine learning model is used to classify the images into the 5 DR severity categories. To reduce the dimension of these features, a principal component analysis (PCA) is performed.

Inception-v3 [33] is a pretrained models by TensorFlow [1]. A pretrained model is a network that has been previously trained on a large dataset, usually on a large-scale image classification task. Transfer learning [34] is used to customize this pretrained model for a given task (Fig. 3).

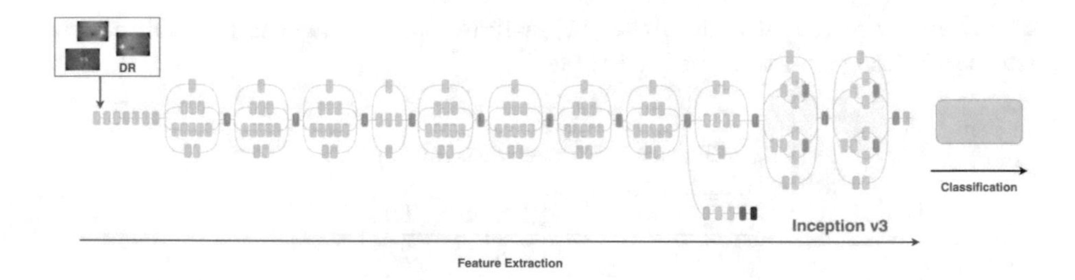

Fig. 3. Feature extraction with Inception v3

The feature that feeds the last classification layer is also called the bottleneck feature.

The features extracted with Inception v3 are processed to form a classifier on the images. Subsequently, an SVM and KNN model are used to classify the images into the 5 classes.

4.3 Results and Discussion

The Table 2 shows a summary of the results obtained in the three methods that are expressed in terms of classification accuracy, execution time and finally also shows the basic size uses for each method.

Inception V3 + SVM model gives better results, however the execution time of this model is relatively long. This is due to the large size of the dataset and hardware limitations (Fig. 4).

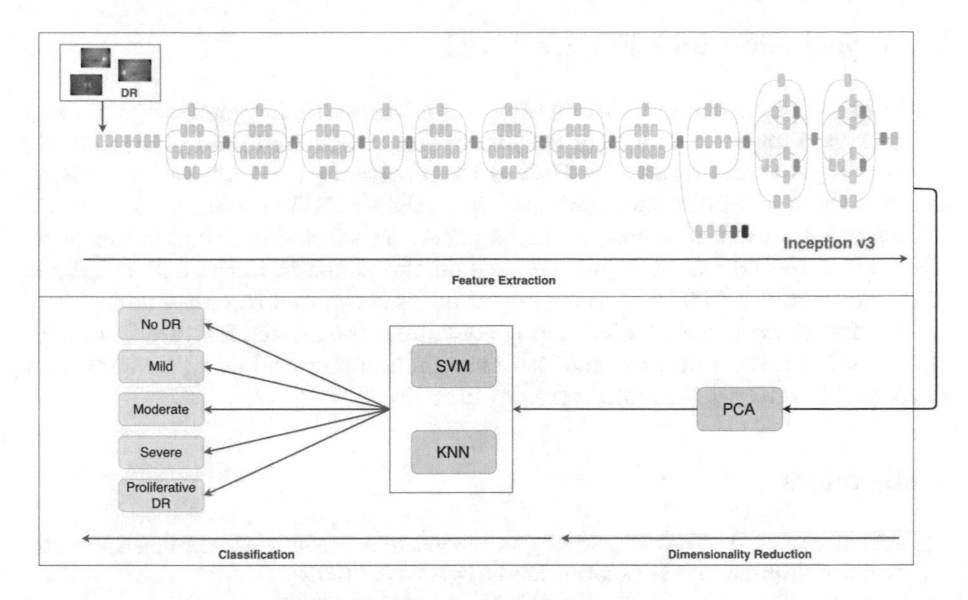

Fig. 4. Hybrid CNN-KNN/SVM model

Table 2. Results in non distributed environment

	Training	Test	Prediction	Execution time
Inception V3 + SVM	35126	53576	80.7%	35 min
Inception V3 + KNN	35126	53576	79%	15 min

To address this problem, we studied how to reduce execution time by distributing all tasks in a big data environment using Hadoop and Apache Spark technologies (Table 3).

Table 3. Results in distributed environment

	Training	Test	Accuracy	Execution time
Inception V3 + SVM	35126	53576	80.7%	35 min
Inception V3 + KNN	35126	53576	79%	15 min
Inception V3 + SVM in Apache Spark	35126	53576	80.7%	5 min

Execution in a distributed environment has significantly optimized execution time while maintaining performance. This confirms the relevance of considering a Big Data environment for medical image analysis.

5 Conclusion and Future Work

Early detection of diabetic retinopathy is a critical task for good diagnosis and efficient care. In order to satisfy this obligation, automated models for diabetic retinopathy diagnoses, specially Hybrid Deep Learning ones, are very promising and are the subject of growing interest from the scientific community.

Among the models discussed in this paper, transferred distributed Inception V3/SVM achieved the best performance on the Diabetic Retinopathy Dataset with an accuracy of 80.7% and an execution time reduced to only 5 min.

For future work, we are considering combining transferred features from multiple CNN to further improve the classification accuracy and using a compressed CNN [21] to optimize the hardware performance.

References

1. Abadi, M., et al.: TensorFlow: large-scale machine learning on heterogeneous distributed systems. arXiv preprint arXiv:1603.04467 (2016)
2. Acharya, U.R., Lim, C.M., Ng, E.Y.K., Chee, C., Tamura, T.: Computer-based detection of diabetes retinopathy stages using digital fundus images. Proc. Inst. Mech. Eng. Part H: J. Eng. Med. **223**(5), 545–553 (2009)
3. Al-Rawi, M., Karajeh, H.: Genetic algorithm matched filter optimization for automated detection of blood vessels from digital retinal images. Comput. Methods Program. Biomed. **87**(3), 248–253 (2007)
4. Alban, M., Gilligan, T.: Automated detection of diabetic retinopathy using fluorescein angiography photographs. Report of Standford Education (2016)
5. Asiri, N., Hussain, M., Al Adel, F., Alzaidi, N.: Deep learning based computer-aided diagnosis systems for diabetic retinopathy: a survey. Artif. Intell. Med. **99**, 101701 (2019)
6. Bravo, M.A., Arbeláez, P.A.: Automatic diabetic retinopathy classification. In: 13th International Conference on Medical Information Processing and Analysis, vol. 10572, p. 105721E. International Society for Optics and Photonics (2017)
7. Carrera, E.V., González, A., Carrera, R.: Automated detection of diabetic retinopathy using SVM. In: 2017 IEEE XXIV International Conference on Electronics, Electrical Engineering and Computing (INTERCON), pp. 1–4. IEEE (2017)
8. Chaudhuri, S., Chatterjee, S., Katz, N., Nelson, M., Goldbaum, M.: Detection of blood vessels in retinal images using two-dimensional matched filters. IEEE Trans. Med. Imaging **8**(3), 263–269 (1989)
9. Chen, Y., Jiang, H., Li, C., Jia, X., Ghamisi, P.: Deep feature extraction and classification of hyperspectral images based on convolutional neural networks. IEEE Trans. Geosci. Remote Sens. **54**(10), 6232–6251 (2016)
10. Cho, N., et al.: IDF diabetes atlas: global estimates of diabetes prevalence for 2017 and projections for 2045. Diabetes Res. Clin. Pract. **138**, 271–281 (2018)
11. Cuadros, J., Bresnick, G.: EyePACS: an adaptable telemedicine system for diabetic retinopathy screening. J. Diab. Sci. Technol. **3**(3), 509–516 (2009)
12. Fleming, A.D., Philip, S., Goatman, K.A., Williams, G.J., Olson, J.A., Sharp, P.F.: Automated detection of exudates for diabetic retinopathy screening. Phys. Med. Biol. **52**(24), 7385 (2007)
13. Fong, D.S., et al.: Retinopathy in diabetes. Diab. Care **27**(suppl 1), s84–s87 (2004)

14. Goldberg, Y.: Neural network methods for natural language processing. Synthesis Lect. Hum. Lang. Technol. **10**(1), 1–309 (2017)
15. Goodfellow, I., Bengio, Y., Courville, A., Bengio, Y.: Deep Learning, vol. 1. MIT Press, Cambridge (2016)
16. Gulshan, V., et al.: Development and validation of a deep learning algorithm for detection of diabetic retinopathy in retinal fundus photographs. JAMA **316**(22), 2402–2410 (2016)
17. Guo, Y., Zou, B., Chen, Z., He, Q., Liu, Q., Zhao, R.: Optic cup segmentation using large pixel patch based CNNs (2016)
18. Jayashree, J., Sruthi, R., Ponnamanda Venkata, S., Vijayashree, J.: Knowledge based expert system for predictingdiabetic retinopathy using machine learningalgorithms. Int. J. Eng. Adv. Technol. (IJEAT) **9**(4) (2020)
19. Lim, G., Cheng, Y., Hsu, W., Lee, M.L.: Integrated optic disc and cup segmentation with deep learning. In: 2015 IEEE 27th International Conference on Tools with Artificial Intelligence (ICTAI), pp. 162–169. IEEE (2015)
20. Mansour, R.F.: Deep-learning-based automatic computer-aided diagnosis system for diabetic retinopathy. Biomed. Eng. Lett. **8**(1), 41–57 (2018)
21. Meddad, M., Moujahdi, C., Mikram, M., Rziza, M.: A hybrid face identification system using a compressed CNN in a big data environment for embedded devices. Int. J. Comput. Digit. Syst. **9**(4), 689–701 (2020)
22. Nayak, J., Bhat, P.S., Acharya, R., Lim, C.M., Kagathi, M.: Automated identification of diabetic retinopathy stages using digital fundus images. J. Med. Syst. **32**(2), 107–115 (2008)
23. Paing, M.P., Choomchuay, S., Yodprom, M.R.: Detection of lesions and classification of diabetic retinopathy using fundus images. In: 2016 9th Biomedical Engineering International Conference (BMEiCON), pp. 1–5. IEEE (2016)
24. Qomariah, D.U.N., Tjandrasa, H., Fatichah, C.: Classification of diabetic retinopathy and normal retinal images using CNN and SVM. In: 2019 12th International Conference on Information & Communication Technology and System (ICTS), pp. 152–157. IEEE (2019)
25. Rahim, S.S., Palade, V., Jayne, C., Holzinger, A., Shuttleworth, J.: Detection of diabetic retinopathy and maculopathy in eye fundus images using fuzzy image processing. In: Guo, Y., Friston, K., Aldo, F., Hill, S., Peng, H. (eds.) BIH 2015. LNCS (LNAI), vol. 9250, pp. 379–388. Springer, Cham (2015). https://doi.org/10.1007/978-3-319-23344-4_37
26. Rhanoui, M., Mikram, M., Yousfi, S., Barzali, S.: A CNN-BiLSTM model for document-level sentiment analysis. Mach. Learn. Knowl. Extract. **1**(3), 832–847 (2019)
27. Ronneberger, O., Fischer, P., Brox, T.: U-net: convolutional networks for biomedical image segmentation. In: Navab, N., Hornegger, J., Wells, W.M., Frangi, A.F. (eds.) MICCAI 2015. LNCS, vol. 9351, pp. 234–241. Springer, Cham (2015). https://doi.org/10.1007/978-3-319-24574-4_28
28. Scarpa, G., Gargiulo, M., Mazza, A., Gaetano, R.: A CNN-based fusion method for feature extraction from sentinel data. Remote Sen. **10**(2), 236 (2018)
29. Sevastopolsky, A.: Optic disc and cup segmentation methods for glaucoma detection with modification of U-net convolutional neural network. Pattern Recognit. Image Anal. **27**(3), 618–624 (2017)
30. Sinthanayothin, C., et al.: Automated detection of diabetic retinopathy on digital fundus images. Diabet. Med. **19**(2), 105–112 (2002)

31. Soares, J.V., Leandro, J.J., Cesar, R.M., Jelinek, H.F., Cree, M.J.: Retinal vessel segmentation using the 2-D gabor wavelet and supervised classification. IEEE Trans. Med. Imaging **25**(9), 1214–1222 (2006)
32. Sopharak, A., Uyyanonvara, B., Barman, S., Williamson, T.H.: Automatic detection of diabetic retinopathy exudates from non-dilated retinal images using mathematical morphology methods. Comput. Med. Imaging Graph. **32**(8), 720–727 (2008)
33. Szegedy, C., Vanhoucke, V., Ioffe, S., Shlens, J., Wojna, Z.: Rethinking the inception architecture for computer vision. In: Proceedings of the IEEE Conference on Computer Vision and Pattern Recognition, pp. 2818–2826 (2016)
34. Tan, J.H., Acharya, U.R., Bhandary, S.V., Chua, K.C., Sivaprasad, S.: Segmentation of optic disc, fovea and retinal vasculature using a single convolutional neural network. J. Comput. Sci. **20**, 70–79 (2017)
35. Wang, Z., Yin, Y., Shi, J., Fang, W., Li, H., Wang, X.: Zoom-in-net: deep mining lesions for diabetic retinopathy detection. In: Descoteaux, M., Maier-Hein, L., Franz, A., Jannin, P., Collins, D.L., Duchesne, S. (eds.) MICCAI 2017. LNCS, vol. 10435, pp. 267–275. Springer, Cham (2017). https://doi.org/10.1007/978-3-319-66179-7_31
36. Wu, A., Xu, Z., Gao, M., Buty, M., Mollura, D.J.: Deep vessel tracking: a generalized probabilistic approach via deep learning. In: 2016 IEEE 13th International Symposium on Biomedical Imaging (ISBI), pp. 1363–1367. IEEE (2016)

Detection of Appliance-Level Abnormal Energy Consumption in Buildings Using Autoencoders and Micro-moments

Yassine Himeur[1](✉), Abdullah Alsalemi[1], Faycal Bensaali[1], and Abbes Amira[2,3]

[1] Department of Electrical Engineering, Qatar University, Doha, Qatar
{yassine.himeur,a.alsalemi,f.bensaali}@qu.edu.qa

[2] Department of Computer Science, University of Sharjah, Sharjah, UAE
aamira@sharjah.ac.aea,abbes.amira@dmu.ac.uk

[3] Institute of Artificial Intelligence, De Montfort University, Leicester, UK

Abstract. The detection of anomalous energy usage could help significantly in signaling energy wastage and identifying faulty appliances, especially if the individual power traces are analyzed. To that end, this paper proposes a novel abnormal energy consumption detection approach at the appliance-level using autoencoder and micro-moments. Accordingly, energy usage footprints of different household appliances along with occupancy patterns are analyzed for modeling normal energy consumption behaviors, and on the flip side, detecting abnormal usage. In effect, energy micro-moments occur when end-users reflexively (i) switch on/off an appliance to start/stop an energy consumption action; (ii) increase/reduce energy consumption of a specific appliance; and (iii) enter/leave a specific room. Put differently, energy micro-moments are captured by reference to end-users' daily tasks usually performed to meet their preferences. In this regard, energy micro-moment patterns are extracted from appliance-level consumption fingerprints and occupancy data using an innovative rule-based algorithm to represent the key intent-driven moments of daily energy use, and hence model normal and abnormal behaviors. Moving forward, energy micro-moment patterns are fed into an autoencoder including 48 input/output neurons, and 4 neurons in the intermediate layer aiming at reducing the computational cost and improving the detection performance. This has helped in accurately detecting two kinds of anomalous energy consumption, i.e. "excessive consumption" and "consumption while outside", where up to 0.95 accuracy and F1 score have been achieved, for example, when analyzing microwave energy consumption.

Keywords: Appliance-level energy consumption · Anomaly detection · Autoencoder · Micro-moments · Excessive consumption · Consumption while outside

M. Lazaar et al. (Eds.): BDIoT 2021, LNNS 489, pp. 179–193, 2022.
https://doi.org/10.1007/978-3-031-07969-6_14

1 Introduction

Recently, a great effort has been put to mitigate climate change, which is a principal concern for the world's population. In effect, the building energy sector is responsible for more than 36% of the global CO_2 emissions, which could be translated to more than 40% of the world's energy consumption [1–3]. Moreover, around 74% of the world electricity is generated from non-renewable resources, e.g. coal, gas, nuclear and oil [4,5]. On the other hand, an increase of 65% of energy usage in buildings, which encompasses residential and commercial structures is expected between 2020 and 2050 [6,7]. This is in turn will aggravate climate change and global warming. Accordingly, while researchers are seeking to further expand the contribution of renewable energy resources to the overall world's electricity supply, it is also of significant importance to reduce our energy consumption in buildings [8–10]. One solution to achieve this goal is via the use of the cutting-edge technologies of artificial intelligence and internet of things (IoT) technologies [11,12].

With the current advance of IoT, it becomes simple to automatically collect massive data of different kinds, remotely. This could note only save time and money but also enables developing real-time applications [13,14]. Furthermore, analyzing big data helps significantly in identifying problems and/or abnormalities, i.e. perform a preventive treatment that considerable helps in improving the operational processes [15–17]. Specifically, by using IoT technologies in the building sector, smart meters can gather different types of data (e.g. energy usage footprints, occupancy patterns, humidity, luminosity and temperature) for several objectives and from different end-users [18–20]. In this line, smart meters can play an essential role in changing the energy consumption of end-users in alignment with the increasing cost of electricity [21–23], and on the flip side, can help utility companies in adjusting their model rates with reference to the demand and electricity cost [24–26].

On the other hand, with the rapid proliferation of smart-meters in the building energy sector, detecting abnormal energy consumption behaviors has significantly been advanced in the recent years to boost energy efficiency and reduce carbon dioxide [27–29]. However, most of the existing anomaly detection frameworks rely on the analysis of only energy usage data, which is not reliable in most of the cases since other factors that could seriously affect the detection of anomalous energy usage have not been considered [30,31]. For instance, the absence/presence of end-users in the considered room/building could cause other kinds of anomalous energy consumptions that could not be detected with conventional anomaly detection methods [32,33].

To that end, this paper presents a powerful energy consumption anomaly detection at an appliance-level using autoencoder and micro-moments. In this context, power usage signatures of various household devices along with occupancy patterns are analyzed to model normal usage and load anomalies [34]. Explicitly, the micro-moment paradigm, recently proposed by Google to develop marketing models based on detecting intent-driven moments when consumers decide to buy (learn, discover or watch) something online[1], is investigated in our case to identify energy micro-moments occurring when end-users reflexively decide to make an action that impacts energy consumption via (i) switching on/off an appliance to start/stop an energy consumption action; (ii) increasing/reducing energy consumption of a specific appliance; and (iii) entering/leaving a specific room while an appliance is on (e.g. air conditioner, television, fan, lamp, etc.). Put simply, energy micro-moments are defined with regard to end-users' daily tasks usually performed to meet their preferences. For that purpose, energy micro-moment patterns are extracted from appliance-level power traces and occupancy data using an innovative rule-based algorithm to represent the key intent-driven moments of daily energy use and hence model normal and abnormal behaviors. Following, energy micro-moment patterns are fed into an autoencoder including 48 input/output neurons, and 4 neurons in the intermediate layer, to effectively reduce the computational cost and improve the anomaly detection accuracy and F1 score.

The remainder of this paper is organized as follows. Related works are discussed in Sect. 2. Section 3 deeply explains the proposed anomaly detection approach based on autoencoder and micro-moments. Section 4 presents the obtained empirical results with a set of comparisons between detection performance using autoencoder-only, combined autoencoder and micro-moments and other conventional classifiers. Finally, Sect. 5 draws the main conclusions of this framework and identify future works.

2 Related Works

The problem of anomaly detection of energy consumption in buildings has been tackled drastically in recent years, however, more of the existing frameworks have been focused on analyzing aggregated energy consumption data. Only a very limited number of works have been dealt with individual data gleaned using submetering [35,36]. To this effect, we focus on this section on only describing recent appliance-level anomaly detection frameworks.

In [37], the authors introduce four different techniques to detect anomalous energy consumption, which rely on the percentage change in energy consumption, principal component analysis (PCA), k-nearest neighbors (KNN) and histogram buckets. Accordingly, energy use patterns gleaned from fifty submeters installed in several hostels of IIIT-Delhi were analyzed and anomalies are detected in an unsupervised manner. While in [38], an unsupervised energy consumption anomaly detection method is proposed, which uses an isolated forest algorithm. This framework is built upon three steps defined as feature extraction, feature reduction, and isolated forest computing. Moving forward, in [39], the authors use one-class support vector machine (SVM) to identify energy consumption anomalies of heating, ventilation, and air conditioning (HVAC) installations. Following, in [37], four different machine learning and feature extraction techniques are used for identifying anomalous power usage in various hostels in

[1] https://www.thinkwithgoogle.com/marketing-strategies/micro-moments/.

Delhi, India. Accordingly, percentage change, KNN, histogram buckets and PCA are implemented to analyze power traces gathered from smart meters deployed inside the hostels.

In [33], a deep neural networks (DNN) model is proposed to capture abnormal energy behaviors by investigating appliance-level energy footprints of two different datasets collected from distinct regions. Similarly, in [31], an anomaly detection of energy consumption solution is proposed to capture abnormal energy usage in academic buildings based on the use of improved KNN and micro-moment analysis. It has a low computational cost since it is based on the use of a simple yet effective classifier. Moving forward, energy consumption anomalies are detected using a deep learning model in [40], where seasonality and trend are firstly removed from the energy patterns. This results in residual values that have been then compared to values generated with a predictive analysis based on recurrent neural networks (RNN). While, in [41], the authors uses non-intrusive load monitoring (NILM) [42,43] to split the aggregated load into various appliance-level data without the need to install submeters. Moving forward, a rule-based algorithm is applied to each appliance-level power trace to identify consumption anomalies.

It is worth noting that the main issue with existing anomaly detection schemes is that most of them rely only on investigating energy consumption footprints without considering other relevant data, which could have a significant impact on detecting accurately the anomalies and further identifying novel types of abnormalities [44]. For example, taking into consideration the occupancy data helps in identifying energy anomalies related to energy consumption while the end-user is outside the room/house [45,46]. To that purpose, we propose in this article a novel anomaly detection based on autoencoder and micro-moments, which can detect not only excessive energy consumption but also other abnormalities related to the presence/absence of the end-user [47].

3 Proposed System

3.1 Preprocessing

First of all, energy data and occupancy patterns gleaned by various submeters and smart sensors have been cleaned and invalid recorded values have then been removed. In effect, the gathered patterns represent raw data that could be incomplete, in which some values could be missed and other attributes could also be lost when recording energy and occupancy information. The missing values are usually due to a failure of the hardware and/or software measurement devices. Second, energy usage footprints have been sampled every 30 min before analyzing the way they vary over each day, which makes a 48 inputs for our model.

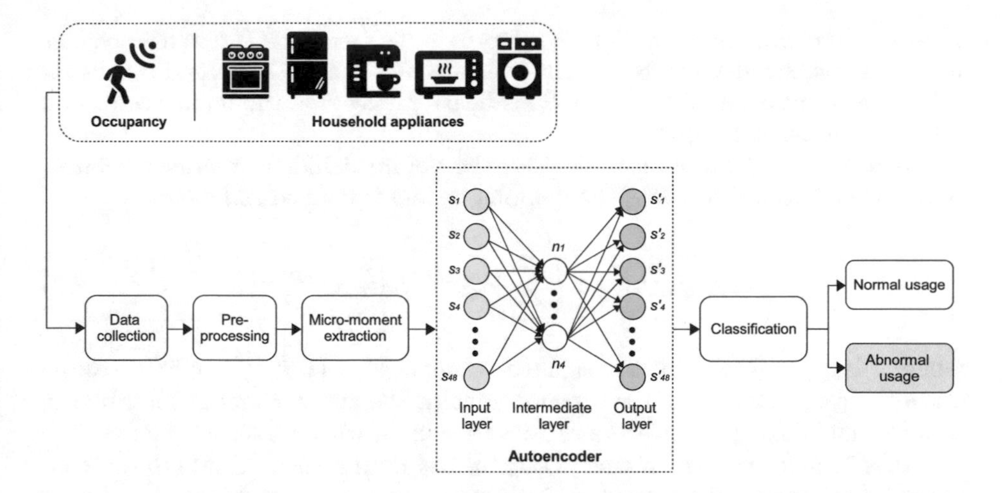

Fig. 1. Flowchart of the proposed anomaly detection approach.

3.2 Autoencoders

Autoencoders are a special kind of neural networks, in which the number of input nodes equals to the number of output nodes. They generally have an intermediate layer having a lower number of neurons. They could be used in both unsupervised and supervised learning although they are more importantly deployed in unsupervised applications, where the labels are not required in the training. Figure 1 presents the flowchart of the proposed anomaly detection scheme, in which the architecture of the implemented autoencoder is portrayed as well.

If we consider a training ensemble of multi-modal data (energy footprints, occupancy, normalized energy, etc.) $\{s_1, s_2, \cdots, s_n\}$, where $s_i \in \mathbf{R}^d$ (d is the dimension of the input vector), the autoencoder aims at constructing a model for projecting the training data (energy footprints, occupancy patterns, normalized energy, etc.) into a lower dimensional sub-ensemble before reproducing the energy data to produce a novel vector $\{s'_1, s'_2, \cdots, s'_n\}$. In this regard, our model is optimized aiming at minimizing the reconstruction error in order to get the optimal sub-ensemble, in which the reconstruction error could be given as:

$$\varepsilon(s_i, s'_i) = \sum_{j=1}^{d} (s_i - s'_i)^2 \tag{1}$$

Accordingly, using the autoencoder model, the dimensionality of the input patterns ($s_i \in R^d$) is reduced into $m (m < d)$ neurons (representing the hidden layer). The activation of the neuron i in the hidden layer is given by:

$$h_i = g_\theta(s) = x\left(\sum_{j=1}^{n} W_{ij}^{input} s_j + a_i^{input}\right) \tag{2}$$

where s represents the input data, θ refers to W^{input} and a^{input} parameters. The former is the encoder weight matrix with size $m \times d$ and the second stands for a bias vector of dimensionality m. This helps in encoding the input vector into a lower dimensional vector.

After that, the autoecoder decodes the obtained hidden representation h_i back to the initial space R^d. The mapping model is defined as follows:

$$s'i = f_{\theta'}(h) = x\left(\sum_{j=1}^{n} W_{ij}^{hidden} h_j + a_i^{hidden}\right) \tag{3}$$

where x is typically the logistic sigmoid function, $\theta' = \{W^{hidden}, a^{hidden}\}$ represents the parameter set of the decoder. Finally, the autoencoder is optimized in order to minimizing the average construction error with reference to θ and θ'.

After completing the training phase, the test data is then fed into the autoencoder for computing the reconstruction errors for every ensemble of data. Moving forward, the anomalies could be then determined using the following equation:

$$c(s_i) = \begin{cases} \text{normal} & \varepsilon_i < \theta \\ \text{anomalous} & \varepsilon_i > \theta \end{cases} \tag{4}$$

where ε represents the reconstruction error that is estimated using Eq. (1). This means that normal energy usage footprints in the test dataset refer to the normal profile, which is built in the training phase with a small reconstruction error. In contrast, anomalous patterns are those having a relatively higher reconstruction error.

Overall, in this article, we adopt an autoencoder architecture including 48 input nodes and the same number of output nodes. Each node represents energy consumption gathered every 30 min. The intermediate layer has only four nodes. This kind of autoeconders adopting this architecture has been widely used because in the intermediate layer, input data are represented in a smaller dimension. Put differently, autoencoders act as a dimensionality reduction scheme, which makes the classification task simple with less computation time.

3.3 Micro-moment Clustering

Aiming at detecting different scenarios of energy consumption anomalies, the (EM)^{3}[2] platform has been developed, which helps in clustering energy usage footprints into five principal categories [48]. They are defined as "good usage", "turn on". "turn off", "excessive usage" and "consumption while outside". In this regard, a rule-based micro-moment extraction algorithm is introduced before applying the autoencoder classifier that is employed for automatic identification of the energy anomalies. Figure 2 outlines the micro-moment classes extracted

[2] (EM)^3: Consumer Engagement Towards Energy Saving Behavior by means of Exploiting Micro Moments and Mobile Recommendation Systems (http://em3.qu.edu.qa/).

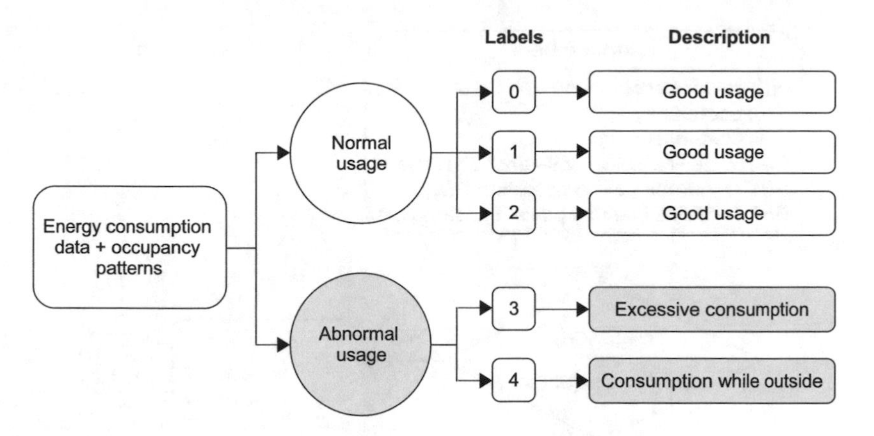

Fig. 2. Micro-moments assumption and labeling.

using the (EM)^3 platform and their description. As it is illustrated, two categories of anomalies are detected, "excessive usage: class 3" and "consumption while outside: class 4". The latter is reserved to detect anomalies that are related to some specific domestic devices, e.g. a television, an air conditioner (AC), a light lamp, a desktop/laptop or a fan, where the end-user must be present during the operation (normal usage) of this kind of appliances [49]. Otherwise the power usage is considered as anomalous [50,51].

By analyzing occupancy profile (O), consumption time (CT) and power consumption (p) of every appliance, the five micro-moment classes have been extracted with reference to each device active consumption range (DACR), normal device operation time (NDOT) and device standby power consumption (DSPC). The proposed methodology utilized for extracting the micro-moment features (MF) through the timestamps is explained in Fig. 3.

Specifically, power consumption readings p of an appliance and occupancy patterns O collected at a sampling rate t are recorded and stored in a dataset backend. Then, the appliance operation parameters are called, including $DACR$, $NDOT$ and $DSPC$. Table 1 presents an example of different appliance parameter specifications that are used in the rule-based algorithm to extract power consumption micro-moments.

4 Experimental Results

Many countries have already started to roll out smart energy meters that can record aggregated and appliance-level electricity data of each household. In this regard, different datasets have been launched recently to help the energy research community in validating their developed energy efficiency algorithms. To that end, a comprehensive energy consumption dataset, named UK-DALE [52], is considered to evaluate the performance of the proposed anomaly detection solution. This dataset is an open-access repository that includes electricity consumption

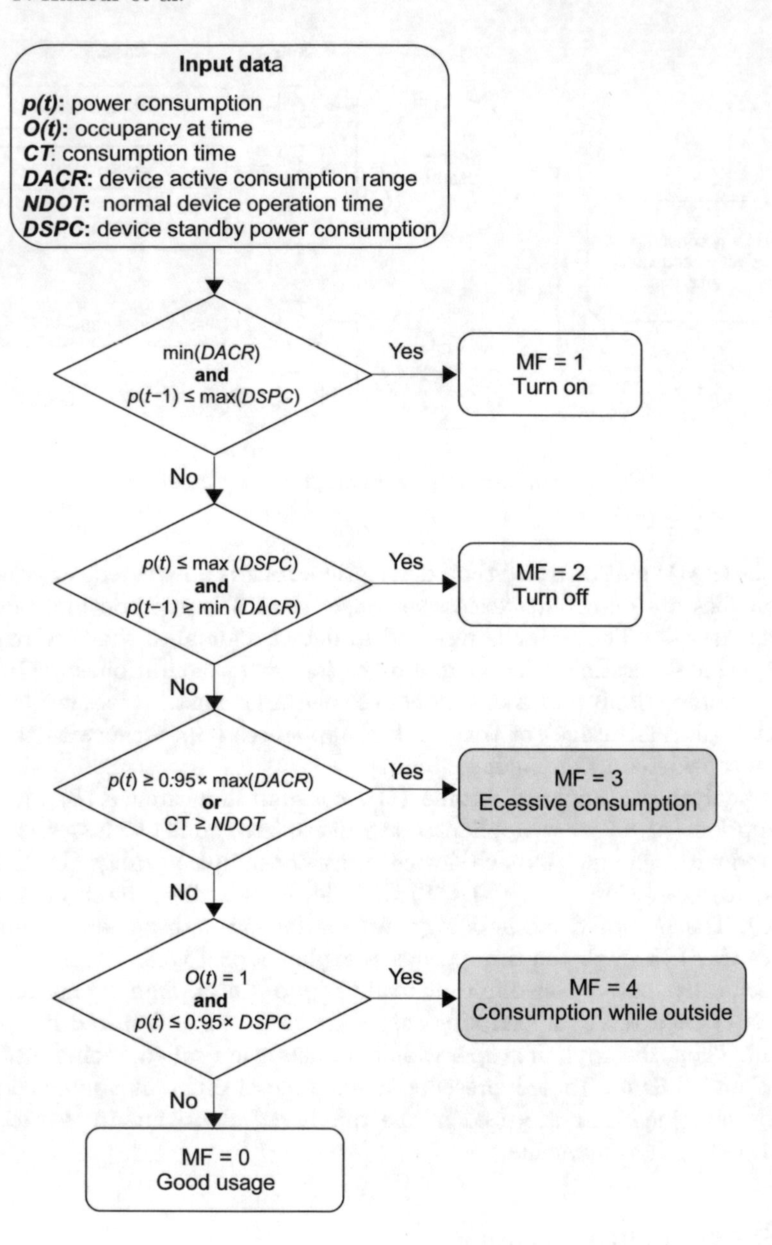

Fig. 3. Energy observation clustering process.

footprints of different appliances registered in a UK household. A sampling rate of 16 kHz is used to glean energy data at the aggregated level, while 1/6 Hz sampling frequency is employed for individual appliances. In this study, we only use individual appliance power traces because our approach has been proposed to identify appliance-level anomalies.

Table 1. Electricity consumption specifications of various electrical devices.

Appliance	NDOT	DACR (watts)	DSPC (watts)
Air conditioner	15 h 30 min	1000	4
Microwave	1 h	1200	7
Oven	3 h	2400	6
Dishwasher	1 h 45 min	1800	3
Laptop	12 h 42 min	100	20
Washing machine	1h	500	6
Light	8 h	60	0
Television	12 h 42 min	65	6
Refrigerator	17 h 30 min	180	0
Desktop	12 h 42 min	250	12

Moving forward, an empirical evaluation of autoencoder-only combined autoencoder and micro-moments has been conducted in comparison with other existing solutions, i.e. KNN and SVM. The obtained results have demonstrated the promising performance of our approach, especially when autoencoder is combined with the micro-moment features. Table 2 summarizes the accuracy and F1 score results obtained for KNN, SVM, autoencoder-only and combined autoencoder and micro-moments. Therefore, this has also proved the efficiency of considering occupancy data and micro-moments to identify abnormal energy consumption. The results of the experiment have shown that the anomaly detection performance varies from an appliance to another. The best results have been obtained for the case of the microwave while the worst ones have been achieved with the dish washer. The difference of the anomaly detection accuracy is mainly due to the imbalance characteristic of collected data, in which the imbalance rate varies from an appliance to another. Moreover, it has been appeared that the autoencoder outperforms both KNN and SMV for all the appliances, and thereby combining autoencoder and micro-moments has significantly improved the anomaly detection performance.

Table 2. Performance of anomaly detection for each appliance.

Appliance	KNN		SVM		Autoencoder		Autoencoder + micro-moments	
	Accuracy	F1 score	Accuracy	F1 score	Accuracy	F1 score	Accuracy	F1 score
Television	0.67	0.66	0.69	0.67	0.74	0.72	0.89	0.88
Microwave	0.75	0.76	0.77	0.77	0.85	0.87	0.95	0.95
Coffee machine	0.65	0.65	0.68	0.67	0.74	0.74	0.87	0.86
Dish washer	0.62	0.63	0.60	0.61	0.63	0.66	0.85	0.87
Toaster	0.67	0.65	0.68	0.67	0.71	0.73	0.91	0.91

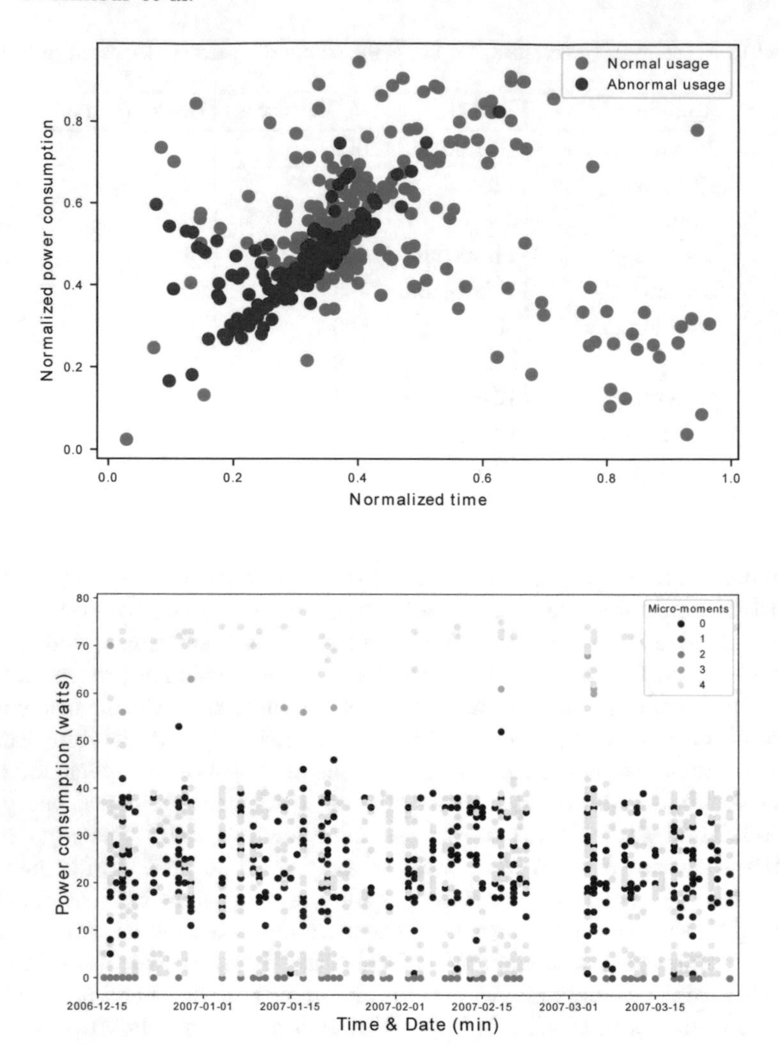

Fig. 4. Example of power consumption anomalies of a television energy footprints gathered in UK-DALE for a period of three months and 10 days: a) scatter plot of autoencoder-only, and b) scatter plot of autoencoder and micro-moments.

Figure 4 illustrates normal and abnormal consumption data collected from UK-DALE dataset [53], which are obtained for both scenarios; (a) when autoencoder-only is considered, and (b) when autoencoder is combined with micro-moment features. It can clearly been seen that in the first scenario, only one type of anomalous consumption patterns are detected, which is related to excessive consumption. Conversely, in the second scenario, two kinds of anomalies are identified that are defined as "excessive consumption" and "consumption while outside". In addition, the normal energy consumption has been categorized into three other different classes. All in all, it is obvious that by combining autoencoder and micro-moments, an accurate analysis of the energy usage has been performed, which provides much more useful information than using autoencoder-only. On the other hand, it is worth noting that the visualization of energy consumption is an important step in developing powerful energy saving systems, where the end-user could easily visualize anomalous energy consumption [54,55].

5 Conclusion

This paper presents an effective anomaly detection scheme using an autoencoder classifier in combination with micro-moment features. Our system enables accurate detection of anomalous energy consumption through detecting "excessive consumption" and "consumption while outside" although the performance varies slightly from an appliance to another. To that end, we have proceeded with analyzing energy consumption observations and occupancy data to firstly cluster energy patterns into five different categories. Following, an autcoencoder classifier has been used to automatically classify test data into normal and abnormal consumption footprints. The results of the empirical evaluation conducted on five different appliances were very promising. They showed that using the autoencoder and micro-moment clustering can not only improve the accuracy of the anomaly detection, but this combination helps as well in detecting other kinds of anomalies, which is impossible to identify by conventional anomaly detection schemes.

The promising results obtained in this study will motivate us for investigating other variations of autoencoders in our future work, such as variational autoencoders. The latter could be a good alternative to overcome the data imbalance issue since they are a special kind of deep generative models. Therefore, they could improve further the anomaly detection accuracy, especially if they are combined with the micro-moment analysis. Moreover, it will part of our future work to use Home Assistant, which is an open-source platform that enables to control and monitor the smart-meters, smart sensors and appliances remotely [56]. It also provides the end-users with the possibility to receive mobile notifications and warnings regarding their energy consumption footprints.

Acknowledgements. This paper was made possible by National Priorities Research Program (NPRP) grant No. 10-0130-170288 from the Qatar National Research Fund (a member of Qatar Foundation). The statements made herein are solely the responsibility of the authors.

References

1. Himeur, Y., Alsalemi, A., Bensaali, F., Amira, A.: Effective non-intrusive load monitoring of buildings based on a novel multi-descriptor fusion with dimensionality reduction. Appl. Energy **279**, 115872 (2020)
2. Himeur, Y., Alsalemi, A., Bensaali, F., Amira, A.: Building power consumption datasets: survey, taxonomy and future directions. Energy Build. **227**, 110404 (2020)

3. Elsalemi, A., Al-kababji, A., Himeur, Y., Bensaali, F., Amira, A.: Cloud energy micro-moment data classification: a platform study. In: IEEE SmartWorld, Ubiquitous Intelligence Computing, Advanced Trusted Computing, Scalable Computing Communications, Cloud Big Data Computing, Internet of People and Smart City Innovation (SmartWorld/SCALCOM/UIC/ATC/CBDCom/IOP/SCI), pp. 1–6 (2020)
4. Sardianos, C., et al.: REHAB-C: recommendations for energy habits change. Futur. Gener. Comput. Syst. **112**, 394–407 (2020)
5. Bedi, G., Venayagamoorthy, G.K., Singh, R.: Development of an IoT-driven building environment for prediction of electric energy consumption. IEEE Internet Things J. **7**(6), 4912–4921 (2020)
6. Zhu, W., Feng, W., Li, X., Zhang, Z.: Analysis of the embodied carbon dioxide in the building sector: a case of China. J. Clean. Prod. **269**, 122438 (2020)
7. Zanchini, E., Naldi, C.: Energy saving obtainable by applying a commercially available m-cycle evaporative cooling system to the air conditioning of an office building in north italy. Energy **179**, 975–988 (2019)
8. Whitney, S., Dreyer, B.C., Riemer, M.: Motivations, barriers and leverage points: exploring pathways for energy consumption reduction in Canadian commercial office buildings. Energy Res. Soc. Sci. **70**, 101687 (2020)
9. Opoku, R., Edwin, I.A., Agyarko, K.A.: Energy efficiency and cost saving opportunities in public and commercial buildings in developing countries-the case of air-conditioners in Ghana. J. Clean. Prod. **230**, 937–944 (2019)
10. Himeur, Y., et al.: On the applicability of 2D local binary patterns for identifying electrical appliances in non-intrusive load monitoring. In: Arai, K., Kapoor, S., Bhatia, R. (eds.) IntelliSys 2020. AISC, vol. 1252, pp. 188–205. Springer, Cham (2021). https://doi.org/10.1007/978-3-030-55190-2_15
11. Mehmood, M.U., Chun, D., Han, H., Jeon, G., Chen, K., et al.: A review of the applications of artificial intelligence and big data to buildings for energy-efficiency and a comfortable indoor living environment. Energy Build. **202**, 109383 (2019)
12. Sardianos, C., et al.: The emergence of explainability of intelligent systems: delivering explainable and personalised recommendations for energy efficiency. Int. J. Intell. Syst. **36**(02), 656–680 (2020)
13. Nižetić, S., Šolić, P., González-de, D.L.-D.-I., Patrono, L., et al.: Internet of things (IoT): opportunities, issues and challenges towards a smart and sustainable future. J. Clean. Prod. **274**, 122877 (2020)
14. Muralidhara, S., Hegde, N., Rekha, P.: An internet of things-based smart energy meter for monitoring device-level consumption of energy. Comput. Electr. Eng. **87**, 106772 (2020)
15. Abate, F., Carratù, M., Liguori, C., Paciello, V.: A low cost smart power meter for IoT. Measurement **136**, 59–66 (2019)
16. Abbas, A.M., Youssef, K.Y., Mahmoud, I.I., Zekry, A.: NB-IoT optimization for smart meters networks of smart cities: case study. Alex. Eng. J. 1–15 (2020). https://doi.org/10.1016/j.aej.2020.07.030
17. Sardianos, C., et al.: Data analytics, automations, and micro-moment based recommendations for energy efficiency. In: 2020 IEEE Sixth International Conference on Big Data Computing Service and Applications (BigDataService), pp. 96–103. IEEE (2020)
18. Abate, F., Carratù, M., Liguori, C., Ferro, M., Paciello, V.: Smart meter for the IoT. In: 2018 IEEE International Instrumentation and Measurement Technology Conference (I2MTC), pp. 1–6. IEEE (2018)

19. Himeur, Y., Alsalemi, A., Al-Kababji, A., Bensaali, F., Amira, A.: Data fusion strategies for energy efficiency in buildings: overview, challenges and novel orientations. Inf. Fusion **64**, 99–120 (2020)
20. Aboelmaged, M., Abdelghani, Y., Abd El Ghany, M.A.: Wireless IoT based metering system for energy efficient smart cites. In: 2017 29th International Conference on Microelectronics (ICM), pp. 1–4. IEEE (2017)
21. Himeur, Y., Alsalemi, A., Bensaali, F., Amira, A.: Efficient multi-descriptor fusion for non-intrusive appliance recognition. In: 2020 IEEE International Symposium on Circuits and Systems (ISCAS), pp. 1–5. IEEE (2020)
22. Himeur, Y., Alsalemi, A., Bensaali, F., Amira, A.: Improving in-home appliance identification using fuzzy-neighbors-preserving analysis based QR-decomposition. In: Yang, X.-S., Sherratt, R.S., Dey, N., Joshi, A. (eds.) ICICT 2020. AISC, vol. 1183, pp. 303–311. Springer, Singapore (2021). https://doi.org/10.1007/978-981-15-5856-6_30
23. Himeur, Y.,Elsalemi, A., Bensaali, F., Amira, A.: Appliance identification using a histogram post-processing of 2D local binary patterns for smart grid applications. In: Proceedings of 25th International Conference on Pattern Recognition (ICPR), pp. 1–8 (2020)
24. Himeur, Y., Alsalemi, A., Bensaali, F., Amira, A.: Robust event-based non-intrusive appliance recognition using multi-scale wavelet packet tree and ensemble bagging tree. Appl. Energy **267**, 114877 (2020)
25. Carratù, M., Ferro, M., Pietrosanto, A., Paciello, V.: Smart power meter for the IoT. In: 2018 IEEE 16th International Conference on Industrial Informatics (INDIN), pp. 514–519. IEEE (2018)
26. Al-Kababji, A., et al.: Energy data visualizations on smartphones for triggering behavioral change: novel vs. conventional. In: 2020 2nd Global Power, Energy and Communication Conference (GPECOM), pp. 312–317 (2020). https://doi.org/10.1109/GPECOM49333.2020.9247901
27. Liu, Y., Pang, Z., Karlsson, M., Gong, S.: Anomaly detection based on machine learning in IoT-based vertical plant wall for indoor climate control. Build. Environ. **183**, 107212 (2020)
28. Yip, S.-C., Tan, W.-N., Tan, C., Gan, M.-T., Wong, K.: An anomaly detection framework for identifying energy theft and defective meters in smart grids. Int. J. Electr. Power Energy Syst. **101**, 189–203 (2018)
29. Araya, D.B., Grolinger, K., ElYamany, H.F., Capretz, M.A., Bitsuamlak, G.: An ensemble learning framework for anomaly detection in building energy consumption. Energy Build. **144**, 191–206 (2017)
30. Wang, X., Ahn, S.-H.: Real-time prediction and anomaly detection of electrical load in a residential community. Appl. Energy **259**, 114145 (2020)
31. Himeur, Y., Alsalemi, A., Bensaali, F., Amira, A.: Smart power consumption abnormality detection in buildings using micromoments and improved k-nearest neighbors. Int. J. Intell. Syst. **36**(6), 2865–2894 (2021)
32. Himeur, Y., Ghanem, K., Alsalemi, A., Bensaali, F., Amira, A.: Artificial intelligence based anomaly detection of energy consumption in buildings: a review, current trends and new perspectives. Appl. Energy **287**, 116601 (2021)
33. Himeur, Y., Alsalemi, A., Bensaali, F., Amira, A.: A novel approach for detecting anomalous energy consumption based on micro-moments and deep neural networks. Cogn. Comput. **12**(6), 1381–1401 (2020)
34. Sardianos, C., et al.: A model for predicting room occupancy based on motion sensor data. In: 2020 IEEE International Conference on Informatics, IoT, and Enabling Technologies (ICIoT), pp. 394–399. IEEE (2020)

35. Himeur, Y., Elsalemi, A., Bensaali, F., Amira, A.: An intelligent non-intrusive load monitoring scheme based on 2D phase encoding of power signals. Int. J. Intell. Syst. **36**(1), 72–93 (2021)
36. Alsalemi, A., et al.: Achieving domestic energy efficiency using micro-moments and intelligent recommendations. IEEE Access **8**, 15047–15055 (2020)
37. Sial, A., Singh, A., Mahanti, A.: Detecting anomalous energy consumption using contextual analysis of smart meter data. Wirel. Netw. **27**, 1–18 (2019)
38. Mao, W., Cao, X., zhou, Q., Yan, T., Zhang, Y.: Anomaly detection for power consumption data based on isolated forest. In: 2018 International Conference on Power System Technology (POWERCON), pp. 4169–4174 (2018)
39. Beghi, A., Cecchinato, L., Corazzol, C., Rampazzo, M., Simmini, F., Susto, G.A.: A one-class SVM based tool for machine learning novelty detection in HVAC chiller systems. IFAC Proc. Vol. **47**(3), 1953–1958 (2014)
40. Hollingsworth, K., et al.: Energy anomaly detection with forecasting and deep learning. In: 2018 IEEE International Conference on Big Data (Big Data), pp. 4921–4925 (2018)
41. Rashid, H., Singh, P., Stankovic, V., Stankovic, L.: Can non-intrusive load monitoring be used for identifying an appliance's anomalous behaviour? Appl. Energy **238**, 796–805 (2019)
42. Himeur, Y., Elsalemi, A., Bensaali, F., Amira, A.: Recent trends of smart non-intrusive load monitoring in buildings: a review, open challenges and future directions. Int. J. Intell. Syst. 1–28 (2020)
43. Himeur, Y., Alsalemi, A., Bensaali, F., Amira, A.: Smart non-intrusive appliance identification using a novel local power histogramming descriptor with an improved k-nearest neighbors classifier. Sustain. Urban Areas **67**, 102764 (2021)
44. Alsalemi, A., Himeur, Y., Bensaali, F., Amira, A.: An innovative edge-based internet of energy solution for promoting energy saving in buildings. Sustain. Urban Areas **78**, 103571 (2022)
45. Himeur, Y., et al.: Techno-economic assessment of building energy efficiency systems using behavioral change: a case study of an edge-based micro-moments solution. J. Clean. Prod. **331**, 129786 (2022)
46. Himeur, Y., Alsalemi, A., Bensaali, F., Amira, A.: The emergence of hybrid edge-cloud computing for energy efficiency in buildings. In: Arai, K. (ed.) IntelliSys 2021. LNNS, vol. 295, pp. 70–83. Springer, Cham (2022). https://doi.org/10.1007/978-3-030-82196-8_6
47. Alsalemi, A., Himeur, Y., Bensaali, F., Amira, A.: Smart sensing and end-user behavioral change in residential buildings: an edge internet of energy perspective. IEEE Sen. J. **21**, 27623–27631 (2021)
48. Sardianos, C., et al.: Real-time personalised energy saving recommendations, in: 2020 International Conferences on Internet of Things (iThings) and IEEE Green Computing and Communications (GreenCom) and IEEE Cyber, Physical and Social Computing (CPSCom) and IEEE Smart Data (SmartData) and IEEE Congress on Cybermatics (Cybermatics), pp. 366–371 (2020)
49. Himeur, Y., et al.: A survey of recommender systems for energy efficiency in buildings: principles, challenges and prospects. Inf. Fusion **72**, 1–21 (2021)
50. Sardianos, C., et al.: Reshaping consumption habits by exploiting energy-related micro-moment recommendations: a case study. In: Helfert, M., Klein, C., Donnellan, B., Gusikhin, O. (eds.) SMARTGREENS/VEHITS -2019. CCIS, vol. 1217, pp. 65–84. Springer, Cham (2021). https://doi.org/10.1007/978-3-030-68028-2_4

51. Alsalemi, A., Himeur, Y., Bensaali, F., Amira, A.: Appliance-level monitoring with micro-moment smart plugs. In: Ben Ahmed, M., Rakıp Karaş, İ, Santos, D., Sergeyeva, O., Boudhir, A.A. (eds.) SCA 2020. LNNS, vol. 183, pp. 942–953. Springer, Cham (2021). https://doi.org/10.1007/978-3-030-66840-2_71
52. Kelly, J., Knottenbelti, W.: The UK-dale dataset, domestic appliance-level electricity demand and whole-house demand from five UK homes. Sci. Data **2**(150007), 1–14 (2015)
53. Kelly, J., Knottenbelt, W.: The UK-dale dataset, domestic appliance-level electricity demand and whole-house demand from five UK homes. Sci. Data **2**(1), 1–14 (2015)
54. Sayed, A., Himeur, Y., Alsalemi, A., Bensaali, F., Amira, A.: Intelligent edge-based recommender system for internet of energy applications. IEEE Syst. J. (2021)
55. Alsalemi, A., et al.: A micro-moment system for domestic energy efficiency analysis. IEEE Syst. J. **15**(1), 1256–1263 (2020)
56. Sayed, A., Alsalemi, A., Himeur, Y., Bensaali, F., Amira, A.: Endorsing energy efficiency through accurate appliance-level power monitoring, automation and data visualization. In: Ben Ahmed, M., Teodorescu, H.-N.L., Mazri, T., Subashini, P., Boudhir, A.A. (eds.) Networking, Intelligent Systems and Security. SIST, vol. 237, pp. 603–617. Springer, Singapore (2022). https://doi.org/10.1007/978-981-16-3637-0_43

Predicting the Mode of Transport from GPS Trajectories

Hichame Kabiri[✉] and Youssef Ghanou

Moulay Ismail University, Meknes, Morocco
hichamme@outlook.fr

Abstract. In this paper, we present a framework that can automatically classify the transportation mode based only on the GPS trajectory of an individual. We intend to show that the extraction of extra features besides speed, acceleration, and the bearing rate [35,36] enables many classifiers to achieve very efficient generalization. We apply machine learning algorithms, Recurrent Neural Network and Convolutional Neural Network. Finally, we compare our approach with state-of-art transportation mode prediction strategies.

Keywords: Machine learning · Artificial neural networks · GPS trajectory · Supervised learning · Preprocessing · Transport mode classification

1 Introduction

With the rapid increase in the number of vehicles and the accelerating growth of urbanization, the cost of annual traffic congestion in cities is rising rapidly, leading to a reduction in the performance of mobility circuits, slower operating times, slower economic growth, and a reduction in air quality. All these problems lead us to create intelligent circuits and construct smart cities using artificial intelligence, but to solve the sophisticated structures of cities, we need to explore and integrate information on the temporal, spatial, socio-demographic characteristics of users, and related behavioral information as well [2].

Therefore, the research on urban traffic forecasting is of great importance and it is an essential element of the urban development and environmental protection strategy. One of the most important and essential challenges to know in the field of traffic flows, which consists of several interdependent and interrelated elements, is to classify and determine the exact modes of transport used in a given route or period.

Many researchers have been working on this topic, transport mode classification, in an accelerated way in recent years using various data sources such as GPS coordinates [7,35], others used accelerometer of mobile phones [15,18] and some other researchers have used GIS data [20]. [24,26,28,35] These researchers used traditional supervised learning algorithms, including the dynamic Bayesian

M. Lazaar et al. (Eds.): BDIoT 2021, LNNS 489, pp. 194–207, 2022.
https://doi.org/10.1007/978-3-031-07969-6_15

network, support vector machine, decision tree, multilayer perception and random forest algorithms. Besides, much research papers have increased GPS data and combined it with other data to improve mode detection. For instance, [12,22] have mixed GPS data with GIS data while other researchers have combined GPS and accelerometer data to create a more relevant and efficient model [19]. In parallel, other sensors have been used for vehicle classification, such as a gyroscope, rotation vector and magnetometer [5,13]. Other works such as that of [1,2] have exploited data based on the temporal, spatial and socio-demographic characteristics of users, which allowed them to generate highly relevant models with greater accuracy.

However [24] carried out a work containing many works done in this field of classification where they examined and compared many approaches used in the literature based on GPS Trajectories. This document provided us with a complete study on the three fundamental aspects of GPS trajectory-based mode detection: segmentation, pre-processing and classifier. In addition, [34] conducted a very comprehensive study of automatic mode detection and developed a segmentation algorithm based on the points of change between modes of transport and a time threshold between points successive to divide the trajectory into a set of segments. Then, they extracted the characteristics of each segment, including the mean and variance of speed, the three highest speeds and accelerations. At the end, all the instances characterized by these features are created and used as input for a set of machine learning algorithms (SVM, MLPC, KNN, DT, RF). A study by the same author is made after improving their previous result by adding more robust features and also applying graph-based pre-processing and filtering algorithms to improve the data distribution [36]. Another contribution made by [25,26] was to increase the number of features using statistical methods (global and local features).

Further works as [6,23] used deep neural networks (DNN) and [3,4] worked with convolutional neural networks (CNN) to get high-quality features automatically. Their fundamental idea was to transform the feature extracted from a GPS trajectory into a 2-D image structure, which will be considered as an input for a NN model. While [21] applied Long Short-Term Memory (LSTM) recurrent neural networks that learned effective hierarchical and stateful representations for temporal sequences. In addition [30] used deep multi-scale learning for classifying transportation mode and speed from trajectory, and [29] carried out a work called Travel Mode Identification with GPS Trajectories Using Wavelet Transform and Deep Learning.

However, a simple and efficient method using a single sensor may be more applicable and very useful, since collecting human mobility information and accessing several and different pieces of information for some location would be so complex, and very expensive. Our work then focuses on applying different aspects of artificial intelligence, such as pre-processing, machine learning and deep learning algorithms to classify the mode of transport using GPS coordinates

[9,14]. To be more precise, we will therefore use the GPS data-set collected by Microsoft in many places, including Asia and some places in Europe and the United States, to deduce the mode of transport used by an individual from his GPS trajectory using deep and machine learning algorithms [35].

Our contribution is to represent new movements and features, we combine the features used in the literature with the new one we create to achieve greater accuracy and generalization. We applied machine learning algorithms and try to build advanced architectures that can be better adapted either to the convolutional neural network (CNN) or the recurrent neural network (RNN) [10,11].

The structure of this article is described: In Sect.2, we describe our methodology we have used and we focus on the most important part of this work, which is the pre-processing stage. In Sect. 3, we first describe the data, then experiment with our results using machine learning and deep learning algorithms, and then compare our work with the state-of-the art. The last section contains conclusions and suggestions for future research.

2 Methodology and Processing the Input

With the advantage of GPS and mobile devices, a better approach to detect the mode of transport is to calculate the main motions of the trajectory. Our method provides to increase the number of features by adding new ones to the ones used in the literature, and using statistical methods for more increasing on features of the data, then applying a filter to enhance more the data.

2.1 Preparing the Data

After obtaining matches between the GPS trajectory and the corresponding label for each user. We cluster our raw GPS data into shape (longitude, latitude, time, mode) [34,35]. We describe our experiment in the following 4 steps:

First Preprocessing. Naturally, we identified and corrected the errors generated from the data source, due to several sources of error; GPS points whose time record is larger than its following point are detected and deleted (The initial data entry order is taken into consideration) [34,35].

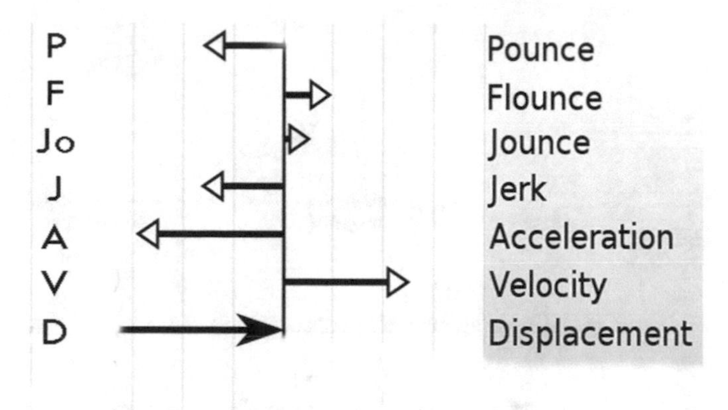

Fig. 1. Derivatives of displacement

Segmentation. In the literature there are several subjects which have to deal with the different segmentation techniques and methods [2,12,17,22,27,35]. In this work, we use the method described in [35] which is for each set at most equal to 200 points and belonging to the same mode of transport are stored as one segment then labelled to the same mode of transport. We do then the same for all points of our trajectories. We note that we do not exceed a time threshold (20 min) between two consecutive GPS points and we remove the segments that have less than 10 points, as a second preprocessing step, because it affects the precision of our model [33–35].

Calculate Features and Motions. For each segment, we calculate many features and movements (characteristics); we believe that these features are unique for each mode of transport. For example, for a given distance, there are differences between the average speed of a car and the average speed of a train and the average speed of a bicycle, etc. Therefore, we will extract a set of such characteristics that will be used as input data for our machine and deep learning algorithms. The figure above, (see Fig. 1), shows the set of movements (features) used in this work.

To calculate the geographical distance between consecutive points P_1 (long, lat) and P_2 (long, lat) of each segment, we use Haversine formula that is based on the spherical Earth (ignoring the ellipsoidal effect). We calculate the different movements as it is presented below, where we note the time between two successive points $\Delta t = t_{P_2} - t_{P_1}$ [3,31].

- $D_{P_1} = Distance_{P_1} = Haversine(P_1, P_2)$
- $S_{P_1} = Speed_{P_1} = (D_{P_2} - D_{P_1})/\Delta t$
- $A_{P_1} = Acceleration_{P_1} = (S_{P_2} - S_{P_1})/\Delta t$
- $J_{P_1} = Jerk_{P_1} = (A_{P_2} - A_{P_1})/\Delta t$
- and Bearing [3,35]

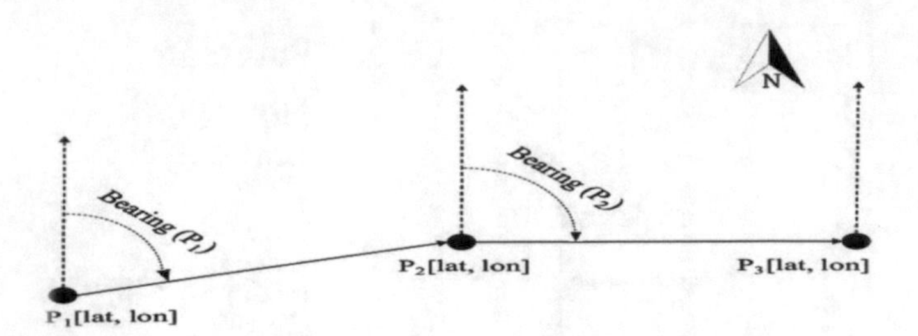

Fig. 2. Bearing between consecutive points [3]

In the following is a set of features that will be used for the first time in the literature of transport mode detection, and they are the fourth, fifth and sixth derivatives of position respectively.

- $Jo_{P_1} = Jounce_{P_1} = (J_{P_2} - J_{P_1})/\Delta t$
- $F_{P_1} = Flounce_{P_1} = (Jo_{P_2} - Jo_{P_1})/\Delta t$
- $P_{P_1} = Pounce_{P_1} = (Fl_{P_2} - Fl_{P_1})/\Delta t.$

The bearing or the rate of change in the heading direction [3] is the horizontal angle between the direction of an object and another object, or between that direction and the true north. This feature is for making more distinction between the modes of transport, cycling and walking, which have more fluidity and flexibility in the deviation of path, contrarily to the other mode of transport which have an alignment in their path trajectory (see Fig. 2) [3].

After applying the data cleaning in the first and second steps, we execute another important preprocessing step to our movements.

Second Pre-processing. To improve more the quality of GPS recording, we define for each mode of transport a derivative of displacement threshold as speed threshold, acceleration threshold, and jerk threshold with the help of experts in transport field [3,35,36]. Then, for all points for which speed, acceleration, and jerk have exceeded its threshold are deleted. Then, we applied the Savitsky-Golay filter to enhance more our data [3,16].

We synthesize our pre-process: First of all, the points whose time record is greater than the point that follows it are detected and deleted. Second, we divide all trajectory data of each user into several segments containing an estimated average number of 200 GPS points that already belong to the same mode of transport, then we reject sample trajectories (segments) containing less than 10 points and mark each segment with its mode of transport. Then, we compute the speed, acceleration, jerk and bearing features as described in the preceding section, then considering a speed threshold, an acceleration threshold and a jerk threshold depending on each mode of transport, so that if the speed between every two points has exceeded its speed threshold the first point is deleted and similarly for the acceleration and jerk thresholds. Then, the three remaining features Jounce, Flounce and Pounce are calculated. As the last step, we apply the Savitsky-Golay filter. Finally, our Input data are of the shape [number of segments, estimated average number of points, features] and their corresponding label.

3 Experiments Details and Results

This section will give a brief description of the Geolife dataset, which used in this work, and then validate our approach with a set of machine learning algorithms and deep learning algorithms and at the end we make a comparison with other works of the same thematic.

3.1 Data Description

This GPS tracking dataset was collected from 182 users for more than five years (April 2007 to August 2012) under the Geolife project (Microsoft Research Asia). The GPS track within this dataset is displayed as a sequence made up of time-stamped points, providing latitude, longitude, and altitude details. This register has 17,621 tracks with a total length of 1,292,951 km and a total running time of 50,176 h. Those tracks were stored with several GPS devices and GPS telephones and have different Sampling Rates. 91.5% were recorded in a condensed format, for example, every 1 to 5 s or every 5 to 10 m per point. This dataset recorded a vast range and variety of external user patterns, which includes not only life routines like returning home and back to work, as well as some leisure and sports experiences such as shopping, sightseeing, eating, walking, and cycling. These trajectory datasets may be used in numerous research areas, for instance, to extract traffic patterns, track user activity, social networks based on location, privacy, and location recommendation. While this dataset is largely deployed in more than 30 cities in China and even in some cities in the US and Europe, most part of the data has been created in Beijing. The references [32,35] give more details about the Data.

```
optimal parameter value:  {'n_estimators': 115}
CLASSIFICATION REPORT RandomForestClassifier :-
              precision    recall  f1-score   support

           1      0.934     0.986     0.959        72
           2      0.833     0.435     0.571        46
           3      0.750     0.794     0.771        68
           4      0.951     0.983     0.967       414

   micro avg      0.920     0.920     0.920       600
   macro avg      0.867     0.800     0.817       600
weighted avg      0.917     0.920     0.913       600

ACCURACY OF COMPLETE STRUCTURE RandomForestClassifier:-
0.92
```

Fig. 3. Classification report RandomForestClassifier

3.2 Experiments and Performances

In our experiment, we merged the car and taxi sub-trajectories and called them driving then we merged the rail-based classes such as train and subway as a train. Therefore, the target classes are walk, bike, bus, driving and train [35].

We only took about 3000 first segments of the preprocessed data, then we applied a set of machine learning algorithms with 80% for training and 20% for testing of the Input data of a shape [number of segments, features] and their corresponding labels. The features used are Speed, Acceleration, Jerk, Bearing rate, Jounce, Flounce, Pounce with their static preprocessing values (mean, standard deviation, median, max and min).

```
optimal parameter value:  {'max_depth': 5}
CLASSIFICATION REPORT DecisionTreeClassifier :-
              precision    recall  f1-score   support

           1      0.957     0.931     0.944        72
           2      0.514     0.413     0.458        46
           3      0.586     0.750     0.658        68
           4      0.961     0.942     0.951       414

   micro avg      0.878     0.878     0.878       600
   macro avg      0.754     0.759     0.753       600
weighted avg      0.883     0.878     0.879       600

ACCURACY OF COMPLETE STRUCTURE DecisionTreeClassifier:-
0.8783333333333333
```

Fig. 4. Classification report DecisionTreeClassifier

```
optimal parameter value:  {'hidden_layer_sizes': (70,)}
CLASSIFICATION REPORT MLPClassifier :-
              precision    recall  f1-score   support

           1      0.861     0.944     0.901        72
           2      0.833     0.326     0.469        46
           3      0.627     0.544     0.583        68
           4      0.912     0.978     0.944       414

   micro avg      0.875     0.875     0.875       600
   macro avg      0.808     0.698     0.724       600
weighted avg      0.868     0.875     0.861       600

ACCURACY OF COMPLETE STRUCTURE MLPClassifier:-
0.875
```

Fig. 5. Classification report MLPClassifier

We used many machines learning algorithms to classify our Input data such as RandomForestClassifier, DecisionTreeClassifier, KNeighborsClassifier, MLP-Classifier(Multilayer Perception Classifier), and SVMClassifier integrated into the Sklearn library.

We present our classification results and model details in regards to the classification report. The set of measures of the efficiency and performance of machine learning algorithms classifier are given in Fig. 3, Fig. 4, Fig. 5, Fig. 6 and Fig. 7. Note that the modes are noted by numbers in this way: walk = 1, driving = 2, bus = 3, and train = 4. We also analyze the relevance of each feature in the classification process.

```
optimal parameter value:  {'n_neighbors': 10}
CLASSIFICATION REPORT KNeighborsClassifier :-
              precision    recall  f1-score   support

           1      0.779     0.833     0.805        72
           2      0.500     0.239     0.324        46
           3      0.481     0.559     0.517        68
           4      0.927     0.944     0.935       414

   micro avg      0.833     0.833     0.833       600
   macro avg      0.672     0.644     0.645       600
weighted avg      0.826     0.833     0.825       600

ACCURACY OF COMPLETE FLAT STRUCTURE KNeighborsClassifier:-
0.8333333333333334
```

Fig. 6. Classification report KNeighborsClassifier

```
optimal parameter value:  {'C': 16}
CLASSIFICATION REPORT SVM :-
              precision    recall  f1-score   support

           1      0.881     0.722     0.794        72
           2      0.643     0.391     0.486        46
           3      0.549     0.412     0.471        68
           4      0.857     0.957     0.904       414

   micro avg      0.823     0.823     0.823       600
   macro avg      0.733     0.620     0.664       600
weighted avg      0.809     0.823     0.810       600

ACCURACY OF COMPLETE FLAT STRUCTURE SVM:-
0.8233333333333334
```

Fig. 7. Classification report SVM Classifier

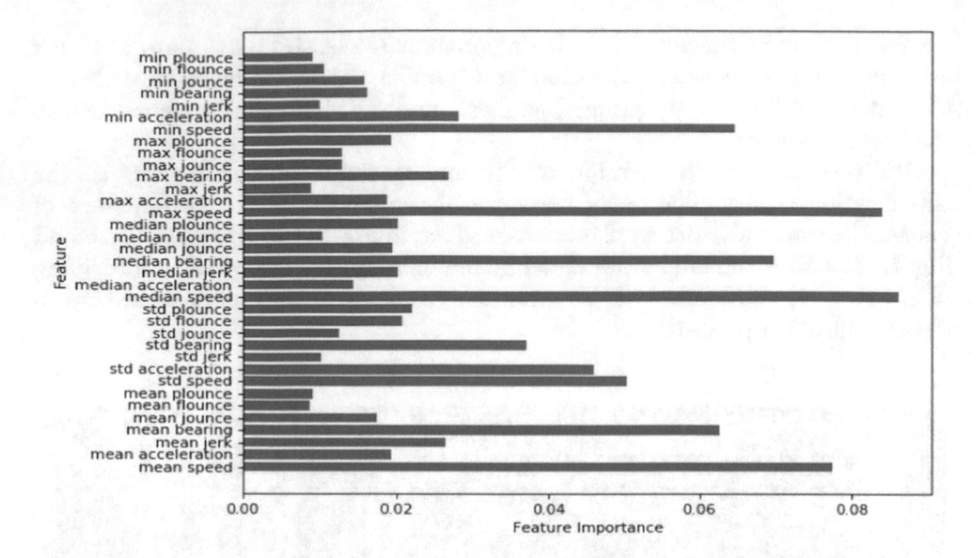

Fig. 8. Feature importance with RandomForestClassifier

To understand the importance of each characteristic in the classification process, we use the feature importance method integrated into the Sklearn library (see Fig. 8). The best characteristics are mean Speed, mean Bearing rate, std Speed, std Acceleration, median Speed, median Bearing rate, max Speed and min Speed.

We encountered many problems in this part where we used deep learning algorithms: Firstly, the difficulty of designing an architecture that is highly compatible with recurrent neural network or convolutional neural network. Secondly, identifying the optimal configuration of each architecture and thirdly, the time spent in the training part using a personal computer.

However an evaluation is made by the RNN, and another one is made with CNN. We give simulation results of the training and the test set (see Fig. 9 and Fig. 10). The target classes are the five 5 classes: walk, bike, bus, driving and train.

We have inspired based on the work of [3] for CNN model architecture. We used the Adam algorithm with the hyper parametrization as [3] and we made our training with the 35 features (see Fig. 8). We used 80% for training and 20% for testing of the Input data [number of segments = 3860, 1, estimated average number of points = 200, features = 35] [3]. We have trained our model with an epoch number equal to 60.

For RNN model we have trained our model with a simple architecture, (see Fig. 11). We used the Adam algorithm with the hyper parametrization as [3] and we made our training with the same 35 features. We used 80% for training and

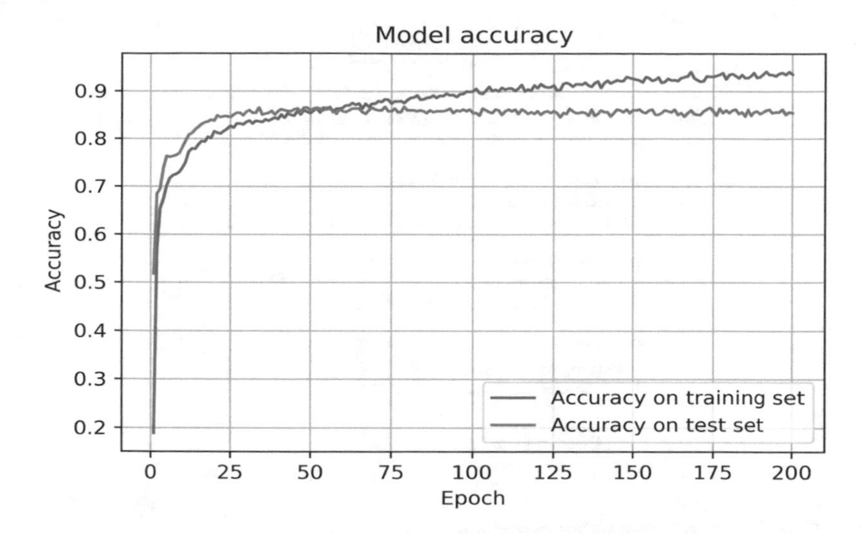

Fig. 9. Train-test accuracy with RNN

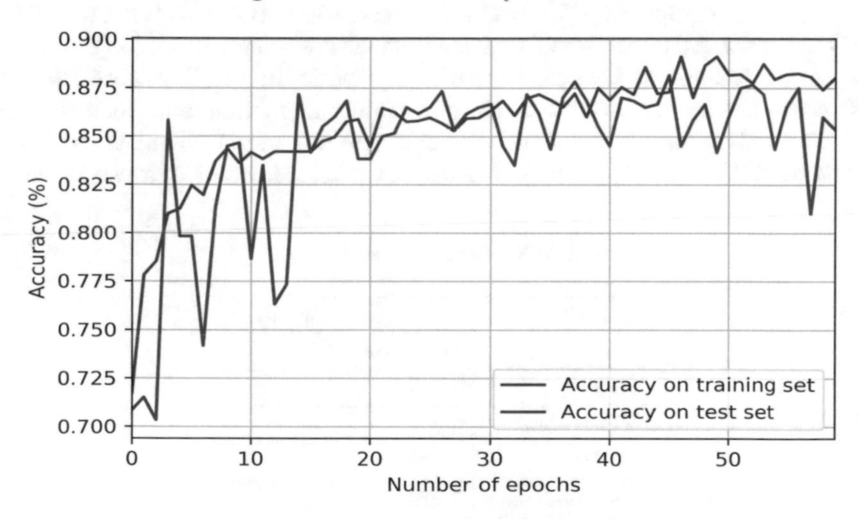

Fig. 10. Train-test accuracy with CNN

20% for testing of the Input data [number of segments = 3860, features = 35]. We have trained our model with an epoch number equal to 200.

Even if there is not much difference in accuracy between the two models, the CNN model is very complicated both in terms of architecture and training time, and moreover the RNN model is more stable during the training process.

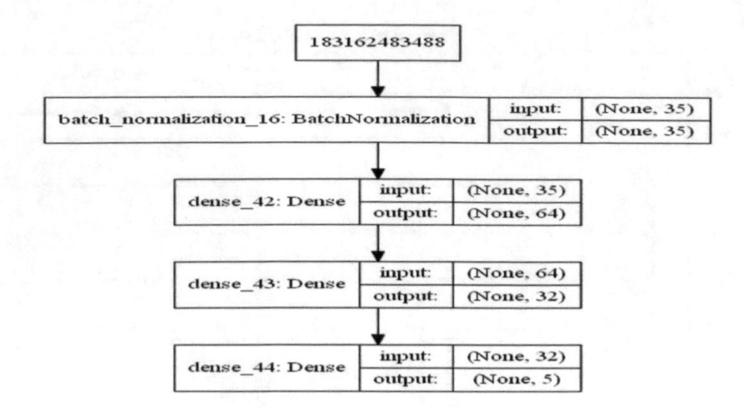

Fig. 11. Visualize RNN model

3.3 Comparison and Discussion

With beginner equipment, We used a personal computer (PC) with a 1.40 GHz AMD E1-2500 APU and 4 GB of random access memory (RAM), We were able to get accuracies that surpassed many works in the literature. We used 3000 segments for traditional machine learning algorithms and 3680 segments for deep learning algorithms, which is the lowest number of segments used in the literature of detection the mode of transport from GPS trajectories (Table 1).

Table 1. Summary of our work

Classifiers	Accuracy of classifiers
	Our accuracy
SVM	0.823
KNeighbors	0.833
MLP	0.875
DecisionTree	0.878
RandomForest	0.92
RNN	0.8488
CNN	0.8564

It is always interesting to study the performance of the proposed process compared to previous results in the literature. To demonstrate this, we present the accuracy results of a selection of various methodologies for detecting the mode of transport but not necessarily from GPS trajectories, namely [3] which has 84.84% as very high precision with their CNN model using the Geolife dataset, while the precision achieved by [2] was 90% using exploited data based on the temporal, spatial and socio-demographic characteristics of users, and the model of [13] which is developed by a support vector machine classifier reached 90% for the precision using combined mobile phone sensors data, also [30] obtained with their model a precision of 75.62% using Geolife dataset, and [26] have achieved a precision of 80% using raw GPS data collected by smartphone-based travel surveys, while [6] has obtained 84.4% as accuracy for their RNN model using two different datasets, and [5] achieved 92.74% as highest accuracy using a combination of smart phone sensors and [8] achieved an accuracy of 92.4% using Geolife dataset and the proposed mechanism of [29] outperforms all the approaches by achieving 92.7% of accuracy using only GPS trajectories.

In general, we can see that our models outperform many of these works with just a few amounts of the public Geolife dataset although many researchers have worked on personal dataset or companies dataset (private) or combined dataset (GPS, GIS, ..) [1,2,5,6,13,25]. It is therefore important to develop a "reference" database accessible to the public, allowing different research laboratories to evaluate and compare their methods.

4 Conclusion

The aim of this work was to detect the mode of transport used by an individual from his GPS trajectory. We expert many new features and applied many algorithms. We achieved very efficient accuracy with the Random Forest algorithm that surpassed many works in the literature.

As a complement to this topic, we can configure and identify the optimal RNN and CNN configurations and use GPU to train our models with all the data to fairly compared with the state of art. As future work we could combine GPS data with accelerometer data to predict the exact human activity: is he running or going downstairs or climbing, or to predict exactly how many steps used for a day or for any length of time.

References

1. Anda, C., Erath, A., Fourie, P.J.: Transport modelling in the age of big data. Int. J. Urban Sci. **21**(June), 19–42 (2017)
2. Bantis, T., Haworth, J.: Who you are is how you travel: a framework for transportation mode detection using individual and environmental characteristics. Transp. Res. Part C Emerg. Technol. **80**, 286–309 (2017)
3. Dabiri, S., Heaslip, K.: Inferring transportation modes from GPS trajectories using a convolutional neural network. Transp. Res. Part C Emerg. Technol. **86**(August 2017), 360–371 (2018). https://doi.org/10.1016/j.trc.2017.11.021
4. Dabiri, S., Lu, C.T., Heaslip, K., Reddy, C.K.: Semi-supervised deep learning approach for transportation mode identification using GPS trajectory data. IEEE Trans. Knowl. Data Eng. **32**, 1010–1023 (2020)
5. Eftekhari, H.R., Ghatee, M.: An inference engine for smartphones to preprocess data and detect stationary and transportation modes. Transp. Res. Part C Emerg. Technol. **69**, 313–327 (2016)

6. Endo, Y., Toda, H., Nishida, K., Kawanobe, A.: Deep feature extraction from trajectories for transportation mode estimation. In: Bailey, J., Khan, L., Washio, T., Dobbie, G., Huang, J.Z., Wang, R. (eds.) PAKDD 2016. LNCS (LNAI), vol. 9652, pp. 54–66. Springer, Cham (2016). https://doi.org/10.1007/978-3-319-31750-2_5
7. Endo, Y., Toda, H., Nishida, K., Kawanobe, A.: Deep feature extraction from trajectories for transportation mode estimation. In: Bailey, J., Khan, L., Washio, T., Dobbie, G., Huang, J.Z., Wang, R. (eds.) PAKDD 2016. LNCS (LNAI and LNB), vol. 9652, pp. 54–66. Springer, Cham (2016). https://doi.org/10.1007/978-3-319-31750-2_5
8. Etemad, M., Soares Júnior, A., Matwin, S.: Predicting transportation modes of GPS trajectories using feature engineering and noise removal. In: Bagheri, E., Cheung, J.C.K. (eds.) Canadian AI 2018. LNCS (LNAI and LNB), vol. 10832, pp. 259–264. Springer, Cham (2018). https://doi.org/10.1007/978-3-319-89656-4_24
9. Ettaouil, M., Ghanou, Y.: Neural architectures optimization and genetic algorithms. WSEAS Trans. Comput. **8**, 526–537 (2009)
10. Ettaouil, M., Ghanou, Y., Elmoutaouakil, K.: A new architecture optimization model for the Kohonen networks and clustering (2011)
11. Ghanou, Y., Bencheikh, G.: Architecture optimization and training for the multilayer perceptron using ant system (2016)
12. Gong, H., Chen, C., Bialostozky, E., Lawson, C.T.: A GPS/GIS method for travel mode detection in New York City. Comput. Environ. Urban Syst. **36**, 131–139 (2012)
13. Jahangiri, A., Rakha, H.: Developing a support vector machine (SVM) classifier for transportation mode identification using mobile phone sensor data. In: 2014 TRB Annual Meeting Compendium of Paper (2014)
14. Khalifi, H., Elqadi, A., Ghanou, Y.: Support vector machines for a new hybrid information retrieval system. Proc. Comput. Sci. **127**, 139–145 (2018)
15. Nham, B., Siangliulue, K., Yeung, S.: Predicting mode of transport from iPhone accelerometer data. Stanford University (2011)
16. Schafer, R.W.: What is a Savitzky-Golay filter? IEEE Signal Process. Mag. **28**, 111–117 (2011)
17. Schuessler, N., Axhausen, K.: Processing raw data from global positioning systems without additional information. Transp. Res. Rec. **2105**, 28–36 (2009)
18. Shafique, M.A., Hato, E.: A comparison among various classification algorithms for travel mode detection using sensors' data collected by smartphones. In: CUPUM 2015 - 14th International Conference on Computers in Urban Planning and Urban Management (2015)
19. Shafique, M.A., Hato, E.: Classification of travel data with multiple sensor information using random forest. Transp. Res. Proc. **22**, 144–153 (2017)
20. Shah, R.C., Wan, C.Y., Lu, H., Nachman, L.: Classifying the mode of transportation on mobile phones using GIS information. In: UbiComp 2014 - Proceedings of 2014 ACM International Joint Conference on Pervasive Ubiquitous Computing (2014)
21. Simoncini, M., Taccari, L., Sambo, F., Bravi, L., Salti, S., Lori, A.: Vehicle classification from low-frequency GPS data with recurrent neural networks. Transp. Res. Part C Emerg. Technol. **91**, 176–191 (2018)
22. Stenneth, L., Wolfson, O., Yu, P.S., Xu, B.: Transportation mode detection using mobile phones and GIS information. In: GIS Proceeding of ACM International Symposium on Advances in Geographic Information Systems (2011)
23. Wang, H., Liu, G., Duan, J., Zhang, L.: Detecting transportation modes using deep neural network. IEICE Trans. Inf. Syst. **100**, 1132–1135 (2017)

24. Wu, L., Yang, B., Jing, P.: Travel mode detection based on GPS raw data collected by smartphones: a systematic review of the existing methodologies (2016)
25. Xiao, G., Juan, Z., Zhang, C.: Travel mode detection based on GPS track data and Bayesian networks. Comput. Environ. Urban Syst. **54**, 14–22 (2015)
26. Xiao, Z., Wang, Y., Fu, K., Wu, F.: Identifying different transportation modes from trajectory data using tree-based ensemble classifiers. ISPRS Int. J. Geo-Inf. **6**, 57 (2017)
27. Yang, X., Stewart, K., Tang, L., Xie, Z., Li, Q.: A review of GPS trajectories classification based on transportation mode. Sens. (Switz.) **18**(11), 1–20 (2018)
28. Yazdizadeh, A., Patterson, Z., Farooq, B.: An automated approach from GPS traces to complete trip information. Int. J. Transp. Sci. Technol. **8**(1), 82–100 (2019). https://doi.org/10.1016/j.ijtst.2018.08.003
29. Yu, J.J.: Travel mode identification with GPS trajectories using wavelet transform and deep learning. IEEE Trans. Intell. Transp. Syst. **22**, 1093–1103 (2019)
30. Zhang, R., Xie, P., Wang, C., Liu, G., Wan, S.: Classifying transportation mode and speed from trajectory data via deep multi-Scale learning. Comput. Netw. **162**, 106861 (2019). https://doi.org/10.1016/j.comnet.2019.106861
31. Zheng, Y.: GPS Trajectories with transportation mode labels (2010)
32. Zheng, Y., Fu, H.: Geolife GPS trajectory dataset - user guide (2011)
33. Zheng, Y., Li, Q., Chen, Y., Xie, X., Ma, W.Y.: Understanding mobility based on GPS data. In: UbiComp 2008 - Proceedings of 10th International Conference on Ubiquitous Computing (2008)
34. Zheng, Y., Liu, L., Wang, L., Xie, X.: Discovering regions of different functions in a city using human mobility and POIs Jing. In: Proceedings of 18th ACM SIGKDD international conference on Knowledge discovery and data mining - KDD 2012 (2008)
35. Zheng, Y., Liu, L., Wang, L., Xie, X.: Learning transportation mode from raw GPS data. In: WWW (2008)
36. Zheng, Y., Xie, X., Ma, W.: GeoLife: a collaborative social networking service among user, location and trajectory. IEEE Data Eng. Bull. **33**(2), 32–40 (2010). http://scholar.google.com/scholar?hl=en&btnG=Search&q=intitle:GeoLife+:+A+Collaborative+Social+Networking+Service+among+User+,+Location+and+Trajectory#0

Application of Artificial Intelligence to X-ray Image-Based Coronavirus Diseases (COVID-19) for Automatic Detection

El Idrissi El-Bouzaidi Youssra[1(✉)] and Abdoun Otman[2]

[1] Laboratory of Advanced Sciences and Technologies, Polydisciplinary Faculty, Abdelmalek Essaadi University, Larache, Morocco
youssra.elidrissi.elbouzaidi@gmail.com

[2] Computer Science Department, Faculty of Science, Abdelmalek Essaadi University, Tetouan, Morocco

Abstract. The automatic diagnosis of the Coronavirus has become essential to minimize the workload of the health system in the face of the epidemic. Today, the coronavirus is spreading at an accelerated rate. It is Therefore essential to identify infected persons in order to avoid any contamination. This virus can be detected using real-time polymerase chain reaction (RT-PCR) kits. These kits are expensive and it takes several hours to confirm infection; another weakness is that in the early stages of the disease, the positivity rate of this test was considered to be very low, however, high international demand has revealed a gap in this kit. Therefore, an alternative strategy is advocated to assist in effective diagnosis. This study examines the issue of automatic classification systems in lung disease, including the new COVID-19. To accurately track this virus, Artificial Intelligence (AI) combined with radiography that can detect radiological symptoms, represents an ideal alternative solution. In this sense, we first propose a deep-transfer learning approach to extract high dimensional features from radiological images and then we apply a machine learning technique to classify the obtained features. The proposed classification model, i.e. DenseNet121 plus SVM, achieved an accuracy and f1 score of 96.09% and 97% respectively, for COVID-19 detection based on three classes. This model is superior to other models in the literature. We implemented 10 CNNs to compare our methodology. The data available in GitHub and Kaggle repository are used as a basis to produce the result.

Keywords: COVID-19 · Artificial intelligence · Hybrid systems · Deep learning · Convolutional neural networks · Transfer learning

1 Introduction

In 2019, emerging coronavirus disease (COVID-19) was officially pronounced by the World Health Organization (WHO) as a pandemic. As announced in its March 11, 2020, press release, this lung disease is caused by the respiratory Coronavirus-2 (SARS-CoV-2), a member of the Coronavirdeae family. Today's press release provides an update on

M. Lazaar et al. (Eds.): BDIoT 2021, LNNS 489, pp. 208–220, 2022.
https://doi.org/10.1007/978-3-031-07969-6_16

the epidemic health problem that has been encountered and shared around the world in recent months. To date, no drug has been developed or officially approved for use in COVID-19 patients worldwide. Approximately 82% of COVID-19 patients suffer from mild symptoms, but the rest are severe or critical [1]. Symptoms associated with infection include respiratory problems, febrile condition, and even dyspnea with cough. It can lead to aggravated pneumonia or death.

The mechanism currently used by the international health entities to confirm the presence of COVID-19 in the patient is the reverse transcription-polymerase chain reaction (RT-PCR). Because RT-PCR is not very sensitive, this means that many patients with COVID-19 will not be identified quickly or treated appropriately. Given the lack of diagnostic resources worldwide, RT-PCR test kits for COVID-19 seem to be a major challenge, causing panic. Effective early detection and rapid diagnostic testing will help patients affected by COVID-19 access appropriate healthcare, helping to save many lives worldwide. The development of a rapid and effective diagnostic tool becomes one of the first priorities to be addressed.

One of two tests justify their effectiveness for the diagnosis of COVID-19: computed tomography (CT) imaging and radiographic (X-rays) imaging. Radiographic imaging has several advantages: it is inexpensive, requires only a low dose of radiation, is widely available, and is easy to use in general or community hospitals. However, X-rays pose a problem: sometimes, because of the properties that describe the existence of COVID-19, radiologists have difficulty diagnosing this disease due to its admixture with other respiratory diseases such as viral or bacterial pneumonia. Artificial intelligence (AI) solutions can help solve these problems [2]. Convolutional neural networks (CNNs), in particular, have proven to be a very promising deep learning technique for image classification [3].

This paper presents a model based on the application of deep learning and machine learning techniques that can automatically predict the presence of the COVID-19 virus. We propose a methodology based on a deep transfer learning technique that extracts features from radiographic images, and an SVM to automatically classify the COVID-19 case. For this reason, we implemented DensNet121, DensNet169, DensNet201, MobileNetV2, Xception, ResNet101, ResNet152, ResNet50, InceptionV3, NASNet-Large, and NASNetMobile pre-trained models to achieve better prediction accuracy for three different classes including COVID-19 X-ray images, patients with bacterial and viral pneumonia and, Normal.

The main sections were structured as shown below: Sect. 3 provides a detailed overview of the proposed model. Section 4 describes the experimental setting, including performance measures. Sections 4.2 and 4.3, respectively, outline and report the results obtained from the proposed model. Finally, Sect. 5 draws a conclusion.

2 Related Work

Researchers have proven, through computer vision and deep learning, their advantage over medical images analysis, specifically using reliable chest X-rays for COVID-19 detection. Advances in these technologies are being used to develop computer-aided detection or diagnostic tool against COVID-19. In addition, the application of artificial intelligence offers significant advantages in detecting disease and measuring infection rates, offering very encouraging and prospective results.

Ozturk et al. [4] developed a model using The DarkNet architecture. They obtained 98.08% and 87.02% rates respectively on two and three classes of their model.

Khan et al. [5] suggested the CoroNet, a proposed classification system using the Xception architecture, which classifies chest X-ray images in four classes: normal, bacterial, viral, and COVID-19 pneumonia. This resulted in overall performance of 89.6%.

Wang and Wong [6] introduced in their paper a deep model for the detection of COVID19, an architecture known as COVID-Net. However, a convolutional neural network (CNN) model using chest X-rays as inputs is the basis of the COVID-Net architecture. They achieved a test accuracy of 92.4% in the classification of healthy, pneumonia, and COVID-19 classes.

Sethy and Behera [7] employed several different pre-trained transfer techniques (AlexNet, VGG16, VGG19, GoogleNet, ResNet18, ResNet50, ResNet101, InceptionV3, InceptionRes-NetV2, DenseNet201, and XceptionNet), and are trained in radiographs extracted by contaminated patients for the extraction of imaging features. For classification, these features were applied to Support Vector Machines (SVM). As a result, the accuracy of the classification achieved in the developed model indicated that ResNet50 with SVM had the best results with an average accuracy of 95.38%.

The study by Alqudah et al. [8] applied for the diagnosis of COVID-19 symptoms in patients with chest X-ray images two different machine learning techniques. Initially, they used the CNNs model for feature extraction such as AOCTNet, MobileNet, and ShuffleNet. Secondly, the features of these images were classified using the support vector machine (SVM) and random forest algorithms (RF). The resulting accuracies from the SVM and RF were 90.5% and 81% respectively.

Elaziz et al. [9] included the following steps: image pre-processing, segmentation of regions of interest (ROI) related to COVID-19 diseases, calculation and identification of 961 effective features of images representing chest X-rays, Then, a KNN classification model was developed based on the fusion of the different features to detect and classify cases. The model resulted in an accuracy of 96.1% to classify COVID-19 and non-COVID-19 cases.

In his paper, Narin et al. [10] presented three different deep-transfer learning models such as ResNet-50, Inception-ResNetV2, and InceptionV3 were presented for a binary classifier to discriminate covid-19 from healthy patients based on chest X-rays. Evaluation results show that ResNet50 exhibited the best classification performance with 98.0% accuracy, 96% recall and 100% accuracy, versus 97.0% for InceptionV3 and 87% for Inception-ResNetV2.

3 Proposed Image-based COVID-19 Disease Detection Model

3.1 Dataset

To create the dataset, the investigation focused on obtaining radiographic images corresponding to three classes such as COVID-19 (+), pneumonia (+), and Normal X-ray images. A detailed search yielded a total collection of 180 X-ray images with COVID-19, in addition, 180 X-rays representing a class that includes bacteria and viruses, called "Pneumonia", and a total of 180 X-rays of normal individuals. This collection includes a selection of radiographs from a GitHub repository created by Dr. Cohen [11], and a selection of radiographs from the Kaggle repository called "Chest XRay Images (Pneumonia)" [12]. Sample images of both datasets shown in Fig. 1.

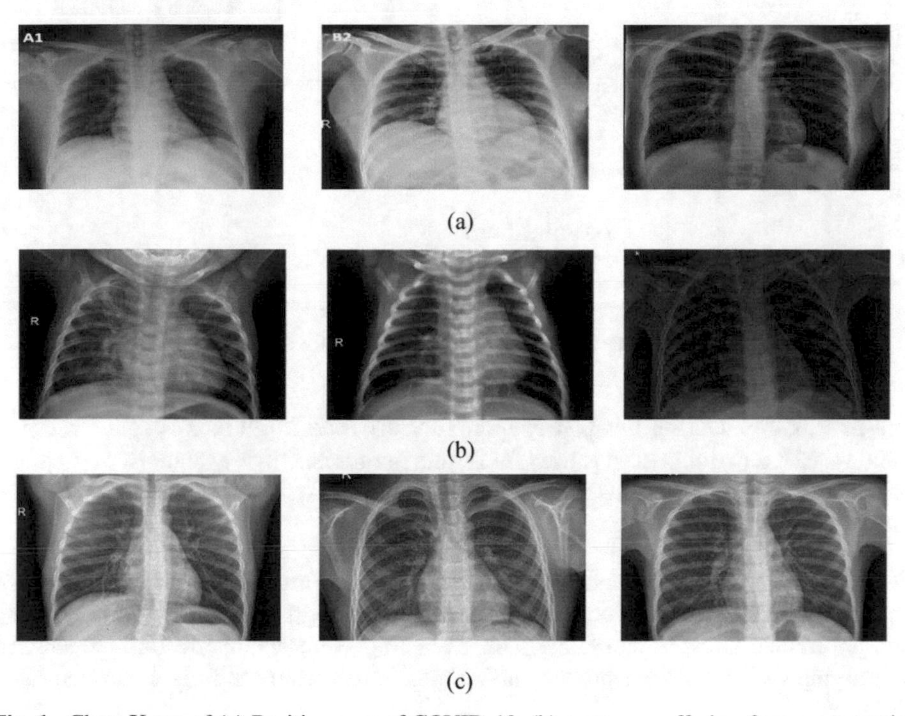

Fig. 1. Chest X-ray of (a) Positive case of COVID-19, (b) a person suffering from pneumonia, and (c) a healthy person.

3.2 Model Architecture Overview

Computational analysis of radiographic images has the benefit of being performed using Artificial Intelligence approaches, which helps to significantly speed up the analysis time and cost-effectiveness. This section describes each part of the architecture of the proposed COVID-19 disease detection model, presented at the beginning of a pre-processing phase applied to the collected images. Subsequently, a deep-transfer learning model for the Feature Extraction is defined. Finally, the features obtained from the CNN network are used by the SVM, a Machine Learning technique. The classification is then performed, and the performance of the classification model is measured. Figure 2 shows the structure of the model.

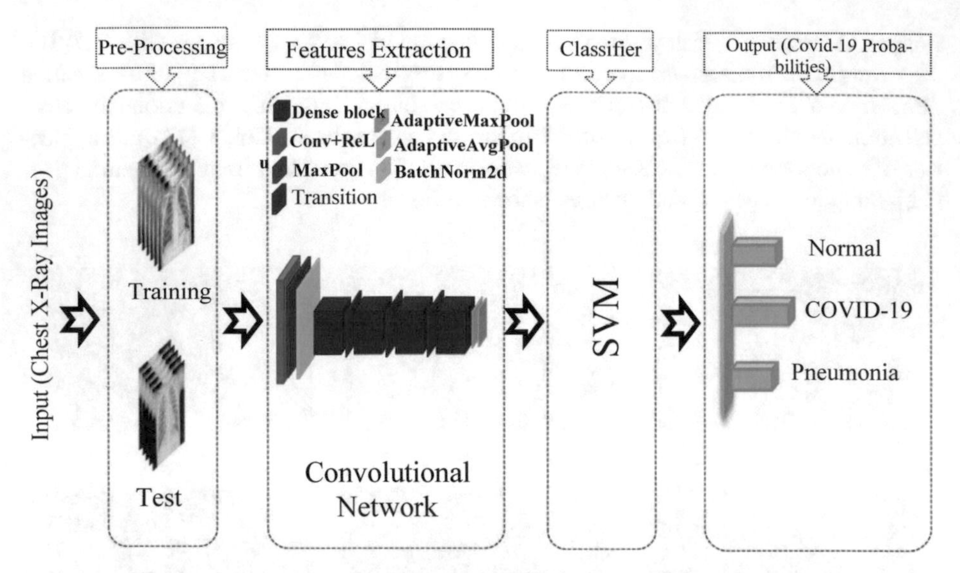

Fig. 2. Architecture of the proposed COVID-19 disease detection model

Pre-processing. During this phase, approximately 80% of all data is split for use in training. The remaining 20% is used for testing purposes. Other operations are applied, such as resizing all images in this dataset to 224 × 224 pixel size. And the normalization (mean and standard deviation) is carried out.

Feature Extraction. In this step, the objective is to maximize the benefits of advanced deep learning architectures on chest X-ray images using transfer learning [13], using this pre-trained network as the basis for extracting representative features capable of identifying COVID-19, Pneumonia, and normal cases from the fully connected layer (FC).

Deep-Transfer Learning. Deep learning is at an elementary level in the field of machine learning. The purpose of using deep learning techniques in the medical field is to strengthen the capacity and quality of the health care system. With recent innovations and advances in the field of medical image analysis, significant success stories are being reported in the field.

The deep learning models for the analysis of medical images often require a lot of data, therefore one of the biggest difficulties encountered is the limited amount of data available, only a limited amount of COVID-19 positive case X-ray image data are available at the moment. This data is both costly and tedious to annotate by experts. An effective way to achieve meaningful results in classification problems with a lack of data in a given domain is to exploit existing knowledge from a similar domain, a technique known as transfer learning, which is used extensively in deep learning, provides the ability to form models with a smaller number of datasets with less cost of calculation.

Transfer Learning. Transfer learning is a method of machine learning where a model for a specific task is developed and exploits existing knowledge in a similar field for another task. There are two categories for performing Transfer Learning, namely Fine-Tuning [14] and Feature Extraction.

The Feature Extraction approach is processed by using the pre-trained convolutional neural network architectures, which is defined from the source task, with no need to retrain the network. This approach only requires a classifier to be added to perform the classification of these features against their importance relative to the particular problem. This makes the procedure generally fast and efficient.

In contrast to the first strategy, the Fine-Tuning strategy involves the use of a pre-trained network and its partial re-training of the weights using our datasets in the training phase. A low learning rate required for the initial layers because they require less tuning and gradually increase on later layers that need more tuning, especially those that are fully connected.

In this study, we relied on the first strategy based on the Feature Extraction acquired from a pre-trained Convolutional Neural Network DenseNet121 architecture as an efficient alternative due to its weights, performed in the ImageNet dataset [15] with fewer parameters, succeeded by the use of the SVM classifier (see Fig. 2) in order to classify COVID-19 X-ray images into COVID-19, Pneumonia and Normal classes. In what follows, a short explanation of the DenseNet121 architecture is characterized.

DenseNet121. Moves the characteristics present per dense block to all the following layers. Therefore, the $l - th$ layer gets as input the features for all preceding layers:

$$\mathrm{x_l = F([x_0, x_1, x_2, \ldots, x_{l-1}])} \tag{1}$$

where, $[x_0, x_1, x_2, \ldots, x_{l-1}]$ corresponds to the elements produced in layers $0, \ldots, l-1$.

The DensNet121 architecture is composed as follows: four dense blocks, each of them having a different set of layers (6, 12, 24, 16 layers). Transition layers are located between each dense layer, in order to reduce the size of the features generated in Eq. (1) by concatenation. The classification layer consists of an overall average pooling layer and a fully connected layer. For further details, please refer to [16]. As part of this work, we have added a sequence of layers in a typical pre-trained DensNet121 network, namely an adaptive Average/Max pooling, Batch normalization, and drop out, which drastically reduces the issue of overfitting (see Fig. 2).

Classification Using SVM. The objective of this model is to use SVM as a classifier instead of Softmax. For the deep features extracted from the model, we have chosen the DensNet121 network, a pre-formed CNN model, as the input element of the classifier, while the output elements of the model represent the probabilities of the class representing these features. A classifier most frequently used for solving classification problems is SVM. In linear type SVM, the separation between data in linear form or even in hyperplane form is performed. The equation describing the linear form is:

$$\mathrm{A(x) = w^T.x + b} \tag{2}$$

where A (x) is the class, w^T is a normal vector to the hyperplane, x is the training data, and b is the Bias.

When the data are not separated linearly, another technique can be applied in mapping incoming features in a higher dimensional space to determine the separation hyperplane. Such a mapping function is called "kernel" and includes polynomial, sigmoid and radial base functions [17]. For the purpose of this paper, we use the kernel Radial Base Function (RBF), which is given by the Eq. (3).

$$\mathrm{B}(\mathrm{x}, \mathrm{y}) = \exp\left(-\frac{\|\mathrm{x} - \mathrm{y}\|}{2\sigma^2}\right) \tag{3}$$

where B (x, y) is the kernel function, x is the training data, y is the classes, and σ is a factor that shapes the width of the Radial Basis Function, with $\sigma > 0$.

Performance Evaluation. To reliably assess the performance of Artificial Intelligent systems, the confusion matrix is the most commonly used indicator. Measures of the confusion matrix include Accuracy, Recall, Precision, and F1 score. The following statistical indices: False Positive, True Positive, False Negative, and True Negative; (FP, TP, FN, TN) are calculated to evaluate each of the measures listed above:

Accuracy. indicates the reconciliation between the measured value and another determined value.

$$\text{Accuracy} = \frac{\text{FN} + \text{TP}}{\text{TN} + \text{TP} + \text{FN} + \text{FP}} \tag{4}$$

Precision (Specificity). Indicates the degree to be achieved by the model in relation to positive predictions.

$$\text{Precision} = \frac{\text{TP}}{\text{TP} + \text{FP}} \tag{5}$$

Recall (Sensitivity). Indicate the number of true positives that can be reported as true positives in the model.

$$\text{Recall} = \frac{\text{TP}}{\text{TP} + \text{FN}} \tag{6}$$

F1 score. provides the balance of precision and recall.

$$\text{F1 score} = 2 \times \frac{\text{Precision} \times \text{Recall}}{\text{Precision} + \text{Recall}} \tag{7}$$

4 Experimental Results and Discussion

This section describes and discusses the experiments with the evaluation criteria used in this paper for test the effectiveness of the proposed model. Based on these results, the optimal strategy is selected. We used the open-source deep learning framework Keras and TensorFlow. All the computational work was performed on a Colab platform.

4.1 Experimental Details

Pre-trained Neural Networks. ten advanced deep learning networks, DenseNet169, Dense-Net201, MobileNetV2, Xception, ResNet50, ResNet101, ResNet152, InceptionV3, NASNetLarge and NASNetMobile, respectively, were included for comparison with DenseNet121 in the Experimental study. Findings are discussed in the next subsection.

Pre-processing. Pre-processing was necessary according to the deep-transfer neural network used for each image. Depending on the architecture, each of these different neural networks requires different image sizes. It was necessary to proceed with re-dimensioning with the standardization process. As an example: InceptionV3 and Xception require images of size 229 × 229, while DenseNet121, NASNetMobile, and MobileNetV2 is able to provide 224 × 224 images, while NASNetLarge supports 331 × 331 images. In the same way, the standardization of all the images was carried out according to the appropriate architectures. The pre-processing is similar to that of the training images and the test images in order to have the same size as requested by the architecture.

4.2 Results

In Table 1, comparative analyses between the proposed model and competing models in terms of Accuracy, F-score, Precision and Recall, over the test dataset, correspond to the three defined classes, (COVID-19, Normal, and Pneumonia).

Table 1. Results of the proposed deep transfer learning and SVM based COVID-19 model.

Models	Accuracy	F1-score	Recall	Precision
DensNet169	0.9296875	0.93	0.93	0.94
DensNet201	0.9453125	0.94	0.95	0.95
MobileNetV2	0.953125	0.95	0.95	0.96
Xception	0.9375	0.93	0.91	0.94
ResNet101	0.7890625	0.77	0.79	0.82
ResNet152	0.5234375	0.52	0.63	0.64
ResNet50 & SVM	0.4921875	0.50	0.49	0.66
InceptionV3	0.875	0.89	0.88	0.88
NASNetLarge	0.890625	0.89	0.86	0.93
NASNetMobile	0.8906250	0.88	0.86	0.91
DensNet121	**0.9609375**	**0.96**	**0.96**	**0.97**

The best results were obtained with an accuracy of 96.09%, a 96% recall and a precision value of 97% by the DenseNet121 pre-training model. On the other hand, with

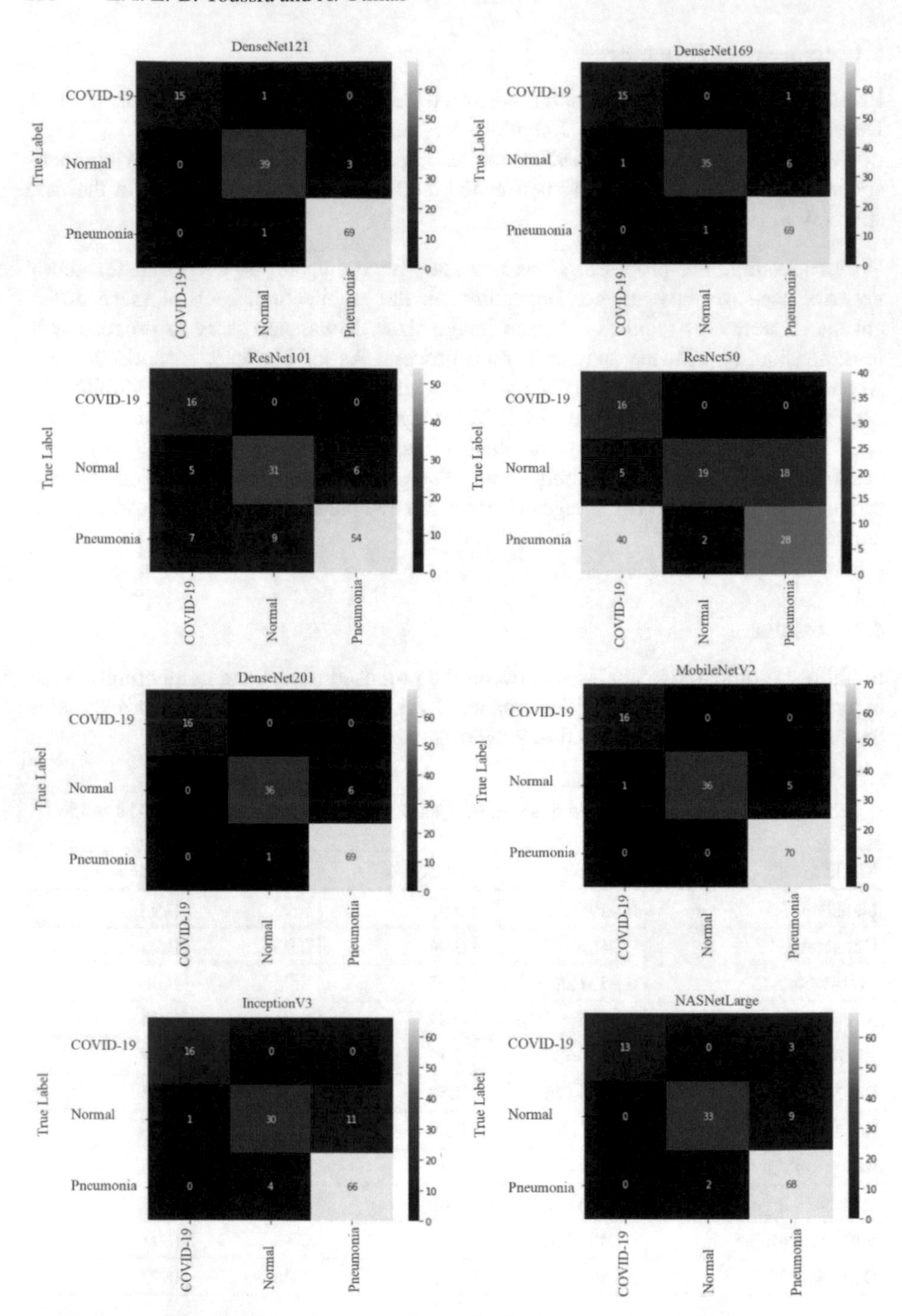

Fig. 3. The confusion matrix obtained using pre-trained models.

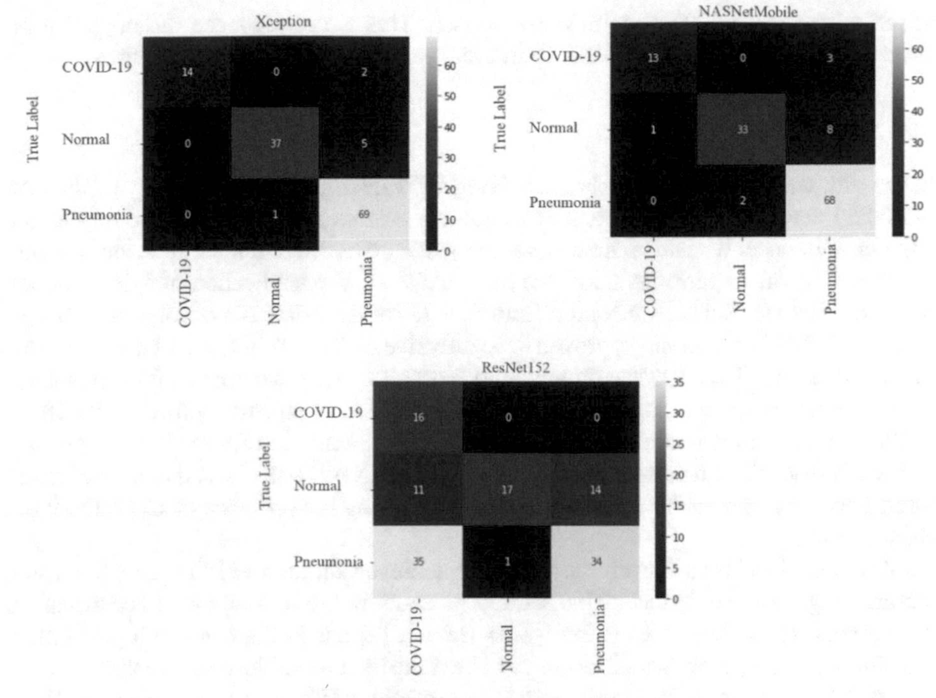

Fig. 3. continued

an Accuracy of 49.21%, a recall of 49% and a precision value of 66% for the ResNet50 model, the lower performances were achieved. These results make the DenseNet121 model superior to other models for both training and testing. The following Table 2 summarizes the detailed classification results of our methodology.

The confusion matrixes of all configurations were obtained to further test the robustness of the proposed methodology (see Fig. 3). The confusion matrix allowed us to measure true positives, true negatives, false positives and false negatives at the same time considering the performance of our model.

Table 2. Precision, Recall, and F1 score, corresponding to different classes, DenseNet121 model.

Class	Precision (%)	Recall (%)	F1 score (%)
COVID-19	100	94	97
Normal	95	93	94
Pneumonia	96	99	97

According to Table 2, the COVID-19 class achieves a good sensitivity, indicating a significant ability of the model to detect COVID-19 cases. An additional and significant result showed a 100% higher positive predictive. That indicates there were no classes misclassified as COVID-19 with other classes. This is critical since the objective of detecting positive cases of COVID-19 in order to reduce the spread of the virus.

4.3 Discussion

In the ultimate goal of effectively classifying COVID-19, the high accuracy of the test (96.09%) demonstrated the potential usefulness of this proposed methodology as an adjunct in medical decisions. Indeed, radiologists are assisted in their decision-making process; only the experts can make the final decision. The application of a deep model based on the DensNet121 CNN architecture, both for the extraction of relevant features and for the SVM as a classifier, proved to be effective in this task in terms of test accuracy for 3 classes (COVID-19, Pneumonia, and Normal). There were more false positives than false negatives, and therefore the misclassification of patients with COVID-19 as healthy or have pneumonia is relatively less, an ideal condition for medical diagnosis, and will replace the much-demand PCR tests for non-COVID-19 cases. Indeed, Artificial Intelligence techniques have been relevant in detecting the presence of COVID-19 on chest X-rays.

A comparison focusing on the accuracy achieved through existing and proposed methodology was performed. This section presents the results obtained according to the authors' respective studies. Sethy and Behera [7] use in their model a preformed transfer technique, such as ResNet-50, to extract the features of the radiographic images. These features were applied to Support vector machine (SVM) for classification. They reported an AUC score of 95.38%. Alqudah et al. [8] propose a radiographic image-based method for the detection of COVID-19 using deep feature and machine learning techniques namely Support vector machine (SVM), K-nearest neighbor (kNN), and random forest (RF). They used AOCTNet, MobileNet, and ShuffleNet CNNs models for feature extraction and fed them to the machine learning techniques to classify COVID-19 and non-COVID-19 cases. They have achieved a validation accuracy of 90.5% and 81% for SVM and RF, respectively. Wang and Wong [6] introduced in their paper COVID-Net, a deep model for the detection of COVID19. They achieved a test accuracy of 92.4% in the classification of normal, Pneumonia, and COVID-19 classes. Table 3 summarizes all of the above results.

Table 3. Transfer learning method, and the Accuracy of the model. Comparison of the proposed methodology with the existing methods.

Transfer learning method	Accuracy (%) 3-class
ResNet50 & SVM	95.38
COVID-Net	92.40
DarkNet	87.02
DenseNet121 & SVM	**96.09**

5 Conclusion

We proposed in the present study a methodology to automatically diagnose cases of COVID-19 on pulmonary radiography images. The proposed method uses deep transfer learning to extract relevant features from COVID-19 radiographic images. Then, the SVM classifier is used to decide whether a given radiographic image is a COVID-19 case, pneumonia or a normal case. The method has achieved comparable performance in terms of accuracy, recall and precision. The proposed approach achieved high performance and low resource consumption. Further applications in the medical field and other relevant areas should be included in our future work.

References

1. Coronavirus is officially a pandemic: Here's why that matters. Science (2020). https://www.nationalgeographic.com/science/2020/02/how-coronavirus-could-become-pandemic-and-why-it-matters/
2. Hosny, A., Parmar, C., Quackenbush, J., Schwartz, L.H., Aerts, H.J.W.L.: Artificial intelligence in radiology. Nat. Rev. Cancer **18**, 500–510 (2018)
3. El Idrissi El-Bouzaidi, Y., Abdoun, O.: DenTcov: Deep Transfer Learning-Based Automatic Detection of Coronavirus Disease (COVID-19) Using Chest X-ray Images. In: Motahhir, S., Bossoufi, B. (eds.) ICDTA 2021. LNNS, vol. 211, pp. 967–977. Springer, Cham (2021). https://doi.org/10.1007/978-3-030-73882-2_88
4. Ozturk, T., et al.: Automated detection of COVID-19 cases using deep neural networks with X-ray images. Comput. Biol. Med. **121**, 103792 (2020)
5. Khan, A.I., Shah, J.L., Bhat, M.: CoroNet: A deep neural network for detection and diagnosis of COVID-19 from chest x-ray images. Comput. Methods Programs Biomed. **196**, 105581 (2020)
6. Wang, L., Wong, A.: COVID-Net: a tailored deep convolutional neural network design for detection of COVID-19 cases from chest X-ray images. (2020) arXiv:2003.09871 [cs, eess]
7. Sethy, P.K., Behera, S.K.: Detection of Coronavirus Disease (COVID-19) Based on Deep Features. (2020). https://doi.org/10.20944/preprints202003.0300.v1
8. Alqudah, A., Qazan, S., Alquran, H., Qasmieh, I., Alqudah, A.: COVID-19 Detection from X-ray images using different artificial intelligence hybrid models. JJEE **6**, 168 (2020)
9. Elaziz, M.A., et al.: New machine learning method for image-based diagnosis of COVID-19. PLoS ONE **15**, e0235187 (2020)
10. Narin, A., Kaya, C., Pamuk, Z.: Automatic Detection of Coronavirus Disease (COVID-19) Using X-ray Images and Deep Convolutional Neural Networks. (2020) arXiv:2003.10849 [cs, eess]
11. Cohen, J.P.: ieee8023/covid-chestxray-dataset (2020)
12. Chest X-Ray Images (Pneumonia) | Kaggle: https://www.kaggle.com/paultimothymooney/chest-xray-pneumonia
13. Shin, H.-C., et al.: Deep convolutional neural networks for computer-aided detection: CNN architectures, dataset characteristics and transfer learning. IEEE Trans Med Imaging **35**, 1285–1298 (2016)
14. Wang, G., et al.: Interactive medical image segmentation using deep learning with image-specific fine tuning. IEEE Trans. Med. Imaging **37**, 1562–1573 (2018)
15. Deng, J., et al.: ImageNet: a large-scale hierarchical image database. In: 2009 IEEE Conference on Computer Vision and Pattern Recognition, pp. 248–255 (2009). https://doi.org/10.1109/CVPR.2009.5206848

16. Huang, G., Liu, Z., Van Der Maaten, L., Weinberger, K.Q.: Densely connected convolutional networks. In: 2017 IEEE Conference on Computer Vision and Pattern Recognition (CVPR), pp. 2261–2269 (2017) https://doi.org/10.1109/CVPR.2017.243
17. Achirul Nanda, M., Boro Seminar, K., Nandika, D., Maddu, A.: A comparison study of kernel functions in the support vector machine and its application for termite detection. Information **9**, 5 (2018)

E-Learning

Personalization Between Pedagogy and Adaptive Hypermedia System

Anoir Lamya(✉), Khaldi Mohamed, and Erradi Mohamed

Research Team in Computer Science and University Pedagogical Engineering, Higher Normal School, Abdelmalek Essaadi University, Tetouan, Morocco
lamya.anoir@uae.ac.ma

Abstract. Adaptive learning is about providing personalized learning adapted to individual needs, both in terms of learning rhythm and also in terms of content. Personalization in adaptive e-Learning has become a very important topic in research in recent years. With the emergence of the Web, Artificial Intelligence, Big Data,…on learning that is based on these technologies and are about to revolutionize teaching/learning. In this article, we discuss the progression of personalization and personalized and adaptive hypermedia systems.

Keywords: Personalization · Pedagogy · Adaptive Hypermedia System (AHS) · Learner profile · Learning styles

1 Introduction

According to Baron and Bruillard in 2008, the rapid progress of Information and Communication Technologies (ICT) has encouraged researchers to develop tools and strategies to lead learning activities such as e-learning and more precisely adapted learning [1]. Modern pedagogical tools require means to facilitate learning for the learner, taking into consideration the choice of the learning situation, the clarification of objectives and the choice of contents, in order to implement adapted teaching–learning procedures, with the aim of developing tools that help the learner to build and appropriate knowledge and know-how according to his profile, his learning preferences and in particular to provide an adaptive and personalized learning system [2].

Adaptive learning is a new approach to education that combines between technology and pedagogy to meet the challenges faced in the world of education [3]. It consists in adapting and personalizing the learning to the needs of each learner, it corresponds to a pedagogical concept whose purpose in order to adapt the pedagogical decisions to the specific skills and needs of each learner. These adaptive pedagogical systems promise to take into consideration the learner's profile (his knowledge, preferences, aptitudes, objectives, etc.) [4]. Our work therefore consists in personalizing the learning process based on the learner's profile, his preferences, his habits, etc., in order to reflect the learner's learning needs in an adaptive hypermedia system (AHS);

In our article, we present the progression of personalization between pedagogy and Adaptive Hypermedia Systems.

M. Lazaar et al. (Eds.): BDIoT 2021, LNNS 489, pp. 223–234, 2022.
https://doi.org/10.1007/978-3-031-07969-6_17

The term personalization is often encountered in the literature and designates the act of providing information specially adapted to the needs of an individual or a group of individuals [5, 6]. While AHS are systems that have the ability to provide some degree of personalization [7].

Personalization can be provided to offer the content or visualization of the system according to the preferences of the learner, personalization is a term which represents a synonym of the terms adaptive and adaptable. These two systems, adaptive and adaptive, can be summarized as custom systems [8]. Today, personalization has been developed at the level of different axes such as: Intelligent Tutor Systems (STI), Big Data, Ontologies, and Artificial Intelligence (IA), Machine learning…

2 Scientific and Problematic Context

Through our article, we focus our work on two essential points. The first point concerns the definition of the concept "personalization and the second concerns the development of personalized educational hypermedia systems".

2.1 An overview of the Concept of Personalization

The notion of the personalization of learning include the individualization of teaching, where the learner can progress at his own rhythm, and that of the differentiation of learning, where he can choose between the different learning methods according to his preferences or specific characteristics [9].

Personalization is different from differentiation and individualization; personalization requires a major change in orientation from a teacher-centered approach to an authentic, learner-centered approach. Personalized forms of learning offer an approach adapted to the diverse abilities, preferences, interests, and other needs of each learner. As such, it gives students more autonomy to develop their own learning paths and with more opportunity for creativity, collaboration, and content creation. Personalization aims to develop individualized learning programs for each student with the intention of involving them in the learning process [10].

Personalized learning is a systematic learning design that is centered on adapting instruction to learners. Personalized learning provides flexibility and supports what, how, when, and where learners learn and make their knowledge proficient. Specifically, these flexibilities and supports are designed in terms of instructional approaches, content, activities, objectives and learning results. Personalized learning systems make often the use of technologies to improve the availability of adaptive and personalized learning for all learners [11].

Personalizing is not individualizing, individualizing relief of a particular course, personalizing is based on the collective and cooperative learning [12]. It is less interested in the unique peculiarities of each individual than taking advantage of the functioning of a group. Personalizing one's teaching means above all seeking to meet the identified needs of groups of learners. The diagram in Fig. 1 shows us the difference between individualizing and personalizing [4, 13].

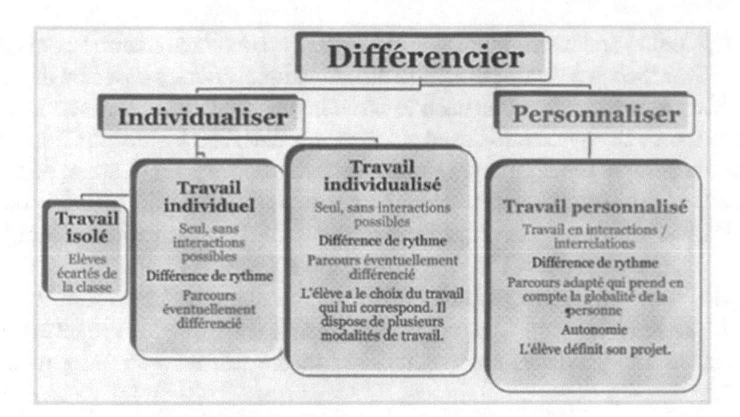

Fig. 1. Personalization and Individualization [4]

In fact, in individualization, educational resources are adapted to the objectives and needs of each learner, according to their characteristics. In personalization, it is the learner who chooses the resources appropriate to him after taking a reflective activity on himself and on his learning process [14].

2.2 Personalized Educational Hypermedia Systems

In recent years, the evolution of the Web has encouraged the development of personalized educational hypermedia systems to support and facilitate learner-based teaching and learning in the exploitation of Adaptive E-Learning.

Adaptive educational hypermedia systems create a new model for the needs, characteristics and knowledge of learners, it identifies and adapts needs based on the interaction of the learner [7]. Adaptive learning environments adapt teaching approaches and techniques to create a personalized learning environments and meet different individual needs [15]. Adaptive learning environments also allow learners to identify their own learning methods, as well as personalized learning opportunities based on their needs [16].

However, most adaptive, personalized systems focus on the preferences, interests and behaviors of the learners, neglecting the importance of the learner's abilities. Indeed, according to Hussain in 2012, some researchers stressed that personalization must take into account the knowledge levels of the learner, especially with regard to learning where the capacities of the learners can be taken into account [17].

Through this section, we offer a chronological overview of the evolution and development of personalized educational hypermedia systems.

In 1982, Sleeman and Brown offer personalized e-learning systems developed in the domains of Intelligent Tutoring Systems (ITS) and Adaptive Hypermedia (AH). These two areas are different in the way they deliver their personalized content offered to different learning preferences and characteristics of learners. The objective is to adaptively deliver content to learners. However, these systems set limits for learners and restrict opportunities to support free exploration. AH provide the most relevant content and navigation paths by tailoring the content to the learning needs of learners [18].

In 2002, Conlan and his collaborators offer the Personalized Learning Service (PLS) which was developed to provide personalized educational courses based on the metadata-driven multi-model approach. It is used to provide personalized online services. Indeed, the architecture of the system is based on three models: the learner model, a narrative model and the content model, it uses three metadata repositories (learner, narration and content) and two information repositories (content and narration) The SAP contains the Adaptive Engine which uses the Java System Shell (JSS) wizard with custom functions as the basis of its rule engine. The role of the rule engines is to produce a personalized course model based on the storytelling and the learner model. The custom course model based on Extensible Markup Language (XML) that encapsulates the structure of the learner's course and produces content tailored to the learner's learning requirements [19].

In 2005, Chen et al., present their personalization of e-learning systems using item response theory which provides personalized learning based on the difficulty settings of course materials and learner responses [20]. Therefore, taking these abilities into consideration can promote learning performance by considering that the ability of the learner has an impact on personalization [21].

In 2008, Sangineto offers the DIOGENE platform who is an adaptive online learning platform that generates lessons by assembling learning materials using static knowledge. This knowledge includes didactic information on a specific area, which is explicitly represented using ontologies. Static information represents both the knowledge of the learners and the learning preferred by the learner. A learner is monitored during interactive testing activities in order to have their knowledge about specific areas [22].

In 2009, according to Hendler, many researchers are adopting the technologies and the Semantic Web to find new methods to design personalized adaptive e-Learning systems based on the description of knowledge using ontologies [23].

In 2013, Yarandi and colleagues propose a system that provides adaptation based on the learning styles of the learner according to the Felder-Solomon approach. This educational approach is a hybridization that merges the ITS and AH approaches. These are adaptive techniques that can be personalized according to the needs of the learner [24].

According to Lin et al., 2013, a web-based education system should offer an intelligent learning environment personalized training tailored to the needs of learners. Data Mining, and in particular with the technical decision has been suggested as an effective way to meet individual needs and improve the effectiveness of learning in an educational environment. Therefore, this study aims to develop a personalized creativity learning system in which the decision however technique is used to provide adaptive learning paths for learners [25].

In 2016, Roberts-mahoney and his collaborators think that personalized learning is formulated as a student interaction with digital screen interfaces and with the interface which is powered by predictive algorithms which learners work on scripted tasks that produce flows, algorithms feed into new tasks to the student based on the data produced through the learning process. Big Data and learning analytics make such user profiling possible in education [26] or analytical study reports indicate that learning needs to be personalized based on the profiles of learners [27].

The Personalized Learning Object Recommender System (PLORS) proposed in 2016 by Imran and other Canadian and Chinese researchers, PLORS provides Recommendations for learning objects in a course. The aim of the system is to enable the Learning Management System (LMS) to provide recommendations to learners, taking into account the learning object they are accessing as well as the learning objects visited by other learners with similar profiles. PLORS uses information on learners such as their learning style, behavior, preferences, by accessing the diary data tracked by the LMS, which includes which learning objects have been visited by each learner and how long they have been spent on each object learning. The general objective of PLORS is to provide learning object recommendations to the learner in a situation where the learner attends the different learning objects [28].

Huang et al. In 2016, a context-sensitive and personalized language learning system was developed for mobile devices, which is easy to use for adults at work. In the future, we believe that as the cost of adaptable/personalized systems for mobile devices is greatly reduced, working adults will be more likely to be selected as research subjects [29, 30].

In 2018, Zhou and his collaborators offer a new comprehensive and personalized learning recommendation model, this model is based on clustering and machine learning techniques it is based on a similarity metric feature on learners, they group together at the beginning a collection of learners to train a model of LSTM Long Short-Term Memory neural networks to predict suitable learning paths and performance for students. Personalized learning paths are then selected from the results of the prediction path, and finally, a complete adapted learning path is recommended specifically for testing the learner [31].

In 2019, according to Schmid and Petko, the term Personal Learning Environment (PLE), for example, was coined to describe an individual combination of digital tools. Some authors view PLE as collections of digital tools that support individual learning processes, while others describe PLE as platforms that aggregate and link information between tools or as environments in which the learner and not the teacher is in control. Empirical results regarding the control of learners in the context of educational technology, however, show inconsistent results and close to zero effects [32].

In 2019, Mwambe and Kamioka proposed an approach based on Brain Computer Interface (BCI), namely the e-learning system for assessing prior knowledge e-learning Prior Knowledge Assessment System (ePKAS) which can detect the history of learner's knowledge profiles and their match score with multimedia content. ePKAS algorithm relies on the biological information of learners acquired using a learner's brain wave sensor, the result has been used to recommend the knowledge profiles of the learners for the online Learning Management System (LMS) to support people with hearing loss [33].

In 2020 Mwambe is its collaborators continued to develop their work carried out in 2019, they proposed an adaptive navigation support based on bioinformatics, namely Adaptive Online prior knowledge assessment system ATdaptive e-learning Prior Knowledge Assessment System (AePKAS). The proposed approach allows AHS to manage the personalization of dynamic content in real time and reduce work overload by reducing the time imposed on navigation links through the integration of dynamic learning styles of learners "Learning Styles" with Learner Objects on a real-time using modification of the learner cognitive processes in e-learning platforms. In This study assumes that learners have prior knowledge in the areas of the subjects tested they know the language of instruction and have basic computer skills to work in an online learning environment. AePKAS delivers mediated learning content, this content is based on a slide size format, and therefore a very long content page is out of scope in this study [34].

According to Lin, 2020 and the other researchers offers a Big Data-based personalized learning approach for education and artificial intelligence to provide learners with intelligent and personalized learning, Big Data creates dynamic and diverse learning resources, by making a collect and analyze learner [35].

The PerLCol framework offered by Al Abri is its collaborators in 2020 is defined as a conceptual model containing the key components to generate personalized learning content with support for social collaboration tools. Adaptation parameters are derived from the interaction with learners with learning activities using collaborative tools. This model is classified according to the learning concept and the learning content, the personalization is done by the adaptation model. This model is associated with three related concepts to generate the personalization, these concepts are: domain model, user model and personalization package, the latter is associated with the learning session to deliver the related personalized learning [36].

In 2020 also we have proposed in our article a general approach to the personalization of the pedagogical scenarios during the different types of pedagogical activities taking into consideration the learning styles of the learners based on the Kolb learning style model. Our research work aimed to developing a learning system personalized and adaptive to the needs of learners and adequate with their preferences and profiles throughout the learning process proposed by the system by making a correspondence with the appropriate pedagogical scenario with each profile and each activity [6].

2.3 The Pedagogical Practices of Personalization

The personalization of learning activities is a fundamental issue in research on adaptive hypermedia systems, the pedagogical practices of personalization constitute the source of the implementation of personalized learning activities or more precisely the personalization of educational scenarios by based on different parameters [37]:

- **Information research** it is a parameter which aims to facilitate the search for information from an information database by using advanced methods for research such as clustering [38].
- **The level of knowledge** of the learner is used to take into account the preferences of the learner when communicating the learning material to the learner.
- **Learning objectives** are used to plan learning and in particular to determine the knowledge to be acquired and the skills to be developed by learners during a learning activity.
- **Educational approaches**
- **The learning style**
- **The level of participation in the educational activity**
- **Progress in the task and Feedbacks**
- **The level of motivation**

- **System interface**

According to Song, 2012 and colleagues Kolb's experiential learning model or experiential learning focuses on experience as the main driving force behind learning because learning is the process by which knowledge is created by the transformation of experience. This model requires that learning scenarios can integrate a series of objectives, activities and outcomes, in order to be integrated into the experiential instructional design [6, 10].

2.4 Learning Styles

In our article, in our approach we propose an adaptive e-Learning system which is an interactive system that personalizes the content, the teaching models, and the interactions between the participants in the environment to meet the individual needs and preferences of the learners when 'they arise.

Personalizing learning means allowing the learner to acquire knowledge and master skills based on the profile of the learner, this profile which will allow us to locate the preferences of the learner. Especially his learning styles to adapt his learning to him and to give an idea about the social dimension of the learner, his relationship with others, and his participation in the construction of this society [4].

User profile provide information about learner to personalize the content and the application defined according to the specific and individual needs of the learner. Content profiles represent descriptive information about search, recommendation, and content management resources. Usage profiles show learner behavior and relationship to content [39]. As well as their learning styles which mainly focus on the characteristics of the learner. When the learning situations are taken into account, which is not the most frequent case, they are mainly described from the learner's point of view, in terms of the content to be acquired and the presentation supports for this content [40].

Therefore several researchers have discussed learning styles and its impact on determining how the learner prefers to learn. In our research, we chose to work with the Kolb model [41], the DUNN and DUNN model [42] and the Felder Silvermen model [43].

The Experiential Learning Theory (ELT) proposed by Kolb is a dynamic learning vision based on the learning cycle, which is driven by the solution of the dual dialectics of action/reflection and experience/abstraction. Learning is defined as "the process of creating knowledge through the transformation of experience.

Knowledge comes from the combination of mastery and transformation experience [44]. Capturing experience refers to the process of acquiring information, while transforming experience is how an individual interprets the information and takes action on it. The ELT model describes two modes of capturing experience related to dialectics:

Experience (EC) and Abstract Conceptualization (AC)-as well as two dialectically related conversion experience modes-reflection observation (RO) and active experiment (AE). Solving the creative tension between these four learning modes is the result of learning [45].

These meanings can be actively tested and can be used as a guide for creating new experiences (Fig. 2).

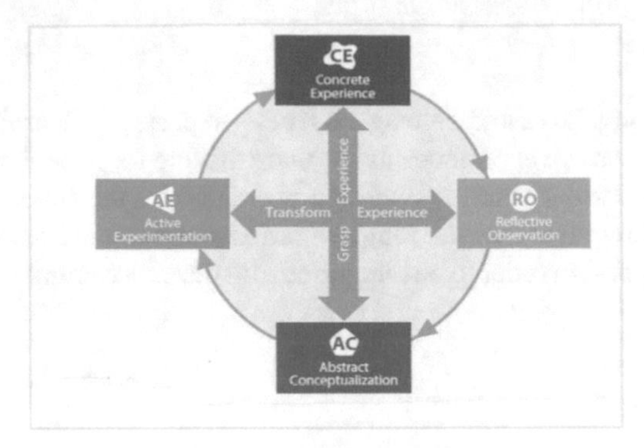

Fig. 2. The experiential learning cycle [45]

The Dunn and Dunn model developed by Dr Rita Dunn and Dr Kenneth Dunn of St John's University in New York. It consists of twenty to twenty one items, which depend on the appropriate age of the assessments administered [46] (Fig. 3).

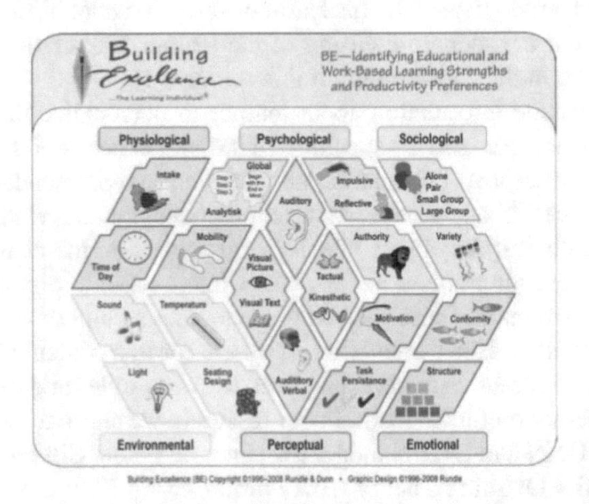

Fig. 3. Learning styles according to Dunn and Dunn model [46]

Felder and Silverman further describes a learner's learning style, is based on learners' preferences and how they receive and process information [43].

Felder and Silverman Learning Styles Model FSLSM has four dimensions with two opposing characteristics. Each learner is characterized by a specific preference for each of these dimensions (Fig. 4).

Indeed the Kolb model focuses on the theoretical and pedagogical aspects of the learning process, and the DUNN and DUNN model will be used to test the context of the learning process through the five dimensions: environment, emotional stimuli, sociological preferences, physiological preferences, psychological processing inclinations, then Felder Silvermen's learning style model which covers four dimensions Active/Reflective, Sensing/Intuitive, Visual-Verbal, Sequential/Global [43].

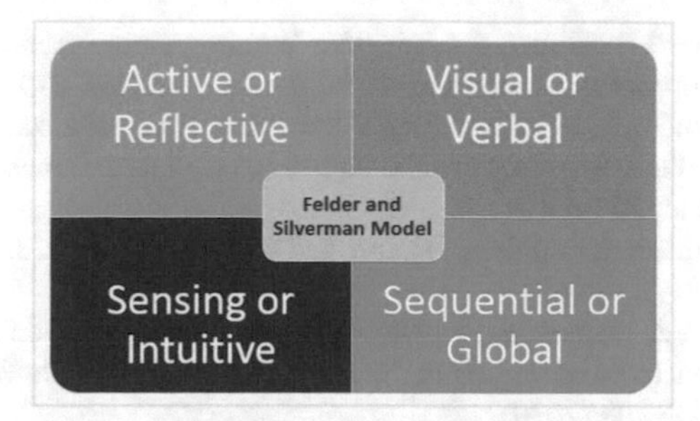

Fig. 4. Felder-Silverman learning style model [43]

The LSI (Learning Styles Instrument) Questionnaire of Kolb) [47], which is an 80-item Questionnaire to identify the learning styles of each learner.

The BE (Building Excellence Survey) [42] based on the Dunn and Dunn model helps identify learning styles according to age appropriate learner administered assessments.

The Index of Learning Styles (ILS), developed by Felder and Soloman, is a 44-item questionnaire to identify learning styles according to the Felder and Soloman, FSLSM, based on individual preferences learner for each dimension of the model [43].

Personalization is based on a set of processes of transformation of the person towards an involvement, progressive and optimized, in the course of his individualized learning journey, these processes are implemented by a so-called personalized pedagogy, focused on support of the learner, based on his motivation and his relations with the trainer, the tutor and the other learners [48, 49].

It is based on three characteristics the preparation of the learner, that is to say this knowledge, understanding and skills linked to a learning sequence. The goal is to improve the knowledge and know-how of the learner and his profile which refers to these learning styles, his knowledge, his preferences … [10]

3 Conclusion

This article is the subject of a presentation of the personalization between pedagogy and AHS in order to develop in our next work a compliant training device based on a pedagogical framework which consists in personalizing the different scenarios of learning according to each learner profile, then create the adaptive scenarios with the aim of developing a personalized and adaptive system in the results that we can publish in future articles.

References

1. Baron, G.-L., Bruillard, E.: Technologies de l'information et de la communication et indigènes numériques: quelle situation?. STICEF (Sciences et Technologies de l'Information et de la Communication pour l'Éducation et la Formation) **15**, 12 (2008)
2. Redecker, C.: Review of learning 2.0 practices: study on the impact of web 2.0 innovations of education and training in Europe (2009)

3. Paramythis, A., Loidl-Reisinger, S.: Adaptive learning environments and E-learning standards. In: Second European Conference on E-Learning, pp. 369–379 (2003)
4. Tadlaoui, M.A., Khaldi, M.: Concepts and interactions of personalization, collaboration, and adaptation in digital learning. In: Personalization and Collaboration in Adaptive E-Learning, pp. 1–33. IGI Global (2020)
5. Kim, W.: Personalization: definition, status, and challenges ahead. J. Object Technol. **1**(1), 29–40 (2002)
6. Lamya, A., Kawtar, Z., Mohamed, E., Mohamed, K.: Personalization of an educational scenario of a learning activity according to the learning styles model David Kolb. Global J. Eng. Tech. Adv. **5**(3), 099–108 (2020)
7. Brusilovsky, P.: Developing adaptive educational hypermedia systems: from design models to authoring tools. In: Authoring tools for advanced technology Learning Environments, pp. 377–409. Springer, Dordrecht (2003)
8. Weibelzahl, S., Weber, G.: Evaluating the inference mechanism of adaptive learning systems. In: International Conference on User Modeling. Springer, pp. 154–162. Berlin, Heidelberg (2003)
9. Bejaoui, R., et al.: Cadre d'analyse de la personnalisation de l'apprentissage dans les cours en ligne ouverts et massifs (CLOM). Revue STICEF (Sciences et Technologies de l'Information et de la Communication pour l'Éducation et la Formation) **24**(2) (2017)
10. Song, Y., Wong, L.-H., Looi, C.-K.: Fostering personalized learning in science inquiry supported by mobile technologies. Edu. Technol. Res. Dev. **60**(4), 679–701 (2012)
11. Walkington, C., Bernacki, M.L.: Appraising research on personalized learning: definitions, theoretical alignment, advancements, and future directions. J. Res. Technol. Educ. **52**(3), 235–252 (2020). https://doi.org/10.1080/15391523.2020.1747757
12. Lamya, A., Mohamed, E., Mohamed, K.: Adaptive E-learning and scenarization tools: the case of personalization. Int. J. Comput. Trends Technol. **69**(6), 28–35 (2021). https://doi.org/10.14445/22312803/IJCTT-V69I6P105
13. Connac, S.: La Personnalisation des Apprentissages. ESF (2017). https://doi.org/10.14375/NP.9782710137887
14. Lefèvre, M., et al.: Personnalisation de l'apprentissage: comparaison des besoins et approches à travers l'étude de quelques dispositifs. Sciences et Technologies de l'Information et de la Communication pour l'Éducation et la Formation **19**(1), 353–387 (2012). https://doi.org/10.3406/stice.2012.1050
15. Inan, F., Grant, M.: Individualized web-based instructional design. In: Instructional Design: Concepts, Methodologies, Tools and Applications, pp. 375–388. IGI Global (2011)
16. Aydoğdu, Y.Ö., Yalçin, N.: A web based system design for creating content in adaptive educational hypermedia and its usability. Malays. Online J. Educ. Technol. **8**(3), 1–24 (2020)
17. Hussain, F.: E-Learning 3.0 = E-Learning 2.0 + Web 3.0?. In: International Association for Development of the Information Society (2012)
18. Sleeman, D., Brown, J.S.: Intelligent Tutoring Systems. Academic Press, London (1982)
19. Conlan, O., et al.: Multi-model, metadata driven approach to adaptive hypermedia services for personalized elearning. In: International Conference on Adaptive Hypermedia and Adaptive Web-Based Systems, pp. 100–111. Springer, Berlin, Heidelberg (2002)
20. Chen, C.-M., Lee, H.-M., Chen, Y.-H.: Personalized e-learning system using item response theory. Comput. Educ. **44**(3), 237–255 (2005)
21. Chen, C.-M.: Intelligent web-based learning system with personalized learning path guidance. Comput. Educ. **51**(2), 787–814 (2008)
22. Sangineto, E.: An adaptive E-learning platform for personalized course generation. In: Architecture Solutions for E-Learning Systems, pp. 262–282. IGI Global (2008)
23. Hendler, J.: Web 3.0 Emerging. Computer **42**(1), 111–113 (2009)

24. Yarandi, M., Jahankhani, H., Tawil, A.: A personalized adaptive E-learning approach based on semantic web technology. Webology **10**(2), 110 (2013)
25. Lin, C.F., et al.: Data mining for providing a personalized learning path in creativity: an application of decision trees. Comput. Educ. **68**, 199–210 (2013)
26. Krippendorff, K.: Content Analysis: An Introduction to its Methodology. Sage Publications (2018)
27. Roberts-Mahoney, H., Means, A.J., Garrison, M.J.: Netflixing human capital development: personalized learning technology and the corporatization of K-12 education. J. Educ.tion Policy **31**(4), 405–420 (2016)
28. Imran, H., et al.: PLORS: a personalized learning object recommender system. Vietnam J. Comput. Sci. **3**(1), 3–13 (2016)
29. Huang, C.S.J., et al.: Effects of situated mobile learning approach on learning motivation and performance of EFL students. J. Educ. Technol. Soc. **19**(1), 263–276 (2016)
30. Xie, H., et al.: Trends and development in technology-enhanced adaptive/personalized learning: a systematic review of journal publications from 2007 to 2017. Comput. Educ. **140**, 103599 (2019)
31. Zhou, Y., et al.: Personalized learning full-path recommendation model based on LSTM neural networks. Inf. Sci. **444**, 135–152 (2018)
32. Schmid, R., Petko, D.: Does the use of educational technology in personalized learning environments correlate with self-reported digital skills and beliefs of secondary-school students? Comput. Educ. **136**, 75–86 (2019)
33. Mwambe, O., Kamioka, E.: Utilization of learners' metacognitive experiences to monitor learners' cognition states in e-learning platforms. Int. J. Inf. Educ. Technol **9**, 362–365 (2019)
34. Mwambe, O.O., Tan, P.X., Kamioka, E.: Bioinformatics-based adaptive system towards real-time dynamic e-learning content personalization. Educ. Sci. **10**(2), 42 (2020)
35. Lin, J., et al.: Personalized learning service based on big data for education. In: 2020 IEEE 2nd International Conference on Computer Science and Educational Informatization (CSEI), pp. 235–238. IEEE (2020)
36. Al Abri, A., et al.: PerLCol: a framework for personalized e-learning with social collaboration support. Int. J. Comput. Digit. Syst. **9**(03) (2020)
37. Essalmi, F., et al.: A fully personalization strategy of E-learning scenarios. Comput. Hum. Behav. **26**(4), 581–591 (2010)
38. Höök, K., et al.: A glass box approach to adaptive hypermedia. In: Adaptive Hypertext and Hypermedia, pp. 143–170. Springer, Dordrecht (1998).
39. Desai, D.: Modeling Personalized E-Learning for Effective Distance Education
40. Chartier, D.: Les styles d'apprentissage: entre flou conceptuel et intérêt pratique. Savoirs (2), 7–28 (2003)
41. Kolb, A.Y., Kolb, D.A.: Learning styles and learning spaces: enhancing experiential learning in higher education. Acad. Manag. Learn. Educ. **4**(2), 193–212 (2005)
42. Dunn, R., et al.: Impact of learningstyle instructional strategies on students' achievement and attitudes: perceptions of educators in diverseinstitutions. Clearing House **82**(3), 135–140 (2008)
43. Graf, S., et al.: In-depth analysis of the Felder-Silverman learning style dimensions. J. Res. Technol. Educ. **40**(1), 79–93 (2007)
44. Kolb, D.A.: Experience as the Source of Learning and Development. Prentice Hall, Upper Sadle River (1984)
45. Kolb, A.Y.: The Kolb Learning Style Inventory-Version 3.1 2005 Technical Specifications, vol. 200, no 72, pp. 166–171. Hay Resource Direct, Boston, MA (2005)
46. Dunn, R., Honigsfeld, A.: Learning styles: what we know and what we need. In: The Educational Forum, pp. 225–232. Taylor and Francis Group (2013)

47. Wintergerst, A.C., DeCapua, A., Itzen, R.C.: The construct validity of one learning styles instrument. System **29**(3), 385–403 (2001). https://doi.org/10.1016/S0346-251X(01)00027-6
48. Vanderspelden, J.: APP: individualiser n'est pas personnaliser ou apprendre à s' autoformer. Actualité de la Formation Permanente (194), 122–129 (2005)
49. Tomlinson, C.A.: Reconcilable differences: Standards-based teaching and differentiation. Educ. Leadersh. **58**(1), 6–13 (2000)

Collaboration in Adaptive E Learning

Zargane Kawtar(✉), Khaldi Mohamed, and Erradi Mohamed

Laboratory of Computer Science and University Pedagogical Engineering (S2ipu), ENS-TETOUAN, Abdelmalek Essaadi University, Tetouan, Morocco
kzargane@uae.ac.ma

Abstract. This article focuses on adaptive e learning between collaboration, which offers a combination of formalizing and well structured for e learning or distance learning, virtual learning environments and adaptive learning systems correspond to distance learning solutions that strive to meet the promise of individualized learning. The collaborative approach combines two approaches: that of the learner and that of the group. Collaborative learning is an active process by which the learner works to build knowledge. The trainer acts as a facilitator of learning while the group participates as a source of information, as a motivating agent, as a means of mutual assistance and support, and as a privileged place of interaction for the collective construction of knowledge. In the collaborative approach, learners collaborate in Group learning and, in turn, the group collaborates with learners.

Keywords: E-learning · Collaboration · Adaptive E-learning · Educational style

1 Introduction

The adaptive system [1] provides a platform that adapts to the learner's behavior. The research question involves the representation of educational content, but the field of pedagogical engineering [1] allows for particular definitions and modifications for specific reorganizations other than equipment using pedagogical technologies, methods, or tools aimed at pedagogical objectives. It will gain knowledge with the audience of learners.

Different communication tools (chat, email, forum, video conferencing, etc.) can help teachers develop collaborative spaces and enable information sharing to use the knowledge and provide learners with new skills. Based on the method of each learner and the ability of the organized world, teaching methods that support the development of learners' knowledge [2].

Learning autonomy is also essential. Belonging it started as a group of people and then turned into a real online community. At the end in the course, learners decide to extend the experience by creating their own experience they invited us to participate in social networks to discuss internships, tools, experience and coaching.

Psychology teacher BF Skinner (BF Skinner) created the learning machine in the mid-1920s, there was already the idea of teaching mechanization. Possibility of feedback based on their answers. It is of course limited, but behaviorist believes this is the future.

To focus on the role and use of learners and teachers in the use of new knowledge tools, we can focus on reflecting our beliefs and values in education.

M. Lazaar et al. (Eds.): BDIoT 2021, LNNS 489, pp. 235–244, 2022.
https://doi.org/10.1007/978-3-031-07969-6_18

2 Collaborative Learning

Collaborative learning [3] is based on a social constructivist approach, which considers knowledge to be a personal psychological construction produced by personal experience gained through interaction with the environment. According to social construction, knowledge corresponds to subjectivity, it is realistic and its value falls within the acceptable range of society.

For constructivists, we need to embrace in such a social environment: sharing, confrontation, and negotiation help learners build knowledge and reach consensus on reality while respecting individual differences. In other words, use the results of a single job through group activities or online courses (Fig. 1).

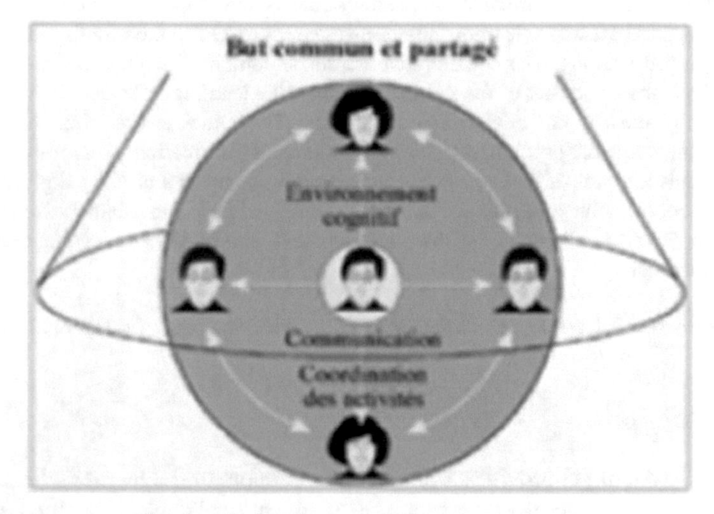

Fig. 1. The E-learning collaboration approach

Collaborative learning is about knowledge sharing. Sharing can take place in face-to-face meetings. However, remote collaboration provides a good learning environment. By providing a collaborative space for the team, everyone can collaborate in real-time or asynchronously, confront each other at any time, negotiate and acquire new knowledge, and ensure diverse communication and coordination for each team member.

Moreover, support tools. In their book on collaborative learning, Henri and Lundgren (2001) [4] any learning activity carried out by a group of learners with a common objective, each being a source of information, motivation, interaction, mutual assistance… [4]. Each of the contributions of the other beneficiaries of the group system as well as a well-presented individual and collective learning tutor's guide.

In addition, to that we can be found in this book [4] it includes the following guiding principles theory of cognitive flexibility, which provides several representations of the same object, and are expressing various ways of thinking to facilitate the acquisition and dissemination of complex knowledge. Therefore, the scope of collaborative online learning is very broad topics and skills, but some concepts or skills may not be suitable for learning.

Table 1. The Electronic forum [4].

Forum electronic	... for the learner
What is it what is it?	- It is a virtual place where the speech of a group, where the learner appropriates new knowledge by conversing with others. - It is an agora of collaboration and socialization.
What does it allow?	- When is it suitable to use it? One can expose ideas, develop one's thinking, build new knowledge, validate and confirm With the support of the group. - It facilitates the learning of complex knowledge belonging to poorly or poorly structured domains. - It promotes a reflective attitude about learning. - It allows a group to live an experience motivating learning and developing a sociocognitive commitment that gives even more learning that is meaning.
Who is participating?	- Human agents who create a social network: The Guardian - Alternatively, the trainer and other learners who work in the within small groups (spontaneous, informal collaboration) or within a large group (supervised collaboration, formal). - Machine agents that make up the environment technology: tools for collaboration, coordination and support cognitive and social processes.
How do we participate?	- Human agents transmit messages to the group; interaction is the semantics of these messages. - Machine agents, grafted to the forum, offer automated support for learning by providing, for example, multiple representations of what is said in messages; this support is immediate, fast, interactive.
When is it suitable to use it?	- When learning involves skills of high level such as analysis, synthesis and evaluation, the forum is an appropriate educational choice. It is the designed as a resolution environment for problems, decision-making. It testifies to the intentionality of the learning situation. - In a formal learning framework, when the learners feel they need others to learn, when a group of 15 to 30 learners want to learn together with the support of a trainer and when they are willing to agree to reach a shared goal. - In an informal setting, when a sufficient number of people want to form a group and work Together to learn.

The navigation between forums reserved for forums is generally transparent [4]. In many systems, one can automatically switch from one forum to another, which encourages members to pay attention to all communications the subject of the group's attention.

Table 1 represents the components, message boards and forums are collaborative tools each of which contributes to the group work, the forum comes from the fact that it can compared to other tools, meet a variety of needs in the collaborative group, as it provides an unprecedented human communication.

3 Adaptive Learning

Adaptive learning [5] raises several questions in the field of online learning in the use of methods, and technologies, and how to provide learners with adaptive content through personalization of personal data, and skills system.

The process of using online technology tools to develop educational content will be divided into several stages: the trainer will create an educational space or environment and manage the process to be followed to obtain and use the process successfully. Information is given to learners, which leads to the personalization of personal files and educational content.

Besides, through teaching and the use of technical tools to develop interactive courses with learners, and adapt to their educational needs, and easy to manage and understand for learners. To respect the learning process of the learner, the teacher will have to propose learning activities. There are rules for establishing learning activities. They are derived from the way the human brain learns. Indeed, we need to build a hierarchy of activities for learners based on what we want to learn [6].

Guild 1997 compared several intelligences, learning styles and key principles of cognitive education, and concluded that these methods overlap a lot in learner-centered methods and emphasize design, deep and extensive learning plans.

The learning style [7] corresponds to the characteristics of a group of learners in multiple dimensions, and each factor can cause individual differences in the learning situation. Each of these elements has its way and at the same time forms a functional whole with other elements. Kolb was the first to adopt an experiential learning style model, and then he influenced the construction of other models. In 1984, Kolb highlighted the principles of learning based on Discovery and experience in his book "Experiential Learning" [8].

Kolb's four learning methods depending on the stage of the learning cycle he prefers, Kolb gives different types of learner names:

Disagreement: He likes the stages of specific experience and reflection on experience. Disagreements are interested in people and emotions. He has he is good at observation and perceiving objects or problems from different angles. He enjoys innovative activities he has a rich imagination and various interests. He is interested in people and values. He likes to learn by doing.

Assimilator: He prefers the stage of reflection on experience and the stage of abs-traction and conceptualization of experience. Assimilators like to create theoretical models and are less interested in the practical application of people and knowledge than others are. It reorganizes information logically, taking into account both thought and theory. He enjoys the theory course.

Convergent: He prefers the abstraction of experience and the conceptualization of theory and the implementation of ideas/actions. Convergent people like to be pragmatic and often have no emotions. Moreover, they prefer to deal with people rather than with people. He likes to solve problems, so the solution is unique. He has abilities in technical tasks and decision-making. He appreciates self-managed projects and activities.

Moderator: He prefers the steps of the specific experience and the implementation of ideas/actions based on this experience. Residents adapt easily to new experiences and tend to find solutions. He learns by manipulating and performing tasks. He likes to participate in the planning and execution of activities, and his work is more trial and error than logical. He tends to rely on other people's thinking rather than his own analysis, and he is willing to take risks. He likes group practice.

Learning style is one way of explaining reasonable dilemmas in the classroom: it is a fact that in the context of a given teaching style, plan or effort, some students learn

Table 2. Learning preferences based on sensory learning styles [13]

Sensory learning style	Learning preferences
Auditive learner	• Hears aloud ; • Listens carefully ; • Discusses what he learns; • Plays vocabulary games ; • Solves and makes puzzles; • Works in collaboration ; • Screams and recites by singing ; • Shares with others what he learns to clarify the lessons etc.
Visual learner	• Memorize by using visual indexes and drawing images in the his head; • Designs diagrams and plots graphs; • Clarifies the ideas through drawings; • Revises by watching videos; • Participates in school activities to live scientific experiences; • Observes how others do • Makes games and puzzles, etc.
Kinesthetic learner	• Seeks to move; • Prefer walking while talking • Writes or draws while listening ; • Elaborates models ; • Marks and highlights the essentials ; • Manipulates material ; • Lives experiences ; • Imagine himself in situations; • Makes interpretations for stories and concepts, etc. [13].

while others do not. Given the key role played by cognitive strengths and weaknesses, learning styles can be explained and understood [9].

The following table expresses that no one can say that one style is better than another is, as most people combine these three styles of sensory learning and prefer one of them [13] (Table 2).

The adaptive learning system resides in an adaptation engine that generates recommendations related topics or concepts to learning material [10].

There are two modules in the adaptive engine "the referent is adaptive engines in adaptive learning suggestions Architecture", as shown in Fig. 2 the difference between this architecture and the convention existence, the adaptive navigation engine can determine the subject or concept that the learner will learn.

The recommendation system for teaching materials, once the adaptive navigation engine recommends a concept, the recommender begins by choosing teaching materials related to the concept and learners.

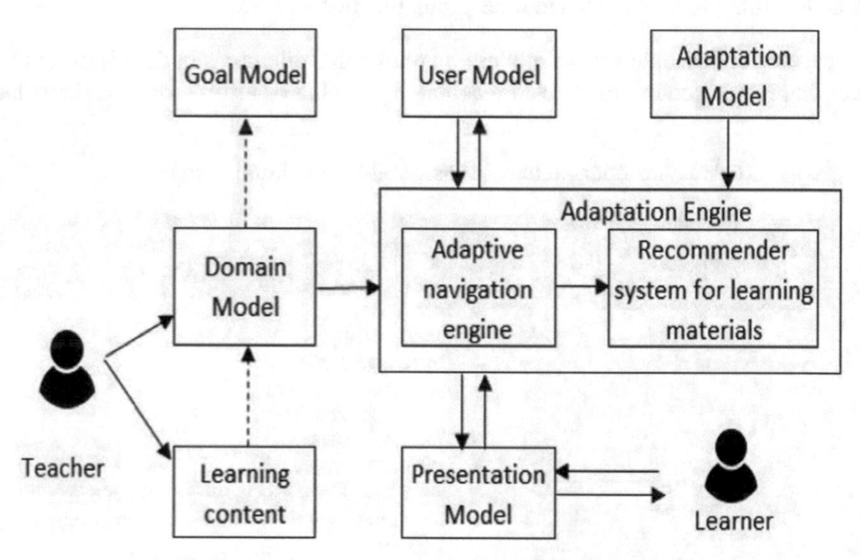

Fig. 2. The architecture of an adaptive learning system [10].

4 Structural Pedagogy

Cognitive psychological theories [11] applied to teaching methods particularly promote the development of teaching methods, prescribe the cognitive path of learners and decide to destroy requirements to create an ideal learning situation.

Therefore, given the prerequisite knowledge, cognitive conflicts, the acquisition of formalism and problem solving are some of the ways and methods that guide learners to insert new knowledge rooted in a cognitive structure and well integrated. The trainer plays a key and irreplaceable role (Bertrand, 1990). No other speaker, not the system can replace the program.

The learners who learn their structure, but the trainer who sets up the learning environment essential to the development of knowledge.

4.1 Educational Strategies

Here the teacher chooses teaching strategies. He wants to know the relevance of different strategies, or teaching strategy [12] strategies for the whole SA, such as direct business simulation, executing projects, solving problems, case studies, virtual meetings, etc. choosing a single educational strategy for the whole SA can be particularly beneficial for collaboration.

Informed learning offers a more inclusive pedagogy because it can understand intelligence in the broad sense and allows students to learn from their own intellectual advantages without being marginalized by traditional learning methods (Barrington 2004) [13].

4.2 Educational Style

The combination of learning style and teaching style [14] can help us improve the teaching process and increase the effectiveness of teachers and learners (Fig. 3).

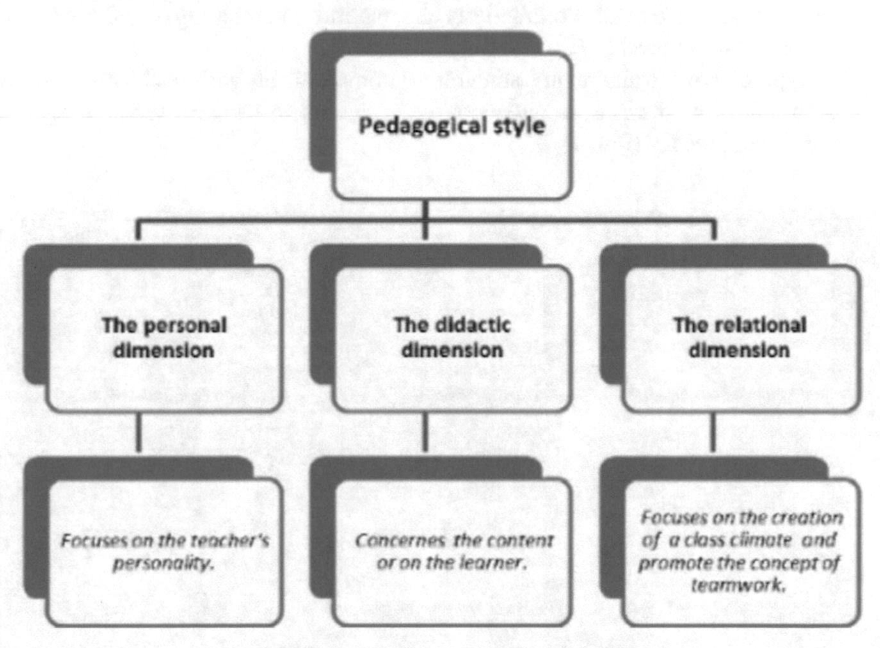

Fig. 3. Educational style dimensions

This figure represents the teaching style is the teaching methodology in the E-learning environment, very simple, it is "teaching method" [14].

5 Collaborative Work in Adaptive Learning

Collaborative learning [15] is an active stage in which learners strive to develop their knowledge. Therefore, at the level of the adaptive system, a spacious and comfortable space must be provided to learn and demonstrate this knowledge. However, the classical representation of basic data and courses is usually found without communication or exchange of information.

In collaborative work [16], there are participants in the group without any division of Labor, and the students will work in this way: build everything in the group and perform each step together, which will create a very coherent and coherent environment. Establish and strengthen group self-confidence among learners. Indeed, collaborative work is based on communication and information sharing, and each step consists of assessing the level of each position. The collaborative work model is a new development in the field of E-learning.

Collaborative work [17] is based on communication and information sharing at each stage, and evaluation of work at each stage. The collaborative work model is a novelty in the field of E-learning. This type of work trains many students to communicate via platforms, forums or videoconferences, and to use new technologies and teaching resources. Therefore, the result of the work is that you can well organize and organize the work and cooperate with stakeholders to perform and complete the same task at the same time, which is also a task work/activity and maximize the sharing of tasks/activities. Information at Group level [17].

This type of work trains many students to communicate with each other through discussion platforms, forums, or videoconferences, and to use new technologies and educational resources… (Fig. 4)

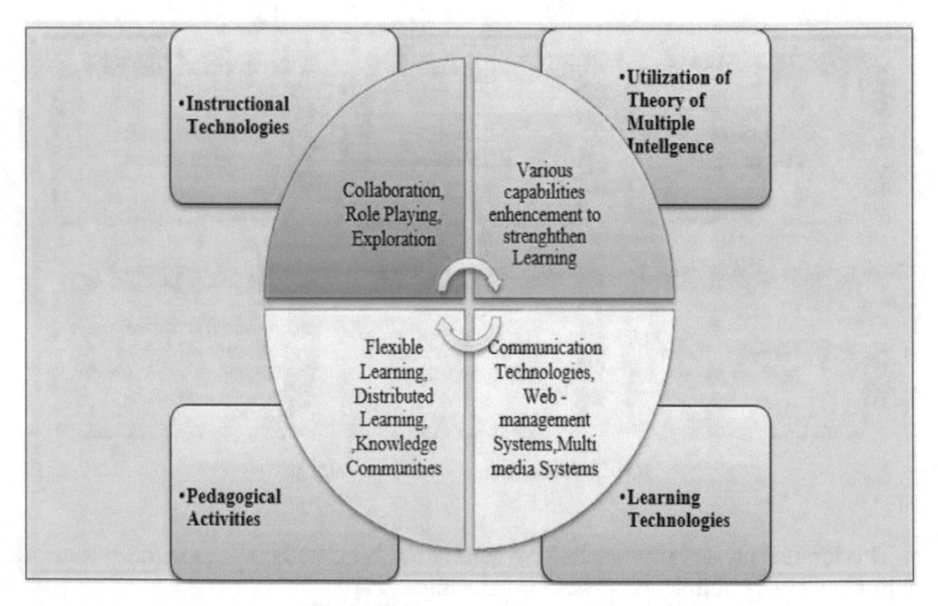

Fig. 4. Collaboration in E-learning

This figure represents a scientific, pedagogical and technological approach, with an emphasis on the role of collaboration in E-learning and adaptation to new technologies [18].

6 Conclusions

The objective of this article is to address collaborative work at the level of adaptive online apprenticeships. Therefore, despite the advanced technological development on a global scale, the role of this module in distance learning. Consider that the educational process is the main issue in the design and development of a new adaptive education system. It is about considering educational activities as educational content. Through formalization and management processes, teaching strategies and methods can be customized.

In short, with the help of the Internet can improve the quality of Social learning, provided that learners and teachers are willing to take on different and even innovative roles: learners must take responsibility for their own study, the teacher should no longer guide the study strictly speaking, but it must be able to guide, encourage and easy.

Learning is a cognitive activity, and one student to another. Analysis of the adaptability of E-learning the system clearly emphasizes modeling the cognitive characteristics of the learner, especially the learning style is the most explored cognitive characteristic. In general, taking into account personal characteristics, allow to improve the abilities of learners within the collaborative group.

References

1. Aubert, S.: Le E-learning adaptatif. Rapport de stage MASTER (2005)
2. Paquette, G.: L'ingénierie Pédagogique. Puq (2002)
3. Tadlaoui, M.A., Khaldi, M.: Concepts and interactions of personalization, collaboration, and adaptation in digital learning. In: Walckiers, M., De Praetere (eds.) Personalization and Collaboration in Adaptive E-Learning, pp. 1–33. IGI Global (2020)
4. Thomas: L'apprentissage collaboratif en ligne, huit avantages qui en font un must. Distances et Savoirs **2**(1), 53–75 (2004)
5. Henri, F. Lundgren-Cayrol, K.: Apprentissage Collaboratif à Distance. Puq (2001)
6. Maha, K., Mohammed, E., Mohamed, K.: Learning situation: the teacher management and decisions according to the context and the situation. IMPACT: Int. J. Res. Eng. Technol. **7**(5), 25–40 (May 2019). ISSN(P): 2347-4599; ISSN(E): 2321-8843 (2019)
7. Lamya, A., Kawtar, Z., Mohamed, E., Mohamed, K.: Personalization of an educational scenario of a learning activity according to the learning styles model David Kolb. Global J. Eng. Technol. Adv. **5**(3), 099–108 (2020). https://doi.org/10.30574/gjeta.2020.5.3.0114
8. Fernandez-Manjon, B., Sancho, P.: Creating cost-effective adaptative educational hypermedia based on markup technologies and E-learning standards. Interact. Educ. Multimedia (4), 1–11 (2002)
9. Kolb, D.A.: Experiential Learning: Experience as the Source of Learning and Development. PrenticeHall Inc., New Jersey (1984)
10. LearningRx Corp: Types of Learning Styles. [Online]. Available: http://www.learningrx.com/types-of-learning-styles-faq.htm#sthash.W0cCt8Ql.dpuf. Accessed 26 Jun 2015
11. https://www.researchgate.net/publication/304020637_Good_and_Similar_Learners%27_Recommendation_in_Adaptive_Learning_Systems

12. Henri, F., Peraya, D., Charlier, B.: La recherche sur les forums de discussion en milieu éducatif: critères de qualité et qualité des pratiques. In: Revue STICEF, vol. 14, pp. 155–192. https://archive-ouverte.unige.ch/unige:17650 (2007)
13. Henri, F., Basque, J.: Conception d'activités d'apprentissage collabo-ratif en mode virtuel. In: Dans Deaudelin, C., Nault, T. (eds.), Collaborer pour apprendre et faire apprendre: La place des outils technologiques, pp. 29–53. Presses de l'Université du Québec, coll. « Éducation - Recherche », Sainte-Foy, Canada (2003). ISBN 978-2-7605-1228-3
14. Arrington, E.: Teaching to student diversity in higher education: how multiple intelligences theory can help. Teach. High. Educ. **9** (2004)
15. El Emrani, S., El Merzouqi, A., Khaldi, M.: The MOOCs in face of pedagogical constraints. Challenge **4**(5) (2015)
16. Larivée, S.J., Kalubi, J.-C., Terrisse, B.: La collaboration école-famille en contexte d'inclusion: entre obstacles, risques et facteurs de réussite. Rev. Des Sci. De L'édu. **32**(3), 525–543 (2006)
17. Kawtar, Z., Lamya, A., Mohamed, E., Mohamed, K.: CDESACL conception and development of educational scenarios for an adaptive online training device based on collaborative/cooperative learning: work methodology, Rajar **7**(01), 2814–2819 (Jan 2021)
18. Mankad, K.B.: The role of multiple intelligence in E-learning. Int. J. Sci. Res. Dev. **3**(5), 1076–1081 (2015)

A Review of the State of Higher Education in MOROCCO at the Time of Covid-19

Kaouni Mouna(✉), Fatima Lakrami, and Ouidad Labouidya

STIC Laboratory, Department of Physics, Faculty of Science, Chouaib Doukkali University, El Jadida, Morocco
mouna_kaouni@outlook.fr, Labouidya.o@ucd.ac.ma

Abstract. This paper describes the current situation of the use of E-learning in Moroccan higher education at the time of COVID-19. We present a comparative study of the most well-known E-learning platforms widely used in universities and higher education for E-learning while focusing on analyzing the solutions adopted by Moroccan establishments during the current crisis. The objective is to highlight the limitations and describe the challenges facing the development of an appropriate E-learning system in Morocco.

Keywords: Online learning · Covid-19 · Distance learning · Moroccan university · E-learning platforms · NICT

1 Introduction

In 2020, the coronavirus pandemic (Covid-19) ravages the world, and it was first reported in Morocco on March 2, 2020. Since the detection of the first infected case, the Ministry of National Education, Vocational Training, Higher Education and Scientific Research has taken critical preventive measures to limit the spread of the Covid-19 pandemic and ensure educational continuity. It has been decided to suspend face-to-face courses in all schools and universities in the public and private sectors and for an undetermined duration. Following this decision, all face-to-face courses have been replaced by distance learning.

Under this crisis, schools and universities were brought to prepare their systems for a global digital transformation. In this regard, some universities have designed their distance learning platforms such as the one launched by Mohammed VI Polytechnic University (UM6P) for its students and Moroccan engineering schools [1]. Other universities have used interactive tools for distance teaching, namely Zoom, Microsoft Teams, Google Classroom, etc.

Under the current circumstances, the concept of "virtual university" has become more pronounced. However, the critical situation caused by the pandemic brings to light many problems. On one hand, this crisis has identified several student constraints related to the lack of a variety of resources such as internet access, availability of computers, tablets, or smartphones, etc. On the second hand, several issues related to trainers' expertise in distance learning were revealed, in terms of "digital" pedagogy and NICT tools.

M. Lazaar et al. (Eds.): BDIoT 2021, LNNS 489, pp. 245–260, 2022.
https://doi.org/10.1007/978-3-031-07969-6_19

The transition to digital in Morocco is not recent, it was initiated by several programs launched by the state in concert with public educational institutions. In 2008, the Ministry of National Education (MEN) in MOROCCO launched a program for the generalization of information and communication technologies in education [2]. This program aims to train over 230,000 people (teachers, inspectors, technicians, school heads, etc.) in the field of ICTE. A project under the name of "INJAZ" was launched in 2013, which aims to distribute laptops and desktop computers at a favorable price as well as a free internet pass available for a year. In 2015, the Minister of Higher Education, Lahcen Daoudi, officially launched the "Lawhati" project. This project is destined for students enrolled in higher education and vocational training establishments. The main goal of "Lawhati" project is to strengthen knowledge sharing and extend the use of ICT in higher education in Morocco, through the distribution of tablets at preferential prices with the integration of content provided by the technological partners of the program. All the cited projects were destined to generalize and encourage the use of ICT in the education field. And thanks to such initiatives, MOROCCO was able to have a solid base to proceed to distance education since the lockdown of its educational components due to the spread of COVID-19.

This paper analyses the deployment of distance learning in Morocco at the time of COVID-19. It gives a review of different platforms used by Moroccan establishments to ensure the continuity of face-to-face learning. The proposed study aims to present a comparative review of the features and limitations of the most relevant online-learning systems and is organized as follows: the first section describes what the literature recalls of distance education, namely the historical background of E-learning and it's different types in addition to the evolution of distance learning platforms. The second section presents the current situation of the use of ICT in Moroccan higher education while the third section presents a comparative study of platforms widely used in distance education as well as challenges in the development of an E-learning system, the last section concludes the paper.

2 Review of Distance Education

2.1 History

In 1840 and before the advent of the Internet, the British inventor Isaac Pitman of the shorthand system created the first correspondence course for distance education in London [3].

In 1960, the electrical engineer and the American computer scientist Donald Bitzer introduced the first training program on the computer called PLATO (Programmed Logic for Automated Teaching Operations), this aims at the diffusion of courses on computers, it was created initially for students at the University of Illinois, but it was then used by all schools of the region [4].

Before the 1970s, distance learning was designed to provide information to students. In contrast, in the early 1970s, online learning started to become more and more interactive. The concept of E-Learning or distance learning started effectively in 1999 during a seminar on online training systems. The rise of this very concept began in Anglo-Saxon countries and spread quickly in France.

In 2003, and because of the explosion in the numbers of digital documents as specified in [5] "the number of digital documents for education is inexorably increasing", new requirements for providing information have evolved. These requirements have given rise to a new form of education, which necessitates the implementation of new technologies of systems and services to collect process, and store relevant information.

Today, thanks to the spread of the internet and the introduction of new information and communication technologies (NICT), as well as the decrease in the costs of computer software. E-Learning has become an essential element of professional training, whether for company employees or students.

2.2 Types of Distance Learning

Distance learning requires the use of a set of technologies to design training that is adaptable to the diverse nature of participants' needs and with relevant content. In alignment with this context, several types of e-learning platforms have been designed, there are:

LMS (Learning Management System): a learning management system used to provide and manage distance learning through resources downloaded into platforms. It allows the planning, monitoring, surveillance, and management of participants' journeys through test results, skill levels, time spent on training, etc.

MOOC (Massive Open Online Course): The MOOC or open and massive online courses represent an online and open access model of education, which consists of creating and diffusing educational courses to the general public, these courses are supervised by educational teams. The MOOC is enriched with exercises, tests, and certificates of achievement at the end of each session.

There are several types of MOOCs: aMOOC or Adaptive MOOC, SPOC (Small Private Online Course, COOC (Corporate Open Online Course), etc. [4, 6–8].

2.3 Evolution of Distance Learning Platforms

The revolution in New Information and Communication Technologies (NICT) at an accelerated rate has revolutionized the culture of the society in several sectors, notably the education sector. Thanks to the development of NICTs and scientific research, several learning solutions have emerged, as well as many E-learning platforms have been created, more than 1,000 solutions on the e-Learning market, allowing information sharing, content creation, and diffusion, resource management, a collaboration between users are now available. Indeed, "this is the era of e-learning" [9].

The emergence of new technological concepts such as Artificial Intelligence, big data, data mining, machine learning, have made it possible to move from classic platforms to dynamic and adaptive platforms, whose content and navigation are personalized according to learner's preferences. Adaptive training systems focus on learners, in other words, their way of learning, their preferences, and their unique characteristics that distinguish one learner from another. This creates a unique and personalized learning path.

One of the previous studies for adaptive e-learning systems was conducted by [10] where the authors have suggested an approach based on generic rules, which are produced automatically. This, to offer content and a user interface adapted to the learning styles of each learner.

Another approach has been presented by [11], in order to have a unique, personalized, and optimal learning path for each learner. This approach consists of the use of ontologies with artificial agents while being based on the ACO (Ant Colony Optimization) metaheuristic.

Based on probabilistic models, [12] presented an adaptive learning system "ALS_CORR [LP]" based on the Bayesian network as well as the learning styles defined by Felder and Silverman.

In contrast to adaptive learning, today, in the era of speed, we talk about platforms based on the Micro-learning approach. This learning approach consists of transmitting information on a single unique idea in a targeted way [13], where the course sessions are short (not even 5 min).

The e-learning platforms currently being designed are based on complex systems that rely on the technological aspect and are mainly based on the learners' needs in terms of content; the pedagogical side is partially taken into account while the adaptive side is completely absent.

3 Current Situation of the Use of ICT in Moroccan Higher Education

3.1 Analysis of the Use of ICT in Moroccan Higher Education

International reports on the problem of higher education in MOROCCO remain systematically alarmist, regarding the recent international rankings of Moroccan universities. Over the past two decades, MOROCCO has undertaken a series of reforms to its education system. From a reformed perspective, the Moroccan University has adopted, for eight years, the "License-Master-Doctorate" (LMD) system. This system had the objective of standardizing training at a national level and promoting student mobility outside the country as well as allowing it to align with international standards, notably European. However, it has shown several limits and inadequacies, especially relating to administrative and educational constraints, which manifest themselves in the form of two series of major blockages: civil and human infrastructure.

The problem of opening up MOROCCO to European renovations is mainly due to the major handicap caused by these two blockages. Such a decision must be accompanied by the necessary financial and human resources, rather than being rushed. The situation is very worrying in public Moroccan establishments. The deterioration of the university education system and that of the skills of university graduates, not to mention the job market, which only receives a minimum percentage of these graduates, has pushed the Moroccan state to reflect in the future of higher education and specifically public institutions. To solve the multitude of problems encountered at the level of the Moroccan university, the Ministry of Education and Higher Education decided to switch back to the bachelor system starting from the 2020/2021 academic year.

In this new model, the emphasis will be on languages and soft skills during the first two years, in parallel with basic learning. This system is also distinguished by an approach based on the integration of personal work and para-university activities. With the introduction of this new approach, students will be able to decide, regardless of the years of the cycle, the hourly volume of their validation of all credits. This is capped in four years, which represents a gain of half a year. This approach aims to encourage the involvement of market jobs in university training and therefore to promote the integration of graduates in companies, such as schools and specialized establishments.

In another rescue attempt, the various universities in MOROCCO recently decided to take their place in the digital space, by increasingly introducing ICTs into teaching pedagogy. This vision of establishment is not limited to the use of IT tools and the availability of practitioners in the field of new information and communication technologies but straightforwardly allows the university to create its own identity by making digitalization a methodological framework for these pedagogies and a new orientation to face several problems.

MOOC/SPOC is a new form of digital learning; it is a complete educational program, which is followed according to a predefined schedule. The invasion of MOOC, as a distance-learning platform and concrete form of ICT, is becoming more pronounced. Their arrival has turned the structure of traditional, launched in 2017 through a call for tenders, which concerned all universities, requested innovative SPOC/MOOC proposals with varied contents, and intended for Moroccan graduate students. It aimed to allow Moroccan universities to overcome certain problems, through a modern pedagogy that offers students the opportunity to acquire new skills and better prepare for their integration into education upside down by proposing new forms of integration.

In recent years, Moroccan universities have started to follow this trend by offering training projects based on MOOC/SPOC. Decisions about their integration into traditional education systems are the subject of much debate. MOROCCO launched in 2017 The national platform « MOROCCO Digital University» (MUN), dedicated to open and massive online courses, The MUN platform offers content in various fields: education and training, science engineer, computer science, economics and finance, fundamental sciences, health, languages, management and entrepreneurship, environment, humanities, and law. This national project aims to help contain massification within Moroccan universities. The MUN project the job market.

At the time of the Covid-19 crisis, the Minister of National Education, Professional Training, Higher Education and Scientific Research mobilized to ensure educational continuity through the launch of several online educational platforms (TelmidTICE, Taalimtice). For students who do not have an Internet connection or computers, the Minister launched the courses via channels such as Athaqafiya, Al Amazighiya and Arriyadia. The minister said on Tuesday, May 12 that nearly 6,000 digital content has been made available to students for all grades and 600,000 students use the distance-learning platform daily. The minister also mobilized to launch content related to children with disabilities. Besides, the Men launched on March 23, through the Taalim.ma platform, the integration of Microsoft's "Teams" collaborative service into the Massar school management system. This is to ensure continuity of lessons via virtual classes and to allow learners to communicate directly with their teachers. Regarding higher education, each university has its electronic platform such as Moodle and Google Classroom in addition to communication and interaction tools. The solutions adopted for EAD at the time of COVID-19 for the 12 Moroccan universities.

To adapt to the current situation, the universities of the Kingdom have opted for the digitization of their education system. The University Mohamed V Rabat filled its electronic sites with about 6,000-course materials for all levels, this, in addition to the use of interactive platforms as a communication tool, namely: Microsoft Teams, ZOOM, Google Classroom, and Moodle (UM5Moodle).

Chouaib Doukkali University also took advantage of a partnership between the ministry and Microsoft to use the Microsoft Team platform. Like other universities, Ibn Tofail University has increasingly put courses on their sites, as well as links to educational spaces and free libraries, it also uses the Moodle platform and Google Classroom. Teachers at this university have a recording studio for courses in the form of MOOCs (Massives open online courses). Similarly, the Hassan II University of Casablanca has strengthened its online learning system by filling its Moodle platform with courses as well as providing free access to large digital libraries. To ensure educational continuity at the time of Covid-19, Hassan 1er University in Settat has deployed the Moodle platform. Similarly, Ibn Zohr University launched on April 21, 2020, a MOOC "Pedagogy and E-learning". This MOOC is intended for teachers for different levels of study and to ensure continuity of courses, other various services offered by Google namely Google Drive were used for sharing all types of files. A large number of measures have been implemented by the Cadi Ayyad University that has developed a digital platform named "UCA Digital Campus" that includes more than 9000 educational resources in different formats and supports (PDF, PowerPoint, Videos, and Audio, etc.). Added to this, the use of interactive platforms such as Teams, Google Classroom, Zoom, and Google Meet, enabled teachers to follow students closely as well as in face-to-face classes. Moreover, to ensure equality between students, the University has made available to the channel "Arriyadia" 124 university courses, in addition to 120 courses broadcast on regional radio in Marrakech.

Besides, Sultan Moulay Slimane University filled its sites with around 3,639-course materials for all levels. This, in addition to the use of interactive platforms, namely Teams and Moodle. Moulay Ismail University has set up a personalized distance education system. UMI launched distance-learning platforms on March 20, 2020, where courses are shared in different formats, namely (PowerPoint, Word, and PDF) with daily production of audiovisual capsules. The university currently uses two platforms, namely: Moodle and the Open EdX platform of the MarMOOC project. Sidi Mohamed Ben Abdellah University of Fez has broadcast 410 courses on regional radio stations for 26 channels in plus of deploying platforms and interactive tools, namely: Moodle and Google Classroom, Google Meet, ZOOM, Google drive, etc. Abdelmalek Essaâdi University filled its platforms (Moodle) and its sites with 4,000 courses, these courses were broadcasted on television channels and regional radios. To ensure pedagogical continuity, Mohammed Premier Oujda University broadcasts its courses through the regional radio station of Oujda daily to guarantee equality between students. In addition to the Moodle platform, which contains more than 1,400 digital courses and 120 videos.

For students without access to technology or connectivity, television and radio programming are supplemented. To ensure that the school closure does not exacerbate inequity in access to these learning resources. Some schools in the country are using online tools such as Google classrooms and G-Suite to continue the learning process from home. All schools have broadcasted accounts to work on the Microsoft Teams platform, which is free of charge. Profiles have been created for all students and teachers. A helpdesk and a phone for questions at the MES are provided.

Each higher education institution organizes independently the distance learning process of its students by using various online platforms and video conferencing software through which distance learning sessions are conducted, as well as other communication channels. Programs such as Office 365, Skype, and Blackboard are used. Free Access and cloud services were provided. On the website of the Ministry of Education and Science, there is information about a "telephone e-education" service to answer questions and provide suggestions related to the educational process.

The government is updating legislation concerning grading and assessment to take into account the distance-learning context. The Ministry of Education has been broadcasting lessons on television since February 2020 for every class and subject until secondary school. The ministry is making all of these lessons available.

At the time of Covid-19, the "TelmidTICE" platform provides students with online digital content (filmed courses), these are lessons grouped according to categories, namely (grade level, student branch, and subject), the grade levels concerned are from the first year of primary school until the second year of the baccalaureate. This portal is accessible via the website «https://soutiensco.men.gov.ma/», and without having to create an account for the student. Therefore, it is the same content broadcast for different learners. To solve the problems of inequality between students, the minister mobilized three channels: Athaqafiya, Al Amazighiya, and Arriyadia. These are the visual lessons given to students (around 59 lessons per day) according to government spokesman Amzazi. On May 31, 2020, the Ministry of Education declared that access to the TelmidTice platform is now free. This is intended to promote equal opportunities between learners so that they can access the platform and download course content without internet pay.

4 Comparative Study of the Various Current EAD Platforms

Distance learning platforms provide learners with a set of tools, features, content, and resources to facilitate communication between the trainer and the learner.

To analyze the different characteristics of e-learning platforms and thus develop a comparative study, we looked for more than 30 distance learning platforms, and our choice was made on 7 learning platforms under a free license. The selected platforms are the most popular and most used in distance education during the last 10 years namely [14–16]: Moodle, Claroline Connect, Google Classroom, Ilias, Canvas Network, Sakai and Atutor.

To proceed with our study, we will define a set of criteria [17, 18] summarized by Fig. 1, the comparative metrics are then divided into four main categories (description of the platform, functionalities, techniques, standards). These characteristics result mainly from the actions carried out by the different actors in an e-learning platform. In addition, these criteria are adapted to institutional needs as well as to student needs, in terms of interface, navigation, functionalities, software, administration, communication and content. In other words, these criteria are used to allow the different actors to maintain an autonomy on the platform. These characteristics are adapted to institutional needs in terms of interface, navigation, functionalities, software and content.

Tables 1, 2 and 3 [19, 20] presents a comparative analysis of the various platforms most used in education.

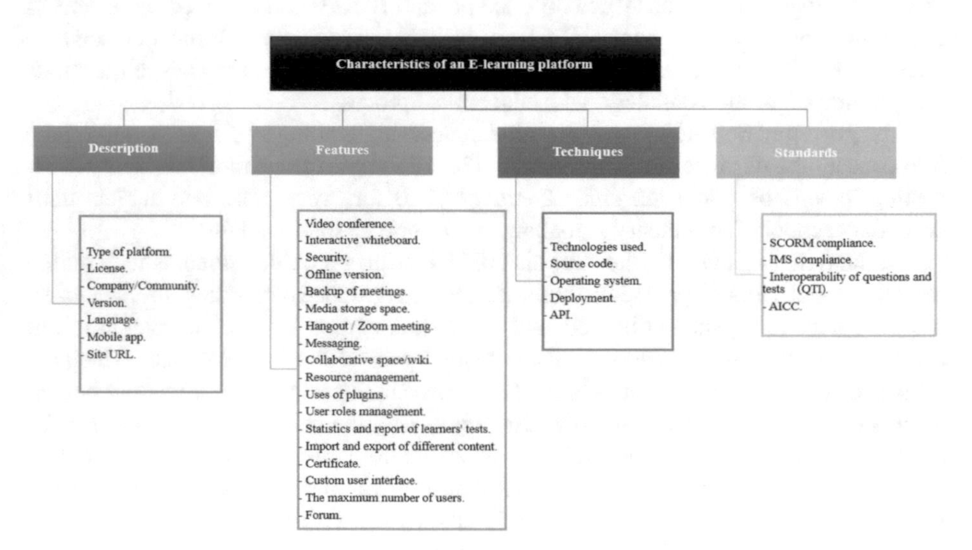

Fig. 1. Characteristics of an E-learning platform.

Table 1. Description of platforms.

Platform name / Characteristics	Moodle	Claroline Connect	Google Classroom	Sakai	Ilias	Canvas Network	Atutor
Description of the platform							
Type of platform	Course management system	Learning management system	Web Service	Learning management system	Learning management system	Learning management system	Learning content management system
License	General Public License (GPL3 +)	General Public License (GPL)	------------	Educational Community License (open source license)	General Public License (GPL)	AGPLv3	General public license version 2
Company / community	Martin Dougiamas; Moodle HQ; Moodle Community	Consortium Claroline / Forma-Libre	Google	Academic institutions, business organizations, and individuals	ILIAS open source e-Learning e. V	Christopher "moot" Poole	Inclusive Design Research Centre, OCAD, University, ATutor-Spaces
Version	Moodle 3.8 (18 November 2019)	Claroline 12.0.13 (15 September 2018)	------------	Sakai 19.3 (11 October 2019)	Ilyas 5.4 .2 (09 Mai 2019)	------------	Atutor 2.2.4 18/06/2018
Language	Multi-language (Fr, en, ar) Over 135 languages	Multi-language 35 languages	Multi-language	Multi-language 19 languages	Multi-language 28 languages	Multi-language	Multi-language Over de 38 languages
Mobile app	Yes	No	Yes	Yes (G-OpenLMS)	Yes	Yes	No
Site URL	https://moodle.org/	https://claroline.net/	https://classroom.google.com/u/0/	https://www.sakailms.org/	https://www.ilias.de/	https://www.canvas.net/	https://atutor.github.io/

Table 2. Platform features.

Characteristics	Platform name						
	Moodle	Claroline connect	Google classroom	Sakai	Ilias	Canvas network	Atutor
Platform features							
Video conference	Yes (Plugins)	No	No	No	Yes	Yes	Yes
Interactive whiteboard	Yes	No	Yes (Google Jamboard)	Yes	No	Yes (Awwapp)	Yes
Security	Yes	Yes	Yes	Yes	Yes	Yes	Yes
Offline version	Yes	No	No	No	No	No	--------------
Backup of meetings	Yes (Plugin)	Yes	Yes	Yes	--------------	Yes	Yes
Media storage space	Limited	Limited (Cloud basic: 50 GO)	Limited	Limited	-------------	Limited	Limited
Hangout/Zoom meeting	Yes (Plugins)	No	Yes	Yes	--------------	Yes (Zoom)	--------------

(*continued*)

Table 2. (*continued*)

Characteristics	Platform name						
	Moodle	Claroline connect	Google classroom	Sakai	Ilias	Canvas network	Atutor
Messaging	Yes (Depuis la version 1.6)	Yes	Yes	Yes	Yes	Yes (Inbox (conversations)	Yes
Collaborative space/wiki	Yes	Yes	Yes	Yes	Yes	Yes	Yes
Resource management	Yes	Yes	Yes	Yes	Yes	Yes	Yes
Uses of plugins	Yes	Yes	Yes	Yes	Yes	Yes	Yes
User roles management	Yes	Yes	Yes	Yes	Yes	Yes	Yes
Learner test statistics and reports	Yes	No	Yes	Yes	Yes	Yes	Yes
Import and export of different content	Yes	Yes	Yes	Yes	Yes	Yes	Yes
Certificate	Yes (Plugins)	Yes	Yes	Yes	Yes (Plugins)	Yes	Yes
Custom user interface	Yes	Yes	No	--------------	--------------	No	--------------
Maximum number of users	Limit of 500 users	250 users (cloud basic)	------------	--------------	--------------	--------------	--------------
Forum	Yes	Yes	Yes	Yes	Yes	Yes	Yes

Table 3. Techniques and standards.

Characteristics	Platform name						
	Moodle	Claroline connect	Google classroom	Sakai	Ilias	Canvas network	Atutor
Techniques & Normes							
Technologies used	PHP 7.1	PHP 5	------------	Java	PHP 7.2	Ruby on Rails	PHP 7.2
Source code	Yes	Yes	No	Yes	Yes	Yes	Yes
Operating system	Windows, macOS, Linux	Windows, macOS, Linux	------------	Microsoft Windows, Linux, and macOS	Linux, Unix, macOS, and Windows	Windows, Linux, Mac	Microsoft Windows, Linux, and macOS

(*continued*)

Table 3. (*continued*)

Characteristics	Platform name						
	Moodle	Claroline connect	Google classroom	Sakai	Ilias	Canvas network	Atutor
Cloud deployment	Cloud	Cloud	Cloud	Cloud	Web, Cloud, SaaS	Cloud	Cloud
API	Yes	Yes	Yes	Yes	Yes	Yes	Yes
SCORM compliance	Yes	Yes	-------------	Yes	Yes	Yes	Yes
AICC	Yes	No	-------------	No	Yes	No	No

4.1 Comparative Technical and Educational Study of the Selected Platforms

4.2 Comparative Analysis of Platforms

The studied platforms guarantee access to rich and operational educational resources, by enabling continuous improvement in teaching. Despite the relevance of these platforms, their analysis shows that each one of them has advantages and limits.

In the table below, we will present the strengths as well as the weaknesses of each platform (Table 4).

4.3 Challenges in the Development of an E-learning System

Distance learning faces a set of constraints, given the number of participants, the resources necessary to set up, the diversity of themes, fields, and the difference in levels between participants. To manage its constraints, it is, therefore, necessary to use a system, which has the role of creating, managing, and storing content, as well as managing the exchange of information between participants, time management, levels, and skills of different learners.

According to Monique Linard [21], a system is defined as follows: "the organization of space, time, objects and actors of a situation with a view to precise objectives". According to this definition, Linard emphasizes the importance of the combination of technical instruments, the organization of places and time as well as the main goal to be achieved.

According to Grace Blanche and Asatsop-Nganmini, an E-Learning system is defined as "a complex and dynamic set of educational resources on digital support, technological, financial resources, and services, which mobilize actors of various skills to enable the learner to carry out his journey in a personalized and responsible autonomous way" [22].

In other words, an e-learning system brings together trainers as well as institutions, to support learners in their training, along with using a set of tools and resources to achieve the objectives sought by institutions by relying on a set of tools and technologies.

Table 4. Strengths and weaknesses of platforms.

Platform	Strong points	Weakness
Moodle	Moodle is an open-source learning platform widely used by institutions. This is thanks to the many features it offers Moodle makes it possible to manage communication and exchanges between the various actors through messaging forums, chats, and collaborative spaces. [14–17] The platform offers the possibility to create quizzes and surveys and to evaluate the learner's results through reports and statistics. It also offers the possibility of adding plugin modules as well as an offline version to solve problems of Internet access	The size of the files to import, the limited number of users, as well as the limited storage space, are considered an obstacle in the use of Moodle by the users. Ergonomics, readability, and accessibility must be revisited to facilitate navigation for learners as well as for teachers
Caroline Connect	Claroline is easy to use by learners as well as teachers; the platform offers a multitude of collaboration and resource management tools. It stands out for its simple installation and compatibility with all operating systems [15]	Lack of interactivity tools such as interactive whiteboards and video conferencing. The ergonomics of the platform must be improved to facilitate learning for students
Google Classroom	Google Classroom is an accessible and easy-to-use web, sharing and saving of documents is done online, which makes it possible to distribute the content to a large number of users, the interface is simple to manipulate by users Google Classroom guarantees its users a multitude of online features	Lack of student assessment tools such as quizzes and automatic tests. Lack of time management tools such as calendars. Besides, account management in Google Classroom is difficult
Sakai	Sakai is an easy-to-use platform for users, it is distinguished by a set of functionalities that meet the needs of learners as well as teachers such as collaborative spaces, messaging, role and resource management, etc. [14, 17]	Like other platforms, file size and storage space are limited. Besides, Sakai does not have an offline version. Learner tracking modules need to be improved

(continued)

Table 4. (*continued*)

Platform	Strong points	Weakness
Ilias	The Ilias platform very rich in features, namely, wikis, role management, content management, messaging, etc. It allows you to add plugins [17, 18]	Ilias is missing the offline version and the platform does not have an interactive whiteboard
Canvas Network	Canvas Network offers users an intuitive graphical interface, it also offers all the basic functionality of an LMS regarding content management, roles, exchange, etc.	Canvas Network is missing conference videos. Also, Canvas offers classic, non-personalized functionality
Atutor	Atutor offers users an interface and ergonomics that are easy to explore, it also offers a multitude of functionalities [15], as well as a set of plugins	Version development is slow, making the interface and navigation classic and not adaptable to the needs of current learners

Below is a figure that represents a standard E-learning device (Fig. 2):

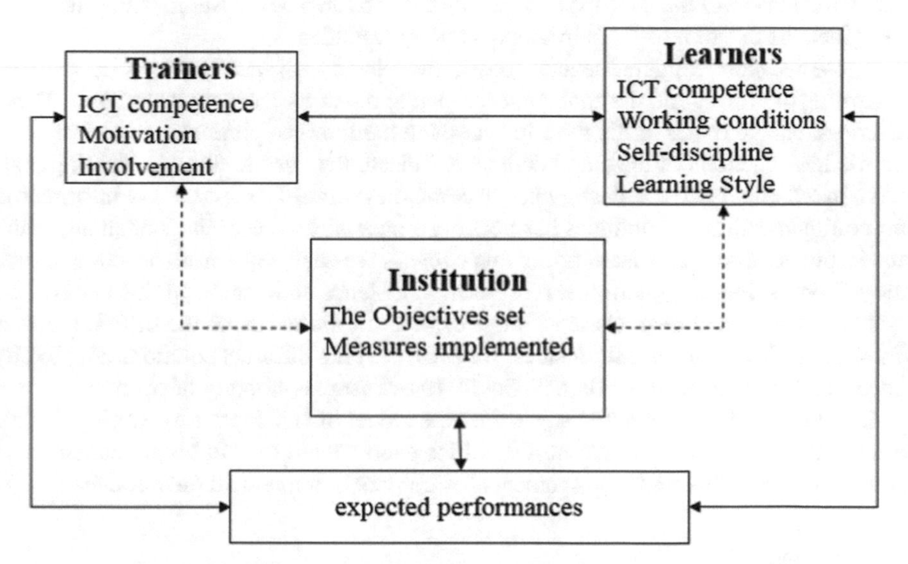

Fig. 2. E-learning system [23]. Source: Houze and Meissonier, s. d.

However, the proposed architecture of the e-learning system above lacks an essential element, which is the adaptability of both: the scientific content/resource and the teaching technique. Indeed, the adaptability of learning style is primordial in both face-to-face and distance learning. Each human has a learning preference and comprehension speed. As an example, a visual student will be more interested in graphical content such as diagrams, images, and illustrations. However, verbal learners favor listening to audio or reading information. Therefore, the presentation of the content is an essential element in any e-learning system.

The criteria of "Adaptability" is missing in all the actual deployed E-learning platforms, which limits them to being management tools for storing and broadcasting content in a static format that may not be appropriate to the needs of both trainers and learners. All platforms were designed a decade ago, which makes them classic in a way, despite the development of new versions that implement several improvements. Some platforms may allow personalized navigation, however, the content remains the same for different learners.

In conclusion, this section served us to focus on the use of ICT in higher education in Morocco.

5 Conclusion

Today, the coronavirus pandemic is an opportunity for a serious and new reflection to review the reform of the education system and to rebuild a new system, which serves to strengthen the use of new ICTs in schools and universities.

These decisions triggered several constraints related to the pedagogical effectiveness of learning platforms and the content developed to replace classroom instruction. These include language issues, difficulties in accessing the Internet, particularly in rural areas and the heterogeneity of the target audience. Indeed, the need to develop learning platforms based on adaptive pedagogy has become inevitable. The use of new information and communication technologies has become a necessity to meet these challenges and thus improve the students learning. In this context, we analyzed, through a comparative study, 7 online learning platforms most used in distance education and this is based on a set of features and characteristics that result from the needs of the different actors of a distance learning system. Besides, we presented the different solutions adopted by Moroccan Universities at the time of Covid-19 to ensure continuity of courses.

In future work, we intend to suggest a new model of an adaptive e-learning device, which takes into account different difficulties encountered by Moroccan students and offers more flexibility and improvements for trainers to implement their content.

References

1. Draissi, Z., ZhanYong, Q.: COVID-19 Outbreak Response Plan: Implementing Distance Education in Moroccan Universities. Social Science Research Network, Rochester, NY (2020). https://doi.org/10.2139/ssrn.3586783
2. Ismaili, J.: Evaluation of information and communication technology in education programs for middle and high schools: GENIE program as a case study. Educ. Inf. Technol. **25**(6), 5067–5086 (2020). https://doi.org/10.1007/s10639-020-10224-1

3. Archibald, D., Worsley, S.: The father of distance learning. TechTrends **63**(2), 100–101 (2019). https://doi.org/10.1007/s11528-019-00373-7
4. Sanchez-Gordon, S., Luján-Mora, S.: Technological innovations in large-scale teaching: five roots of massive open online courses. J. Educ. Comput. Res. **56**, 623–644 (2018). https://doi.org/10.1177/0735633117727597
5. Haffar, N., Maraoui, M., Aljawarneh, S., Bouhorma, M., Alnuaimi, A., Hawashin, B.: pedagogical indexed Arabic text in cloud E-learning system. Int. J. Cloud Appl. Comput. **7**, 32–46 (2017). https://doi.org/10.4018/IJCAC.2017010102
6. Ewais, A., Samra, D.A.: Adaptive MOOCs: a framework for adaptive course based on intended learning outcomes. In: 2017 2nd International Conference on Knowledge Engineering and Applications (ICKEA), pp. 204–209. IEEE, London (2017). https://doi.org/10.1109/ICKEA.2017.8169930
7. Luna, R.R., Neves, M.: MOOC as an innovative tool for design teaching. In: Ahram, T., Taiar, R., Gremeaux-Bader, V., Aminian, K. (eds.) IHIET 2020. AISC, vol. 1152, pp. 258–263. Springer, Cham (2020). https://doi.org/10.1007/978-3-030-44267-5_39
8. Pilli, O., Admiraal, W.: A taxonomy of massive open online courses. Contemp. Educ. Technol. **7** (2016). https://doi.org/10.30935/cedtech/6174
9. Brunel, S., Lamago, M., Girard, P.: Platforms for online teaching: towards a general modeling of their functions. In: 4ème Colloque International du RAIFFET, Marrakech, Morocco (2014)
10. Kolekar, S.V., Pai, R.M., Pai, M.M.M.: Rule based adaptive user interface for adaptive E-learning system. Educ. Inf. Technol. **24**(1), 613–641 (2018). https://doi.org/10.1007/s10639-018-9788-1
11. Lakkah, S.E., Alimam, M.A., Seghiouer, H.: Adaptive E-learning system based on learning style and ant colony optimization. In: 2017 Intelligent Systems and Computer Vision (ISCV), pp. 1–5. IEEE, Fez, Morocco (2017). https://doi.org/10.1109/ISACV.2017.8054963
12. Elghouch, N., En-Naimi, E.M., Seghroucheni, Y.Z., El Mohajir, B.E., Achhab, M.A.: ALS_CORR[LP]: an adaptive learning system based on the learning styles of Felder-Silverman and a Bayesian network. In: 2016 4th IEEE International Colloquium on Information Science and Technology (CiSt), pp. 494–499. IEEE, Tangier, Morocco (2016). https://doi.org/10.1109/CIST.2016.7805098
13. Dolasinski, M.J., Reynolds, J.: Microlearning: a new learning model. J. Hosp. Tour. Res. **44**, 551–561 (2020)
14. Etude comparative de plates-formes de formation à distance. https://www.projet-plume.org/files/Choix_plateforme_a2l.pdf. Last Accessed 21 Jun 2020
15. Ouadoud, M., Chkouri, M.Y., Nejjari, A., El Kadiri, K.E.: Studying and comparing the free E-learning platforms. In: 2016 4th IEEE International Colloquium on Information Science and Technology (CiSt), pp. 581–586. IEEE, Tangier, Morocco (2016)
16. Ajlan, S.: A comparative study between E-learning features. In: Pontes, E. (ed.) Methodologies, Tools and New Developments for E-Learning. InTech (2012). https://doi.org/10.5772/29854
17. Faxén, T.: Improving the outcome of E-learning using new technologies in LMS systems 81
18. Ouadoud, M., Chkouri, M.Y., Nejjari, A., Kadiri, K.E.E.: Studying and analyzing the evaluation dimensions of E-learning platforms relying on a software engineering approach. Int. J. Emerg. Technol. Learn. (iJET) **11**, 11–20 (2016)
19. Cavus, N., Zabadi, T.: A comparison of open source learning management systems. Procedia Soc. Behav. Sci. **143** (2014)
20. Louhab, F.E., Bahnasse, A., Talea, M.: Smart adaptive learning based on Moodle platform. In: Proceedings of the Mediterranean Symposium on Smart City Application - SCAMS '17, pp. 1–5. ACM Press, Tangier, Morocco (2017)
21. Bluteau, M.: Hybridations et alternances, caractéristiques et prescrits de reliances. Centre National Pédagogique et de Ressources des MFR (2020)

22. Asatsop-Nganmini, G.B.: La mesure de la qualité perçue d'un dispositif de e-learning **419**
23. Houze, E., Meissonier, R.: Performance du E-Learning: de l'amélioration des résultats de l'apprenant à la prise en compte des enjeux institutionnels

Internet of Things

Digital Twin-Driven Approach for Smart Industrial Product Design

Mohammed Abadi[1(✉)], Chaimae Abadi[2], Asmae Abadi[3], and Hussain Ben-Azza[1]

[1] Industrial and Manufacturing Engineering Department, ENSAM-Meknes, Moulay Ismail University, Meknes, Morocco
abadi.s.mohammed@gmail.com

[2] Laboratory of Mechanics, Mechatronics and Command, Team of Electrical Energy, Maintenance and Innovation, ENSAM-Meknes, Moulay Ismail University, Meknes, Morocco

[3] Euromed Research Center, INSA Euro-Méditerranée, Euromed University of Fes, Route de Meknès (Rond Point Bensouda), 30 000 Fès, Morocco
as.abadi@insa.ueuromed.org

Abstract. The new emerging technologies, such as Internet of Things (IoT), big data analytics, cloud computing and rapid advances in smart software/hardware systems continue to enhance industrials capabilities for the development of efficient Digital Twins (DT). While this emerging DT is seen as a promising track for achieving smart integrated product design processes, industrials and researchers are still confronted to a set of challenges in DT development related to semantic interoperability, effective integration between the virtual and physical entities and the persistent need of inherent reasoning abilities in the developed design frameworks. In response to this increasing interest and challenges, we explore in this paper the potentialities of using DT-driven approaches in complex industrial product design, we identify the main remaining and future challenges to achieve seamless integration and smart abilities all throughout the product design process and we propose a new DT-driven approach for smart product design that combines the potentialities of the new technologies such as IoT and Big Data Analytics with the potentialities of inference ontologies, particularly their expressiveness and reasoning abilities. An industrial case of study is developed to illustrate the application of the proposed DT-driven design approach.

Keywords: Digital twin · Integrated product design · Information modelling · Inference ontologies · Design rules · Internet of Things (IoT) · Big data analytics · Artificial intelligence

1 Introduction

In the context of the fourth industrial revolution, industrial products design processes are considerably affected by the rapid advances in information management, computer technologies and artificial intelligence emerging tools.

M. Lazaar et al. (Eds.): BDIoT 2021, LNNS 489, pp. 263–273, 2022.
https://doi.org/10.1007/978-3-031-07969-6_20

In fact, today products and processes are becoming more and more complex and their development processes are becoming more and more digitalized. In this context, the Digital Twin concept has emerged as a promising paradigm to achieve the flexibility and the integration needed in nowadays highly competitive industrial context.

The Digital Twin (DT) concept was introduced by Grieves as a digital entity to represent the physical product [1]. It was reduced at the beginning to a digital 3D modeling and system simulation of the developed product. Then later on, with the emergence of new technologies such as Internet of Things, augmented reality, cloud computing and big data analysis, more interest has been undergoing for integrating real-time modeling, data fusion, intelligent industrial services and industrial extrapolations in the Digital Twins models [1].

In fact, Digital Twin is characterized by the two-way interactions between the digital and physical worlds, which can possibly lead to many benefits particularly in the product design phase [2]. Reviewing the existing papers that have explored the use of DT in the product design phase [1, 2], we notice that many challenges are still remaining for the achievement of the intended flexibly and integration in nowadays product design processes. In fact, while the interest of developing more intelligent services is undergoing, few efforts have been devoted to exploring the advantages of combining the potentialities of new emerging artificial intelligence (AI) tools such as inference ontologies in the DT frameworks. The challenges related to semantic interoperability between the DT entities and many challenges were not yet addressed too.

The aim of this paper is then to explore the potentialities of developing Digital Twin models for the optimization of the product design process. Our aim is also to identify the different challenges that industrials and researchers must overcome in order to achieve the demanded flexibility, reactivity, integration and systems intelligence and then to propose a new Digital Twin-driven approach for industrial product design.

To do so, the remainder of this paper is organized as follows. Section 2 characterizes the concept of Digital Twin and reviews its main previous relevant applications in the industrial products development processes. Section 3 identifies the remaining and future challenges to overcome then describes the new proposed DT-driven approach for smart industrial product design. Section 4 presents a case study from the impression-die forging industry to illustrate the application of the proposed DT-driven design approach. Finally, conclusion and future work are drawn in Sect. 5.

2 Digital Twin: Concept and Applications

2.1 Concept of DT

Digital Twin is one of the new emerged promising tools to deal with the previous problems. This concept has been defined in different ways. Actually, according to the NASA, a Digital Twin can be defined as an integrated multi-physics, multiscale, probabilistic simulation of an as-built vehicle or system that uses the best available physical models, sensor updates, fleet history, etc., to mirror the life of its corresponding flying twin [3]. Another definition of DT is the one proposed by Grieves et al. In fact, they define a Digital Twin as "a set of virtual information constructs that fully describes a potential or actual physical manufactured product from the micro atomic level to the macro geometrical level" [4].

According to those definitions, the functioning of DT consists of three principal parts, namely: the physical product, the virtual product, and the connected data that tie and indissolubly connect the physical and virtual product [5–7]. However, a five-dimension DT for a product has been developed [8–10]. It is based on five main elements:

Physical entity: It is the real product which is manufactured from raw materials or by assembling different parts. It can be proceeded and operated by the human. Moreover, the physical entity allows the collection of the different data generated all along the product development cycle in real time, for examples: the production information, the customer usage methods, the maintenance data… All those data can be completed thereafter by the manual product data, the online customer records, the download records and also the evaluation feedbacks. Then, they are stored and treated even by designers or users. In fact, through the evolution of technologies in terms of sensors and IoT, all the previous cited operations, specifically the data collection, the storage and the processing, have become possible.

Virtual entity: It is considered as a mirror image of the physical product. It integrates four models kinds, namely: geometry models, physics models, behavior models and rule models [9]. Thus, the geometry and the physical performances are described through the two first ones. In regards to the behavior model, it is used to describe and analyze behaviors of the product, the users and also the environment. In addition to that, it permits the study of the different interactions between those elements. Finally, the rule models consist of evaluation, optimization, and forecasting models which are constructed basing on several laws of product operation and maintenance, etc. [8]. The virtual entity is built, deployed and modified through the cloud environment that facilitates the access for all interveners. Moreover, and through to the advantages of virtual reality (VR) and augmented reality (AR), the interaction of the virtual entity and the user in a real time is possible either in completely virtual environment or a mixed cyber-physical environment.

Digital twin data: It is the data collected from the physical and the virtual entities. It can be analyzed, integrated, and visualized [8]. Actually, the conversion of the data analytics into concrete information is done. This operation facilitates the designers' decision making. Then, data integration is performed in order to assimilate the different collected data from several sources in the same data source. In this way, many hidden patterns are discovered. After this, an explicit data presentation is done using data visualization technologies. Finally, basing on advanced artificial intelligence techniques, the DT abilities are enhanced.

Services: They are divided into two main types functional service and business service (BS) [10]. The rule of the first one is to convert different data, models, algorithms, etc. to services in order to support the DT functioning. In regards to the second type, it consists on presenting the standard inputs and outputs in a simple way through software interfaces.

Connections: It is about connections between the first fourth elements described previously. Those connections are ensured by different technologies, namely, network communication, IoT, and network security [8].

2.2 Digital Twin Applications

The Digital Twin is not used only for products, but also to represent a complex production system [11]. In fact, this artificial intelligence tool permits increasing spread of inline measurements and operating data. By consequent, the deviations and malfunctions of systems are gone over quickly. In addition to that, it gives the possibility to the system to react themselves by initiating countermeasures as cyber physical production systems [12, 13].

Moreover, the Digital Twin tool makes possible the production with high quality, complex specifications and great robustness in a minimum cost. Thus, the efficacy and the power of products are increased [14]. In brief, Digital Twin is an artificial intelligence tool that offers different opportunities in different domains in terms of efficiency, quality and productivity. It has been applied into different phases and stages of product development cycle.

Digital Twin is used, actually, to develop the production phase in terms of human machine interaction [15], energy consumption management [16] and process planning [17]. In addition to that, this artificial intelligence tool is exploited in the service stage, specifically maintenance [18], recycling [19] and prognostics [9].

Another applications of Digital Twin are the one which concerns the design stage [6, 20, 21]. Actually, in this context, Zhuang et al. [22] have developed some relevant theories and tools to implement the DT-based product design basing on the connotation, architecture, and trends of DT. Moreover, a new DT model have been proposed in regard to 3D product configuration [23]. In this way the interface design/production is reinforced. In addition to that, a new product design DT based approach have been presented in [8]. It aims to facilitate the different stages of the design phase.

Another proposed approach in the context of design phase, based on DT capacities, is the one developed by Schleich et al. [24]. Its principal is about exploring DT in order to manage geometrical variations. Thus, the part deviations of the product are evaluated and solved from the early stages of product life cycle. Furthermore, and in the same context, a DT-based approach has been developed by Zhang et al. [25] in order to design production lines.

In brief, several approaches have been based on DT in different domains. The advantage of this resides in the fact of integrating experimental, actual measurement, calculation and real-test environment parameters at the same time. Thus, it is to deduct that DT is an efficient artificial tool for designers. However, the success of such integrated and complex DT-based approaches is conditioned with the resolution of a set of challenges related to both Big Data analytics, Internet of Things than Digital Twin. We Identify all these different challenges in the next section.

3 Big Data Analytics, IoT and Digital Twin Challenges in the Area of Industrial Product Development

In order to tackle the challenges related to the implementation to Digital Twin technologies, we should first identify the remaining challenges related to Data analytics and Internet of Things. In fact, Digital Twin is facilitated through advanced data analytics and IoT connectivity particularly in the area of industrial product development.

Throughout IoT, manufacturers have the ability to store and use increased volume of data related not only to products, but also to manufacturing resources, process and to all the interactors evolved in the industrial value stream. Coupled with Big Data analytics, IoT offers a rich environment to execute different reasoning tasks providing pertinent resource for fault and anomaly detection, predictive maintenance, traffic management and such more applications [28, 29]. Our aim in this section is then to identify the common challenges for data analytics, Internet of Things and Digital Twins particularly in the area of industrial product development.

Reviewing the literature, we can group them in four main categories which are, IT infrastructure, consistent and quality data, security and modeling challenges:

- **IT infrastructure challenges:** The rapid growth of digitalization and AI integration needs are conditioned by the development of high-performance infrastructure incorporating up to date hardware and software. Related to data analytics, the challenge is down to the cost of implementing these reactive updated systems. Related to IoT infrastructure, the challenge is mainly connecting old machines to the IoT environment. For Digital Twins, without a connected and persistent IT infrastructure, achieving the goals behind Digital Twins implementation will not be possible.
- **Consistent and quality data challenges:** It is important to ensure that only quality data which is noise-free, with a constant data stream, is fed into the AI algorithms. But this challenge becomes more and more complex with the advent of big data. The use of IoT increases the large volumes of unstructured data. For DT, collecting, sorting, organizing, managing and exploiting these amounts of data is a necessity for value creation and reactivity goals achieving.
- **Security challenges:** To remain competitive, it is clear that for industrial users the privacy, security and trust associated to DT technologies are main challenges. To overcome this challenges, IoT and data analytics, which are the key enabling technologies for DT, must respect updates in security regulations. For Trust challenges, manufacturing companies must verify and validate that the implemented Digital Twins are performing as expected. To do so, modeling issues must be tackled.
- **Modeling challenges:** From the earliest design phases to simulation of DT, there are fundamental needs of standardization and integration in the adopted information modeling approaches. The developed models must be generic and integrated enough to ensure the federation between all the component of the DT and IoT environment and at the same time these models must be specific enough to consider the details of each engineering domain involved in the integrated product development.

We conclude then that industrials can effectively realize real benefits throughout the implementation of Digital Twins technologies but the fist step for them to achieve this will be to solve the shared challenges related to the key enabling technologies of Digital Twins, which are Big Data analytics and Internet of Things. In this work, we will particularly focus on the modeling and interoperability challenges and we will explore the benefit of incorporating inference ontologies in Digital Twins environment to do this.

4 The Proposed Digital Twin-Driven Approach for Smart Product Design

4.1 Motivations

As shown previously, the emergent use of Digital Twins offers great opportunities of the optimization of products development processes. In fact, with the advancement of new IT and AI tools, in the virtual world, digital mock-up of the product is developed to model and visualize the product structure, simulate its behavior and predict its performance. Simultaneously, in the physical world, thanks to the emergence of Internet of Things (IoT), cloud and AI tools, products behavior and performance can be captured and analyzed in real time. The advantages linked to the development of successful Digital Twins consist mainly on connecting the virtual and physical products through the implementation of a data-driven product development approach.

Although the development of efficient Digital Twins that allow industrials to achieve the flexibility, integration and synchronization between the physical and virtual product worlds needed in the development of nowadays complex products is not easy to realize. Industrials and researchers have the necessity to deal with some important challenges as explained in the previous section [26, 27]:

- Semantic interoperability between all the involved inter-actors
- Integration of different domains in the Product Engineering Process in Digital Twin
- Standardized Information Exchange between Digital Twins entities
- Efficient Design of Information Flow in order to ensure the continuation and completeness of the information flow all throughout the product development process
- Automated and reactive decision-making process.

The aim of this contribution is then to propose a new Digital Twin-driven approach that overcomes all the identified challenges in order to achieve the smart product design process with all the flexibility, integration and intelligent abilities needed.

4.2 The Approach Outlines

Figure 1 represents the flowchart of the proposed approach. The principal is to develop simultaneously not only two main systems 'the Physical and Virtual Product' but three systems. The newest proposed one is what we call: 'the AI-based information modeling and reasoning system':

- **Physical Product Design:** We base the development of the physical product in our approach on the most commonly used design approach which is the systematic design proposed by (Pahl, 2007). This design approach divides the product design process into four main phases: task clarification, conceptual design, embodiment design and detail design. We conserve the same four steps but we integrate in each step not only costumers' specification but also the requirements of all the inter-actors involved in the product lifecycle (manufacturability, assembly, maintenance, robustness, safety …) in an integrated manner.

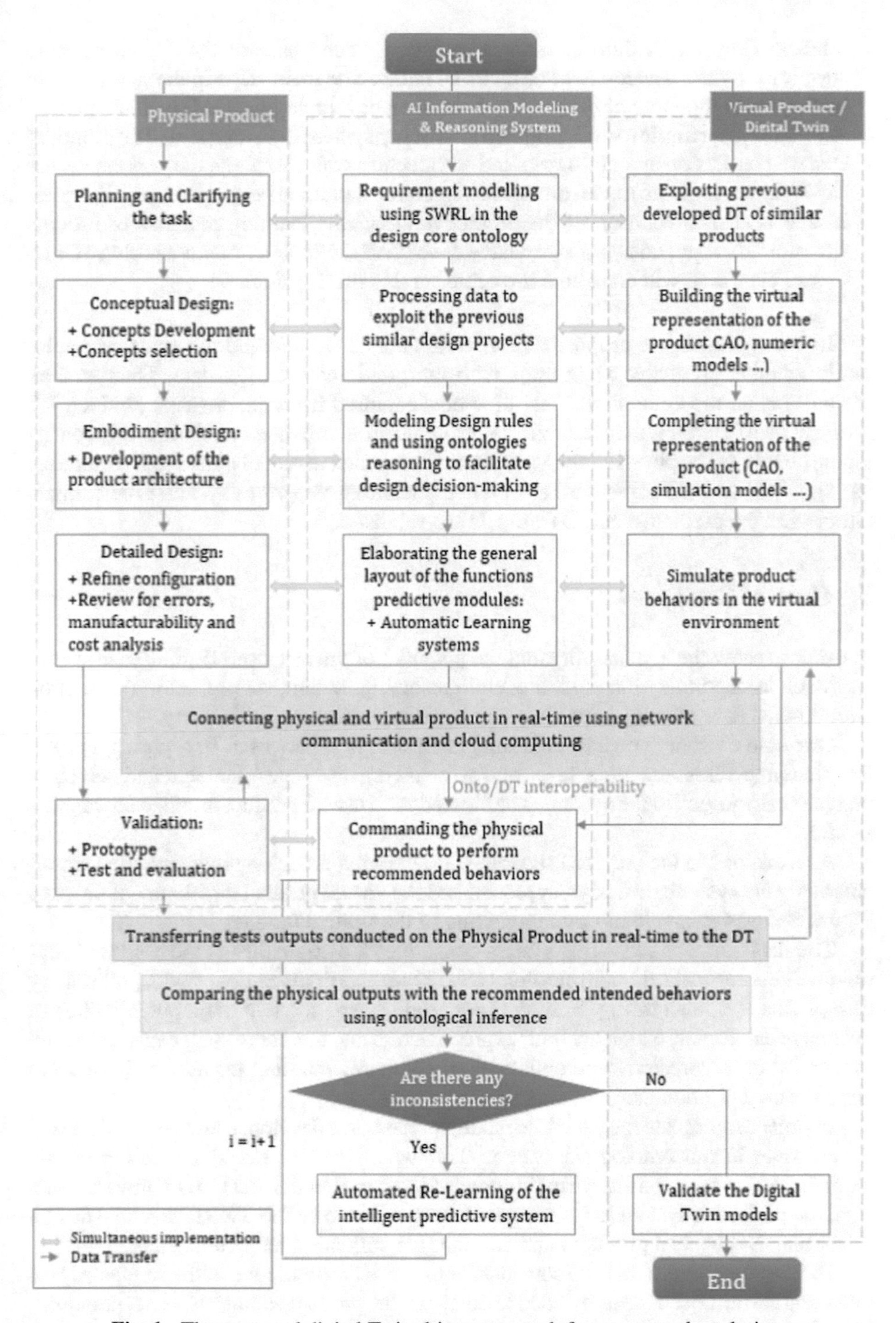

Fig. 1. The proposed digital Twin-driven approach for smart product design.

- **Virtual Product Design:** As shown in Fig. 1, each phase of the design phase is supported by the development of a virtual entity. The main steps in the development of the digital mock-up of the product consist on developing the CAO and 3D models.
- **AI-based information modeling and reasoning system development:** The proposed system consists on an ontology-based architecture which exploits the expressiveness of OWL ontologies to model product design information and engineering design rules. It rule is also to connect all the predictive automatic learning modules of product functions. In a previous work, we have developed the design domain ontology CPD-Onto [27] which will constitute the center of this third system.

In our approach, we propose first to enrich the CPD-Onto and the associated automatic learning predictive algorithms with the preliminary design data. Then second, we will enrich the system with real time data captured from the physical product. We propose then to exploit the ontological reasoning abilities to detect if there any inconsistencies between the physical, the virtual and the predictive models. In case the ontology has shown some inconsistencies, an automatic learning step of the system is performed. Otherwise, the predictive and DT models are validated.

5 Case of Study

In order to show the applicability and the potential of the proposed Digital Twin-driven approach incorporating the modeling and ransoming system, we present in this section an industrial case of study relating to the impression-die forging industry.

Our case of study concerns the integrated design a part used in a piping system, for changing direction, branching or for changing the pipe diameter and which is mechanically joined to the system. The studied mechanical product is called the tapping saddle.

As explained in the previous section, we start our approach by implementing simultaneously the systematic design approach and the virtual product development process. Figure 2 shows a sample of CAO modelling of the studied product.

The modeling and reasoning system constitutes the originality of the proposed approach. The core design domain ontology CPD-Onto was firstly enriched with preliminary design data as shown in Fig. 3. In the proposed system, the two main potentialities of inference ontologies, which are their expressiveness and their reasoning capabilities, are associated to the predictive potentialities of automatic learning techniques in order to enhance our DT abilities.

In order to automate the decision making process, we developed and encoded SWRL design rules in our ontological system. This we allow also the identification of any inconsistencies between the virtual product (developed under the CAO software Catia V5), the predictive system (consisting mainly in this step on the SWRL executed design rules) and the physical product behavior ones the validation tests conducted.

To illustrate this process, we consider for instance the prediction of the forging engine mass and the number of drops needed to obtain all the product geometrical specifications. The virtual system here consists on the simulation model. The predictive system consists on a set of SWRL design rules that we have formalized, for instance to predict the needed:

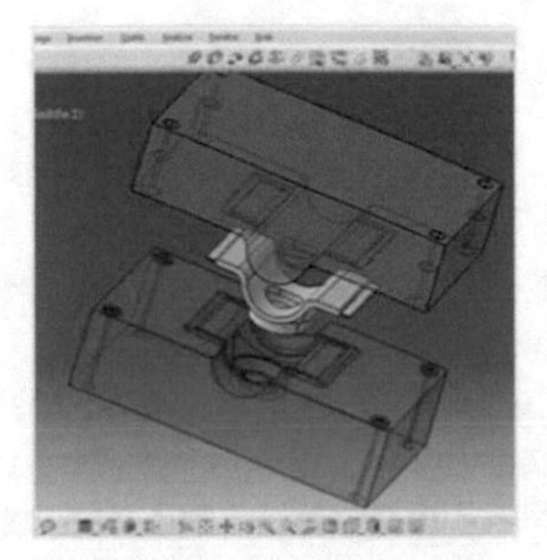

Fig. 2. The studied virtual product in the case of study [27]

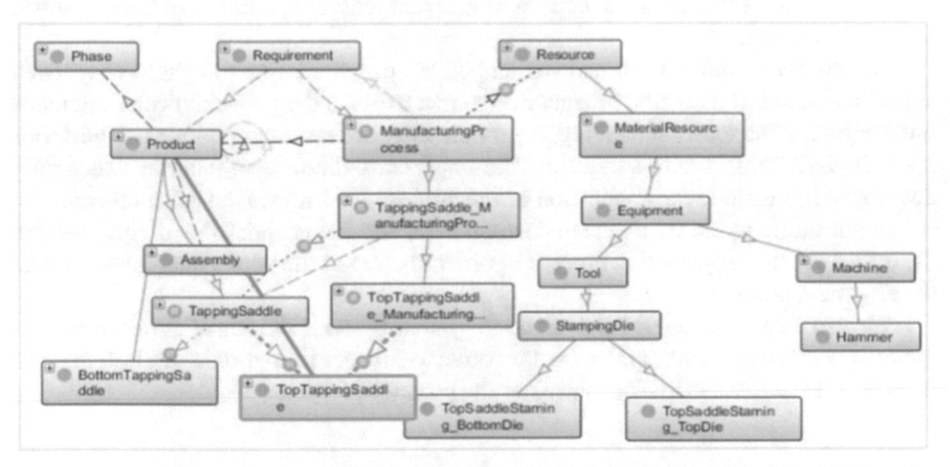

Fig. 3. The modeling components of the Digital Twin-driven architecture for the case of study: The modeling and reasoning system [27].

- Engine Mass: Product(?p) ^ hasSurface_PartWithFlash(?p, ?s) ^ hasCorrectedM-SPU(?p, ?a) ^ swrlb:multiply(?m, ?s, ?a) ^ swrlb:lessThan(?m, 630) -> hasEngine-Mass(?p, ?m) ^ useHammer(?p, "Hammer630T")
- Number of drops: Product(?p) ^ hasWeight(?p, ?w) ^ hasFlashInPer100(?p, 5) ^ swrlb:lessThan(?w, 1) ^ swrlb:greaterThan(?w, 2) -> useNumberOfDrops(?p, 5)

To validate the DT and predictive system efficiency, an experience is conducted in the physical world with the predicted parameters (the predicted engine mass and the number of drops).

The behavior and the geometric specifications of the obtained physical product are then transfer in real time to modeling and reasoning system. The consistency checking of the ontology CPD-Onto after its alimentation with the test experience outputs allows to confirm the efficacy of the DT and predictive models or to readjust and reteach them in order to obtain better systems in term of precision.

6 Conclusion

We explore in this work the potentialities of one of the most relevant emerging paradigms in the industrial context which is the Digital Twin. Since the design process is one of the most strategic phases of industrial products lifecycle, we focused on the potentialities of DT particularly for industrial product design.

The paper characterizes Digital Twin entities, analyses its promising applications and identifies the key challenges that industrials and researchers must overcome to achieve the demanded flexibility, reactivity, integration and systems intelligence in today's industrial products design processes. These challenges are related mainly to semantic interoperability between the involved entities, effective integration between the virtual and physical entities and the persistent need of inherent reasoning abilities in the developed design frameworks.

In response to these needs and challenges, we proposed a new Digital Twin-driven approach for smart industrial product design that exploits the potentialities of inference ontologies to enhance the modelling, interoperability and reasoning abilities of the developed Digital Twin. A case study from the impression-die forging industry was finally developed to illustrate the application of the proposed DT-driven design approach.

In our future work, we intend to develop a holistic framework using digital twins in the entire product engineering process in order to support the whole integrated product development process.

We also plan to complete the work done by incorporating the potentialities of ontologies and other AI tools in our DT models in order to enhance the reasoning and automation of other relevant issues in other stages of the product lifecycle.

References

1. Wang, Y., Liu, L., Liu, A.: Conceptual design driven digital twin configuration. In: Digital Twin Driven Smart Design, pp. 67–107. Elsevier (2020). https://doi.org/10.1016/B978-0-12-818918-4.00003-8
2. Tao, F., et al.: Digital twin-driven product design framework. Int. J. Prod. Res. **57**(12), 3935–3953 (2019)
3. Glaessgen, E., Stargel, D. : The digital twin paradigm for future NASA and U.S. air force vehicles. In: Structures, Structural Dynamics, and Materials and Co-located Conferences, 22267B (2012)
4. Grieves, M., Vickers, J.: Digital twin: mitigating unpredictable, undesirable emergent behavior in complex systems. In: Transdisciplinary Perspectives on Complex Systems: New Findings and Approaches, pp. 85. Springer International Publishing, Cham, s.l. (2017)
5. Schleich, B., Anwer, N., Mathieu, L., Wartzack, S.: Shaping the digital twin for design and production engineering. CIRP Annals **66**, 141 (2017)
6. Boschert, S., Rosen, R.: Digital twin—the simulation aspect. In: Mechatronic Futures: Challenges and Solutions for Mechatronic Systems and their Designers, pp. 59. Springer International Publishing (2016)
7. Rosen, R., von Wichert, G., Lo, G., Bettenhausen, K.D.: about the importance of autonomy and digital twins for the future of manufacturing. IFAC-PapersOnLine **48**, 567 (2015)
8. Tao, F., et al.: Digital twin driven product design framework. Int. J. Prod. Res. **57**(12), 3935–3953 (2019)

9. Tao, F., Zhang, M., Liu, Y., Nee, A.Y.C.: Digital twin driven prognostics and health management for complex equipment. CIRP Ann. Manuf. Technol. **67**, 169–172 (2018)
10. Tao, F., et al.: Five-dimension digital twin model and its ten applications. Comput. Integr. Manuf. Syst. **25**(1). pp. 1–18 (2019)
11. Botkina, D., et al.: Digital twin of a cutting tool. Procedia CIRP **72**, 215 (2018)
12. Monostori, L., et al.: Cyber-physical systems in manufacturing. CIRP Annals **65**, 621 (2016)
13. Lanza, G.: In-line measurement technology and quality control. In: Metrology. Springer (2019)
14. Wartzack, S., Schleich, B., Aschenbrenner, A., Heling, B.: Toleranzmanagement im kontext von industrie 4.0. ZWF Z. füre Wirtsch. Fabr. **112**(3), 170–172 (2017)
15. Bilberg, A., Malik, A.: Digital twin driven human-robot collaborative assembly. CIRP Ann. Manuf. Technol. **68**(1), 499–502 (2019)
16. Zhang, M., Zuo, Y., Tao, F.: Equipment energy consumption management in digital twin shop-floor: a framework and potential applications. In: 2018 IEEE 15th International Conference on Networking, Sensing and Control (ICNSC), Zhuhai, (2018)
17. Liu, J.F., Zhou, H.G., Tian, G.Z., Liu, X.J., Jing, X.W.: Digital twin-based process reuse and evaluation approach for smart process planning. Int. J. Adv. Manuf. Technol. **100**(5–8), 1619–1634 (2019)
18. Liu, Z., Meyendorf, N., Mrad, N.: The role of data fusion in predictive maintenance using digital twin. AIP Conf. Proc. **1949**(1), 020023 (2018)
19. Huang, B.B., Zhang, Y.F., Zhang, G., Ren, S.: A framework for digital twin driven product recycle, disassembly and reassembly. In: Proceedings of International Conference on Computers and Industrial Engineering, Auckland (2018)
20. Tao, F., Zhang, H., Liu, A., Nee, A.Y.C.: Digital twin in industry: state-of-the-art. IEEE Trans. Ind. Inf. **15**(4), 2405–2415 (2019)
21. Tao, F., Zhang, M., Nee, A.Y.C.: Digital Twin Driven Smart Manufacturing. Elsevier (2019)
22. Zhuang, C.B., Liu, J.H., Xiong, H., Ding, X.Y., Liu, S.L., Weng, G.: Connotation, architecture and trends of product digital twin. Comput. Integr. Manuf. Syst. **23**(4), 753–768 (2017)
23. Yu, Y., Fan, S.T., Peng, G.Y., Dai, S., Zhao, G.: Study on application of digital twin model in product configuration management. Aeronaut. Manuf. Technol. **526**(77), 41–45 (2017)
24. Schleich, B., Anwer, N., Mathieu, L., Wartzack, S.: Shaping the digital twin for design and production engineering. CIRP Ann. Manuf. Technol. **66**(1), 141–144 (2017)
25. Zhang, H., Liu, Q., Chen, X., Zhang, D., Leng, J.: A digital twin based approach for designing and multi-objective optimization of hollow glass production line. IEEE Access **5**, 26901–26911 (2017)
26. Wagner, R., Schleich, B., Haefner, B., Kuhnle, A., Wartzack, S., Lanza, G.: Challenges and potentials of digital twins and Industry 4.0 in product design and production for high performance products. In: Procedia CIRP, vol. 84, pp. 88–93 (2019)
27. Abadi, A., Ben-Azza, H., Sekkat, S.: Improving integrated product design using SWRL rules expression and ontology-based reasoning. Procedia Comput. Sci. **127**, 416–425 (2018)
28. Bilberg, A., Malik, A.: Digital twin driven human–robot collaborative assembly. CIRP Ann. **68**(1), 499–502 (2019)
29. Mandolla, C., Petruzzelli, A., Percoco, G., Urbinati, A.: Building a digital twin for additive manufacturing through the exploitation of blockchain: A case analysis of the aircraft industry. Comput. Ind. **109**, 134–152 (2019)

Towards a Generic Architecture of Context-Aware and Intentional System

Imane Choukri(✉), Hatim Guermah, and Mahmoud Nassar

IMS Team, ADMIR Laboratory, Rabat IT Center, ENSIAS, Mohammed V University in Rabat, Rabat, Morocco
imane.choukri@um5s.net.ma, {hatim.guermah, mahmoud.nassar}@um5.ac.ma

Abstract. Context-awareness is a vital character of IoT and pervasive systems, helping to understand user needs with minimum human intervention. Furthermore, the context-awareness must be completed with Intentionality to improve the accuracy of delivering suitable service for a user. Most proposed approaches focus on the context's static nature but don't give enough importance to the user's intention and context's dynamic nature. Then bringing Intentionality to context-awareness methods is an open-ended research topic with many challenges and opportunities for innovation. The key issue of our research is enhancing the relationship between the user's context and intention based on a situation-centered approach. This paper proposes a situation-centric meta model by providing practitioners with architecture to develop a context-sensitive intentional service, improving the relationship between the user's context and the intention. An illustrative example is provided to explain the proposed architecture using E-tourism as a case study.

Keywords: Intentionality · Context-awareness · Situation · Behavior · Service · Pervasive computing · Mobile computing

1 Introduction

Nowadays, mobile phones have evolved beyond simple telecommunication devices into smart gadgets incorporated into our personal daily life. The advent of wireless LANs and the democratization of several mobile devices and wearable computers (like tablets, notebooks, laptops, smartwatches, digital fitness bands) turn mobile computing more convenient and a catalyst for many research efforts.

On the other hand, the new mobile-centric perspective creates an environment saturated with computing and telecommunications technologies and then provides a platform for pervasive computing. Actually, pervasive computing became a significant trend in the 21st century [22], with the challenge of integrating embedding computational capability to work together and yet, to interact gracefully with users within minimum conscious interaction.

Therefore, the design of such Pervasive Information Systems (PIS) must support all practical user situations, which main require more efforts from the IT department on extending designs and models.

M. Lazaar et al. (Eds.): BDIoT 2021, LNNS 489, pp. 274–285, 2022.
https://doi.org/10.1007/978-3-031-07969-6_21

To overcome the complexity of such systems, a pervasive system must be able to adapt its behavior to the surrounding environment and to user preferences and desires in a dynamic manner. This objective could be achieved by raising the system capabilities, ranging from context-awareness to intentionality-awareness. That is why we stand for a user situation centric vision, combining intentionality and context awareness to meet user expectations without distracting users at a conscious level.

To achieve this goal, it is suitable to identify and formalize the contextual knowledge relevant to its runtime reconfiguration; furthermore, the system must observe user requests and user behaviors to understand user intention and targeted goal. Many approaches have been proposed to deal with context-awareness, intentionality, and both of them, but few of them have highlighted the intern relationship linking Context and Intention. Therefore, we promote a situation-centric vision of PIS, bringing a context-aware intentional approach that follows specific interpretation rules or reasoning to detect and compose suitable services able to satisfy the user's ultimate goal. This paper argues the need for a generic architecture for modeling and reasoning on user situations.

The remainder of this paper is structured as follows: The next section presents a motivation scenario that concerns the E-Tourism system. Section 3 presents some related work in context awareness and intentionality. In Sect. 4, we describe the proposed architecture and outline an intentional meta-model. The last section summarizes concluding remarks and plans for future works.

2 Motivating Scenario

Sharing information on social media platforms, especially posts made by online influencers, and bloggers significantly influence online customer purchase behaviour. The integration of new technologies in the tourism industry gives rise to a new business model based on ICT (Information and Communications Technology). Contemporary tourism businesses able to work successfully on online services are able to improve their customization effectiveness and reach new customers. However, in this field, a better knowledge of the user and his preferences is necessary to sell timely and intended/desired service to the target customer. The keys to online customer-centric marketing's success are the comprehension of his situation, intention, and the context he demands a service.

Let's consider a user connected to social networks with his profile (user profiling: traveller, business traveller), executing a sequence of actions (looking at last holidays friends' pictures, chatting in a traveling group, etc.). This user is a potential E-tourism customer as his intention may be "going on holiday or a business trip". This intention may be explicitly expressed on his request or understood from his observed actions. Here, the user situation includes his context information (profile, preference, etc.) and his intention information (his requests, his performed actions.). The system must use this information about the user to infers and obtains a result that he is "planning travel" and then proposing services to meet this user's needs.

The scenarios depicted above illustrate the benefits of context-awareness and intentionality for E-tourism systems. The adopted system example of the following paragraphs is an E-tourism system. Since our research scope is more interested in service-oriented Systems (SOA), we extend this system into a service-based system which also cooperates with external services providers.

3 Background and Related Work

3.1 Context-Awareness

In general, a context refers to "any information that can be used to characterize the situation of entities (i.e., whether a person, place, or object) that are considered relevant to the interaction between a user and an application, including the user and the application themselves." [1].

The same author Dey [5] presents a general definition "A system is context-aware if it uses context to provide relevant information and/or services to the user, where relevancy depends on the user's task".

Therefore, we can state that the context-awareness characterizes the system's ability to perceive incoming context information or gather them itself and then behave accordingly. This context-awareness feature allows the system to react according to the perceived user context and then enhance the system autonomy. It contributes to minimizing system intrusiveness and is becoming one of the most crucial concepts of a pervasive system.

3.2 Intentionality

Intention Definition. Several works have attempted to formalize the meaning of intention in the computing area. However, a universally accepted definition is yet to be agreed.

According to Jackson [10], an intention refers to "an optative" statement, by opposition to an indicative one, expressing a state expected to be reached or maintained in the future. In general terms, the notion of intention can be defined as the desired and intended goal needless to specify how to reach it [17, 18]. Therefore, it can be seen as the requirement that a user desires to be satisfied without specifying how to perform it or care about what service allows him to reach it [14].

On the other hand, the difference between intention and requirement was emphasized by Van Lamsweede [21]. He considers intention as a perspective statement the whole system under consideration should achieve when a requirement is a perspective statement that only the system's software part is considered by.

From another perspective, according to Santos et al. 2009, an intention is a goal that can be achieved by performing a process presented as a sequence of strategy and target intention [20].

Intentional Modeling. Several research works attempted to represent the user's intention, and different kinds of intention representations have been proposed. Nevertheless, we can categorize the significant tendency as follow:

- Task model:

  Tasks are considered as intentional actions performed to reach a goal. The task model represents the relation between the goal and the task, which is supposed to support the goal in a successful execution [19, 20].

- Goal model:

  The intention is expressed using a statement respecting a linguistic pattern; verb, target, and parameters [15, 16].
- Map model:

  Mirbel and Crescenzo [13] captured the system's intentional properties, at a business level, using a map representation introduced by C. Rolland [17]. That map representation was the first to link intention to his strategy terms (manner to achieve the intention) and introduce capturing variability at the goal level. The map is a labelled directed graph with nodes as intentions and strategies as edges.
- GORE models:

  GORE models (Goal-oriented requirements engineering) provide an earlier representation of the system-to-be and its environment at the requirement analysis phase. GORE proposal (GORE, i* [24], Tropos [2], Knowledge Acquisition in Automated Specification (KAOS) [4]) considers that actors and their goals are the reason for requirements. GORE brings useful constructs to goal elicitation, goal refinement, capturing variability [9], and handling conflicts.
- Behavioural approach:

  This approach recognizes intention based on observed user behaviour, logged action sequence [3, 7, 8, 11, 23, 25]. Those studies bring a temporal dimension to intention, observing user behavior, logging his action, and considering a situation's sequence over a period.

  Those models bring understandability and expressiveness, but regarding variability, the Map model outperforms all of them as it proposes a multitude of paths and strategies to gain an intention.

4 Contribution

4.1 Intention and Context Relation

According to the definitions listed above (see III.A), the Context-aware systems are more concerned with service execution conditions (who is, where is, when is, and what is …). It does not look at why a situation occurs; it only uses information about circumstances to execute the corresponding and designed out coming action. However, intention-awareness (intentionality) is more concerned with what the current user wants and why a service is called. Combining intention and context reduces the gap between the user's real needs and what he gets as a service proposition at execution time.

On the other hand, it is essential to highlight the close relationship between the Intention and Context [12]. The intention emerges in a given context; the surrounding context influences the user's intention; the intention is meaningful, using this context. Therefore, the intention is not a coincidence, but it is also not totally or exclusively or only depending on context; a user does not require a service just because he is under a given context (see Fig. 1. Intention and context relationships).

From what is quoted above, combining both concepts enhance the understanding of the user's real need. Intention leads us to capture user's desires, while Context helps us manage the variability and dynamics of the surrounding environment and infer and understand the user's implicit intention.

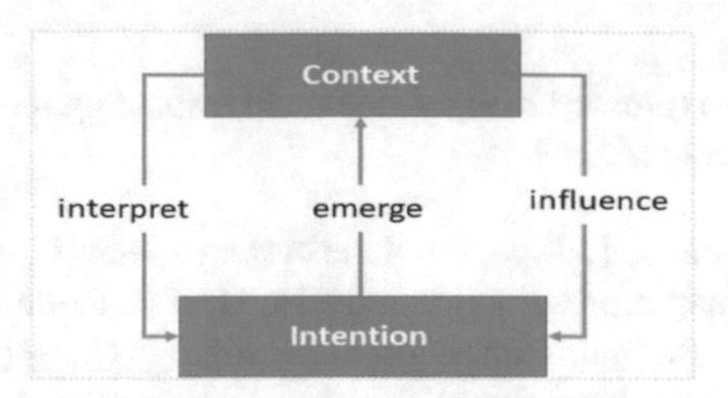

Fig. 1. Intention and context relationships

Intention Concept Understanding. As discussed earlier (paragraph Intention definition), we consider that an intention is an optative statement expressing an abstract description of a contextual situation that is expected to be maintained or to be reached in the future without caring about how to perform it.

The user's implicit intention can be deduced from a sequence of his action. We believe that user behavior is driven by his intention.

According to our understanding of intention, we specify three types of intention (see Fig. 2. Intention type hierarchy.): Declarative intention, Behavioral intention, and Semantic intention.

- Declarative intention: Explicitly expresses the desired results or outcomes of user's requests respecting a predefined model, using an optative statement, or merely a natural language.
- Behavioral intention: Believing that user's actions and behavior are driven by an implicit intention, we consider that the observation of user's participation and interaction takes effect in determining the behavioral intention and taking part in selecting suitable service.
- Semantic intention: The user's implicit intention that is not directly expressed by the user and needs to be inferred from a simple user request, background knowledge, and auxiliary information, such as the user's profile and context data.

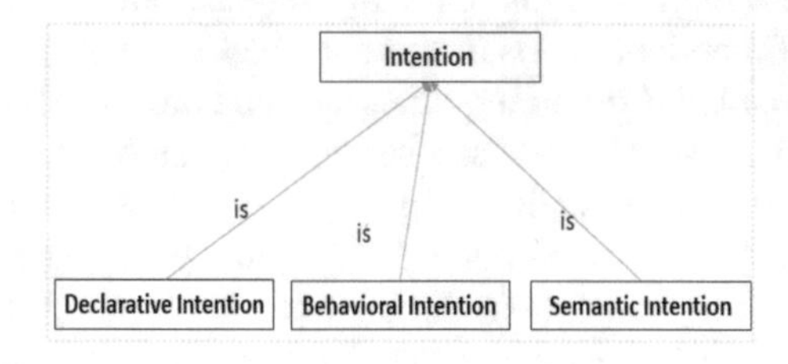

Fig. 2. Intention type hierarchy

Context Concept Understanding. In order to exploit context in a relevant way to our intentional approach, we enumerate three context types (see Fig. 3. Context type hierarchy):

- Personal context: Set of personal information, condition, and states of the user, such as his profile, health status, and physical states.
- Environmental context: The set of user's external factors like the weather and geographical settings. It includes all aspects related to the real surrounding environment in addition to hardware and software systems (such as internet connectivity).
- Behavioral context: Records of a user action such as consuming services like booking a hotel. It also includes the user's passive activity, such as a walking activity be detected by sensors.

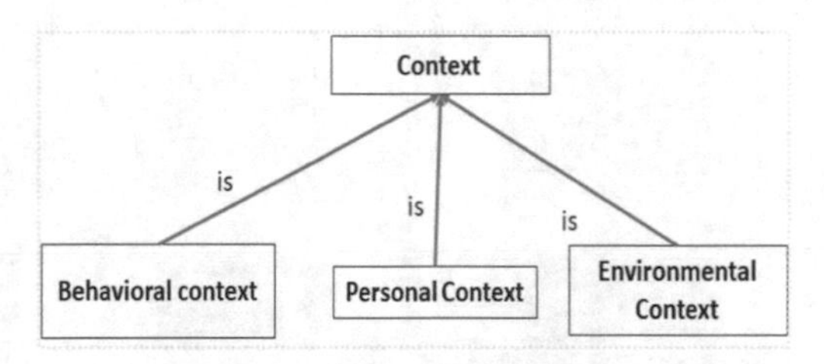

Fig. 3. Context type hierarchy

4.2 Proposal for a Preliminary Architecture

In this section, we propose an architecture to extract and use information from the user's situation to recognize the user's intention and select useful services for the user in different contexts (see Fig. 4 Overview of the proposed architecture). This architecture relies on the proposed intention meta-model, considering that the user's intention and surrounding context are part of the real-time user situation. This architecture and the meta-model are based on the following aspects:

- Relying context and intention to a situation to make them more relevant,
- Considering the relations between the intention and context,
- Understanding user intention from his behavior and not only his declarative intention,
- Extending the semantic meaning of the perceived intention.

The components of the proposed architecture can be divided into four modules:

- **Communication Module:** represents the bridge that lies between the user, internal modules, and external service. It has for role in enabling communication and handling data exchange between the different elements of the system.
- **Context Extraction Module:** analyzes the captured context and extracts Environmental Context, Personal Context, and Behavioral Context. Thus, it builds up the user context.

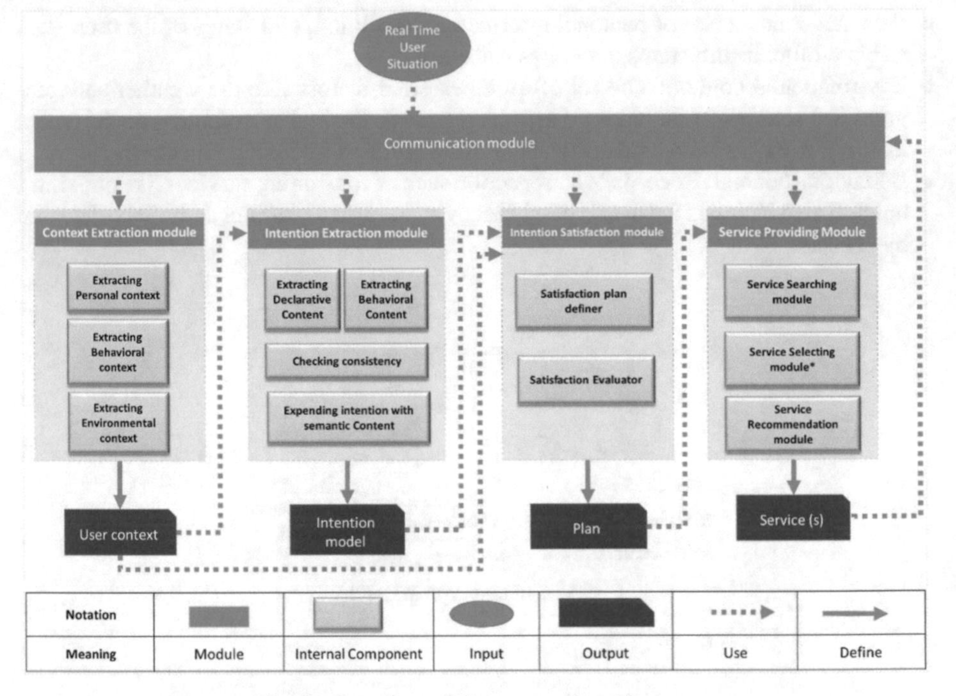

Fig. 4. Overview of the proposed architecture

- **Intention Extraction Module:** composed of three Internal components, in charge of extracting the intention behavioral content, intention declarative content, and extending intention within semantic content using user situation and the extracted user context. The "Checking consistency component" check consistency between different kinds of extracted intentions.
- **Intention Satisfaction Module:** uses the identified intention model and user context to recognize the plan satisfaction. Thus, it determines the sets of intermediate intention compatible with constraints and satisfying the satisfaction indicator.
- **Service Providing Module:** in charge of delivering service (s) that satisfy the generated plan.

4.3 The Proposed Intentional Model

Based on our understanding of intentionality and context-awareness, we propose a model combining both concepts and considering context and intention as part of the user situation (Fig. 5).

Meta-model Presentation. We introduce below the different concepts used in the proposed model and their relationships (Fig. 1. Intentional meta-model).

- **Situation:** We define a situation at time t as a pair of {I, C}t where 'I' represents the user's intention, 'C' stands of Context associated with the situation, on which the intention merged.

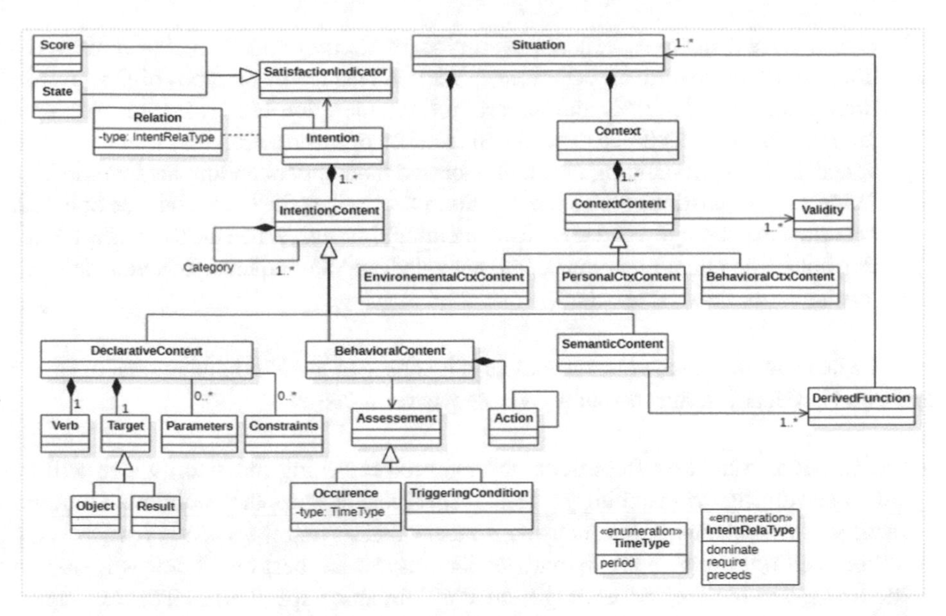

Fig. 5. Intentional meta-model

- **Context:** the context is constituted of "ContextContent" ("BehavioralCtxContent", "PersonalCtxContent", "EnvironementCtxContent") in this work we are not going to present detailed context information.
- **Intention:** the intention content is represented within its "IntentionContent", which is subdivided into three sub-class DeclarativeContent", "BehavioralContent", and "SemanticContent" (described following).
- **Intention Declarative Content structure:** Following the goal template represented by [15], a declarative intention is expressed with a declarative sentence: Verb, Target, and set of parameters and constraints.

  - **Verb**: expresses the action that determined the intention's meaning and allowing the realization of intention.
  - **Target**: is mandatory, denotes either the Object or Result of the action of the intention.
  - **Parameters**: optional, but they are recommended, any characteristic that can help bring precision and explanation to the intention definition, especially the Target.
  - **Constraints**: bring restriction and particular limits to the Target (Object o Result) or the action to be executed (Verb).

- **Intention Behavioral Content structure:** A temporally ordered sequence of named actions that represent an intention to achieve. When this sequence of action is perceived on user behavior, it is not necessarily intentional behavior. Then the assessment of behavior indicates if the intention is activated or not. We define two kinds of behavior assessment as follow:

- **Occurrence time:** the user action is significant, for the comprehension of his intention, according to a given occurrence. We define three basic types of Occurrence time: frequency, duration, and interval (cf. enumeration Fig. 1. Intentional meta-model). This type's choice depends on the kind of information that would be most beneficial to understanding user behavior and then understanding his intention.
- **Triggering condition:** Notes the condition that makes the user sequence of action occurrence meaningful and gives it an intentional aspect. When the triggering condition holds, the intention is considered activated, and the sequence of action defining this intention must be checked.

The behavior assessment is unnecessary; it is used to avoid nonintentional behavior and non-significant occurrence of a given sequence of action.

- **Satisfaction indicator:** Represents the required matching indicator, which will be calculated during the matching process. It specifies the acceptable minimum requirement for the selection of alternatives. It means that during the selection process, the system will not evaluate all alternatives looking for the best one, but it will stop on the first satisfying the "Satisficing indicator". It can be specified as a score, degree, or as a state. It is optional, but it improves the performance and optimizes effort on evaluating all alternatives.
- **Relationships:** following [6], we defined three structural relations between intentions, namely, dominate, precedes, and require.

All that named element extend the superclass "NamedElement".

A Concert Illustration Example. To illustrate our approach, we apply the described meta-model over the following two scenarios presented as tables within the situation's elements:

- **The first scenario:** refers to a user performing a request "book hotel.". Table 1: The first scenario shows our adopted meta-model's instantiation according to the real-time user situation.

Table 1. The first scenario

Elements	Example of instantiation
Intention Declarative Content	**Request**: "Book hotel"
Intention Behavioral Content	**List of Action**: {looking on trip picture}, **Occurrence**: 4 times this day (frequently)
Context Content	**Personal State**: on holiday
Semantic Intention	Searching for booking a hotel

- **The second scenario:** with the same user request of the first scenario, but this time within different values of Intention Behavioral Content, Context Content (cf. Table 2: The second scenario). Thus the inferred semantic intention is different "Searching for books about hotels for rent or purchase". This example shows the importance of using of a consistency checker component; it helps to avoid intention ambiguity.

Table 2. The second scenario

Elements	Example of instantiation
Intention Declarative Content	**Request**: "Book hotel."
Intention Behavioral Content	**List of Action**: {Searching for book}, **triggering condition**: The user is connected to the library system
Context Content	**User schedule**: reading time
Semantic Intention	Searching for **books** about **hotel**s for rent or purchase

5 Conclusion

In this paper, we presented a user situation-based architecture for the development of context-aware and intentional services. This architecture introduces a new approach to understand user intention. It links context and intention to a situation to make them more relevant, considers the relationship between the intention and context. It goes beyond user declarative intention and uses observed user behaviour to understand his intention, and finally, it extends semantic mining of the perceived intention. Actually, this architecture proposes using observed user situations to extract information, formalize and construct contextual and intentional knowledge relevant to the user. And then, our approach uses reasoning methods (which are not in the scope of this paper) on the extracted intention model, taking into account user context to perform service adaptation to the users. Therefore, constructing knowledge for user situation interpretation is the cornerstone of our architecture. We proposed a model combining both concept context and intention, considering them as part of the user situation, and extending the Intention's semantic meaning. This model has the particularity to be generic and extensible; it can be integrated with various domains depending on needs.

Nevertheless, we illustrate our proposition by an E-tourism motivating scenario. We project to provide more details on each module and component in our future work, especially the context extracting module and intention extracting module. For the Intention Satisfaction Module, we will evaluate reasoning mechanisms and specify the used algorithm in each step of our Satisfaction plan definer and Satisfaction Evaluator components.

References

1. Abowd, G.D., Dey, A.K., Brown, P.J., Davies, N., Smith, M., Steggles, P.: Towards a better understanding of context and context-awareness. In: Gellersen, H.-W. (ed.) HUC 1999. LNCS, vol. 1707, pp. 304–307. Springer, Heidelberg (1999). https://doi.org/10.1007/3-540-48157-5_29
2. Bresciani, P., Perini, A., Giorgini, P., Giunchiglia, F., Mylopoulos, J.: Tropos: an agent-oriented software development methodology. Auton. Agent. Multi-Agent Syst. **8**, 203–236 (2004). https://doi.org/10.1023/B:AGNT.0000018806.20944.ef
3. Chang, C.K., Jiang, H., Ming, H., Oyama, K.: Situ: A situation-theoretic approach to context-aware service evolution. IEEE Trans Serv Comput **2**, 261–275 (2009). https://doi.org/10.1109/TSC.2009.21
4. Dardenne, A., van Lamsweerde, A., Fickas, S.: Goal-directed requirements acquisition. Sci. Comput. Program. **20**, 3–50 (1993). https://doi.org/10.1016/0167-6423(93)90021-G
5. Dey, A.K.: Understanding and using context. Pers. Ubiquit. Comput. **5**, 4–7 (2001). https://doi.org/10.1007/s007790170019
6. Fki, E., Tazi, S., Drira, K.: Automated and flexible composition based on abstract services for a better adaptation to user intentions. Futur. Gener. Comput. Syst. **68**, 376–390 (2017). https://doi.org/10.1016/j.future.2016.07.008
7. Ghasemi, M., Amyot, D.: From event logs to goals: a systematic literature review of goal-oriented process mining. Requirements Eng. **25**(1), 67–93 (2019). https://doi.org/10.1007/s00766-018-00308-3
8. Hajer, B., Arwa, B., Lobna, H., Khaled, G.: Intention mining data preprocessing based on multi-agents system. Procedia Comput. Sci. **176**, 888–897 (2020). https://doi.org/10.1016/j.procs.2020.09.084
9. Horkoff, J., et al.: Goal-oriented requirements engineering: an extended systematic mapping study. Requirements Eng. **24**(2), 133–160 (2017). https://doi.org/10.1007/s00766-017-0280-z
10. Jackson, M.J.: Software Requirements and Specifications: a Lexicon of Practice, Principles, and Prejudices. ACM Press ; Addison-Wesley Pub. Co, New York : Wokingham, England ; Reading, Mass (1995)
11. Lee, W.-P., Lee, K.-H.: Making smartphone service recommendations by predicting users' intentions: A context-aware approach. Inf. Sci. **277**, 21–35 (2014). https://doi.org/10.1016/j.ins.2014.04.033
12. Liu, C.: Examining effects of context-awareness on ambient intelligence of logistics service quality: user awareness compatibility as a moderator, **8** (2018)
13. Mirbel, I., Crescenzo, P.: Improving collaborations in neuroscientist community. In: 2009 IEEE/WIC/ACM International Joint Conference on Web Intelligence and Intelligent Agent Technology. IEEE, Milan, Italy, pp 567–570 (2009)
14. Najar, S., Pinheiro, M,K., Souveyet, C.: A context-aware intentional service prediction mechanism in PIS. In: 2014 IEEE International Conference on Web Services. IEEE, Anchorage, pp. 662–669 (2014)
15. Prat, N.: Goal Formalisation and Classification for Requirements Engineering, pp. 145–156. Barcelena, Spain (1997)
16. Prat, N.: Goal formalization and classification for requirements engineering, fifteen years later. In: IEEE 7th International Conference on Research Challenges in Information Science (RCIS). IEEE, Paris, France, pp. 1–12 (2013)
17. Rolland, C.: Capturing system intentionality with maps. In: Krogstie, J., Opdahl, A.L., Brinkkemper, S. (eds.) Conceptual Modelling in Information Systems Engineering, pp. 141–158. Springer, Berlin Heidelberg (2007)

18. Kaabi, R.S.: A methodological approach for modeling and operationalzing intentional services. Paris 1 La Sorbonne University (2007)
19. Shimada, H., Nakagawa, H., Tsuchiya, T.: Constructing a goal model from requirements descriptions based on extraction rules. In: Kamalrudin, M., Ahmad, S., Ikram, N. (eds.) APRES 2017. CCIS, vol. 809, pp. 175–188. Springer, Singapore (2018). https://doi.org/10.1007/978-981-10-7796-8_14
20. da Silva Santos, L.O.B., Guizzardi, G., Pires, L.F., van Sinderen, M.: From user goals to service discovery and composition. In: Heuser, C.A., Pernul, G. (eds.) ER 2009. LNCS, vol. 5833, pp. 265–274. Springer, Heidelberg (2009). https://doi.org/10.1007/978-3-642-04947-7_32
21. Van Lamsweerde, A.: Goal-oriented requirements engineering: a guided tour. In: Proceedings fifth IEEE International Symposium on Requirements Engineering. IEEE, pp 249–262 (2001)
22. Weiser, M.: The Computer for the 21st Century. SIGMOBILE Mob. Comput. Commun. Rev. **3**, 3–11 (1999). https://doi.org/10.1145/329124.329126
23. Xie, H.: Observable context-based user intention specification in context-aware environments. In: 2011 IEEE 35th Annual Computer Software and Applications Conference. IEEE, Munich, Germany, pp. 712–715 (2011)
24. Yu, E.S.K.: Towards modelling and reasoning support for early-phase requirements engineering. In: Proceedings of ISRE 1997: 3rd IEEE International Symposium on Requirements Engineering, pp 226–235. IEEE Computer Science Press, Annapolis (1997)
25. Yu, W., Yu, M., Zhao, T., Jiang, M.: Identifying Referential Intention with Heterogeneous Contexts. New York, pp. 962–972 (2020)

A Smart Healthcare Imbalanced Classes Model Using Multi Conditional-Task GAN

Wayoud Bouzeraib(✉), Afifa Ghenai, and Nadia Zeghib

Constantine2 – Abdelhamid Mehri University, Constantine, Algeria
{wayoud.bouzeraib,afifa.ghenai,
nadia.zeghib}@univ-constantine2.dz

Abstract. The rapid development in the internet of things (IoT) produced a heavenland for attackers due to neglecting the security aspect in manufacturing. Intrusion Detection System has shown a great promise in anomaly detection traffic in the network. However, this lastest had problems because of imbalance data and the minority classes problem. This paper presents a new security model using Machine Learning. In particular, this model is based on a Multi Conditional-Task Generative Adversarial network (MCTGAN) to address the problem of class imbalance in anomaly detection System. The proposed approach use is illustrated by a case study: A Smart Healthcare System-based scenario.

Keywords: Anomaly detection · Generative Adversarial Network (GAN) · Intrusion detection system · Internet of Things (IoT) · Machine learning

1 Introduction

In the era of the Internet of Things (IoT), an enormous amount of sensing devices collects and/or generates various sensory data over time for a wide range of fields and applications. Based on the nature of the application, these devices will result in big or fast/Realtime data streams. The applying analytics over such data streams may lead to discover new information and detect anomalies. Anomalies are system behavior patterns in time steps that do not conform to a well-defined notion of normal behavior [1]. Anomaly detection provides the opportunity to take actions to investigate and resolve the underlying issue before it causes disasters. Closely monitoring these environments is necessary as many of them are deployed for mission-critical tasks.

The Intrusion Detection System (IDS) is an essential part of network security. The IDS aim to automatically identify if there is malicious activity or policy violation by detecting abnormal network patterns. Most previous work implements intrusion detection with classical statistical learning methods, such as Naive Bayes, Decision Tree, Random Forest and Support Vector Machine. Inspired by the remarkable effect of deep learning, several recent studies have used neural networks for intrusion detection, including the Multilayer Perceptron, the Convolutional Neural Network, and the Recurrent Neural Network [2]. Anomaly detection is commonly approached as an unsupervised machine learning task due to the lack of label information [3].

M. Lazaar et al. (Eds.): BDIoT 2021, LNNS 489, pp. 286–299, 2022.
https://doi.org/10.1007/978-3-031-07969-6_22

Though the previous approaches have made profound progress, the class-imbalanced data are still a challenging problem that hinders the performance of most intrusion detection systems [4]. The class imbalance problem occurs when the number of intrusion samples is significantly lower than that of the normal ones.

To address the class imbalance problem in intrusion detection, we propose a novel Multi Conditional Tasks Generative Adversarial Network (GAN) based on Multi-Task learning and Conditional GANs concepts to generating representative samples data for minority classes. This work proposes MCTGAN to generate representative samples for minority classes, countering the class imbalance problem in intrusion detection and anomaly detection.

The remainder of this paper is organized as follows. Section 2 presents most relevant related work. The proposed solution is presented in Sect. 3. Section 4 is dedicated to a case study illustrating the proposed approach. Finally, Sect. 5 concludes this paper and highlights our future work.

2 Related Work

In the aim to highlight the actual contribution of this paper, it is necessary (1) to present class imbalance problem and the use of oversampling method, (2) to underline the limitations of the works which use only Generative Adversarial Network (GANs), (3) to focus on Conditional Generative Adversarial Networks, and (4) Multi-Task Generative Adversarial Networks. This allows situating our work relatively to the most relevant ones in this area.

2.1 Class Imbalance Problem

Class imbalance is a common problem in practical classification tasks (Fraud detection, Prediction of disputed/delayed, Predictive maintenance data sets…). The scarce samples of a minority class make the cost of misclassifying minority class could very high. Therefore, the key to solve class imbalance is to learn an accurate distribution of the minority classes with limited samples. There are multiple ways of handling unbalanced data. Some of them are: collecting more data, trying out different ML algorithms, modifying class weights, penalizing the models, using anomaly detection techniques, oversampling, and under-sampling techniques.

We are focusing mainly on oversampling techniques in this work since the majority does not have direct control over the collection of more data. Also, applying class weights or too much parameter tuning can lead to overfitting. The under-sampling technique can lead to the loss of important information. But that might not be the case with oversampling techniques. Oversampling methods can be easily tried and embedded in many frameworks. Oversampling can improve the accuracy of the learning on the distributions of minority classes by increasing minority samples, which is a common method to tackle the class imbalance problem [5]. Random oversampling, Synthetic Minority Over-Sampling Technique (SMOTE) [6] and Border-line SMOTE [7] are commonly used oversampling methods in classic imbalance problems. However, when dealing with data in high dimensional space, the quality of the synthesized new data points could still be compromised due to noise and poor distance measurement in the high dimensional space [8].

2.2 Generative Adversarial Networks

Generative models aim to learn exact data distribution to generate new samples points with some variations. The most efficient generative approaches are Variational Autoencoders (VAE) [9] and Generative Adversarial Networks (GAN) [10]. A variant of GAN architecture is Adversarial autoencoders (AAE) [11] that use adversarial training to impose an arbitrary prior on the latent code learned within hidden layers of autoencoder are also shown to learn the input distribution effectively. With this ability a several Generative Adversarial Networks-based Anomaly Detection (GAN-AD) frameworks [12–16] proposed to be effective in identifying anomalies on high dimensional and complex datasets. However traditional methods such as K nearest neighbors (KNN) are shown to perform better with lesser number of anomalies when compared to deep generative models [17].

Generative Adversarial Networks (GANs) trained in semi-supervised learning model have shown great promise [18], even if it was use of one class labeled data, can produce considerable performance improvement over unsupervised techniques. However, the fundamental disadvantages of semi-supervised techniques presented by [19] are applicable even in a deep learning context. Furthermore, the hierarchical features extracted within hidden layers may not be representative of fewer anomalous instances hence are prone to the over-fitting problem.

Latest development is anomaly detection models built on the GAN framework. The majority of GAN based methods have been applied in the image domain. A GAN-based method allows learning generative models that can generate realistic images [10]. In [20], the authors use a GAN based unsupervised learning model namely AnoGan to identify anomalies in medical imaging data. AnoGan uses an adversarial network to learn normal anatomical variability. Then it uses an anomaly scoring scheme by mapping images from image space to a latent space and reconstructing the images. The loss in a reconstructed image is used to estimate anomaly score. The anomaly score of an image indicates its deviation from the general data distribution and it can identify anomaly images. This model uses Convolutional Neural Network (CNN) in its generator and discriminator to effectively identify anomalies in images. On the other hand, in [21] LSTM are used in both generator and discriminator of a GAN-based model to generate realistic time series data in medical domain. However, they were not designed to detect anomalies. In [22], the authors propose a novel GAN-based unsupervised method called TAnoGan for detecting anomalies in time series when a small number of data points are available.

2.3 Conditional Generative Adversarial Networks

Conditional GANs apply GANs in the conditional setting. Prior and concurrent works have conditioned GANs on discrete labels [23], text [24], and, indeed, images. The image-conditional models have tackled image prediction from a normal map [25], future frame prediction [26], product photo generation [27], and image generation from sparse annotations [28]. In [29], authors deal with simpler CGANs than most others model with Image-to-Image Translation. The work in [30] uses Deep Conditional Generative Adversarial Nets and Feedforward Convolutional Neural Networks for Occluded Visual Object Recognition. Several other papers have also used GANs for image-to-image mappings, but only applied the GAN unconditionally, relying on other terms (such as L2 regression) to force the output to be conditioned on the input. These papers have achieved impressive results on inpainting [31], future state prediction [32], image manipulation guided by user constraints [33].

2.4 Multi-task Generative Adversarial Networks

In the seminal work [10], the generative adversarial network (GAN) is introduced to generate realistic-looking images from random noise inputs. GANs have achieved impressive results in many domains, leading to impressive and promising results. But SRGAN in [34] assumes that the input is a high-resolution image and contains fine details which are not available for the small objects in the wild. Compared to super resolution on natural images, images of specific objects in the COCO or WIDER FACE benchmarks are full of diversity (e.g. blur, pose and illumination), thus, making the super resolution process on these images much more challenging. In fact, the super-resolution images generated by SRGAN are blurred especially for low-resolution small objects, which is not helpful to train an accurate object classifier.

Consequently, the concept of multi-task in GANs is come to solve this problem as in [35] propose an end-to-end multi-task generative adversarial network (MTGAN) to solve the small object detection problem and [36] propose Sparsely Grouped Generative Adversarial Networks (SG-GAN) that can translate images in sparsely grouped datasets where only a few train samples are labelled in one model. Authors in [37] propose to transfer the style between synthetic and real data using Multi-Task Generative Adversarial Networks (SYN-MTGAN) before training the neural network which conducts the detection of roadside objects.

3 The Proposed Solution

In this section, an enhanced multi conditional-task generative adversarial networks (MCTGAN) is proposed to produce reproduce the distributions of minority classes from scarce samples and provide an effective solution for the class imbalance problem. Strengthen the typical Generative Adversarial Network (GAN) [10] to alleviate the class imbalance problem by generating samples for minority classes.

A typical GAN includes a generative model (Generator, G) and a discriminative model (Discriminator, D). G takes the latent space z as inputs and reconstructs them into synthesized samples G(z), which implicitly defines a generative distribution pg(z). D takes sample x as input and outputs the probability D(x) that x comes from the real distribution pdata(x) rather than pg(x). In other words, G maps z from the latent space distribution pz(z) to pg(z), while D distinguishes pdata(x) from pg(x). G is to maximize the probability that D makes a mistake. The optimization of GAN is performed to increase the similarity between pdata(x) and pg(z) as shown in Fig. 1.

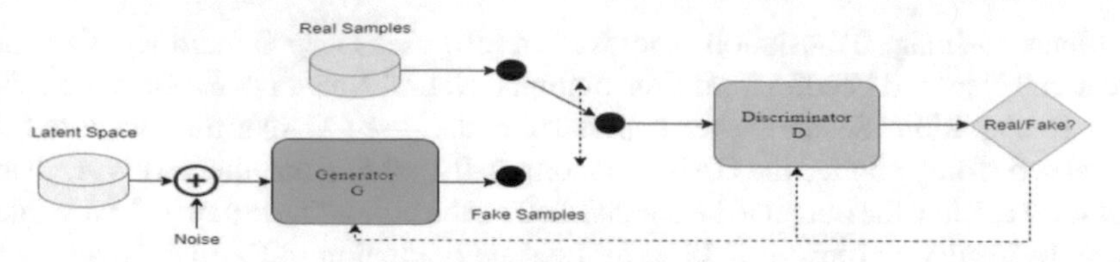

Fig. 1. Generative adversarial network [10]

GANs have achieved great success in image generation [38, 39], image-to-image synthesis [29], image super-resolution [34], and other applications due to its excellent capability of learning the data distributions by providing abundant training samples. In addition, GANs have also shown some potentials to solve class imbalance problems by learning the distributions of minority classes. Although GANs have been successfully applied in many tasks, due to the limited number of samples in the minority class as show in Fig. 1, GANs may only be able to learn part of the minority class distribution at the end of the train step, and therefore could be trapped by the local optimum. Some studies have been done to enable GANs to learn a more accurate distribution during the training process by utilizing improved adversarial learning objectives (e.g. LSGAN [40] energy-based GAN [41] and WGAN [42]). Nevertheless, there still exist limitations when using fixed adversarial training objectives in the training of GANs. More recently, Evolutionary-GAN (E-GAN) [43] was proposed and multiple generators were created by different adversarial objectives to overcome the limitations of the fixed adversarial objectives, and always kept the well-performed generator in the training process. However, the local optimum trapping problem has yet been addressed.

3.1 Conditional GAN (cGAN)

There are two motivations for making use of the class label information in a GAN model: (1) Improve the GAN, (2) Targeted data Generation. As show in Fig. 1 Additional information c that is correlated with the input data z, such as class labels, can be used to improve the GAN. This improvement may come in the form of more stable training, faster training, and/or generated data that have better quality. Class labels c can also be used for the deliberate or targeted generation of data of a given type. A limitation of a GAN model is that it may generate a random data from the domain G(z). There is a relationship between points in the latent space z to the generated data G(z'), but this relationship is complex and hard to map. Alternately, a GAN can be trained in such a way that both the generator G and the discriminator D models are conditioned on the class label. This means that when the trained generator model is used as a standalone model to generate data in the domain, data of a given type G(z'), or class label, can be generated.

3.2 Multi-Task GAN (MTGAN)

Multi-task learning (MTL) has several known mechanisms by which it works. These seem to have been first identified by Caruana [44].

Implicit Data Amplification
MTL can effectively increase the size of the dataset. This can occur in a direct manner, if the tasks are being trained on different datasets. Or it can be indirect, if the tasks are being trained on the same dataset. In this case, data amplification occurs since the tasks have their own data-dependent noise associated with them. Extra tasks mean the model can learn representations that have less data-dependent noise. The effect is similar to adding more data.

Eavesdropping
For some tasks, certain representations are easier to learn than they are for other tasks. This can depend on the quality of the labels, whether they can transmit good training signals through the model. In MTL, tasks can "eavesdrop" on other tasks to learn representations that they are struggling on.

Representation Bias
Hidden representations that benefit one task can often benefit another. MTL biases the training to learn those representations that benefit multiple tasks. This increases the generalization power of the model in facing new tasks.

Attention Focusing
If data is scarce or noisy, it can be difficult for a model to discern between irrelevant and relevant representations. Training with extra tasks provides extra evidence as to which representations are actually useful.

Before applying MTL to GAN architecture and running lots of very time-consuming experiments to explore their hyperparameter space, it is efficient to explore MTL applied to classification of CIFAR-100. This also serves to test the suitability of CIFAR as a dataset to perform MTL on.

For this reason, we split input data to multiple parts and try to generate each part as a task to learn and fusion this learning method to generate a new samples data for minority classes.

3.3 Learning General Data Distribution

A GAN trains the D and G alternately Because of this, at a certain step, a problem may occur whereby the training of the D and the training of the G cancel each other out [45]. By applying the multi-task concepts in generation of samples, as shown in Fig. 2, we attempt to improve the discrimination ability of the D and, accordingly, improve the performance of the G. First, the division of input data X to many inputs covers all the tasks of data coming with taking into consideration the condition in generation of each fake sample. After that, we feed the different inputs to Conditional generative model to try to generate samples that were discriminated against as real by the D. These different

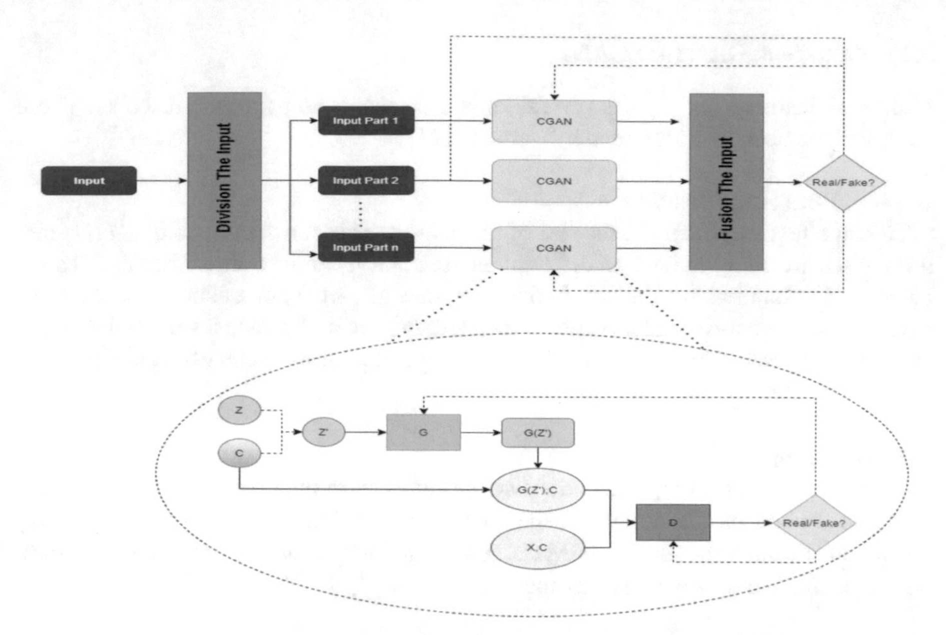

Fig. 2. Multi Conditional-Task Generative Adversarial Network **(MCTGAN)**

samples fusion together and feed another discriminator to check if that looks like the real one or not and give a feedback to discriminator and different CGANs. The D was trained. Through this process, the D should be performed better, in turn causing the G to improve even more. We propose Multi-Objectives Genetec GAN that aims to learn the accurate distribution from the minority class. First, our model uses different adversarial learning tasks to improve the performance of the generator. Second, our model incorporates the mechanism Conditional GAN into the training so that the model can converge to the distribution that is closest to the minority class.

3.4 Anomaly Detection

During anomaly detection in a small sequence data, we evaluate each small x as being a normal (i.e. originating from the general data distribution) or anomalous (i.e. deviated from the general data distribution) observation [22].

In the adversarial training process, the generator G learns the general data distribution p_g of latent space Z and the general data distribution p_{data} of real data space X. That is, each fake small sequence $G'(z)$ is generated based on the general data distribution of X. In every updating iteration λ, the loss function L evaluates the dissimilarity of the fake small sequence $G'(z)$ with the real small sequence x. An anomaly score A(x) expressing the fit of a given x to the general data distribution (i.e. model of normal small sequences) can be directly derived from L.

$$A(x) = (1 - \gamma) \cdot R(x) + \gamma \cdot D(x);$$

where the residual score R(x) and the discrimination score D(x) are defined by the residual loss LR(z) and the discrimination loss LD(z), respectively, at the final updating iteration of the mapping procedure to the latent space. A large anomaly score A(x) indicates an anomalous small sequence data whereas a small anomaly score indicates a small sequence fitting to the general data distribution of X learned by G during adversarial training.

4 Case Study

4.1 Smart Healthcare Domain

The Internet of things is fast growing and in the next few years, the medical sector is expected to see widespread Internet access through new Internet devices and applications. Healthcare devices and services are expected to deal with vital patient information such as personal health care data valuable information and cannot be allowed to change or interfere with the procedures followed with the information. In addition, these smart devices may be connected to global information networks to access them anytime and anywhere, therefore, the area of care in the Internet of things may be a target for attackers.

This technology is set to transform the healthcare industry within the next decade, as it has great potential for multiple applications, from remote monitoring to medical device integration. The next step seems obvious—the IoT will take into account electronic devices that catch or screen information and join them to a personal or public cloud so that these devices can automatically trigger certain events. Given the numerous smart devices in healthcare, such as thermometers, blood-pressure analyzers, glucose meters, smart beds, mobile X-ray machines, and ultrasound units, the IoT in healthcare could transform patient care.

4.2 System Description

As show in Fig. 3, IoT medical devices record data to send it to cloud for analysis and predict. Medical devices used for various healthcare applications can be subdivided into three major groups:

Wearable External Devices
Usually, these are biosensors that monitor physiological data with remote/wireless communication that can be used for telemedicine and inpatient monitoring. These devices monitor blood pressure, electrocardiogram (EKG), temperature, continuous glucose, oxygen level, etc.

Implanted Medical Devices
The implanted devices "replace a missing biological structure or support a damaged biological structure or enhance an existing biological structure." This category includes implantable infusion pumps and other drug-delivery devices, cardiac pacemakers, implantable neurostimulator systems, and glucose monitors.

Stationary Medical Devices
There is a wide scope of stationary restorative hardware that can be utilized in many different applications.

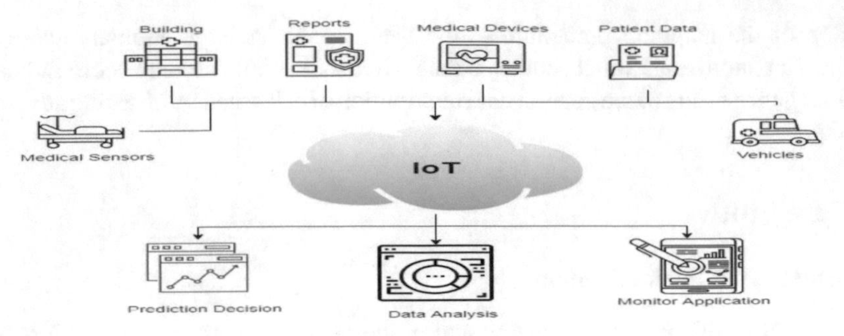

Fig. 3. IoT in Healthcare

In healthcare, the IoT assumes a vital position in numerous applications. The classifications of IoT based personalized healthcare systems are clinical care, remote monitoring, and context-awareness.

Clinical Care
In medical clinics, hospitalized patients, particularly those in critical conditions, need consistent and close consideration to react rather possible in an emergency case, which will increase the chances of saving a patient's life. Thanks to IPbased sensors, the necessary information about a patient's health can be collected remotely and sent to caregivers using the internet for further review and analysis.

Remote Monitoring
Some older and younger patients, as well as chronically sick patients, should be inspected almost every day. Remote monitoring will help these patients avoid making trips to the health center to be checked. In light of their critical status, some changes in their wellbeing will go unnoticed until an infection reaches the point of emergency. Remote access sensors will assist caregivers as they will provide them with analysis and the opportunity to intervene before an infection develops. In this manner, individuals with various degrees of psychological and physical incapacities will be empowered to have an increasingly autonomous and simple life.

Context Awareness
Being able to recognize a patient's condition and their surroundings will significantly help healthcare professionals to understand the variations that can impact the health of a patient. This is because nature has a significant impact on the health of all individuals. The utilization of a few kinds of specific sensors can collect data about the patient's physical condition, such as when they walk, sleep, or run, as well as data about their surroundings and whether it is wet, hot, or cold.

4.3 System Problem

Typically, in a practical scenario, the data sensed by the sensor nodes are either directly transmitted to the cloud or forwarded through the relay nodes so that this data is exposed to risk at every step of its transfer, as shown in Fig. 4. Further, the data sensed by the nodes are vulnerable to attacks and are prone to security threats. The integrity of this data may be reformed or modified by any attacker during the process of data transmission. In such a situation, the decisions generated from these sensor data may be misleading in nature.

The security of networks is necessary and vital to ensure access to sensitive and confidential information for enterprises, after the huge development in networks and communications in the past period was matched by a development in the field of penetration and the manufacture of electronic attacks. Clearly, as these attacks adversely affect the performance of the network where the damage and delay data from causing malfunctions and downtime in the network.

Therefore, the previously presented proposal supports this hypothesis as it allows the provision of rare-appearing data and enables us to complete the actual distribution of the distress data by selecting and modifying the generated data through the multi conditional tasks, which simulates the new attacks derived and modified.

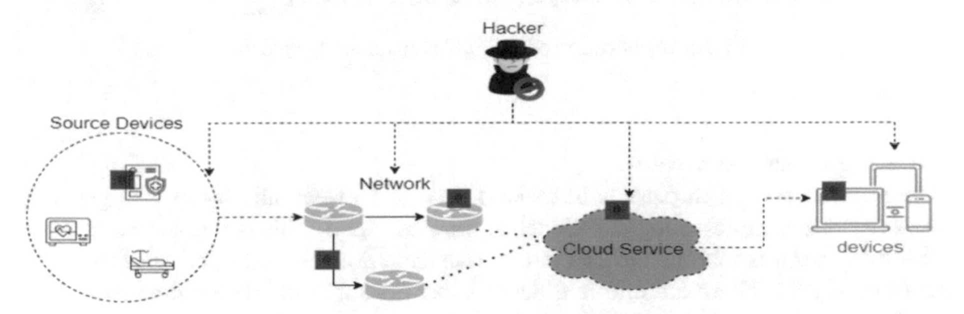

Fig. 4. IoT healthcare threats

4.4 Smart Healthcare Application

One of the most important uses of machine learning in the field of protecting healthcare is the intrusion detection systems (IDS) and Anomaly detection working on the cloud, as shown in Fig. 5, when the system alerts the owner in the event of a risk to the system. In order to improve the performance of smart healthcare in terms of security, many technologies were added to ensure data protection from change and loss, but this was not enough as we mentioned before. This is due to the fact that the healthcare devices contain many heterogeneous devices that rely on different protocols to communicate (such as Bluetooth, zigBee, and Wi-Fi). Therefore, different and unbalanced data saved on the cloud may cause errors in detection leading to multiple false alerts. Therefore, the process of controlling the balance of the data and monitory classes reduces the errors of detection.

In this case, our proposal not only helps generate the data but also contributes to speeding up the process of converging to the globe's minimal data distribution by using the multi conditional tasks mechanisms. This is achieved by generating the shortfall that occurs, which reduces the time taken cost of collect data from n.t + t' to just t. As we mentioned earlier, the process of making data balanced and of equal distribution especially monitoring classes.

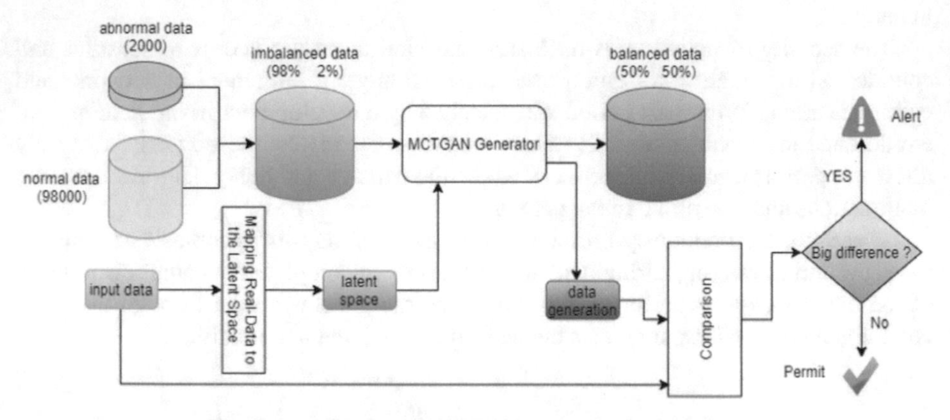

Fig. 5. Application of MCTGAN in anomaly detection

Minority Classes Generative

The generation of each class is to make a balance between all classes. We generate minority data samples using our model as show in Fig. 5. This is due to two reasons. The first reason is that the devices in the healthcare domain do not produce the same amount of data. There are different devices that do not work all the time like alarm devices. The second reason is the lack of some data that rarely occurs, such as attack data, from what causes the model to ignore it and not take it into consideration bias towards the majority (It takes 98,000 samples into account and neglects 2,000 samples). For this reason, making all kinds of data equal (98,000 samples) representation makes the model take into account all classes of data, which increases the accuracy of its attack or anomaly detection, especially new ones because it avoids being overfitting and gets the global minimum.

We consider that we want to identify deviations in the data transmitted by various medical devices, so we note that most of the available data contain a large number of normal data (98,000 samples in our example), while other types of data contain a different personality, such as modified data from attackers and lost and deleted data as well as data of impersonation and hacking. They are present in very small samples (2000 samples), which creates the problem of the anomaly detection model in distinguishing between normal and abnormal behavior of each devices.

Therefore, in order to create a balance between the different categories, we use the proposed model to generate samples that simulate the real distribution by exploiting multiple conditional tasks to create samples close to reality by dividing the input into many parts with different tasks and trying to generate samples that simulate the latter, such as the protocol and destination transmission and ports transport to increase the convergence of the model with the global minimum, as the concepts of multiple tasks give us the ability to simulate the distribution of transmitted data even in the presence of a rare number of samples by combining samples of parts that give us the new information to distribute data that was not available before where it is modified Sample ratios on the one hand and increase the distribution information on the other hand, which increases the accuracy and efficiency of the anomaly detection model.

Anomaly Detection Using Model Generator

As showed in Fig. 5 we produce samples that look like to the original samples for that we can use the generator for anomaly detection field. The generator G learns the general data distribution p_g of latent space Z and the general data distribution p_{data} of real data space X with different parts of data for different tasks. That is, each fake small data $G'(z)$ is generated based on the general data distribution of X. In every updating iteration λ, the loss function L evaluates the dissimilarity of the fake small data $G'(z)$ with the real small data x. An anomaly score value A(x) expressing the fit of a given x to the general data distribution (i.e. model of normal small data) can be directly derived from L. A large anomaly score A(x) indicates an anomalous small data whereas a small anomaly score indicates a small sequence fitting to the general data distribution of X learned by G during adversarial training.

5 Conclusion

In this work, we proposed a Multi Conditional-Tasks GAN generative model to balance the data for Anomaly detection. In this scenario, the main problem is the rarity of the events to identify (i.e. the anomaly). Therefore, a semi-supervised solution is adopted to address this issue. Specifically, the contribution of the paper is twofold: (1) we define a Multi Conditional-Tasks Generative Adversarial Network (MCTGAN) to generate data; (2) we define anomaly detection to identify anomaly distribution of data. We illustrate our approach through a case study: a smart healthcare system-based scenario.

As future work, we plan to study the usage of our model to identify anomaly in complex networks using a multi-Tasks principal. On the other hand, we are interested in defining models that improve network security by detecting anomalies in the network behavior.

References

1. Chandola, V., Mithal, V., Kumar, V.: Comparative evaluation of anomaly detection techniques for sequence data. In: Proceedings - IEEE International Conference on Data Mining, ICDM, pp. 743–748 (2008)
2. Almi'ani, M., Abu Ghazleh, A., Al-rahayfeh, A., Atiewi, S., Razaque, A.: Deep Recurrent Neural Network For IoT Intrusion Detection System. Simulation Modelling Practice and Theory. **101**, 102031 (2019) https://doi.org/10.1016/j.simpat.2019.102031
3. Di Mattia, F., Galeone, P., De Simoni, M., Ghelfi, E.: A Survey on GANs for Anomaly Detection. (2019) http://arxiv.org/abs/1906.11632
4. Yang, Y., Zheng, K., Wu, C., Yang, Y.: Improving the classification effectiveness of intrusion detection by using improved conditional variational autoencoder and deep neural network. Sensors **19**(11), 2528 (2019). https://doi.org/10.3390/s19112528

5. He, H., Garcia, E.A.: Learning from imbalanced data. IEEE Trans. Knowledge Data Eng. **21**(9), 1263–1284 (2009)
6. Chawla, N.V., Bowyer, K.W., Hall, L.O., Philip Kegelmeyer, W.: Smote: synthetic minority over-sampling technique. J. Artificial Intell. Res. **16**, 321–357 (2002)
7. Han, H., Wang, W.-Y., Mao, B.-H.: Borderline-smote: a new oversampling method in imbalanced data sets learning. In: International conference on intelligent computing, pp. 878–887. Springer, (2005) https://doi.org/10.1007/11538059_91
8. Blagus, R., Lusa, L.: Smote for high-dimensional class-imbalanced data. BMC Bioinformatics **14**(1), 106 (2013)
9. Kingma, D.P., Welling, M.: Auto-encoding variational bayes. arXiv preprint arXiv:1312.6114, (2013)
10. Goodfellow, I., et al.: Generative Adversarial Nets. Advances in Neural Information Processing Systems, pp. 2672–2680, 8 [Online]. (2014) http://arxiv.org/abs/1908.08930
11. Makhzani, A., Shlens, J., Jaitly, N., Goodfellow, I., Frey, B.: Adversarial autoencoders. arXiv preprint arXiv:1511.05644, (2015)
12. Li, D., Chen, D., Goh, J., Ng, S.-K.: Anomaly detection with generative adversarial networks for multivariate time series. arXiv preprint arXiv:1809.04758, (2018)
13. Deecke, L., Vandermeulen, R., Ruff, L., Mandt, S., Kloft, M.: Anomaly Detection with Generative Adversarial Networks (2018)
14. Schlegl, T., Seebock, P., Waldstein, S.M., Schmidt-Erfurth, U., Langs, G.: Unsupervised anomaly detection with generative adversarial networks to guide marker discovery. In: International Conference on Information Processing in Medical Imaging, pp. 146–157. Springer, (2017) https://doi.org/10.1007/978-3-319-59050-9_12
15. Ravanbakhsh, M., Nabi, M., Sangineto, E., Marcenaro, L., Regazzoni, C., Sebe, N.: Abnormal event detection in videos using generative adversarial nets. In: Image Processing (ICIP), 2017 IEEE International Conference on, pp. 1577–1581. IEEE, (2017)
16. Eide, A.W.W.: Applying Generative Adversarial Networks For Anomaly Detection in Hyperspectral Remote Sensing Imagery. Master's thesis, NTNU (2018)
17. Škvára, V., Pevný, T., Smidl, V.: Are generative deep models for novelty detection truly better? (2018)
18. Chalapathy, R., Chawla, S.: Deep Learning for Anomaly Detection: A Survey (2019)
19. Tyler, T.L.: Fundamental Limitations of Semi-Supervised Learning. Master's thesis, University of Waterloo (2009)
20. Schlegl, T., Seeböck, P., Waldstein, S.M., Schmidt-Erfurth, U., Langs, G.: Unsupervised anomaly detection with generative adversarial networks to guide marker discovery. In: Niethammer, M., et al. (eds.) IPMI 2017. LNCS, vol. 10265, pp. 146–157. Springer, Cham (2017). https://doi.org/10.1007/978-3-319-59050-9_12
21. Esteban, C., Hyland, S.L., Ratsch, G.: Real-valued (Medical) ¨ Time Series Generation with Recurrent Conditional GANs. (2017) http://arxiv.org/abs/1706.02633
22. Bashar, M.A., Nayak, R.: TAnoGAN: Time Series Anomaly Detection with Generative Adversarial Networks. (2020) https://arxiv.org/abs/2008.09567v1
23. Denton, E., Chintala, S., Szlam, A., Fergus, R.: Deep generative image models using a laplacian pyramid of adversarial networks. In: NIPS, (2015)
24. Reed, S., Akata, Z., Yan, X., Logeswaran, L., Schiele, B., Lee, H.: Generative adversarial text to image synthesis. In: ICML, (2016)
25. Wang, X., Gupta, A.: Generative image modeling using style and structure adversarial networks. In: Leibe, B., Matas, J., Sebe, N., Welling, M. (eds.) ECCV 2016. LNCS, vol. 9908, pp. 318–335. Springer, Cham (2016). https://doi.org/10.1007/978-3-319-46493-0_20
26. Mathieu, M., Couprie, C., LeCun, Y.: Deep multi-scale video prediction beyond mean square error. ICLR, (2016)

27. Yoo, D., Kim, N., Park, S., Paek, A.S., Kweon, I.S.: Pixellevel domain transfer. ECCV (2016)
28. Reed, S.E., Akata, Z., Mohan, S., Tenka, S., Schiele, B., Lee, H.: Learning what and where to draw. In: NIPS (2016)
29. Isola, P., Zhu, J.-Y., Zhou, T., Efros, A.: Image-to-Image Translation with Conditional Adversarial Networks. 5967–5976 (2017) https://doi.org/10.1109/CVPR.2017.632
30. Khazaie, V.R., Akhavan P.A., Ebrahimpour, R.: Occluded Visual Object Recognition Using Deep Conditional Generative Adversarial Nets and Feedforward Convolutional Neural Networks. 1–6. (2020) https://doi.org/10.1109/MVIP49855.2020.9116887
31. Pathak, D., Krahenbuhl, P., Donahue, J., Darrell, T., Efros, A.A.: Context encoders: Feature learning by inpainting. In: CVPR (2016)
32. Zhou, Y., Berg, T.L.: Learning temporal transformations from time-lapse videos. In: Leibe, B., Matas, J., Sebe, N., Welling, M. (eds.) ECCV 2016. LNCS, vol. 9912, pp. 262–277. Springer, Cham (2016). https://doi.org/10.1007/978-3-319-46484-8_16
33. Zhu, J.-Y., Krähenbühl, P., Shechtman, E., Efros, A.A.: Generative visual manipulation on the natural image manifold. In: Leibe, B., Matas, J., Sebe, N., Welling, M. (eds.) ECCV 2016. LNCS, vol. 9909, pp. 597–613. Springer, Cham (2016). https://doi.org/10.1007/978-3-319-46454-1_36
34. Ledig, C., Theis, L., Huszar, F., Caballero, J., Cunningham, A., Acosta, A., et al.: Photo-realistic single image super-resolution using a generative adversarial network. In: CVPR, pp. 4681–4690 (2017)
35. Zhang, Y., Bai, Y., Ding, M., Ghanem, B.: Multi-task generative adversarial network for detecting small objects in the wild. Int. J. Comput. Vision **128**(6), 1810–1828 (2020). https://doi.org/10.1007/s11263-020-01301-6
36. Zhang, J., Shu, Y., Xu, S., Cao, G., Zhong, F., Liu, M., Qin, X.: Sparsely Grouped Multi-Task Generative Adversarial Networks for Facial Attribute Manipulation. 392–401. (2018) https://doi.org/10.1145/3240508.3240594
37. Lin, Y., Suzuki, K., Takeda, H., Nakamura, K.: Generating synthetic training data for object detection using multi-task generative adversarial networks. ISPRS Annals of Photogrammetry, Remote Sensing and Spatial Information Sciences. V-2–2020. 443–449 (2020) https://doi.org/10.5194/isprs-annals-V-2-2020-443-2020
38. Han, Z., Tao, X., Hongsheng, L, Shaoting, Z., Xiaogang, W., Xiaolei, H., Dimitris, N.M.: Stackgan: Text to photo realistic image synthesis with stacked generative adversarial networks. In: Proceedings of the IEEE International Conference on Computer Vision, pp. 5907–5915 (2017)
39. Qi, M., Hsin-Ying, L., Hung-Yu, T., Siwei, M., Ming-Hsuan, Y.: Mode seeking generative adversarial networks for diverse image synthesis. In: Proceedings of the IEEE/CVF Conference on Computer Vision and Pattern Recognition (CVPR) (2019)
40. Xudong, M., Qing, L., Haoran, X., Raymond, Y.K.L., Zhenm W., Stephen, P.S.: Least squares generative adversarial networks. In: Proceedings of the IEEE International Conference on Computer Vision, pp. 2794–2802 (2017)
41. Junbo, Z., Michael, M., Yann, L.: Energy-based generative adversarial network. arXiv preprint arXiv:1609.03126, 2016.] and WGAN [Martin Arjovsky, Soumith Chintala, and Léon Bottou. Wasserstein gan. arXiv preprint arXiv:1701.07875, (2017)
42. Chaoyue, W., Chang, X., Xin, Y., Dacheng, T.: Evolutionary generative adversarial networks. IEEE Transactions on Evolutionary Computation (2019)
43. Martin, A., Soumith, C., Léon, B.: Wasserstein gan. arXiv preprint arXiv:1701.07875 (2017)
44. Caruana, R.: Multitask Learning. Machine Learning, 28, vol. 75, no. September, pp. 41–75, (1997) http://www.cs.cornell.edu/~caruana/mlj97.pdf

Machine Learning

Prediction of Risks in Intelligent Transport Systems

Soukaina Bouhsissin(✉), Nawal Sael, and Faouzia Benabbou

Laboratory of Information Technology and Modeling, Faculty of Sciences Ben M'Sik, Hassan II University, Casablanca, Morocco
bouhsissin.soukaina@gmail.com, saelnawal@hotmail.com, faouzia.benabbou@univh2c.ma

Abstract. With the advancement of technology and the improvement of the standard of living, the number of vehicles gradually increases, and at the same time, the road problems worsen. This paper aims to synthesize the current researches regarding the prediction of dangers in traffic roads. A comparative study was therefore conducted to explore the various solutions and strategies employed to classify dangers and to address the most important techniques used. An experiment was also developed to classify the most used techniques in the same context and over a unique dataset. In addition, to synthesize several key points in the prediction of the dangers (in traffic road), this research allowed us to conclude that ML algorithms can predict the severity of road accidents with over 90.46% accuracy and 0.94 AUC. Random forest and XGboost are the best models for predicting freeway crashes.

Keywords: ITS · Accident severity · Accident risk prediction · Machine learning

1 Introduction

Intelligent transport systems ITS refer to the use of data and technologies advances for the administration of issues in traditional vehicle frameworks. They use innovative technologies in these areas to improve safety, effectiveness, efficiency, accessibility, and sustainability of the transport network. Nowadays, the increase in traffic and the number of vehicles creates new needs in terms of intelligent management [1]. This area (intelligent traffic management) is a very important role of ITS witch interests us. In fact, intelligent Transportation Systems can make a significant contribution to improving the management of road traffic (urban, inter-urban and peri-urban) and more generally of the various modes of transport.

In many cases, it is difficult to fully understand the characteristics of the transport system and the relationships between them; AI allowed the development of many intelligent solutions for complex problems like complex systems that cannot be managed using traditional methods. Numerous scientists have exhibited the benefits of AI in transport. An example of that includes:

M. Lazaar et al. (Eds.): BDIoT 2021, LNNS 489, pp. 303–316, 2022.
https://doi.org/10.1007/978-3-031-07969-6_23

- Accident forecasting.
- Identification of the factors contributing to accidents: human, environmental, traffic, and pavement related factors.
- Traffic management by forecasting traffic flow, traffic conflicts, and speed.
- Prediction of accident severity.
- Detection and identification of obstacles, crashes, cracks, imperfections, and aberrations.
- Classification of driver behavior.

Road traffic accidents are among the most critical problems facing the world as they cause many deaths and injuries every year as well as economic losses. Accurate models for predicting accident risks such as the severity of traffic accidents are an essential task for transportation systems.

The objective of this works is to investigate the potentials of AI to identify and predict the types of risks in traffic to reduce their incidence.

The rest of the paper is organized as follows. Section 2 exposes the related works in the field of accident risk prediction. Section 3 presents a comparative study between those works and details their analysis and discusses their most important results. Section 4 develops an experimental study between the most relevant AI techniques used in this area over the same dataset and context. Finally, conclusions are drawn and future research directions are indicated in Sect. 5.

2 Related Work

Highway safety remains a challenge for researchers and practitioners. Therefore, accident risk prediction helps to investigate and resolve these issues, and to improve operations and safe. In this section, we investigate variants research papers conducted in this context. In [2] authors aimed to predict the risk of accidents on highways taking into account both the impact of critical events for safety and traffic conditions. Based on the data from the simulation experiment, a new accident risk forecasting model based on traffic conflict technique TCT (collision time) and BPNN back propagation neural network is introduced. The results are the case of risk and risk-free status; we have the TPR, AUC and accuracy 70.79%, 0.75 and 67.79%, and in the high and low risk case, the values of TPR, AUC and precision are 82.33%, 0.93 and 86.62%. More [3] proposes a BN-RF model to predict the risk of an accident in real time using data collected on the highway. The proposed method uses random forest (RF) to rank the importance of explanatory variables according to the Gini index and Bayesian network (BN) is built to predict the crash in real time. The corresponding performance is evaluated by the ROC curve, where they found 88.35%. [4] addresses the problem of vehicle accident risk prediction, and the Adaboost trichotomy algorithm using in the field of VANET-Big Data extraction, with SMOTE and One-Hot encoding (AdaBoost-SO) to achieve the vehicle accident risk prediction model. Article [5] attempts to create a new type of real-time traffic accident prediction model to solve the problem of unbalanced data. The unbalanced classification algorithm is optimized in three aspects: At the output level, at the data level, and at the MLP algorithm level. Study [6] aimed to develop a model for predicting accident risks in real time on highways in foggy conditions. Bayesian logistic regression models of three time slices before crashes were established to estimate the relationship between the binary response variable (crash or not crash) and the input variables. In [24] proposes an LSTM-based framework considering traffic data of different temporal resolutions (LSTMDTR) for crash detection on freeways. The traffic conditions before an accident are defined as preconditions for the accident, and the traffic conditions according to specific criteria, which do not lead to an accident, are defined as normal traffic conditions. The detection model is designed to rank the pre-crash conditions against normal traffic conditions. Therefore, collision samples (i.e., pre-crash conditions) and non-collision samples (i.e., normal traffic conditions). The article [7] proposes a model for predicting the risk of accident in real time on arteries more specifically on road segments, using a long-term convolutional neural network (LSTM-CNN). In addition, the objective of research [8] is to use deep learning models to detect the occurrence of an accident. The work focuses on designing deep neural networks and performing experiments on parameter tuning, choice of data sets, to compare with traditional machine learning models. In [25] Logistic Regression (LR), Decision Tree (DT), Naive Bayesian Classifiers (NB), Random Forest (RF), K-Nearest Neighbor Algorithm (k-NN), Bayesian Logistic Regression (BLR) and Vector Machine support (SVM) are used to predict road accidents based on driving simulation. Then select the most accurate predictive model that will help reduce these highway crashes. Based on the area under the ROC curve, the random forest was approximately 0.826. In [26] the study aims to develop a model for detecting the risks of driving in the deceleration zone on a motorway with generalized regression neural network (GRNN) and ANOVA as a feature selection method.

In addition to all these works, other researchers take into account, inter-domain data, and spatiotemporal dependencies. In [9], the objective is to study how the deep learning approach spatiotemporal convolutional long short-term memory network (STCL-Net) contributes to the prediction of short-term collision risks at the city scale by exploiting multi-source data sets. The data collected was divided into three types of data: The type of variables is only spatially varied but temporally static during the study period, the type of variables varies only temporally but spatially static during the study period and ultimately type of variables varies both in space and in time during the study period. In [10] applies and compares two statistical methods, the negative binomial (NB) model and a random negative binomial (RENB) model, in predicting the number of accidents by considering the unobserved heterogeneity in accident data, and identifies the key factors involved in the accident to improve road safety. The article [11] proposes a multi-view spatiotemporal deep learning framework to merge inter-domain urban data. In addition, a ResNet based multitasking learning framework with a speed inference model to realize the prediction of the accident risk distribution in the near future. [12] Presents an important feature of traffic accidents the spatiotemporal correlation, and then built a deep learning model (LSTM) for the prediction of traffic accident risk based on the spatiotemporal correlation model.

The behavior of drivers and their emotional states strongly influence the efficiency of driving and can contribute to the increase of the various transport problems. Several works have studied these aspects. Taxi drivers around the world often have very long driving hours and experience frequent fatigue. These conditions are associated with a high prevalence of fatigue and accidents. The objective of article [13] is to provide a validated prediction model that helps to understand the association of taxi driver fatigue-related accident risk (FRAR) and related factors. They used self-reported data on fatigue-related accidents; the validity of these data was questioned in the questionnaires. They adopted logistic ridge regression (LRR), logistic regression (LR), and support vector machine (SVM) methods to fit and validate the models.

Speed has a direct influence on the frequency and severity of accidents. In [14] bidirectional recurrent neural network BRNN is good at processing time series data for speed prediction and comparing with long short-term memory (LSTM) model and gated recurrent unit (GRU) model.

Accurate and real-time forecasting of traffic flows plays an important role in the construction of intelligent transport systems and in the control and induction of traffic. Article [19] focuses on forecasting short-term traffic flows for the next given period (5 min). They used (k-means) for grouping historical data according to their models. Subsequently the predictor is formed for each group based on a CNN and a LSTM. The objective of the article [20] is to try to build a deep architecture for the prediction of traffic flow (15 min) which can solve the problem of incomplete data and learn the spatiotemporal correlations on the traffic network from the deep hierarchical representation of the characteristics. Article [21] presents a residual deconvolution based deep generative network (RDBDGN) to manage the problem of predicting long-term traffic flows on elevated highways. [22] Attempts to take into account the impact of precipitation when predicting traffic flow with deep bi-directional long short-term memory (DBL) model.

Predicting the severity of traffic accidents is an essential task for transportation systems. The main objective of this study [15] is to build a model to classify the severity of injuries and to select a set of factors influencing the severity of road accidents also in [16]; three data mining models were applied to provide a comprehensive analysis of the risk factors related to the severity of road accidents.

3 Comparative Study and Discussion

3.1 Comparative Study

In this section, the solutions discussed above are compared in Table 1. Our objective is to compare the proposed solution depending of many characteristics, which are:

- Year.
- Task: the paper objective.
- Dataset: which cover the data source and type, its size, even if the dataset is static or dynamic (in this case we specify if the stored data change depending on time, space or both of them).
- Technical: the approach synthesis.
- Comparison: the ML or DL technique compared.
- Validation: the model performance.

Table 1. Comparison between articles.

N°	Index	Year	Task	Data						Technical	Comparison	Validation
				Type and source	Size	Static	Time	Space	Time and space			
[2]	Science Direct	2019	Prediction of accident risk on motorways	Aimsun simulation: Traffic conditions, Traffic information,vehicle trajectory data	1800	Yes	No	No	No	BPNN was applied to model and determine the state of accident risk on road		TPR = 82.33%, AUC = 0.93 Precision = 86.62%
[3]	IEEE	2019	Predict the risk of an accident on the highway	Data from: Traffic and accidents	1105	Yes	No	No	No	BN-RF: RF to rank the importance of the explanatory variables and BN to predict the crash in real time	SVM LR KNN	Mean sensitivity = 70.46% Accuracy = 83.18% False alarm rate = 16,07%
[4]	IEEE	2019	Prediction of vehicle accident risk	Data from: Traffic and vehicle	561659	Yes	No	No	No	AdaBoost-SO trichotomy used to train a series of weak classifiers from the data set and then combine	Trichotomie Adaboost SVM DT kNN	AUC = 0.77 and in real time
[5]	Science Direct	2020	Prediction of traffic accidents to solve the problem of unbalanced data	Data collected by a loop detector	7559352	Yes	No	No	No	Cost-sensitive MLP and Adaboost.M2		Rusboost: AUC = 0.892 Sensitivity = 84.2% Specificity = 81.6%
[6]	Science Direct	2020	Prediction of accident risks on highways in foggy conditions	Data from: traffic, flux, fog-related accidents, goad geometry, meteorological historical visibility	224	Yes	Yes	Yes	No	BLR used to develop the real-time accident risk prediction model		Accuracy = 0.76% False alarm rate = 0.304%
[7]	Science Direct	2020	Prediction of the risk of accident on the arteries	Data from: crash, traffic volume, speed, meteorological	7098701	Yes	No	No	No	LSTM-CNN parallel: Captures long-term dependency while and extracts time invariant features	XGBoost BLR LSTM CNN Sequential LSTMCNN	AUC = 0.932 False Alarm Rate = 0.13% Sensitivity = 0.86%
[8]	Science Direct	2020	Predict the risk of an accident 1 min, 5 min, 10 min before the occurrence of the accident	Image radar Data from: Traffic volume, speed, occupation of sensor time	–	Yes	No	No	No	Prediction of accident risk: CNN + GAP	LR DT RF SVC KNN	Prediction of accident risk: Case (5min): Accuracy = 0.802% F1-score = 0.793%

(continued)

Table 1. (*continued*)

N°	Index	Year	Task	Data						Technical	Comparison	Validation
				Type and source	Size	Static	Time	Space	Time and space			
[9]	Science Direct	2019	City-wide short-term collision risk prediction	Data from: Accidents, GPS, road network attributes, population, meteorological	53354	No	Yes	Yes	Yes	STCL-Net includes 3 components CNN, LSTM, ConvLSTM The 3 components are then merged	ARIMA GWR CNN LSTM ANN GBRT	The prediction performance of the proposed model decreases as the spatiotemporal resolution of the prediction task increases
[10]	Science Direct	2019	Prediction of the number of accidents, and identify the key factors involved in the accident	Data from: Accidents collisions, geometric, traffic, meteorological	48154	Yes	No	Yes	No	The random negative binomial (RENB) and negative binomial (NB) models were explored respectively to predict the number of accidents		RENB: RE = 7.92 MAE = 1.28 SSE = 505 RMSE = 0.15 NB: RE = 12.37 MAE = 1.29 SSE = 589 RMSE = 0.16
[11]	IEEE	2019	Predicting the distribution of accident risk in the near future	Data from: Accidents, social, calendar	191	No	Yes	Yes	Yes	ResNet: The spatio-temporal models of traffic accidents are used in three ResNet modules	Arima ConvLSTM	MSE = 0.160 RMSE = 0.400 Accuracy = 88.89% Detection rate = 23.38%, Training time (s) 116
[12]	IEEE	2018	Prediction of traffic accident risk	Accident data (sensors)		No	Yes	Yes	Yes	LSTM	Lasso Ridge SVR DTR RFR MLP ARMA	MAE:0.014, MSE:0.001 RMSE:0.034
[13]	Science Direct	2019	Prediction of a model that helps to understand the association of accident risk related to taxi driver fatigue and accident risk factors	Self-reported data on fatigue-related accidents, the validity of these data was questioned in the questionnaires	269	Yes	No	No	No	LRR, LR and SVM to adjust and validate models		Precision: LRR = 93.4% LR = 91.3% SVM = 93.0%
[14]	IEEE	2019	Speed prediction	GPS data from buses		Yes	No	No	No	BRNN consists of two RNNs	LSTM GRU	BRNN predicts the speed at 3,6,12, time steps
[15]	IEEE	2019	Classify injury severity	Accident data	27563	Yes	No	No	No	AdaBoost, LR, NB and RF		Precision of RF = 75.5% LR = 74.5%, NB = 73.1% AdaBoost = 74.5%
[16]	Science Direct	2020	Identify the risk factors linked to the severity of accidents	Police service data	5740	Yes	No	No	No	DT, NB and SVM		Accuracy: DT = 66.06% BN = 66.18% SVM = 65.19%

(*continued*)

Table 1. (*continued*)

N°	Index	Year	Task	Data						Technical	Comparison	Validation
				Type and source	Size	Static	Time	Space	Time and space			
[19]	IEEE	2018	Forecasting short-term traffic flows	Trafficdata (DRIVENET)		No	Yes	Yes	No	Cnn: extract spatial characteristics LSTM: characteristics of time series	Arima KNN LSTM GRU SAEs	he average error over the ten target days is 7.66%
[20]	IEEE	2018	Prediction of traffic flow	Traffic data (PeMS)		Yes	Yes	Yes	No	RSCNN: The CNN part is used as a feature extractor	SVR NN Stacked Autoencoder DBN	RSCNN intuitively shows acceptable adjustment capability
[21]	IEEE	2019	Long-term traffic flow prediction	Loop Detector Data		Yes	No	No	No	RDBDGN consists of: Generator is a CNN based RDNN Discriminator is a multichannel CNN	ANN, CNN, STResNet, ConvLSTM	The model is robust
[22]	IEEE	2018	Forecast traffic flow and take into account the impact of precipitation	Traffic data (PeMS, KDD)	312139	Yes	Yes	No	No	DBL deep bi-directional long short-term memory	Arima, Sarima, SVR, RF, DBN, SAE, LSTM, DBL, P-DBL	• PMES (30 min) MAPE = 3,23, RMSE = 110,43 • KDD (20 min) MAPE + 13,75, RMSE = 7,43
[23]	Science direct	2020	Predict traffic conflicts	Traffic data (Sensors, SSM)	1200000	Yes	No	No	No	R-CNN: Identification of conflicts DNN: Conflict detection		DNN can predict 71% and 78% of traffic conflicts at the cost of 5% and 10% false alarm respectively Precision = 94%
[25]	IEEE	2019	Predict highway traffic accidents	Driving simulator		Yes	No	No	No	Six classification algorithms were used to classify traffic accidents, (BLR), (DT), (RF), (KNN), (NBC) and (SVM)		AUC: BLR = 0,79, DT = 0,81, RF = 0,82, SVM = 0,61, NBC = 0,60, KNN = 0,65
[26]	Journal of Advanced Transportation	2018	Detect the risks of driving in the deceleration zone on a motorway	Sensors	399	Yes	No	No	No	Generalized regression neural network (GRNN)		Accuracy 86,4%

3.2 Discussion

This researches analysis shows that, the risk in traffic takes different contexts. A large number of works focus on the forecasting and identification of accident risks, classification of driver behavior and accident severity, the identification of factors contributing to accidents, the conflict prediction and traffic flow.

For the traffic flow, this study shows that, the purpose of predicting traffic flow is to predict the number of vehicles in a given time interval based on historical traffic information and numerous researches suggest that 15 min are considered a typical time interval.

The traffic conditions at five-minute intervals before an accident are very sensitive to the prediction of accident risk.

Prediction of road safety risks generally belongs to the category of supervised learning objects. In addition, the problem of creating an accident risk prediction model is a typical problem of classifying unbalanced data, which means that the number of majority class samples is much higher than that of the minority class samples.

The traffic data of a single time resolution cannot fully represent the traffic trend and dynamic transitions at different time intervals, to solve this problem; we can take into account the traffic data of different spatiotemporal resolutions.

Even if, the prediction of the occurrence of an accident may not only be associated with the characteristics of the traffic, other factors such as weather conditions or human behavior may also contribute to an accident. However, Collision cases should contain information about the date and time of the accident, duration, location, severity of the accident (collected at the same time, in the same place, on the same day of the week, and in the same season as the collision event), the effect of location and weather on traffic conditions can be effectively eliminated.

Traffic data analysis can help capture, the spatial dependencies and extract the spatial characteristics from variables that are only spatially varied but temporally static during the study period. The temporal dependencies and extract the temporal characteristic from the type of variable varies only temporally but spatially static during the study period. And also, the spatiotemporal characteristics from the variables vary both spatially and in time over the period of study.

At the algorithm level, we can conclude that:

Unlike the static data, which can be efficiently trained using Machine Learning algorithms, dynamic data (spatially and or temporarily dependent) need more sophisticated algorithms such as Deep Learning.

Predicting collision risks involves matching each type of data to the appropriate algorithm and then merging the results.

Training learning algorithms parallelly may work better than sequentially, other interesting alternative can be testing different models and then combining the models to improve performance.

The most widely used algorithm for risk prediction is LSTM. Figure (see Fig. 1) shows the eight most used algorithms. The most important performance found was 0.93 AUC.

At the technical level, the dropout technique can reduce overfitting and improve the generalizability of the model. Besides, release layers are added to avoid overfitting.

Furthermore, the most used performance measures for risk prediction are accuracy, AUC, and RMSE.

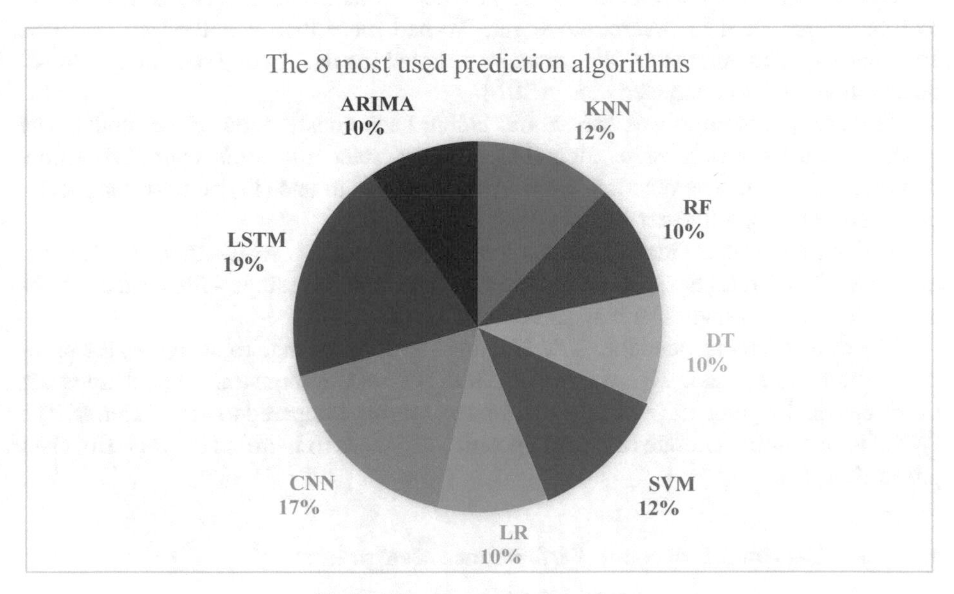

Fig. 1. The most used algorithms.

In the next section, we develop an experimental study to explore the different solutions and strategies employed to classify risk severalty and to address the most important techniques used.

4 Experiments

The purpose of this experiment is to detect the most performant technique used to predict accident risk; our goal is to build the classification rules for prediction of the best performing model.

This section discusses the methods used in this research study, including data description, preprocessing, building the classification models, and extracting the required knowledge.

4.1 Dataset and Preprocessing

The UK government[1] gathers and distributes (normally on a yearly premise) nitty-gritty data on road accidents the nation over. This information includes, but is not limited to, geographic locations, weather conditions, vehicle type, number of casualties, and vehicle maneuvers, making it a very interesting and comprehensive data set for analysis

[1] Dataset: https://roadtraffic.dft.gov.uk/custom-downloads.

and research. The data comes from the UK government's Open Data website, where the Department for Transport published it.

Our objective is to predict the severity of the accident. This severity is divided into two categories, fatal or severe and slight. We had more than 2 million observations and nearly 60 characteristics. We therefore sampled the data in approximately 146342 observations and 34 characteristics in 2014.

The data preparation was carried out before each construction of the model. The process includes various steps including cleaning, standardization, feature selection, and transformation. The reaction variable was paired with one (1) demonstrating lethal or extreme mishap while zero (0) showing minor mishap.

Major causes of accidents include speed limit, weather, road surface conditions, light conditions, road hazards, type of roads, pedestrian crossings - Physical facilities and pedestrian crossings - Human and other controls.

The data was partitioned into 70% learning and 30% testing. To overcome the problem of unbalanced data, we have used the SMOTE resampling strategy as there was a imbalance in the percentage of fatal and severe injuries compared to other injuries. The SMOTE algorithm generates synthetic positive instances to increase the minority class proportion [17].

4.2 Classification Models and Performance Measurement

To develop the injury severity prediction model, we have used the most used and performant techniques founded in our literature review. The following classification techniques were studied:

- Machine learning: Decision tree, logistic regression, random forest, XGboost and Naive Gaussian Bayesian.
- Deep learning: LSTM, bidirectional LSTM, GRU, Bi-GRU.

To assess the performance of our models for classification problems and detect the most important ones, many metrics exist. In this study, we used: Accuracy, precision, sensitivity or recall, F1_score, ROC curve, and area under the receiver operating characteristic curve (AUC) to better perform this comparison. The following Table 2 shows the results for each model.

Precision is a measure of accuracy or quality, while recall is a measure of completeness or quantity. Bi-LSTM achieves the highest precision among all classifiers of 86.26%, and the XGboost classifier achieves the greatest recall of 99.51%. The Random Forest classifier achieves the highest 90.46% accuracy among all classifiers for the test set with Smote.

AUC is the area under the ROC curve and is a ratio between 0 and 1, where a value of 1 indicates a perfect classifier, while a value close to 0.5 indicates a bad model, since this is equivalent to a random classification [18]. RF and XGboost algorithms are the best classification models with 0.94. These encouraging AUCs provide statistical evidence for the excellent classification ability of Random Forest and XGboost in this study. All the results are illustrated in the following figures (see Fig. 2 and Fig. 3).

Table 2. Models evaluation.

Algorithms	Accuracy	Precision	Recall	F1-score	AUC
DT	73.91	69.68	85.15	76.64	0.83
RL	52.85	54.61	36.54	43.79	0.54
RF	90.46	85.59	97.42	91.72	0.94
BN	61.48	66.56	46.93	55.05	0.67
XGboost	89.64	83.16	99.51	90.662	0.94
LSTM	82.28	78.47	88.42	83.33	0.64
Bi-LSTM	78.24	86.26	88.39	87.31	0.65
GRU	84.63	78.29	95.7	86.13	0.91
Bi-GRU	84.27	84.87	99.09	91.43	0.67

The test result shows that based on the confusion matrix, RF and XGboost seem to perform better than the other models. Based on the area under the ROC curve, RF and XGboost had an area of approximately 0.94.

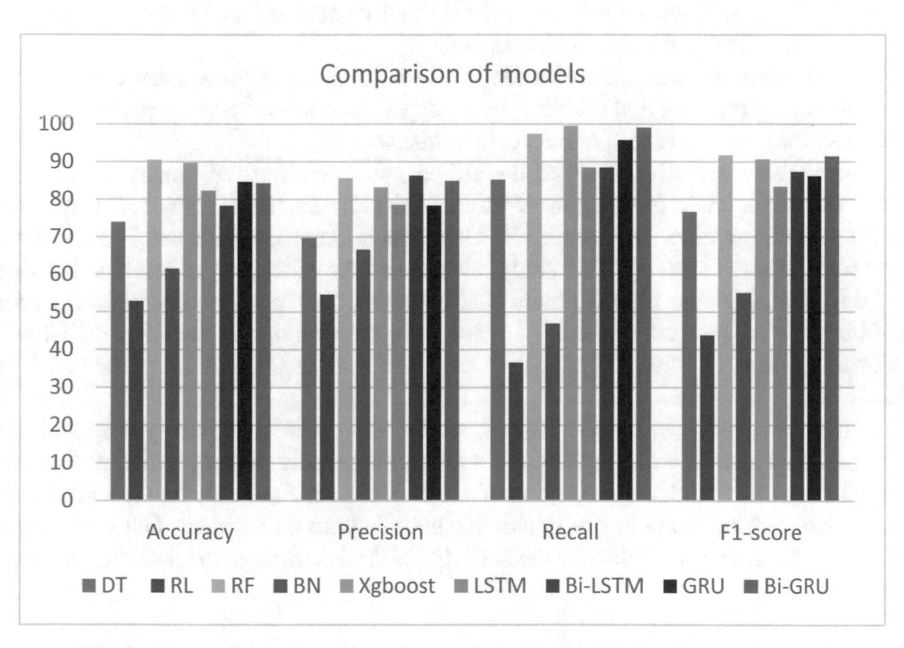

Fig. 2. Comparison of models using confusion matrix performance measures.

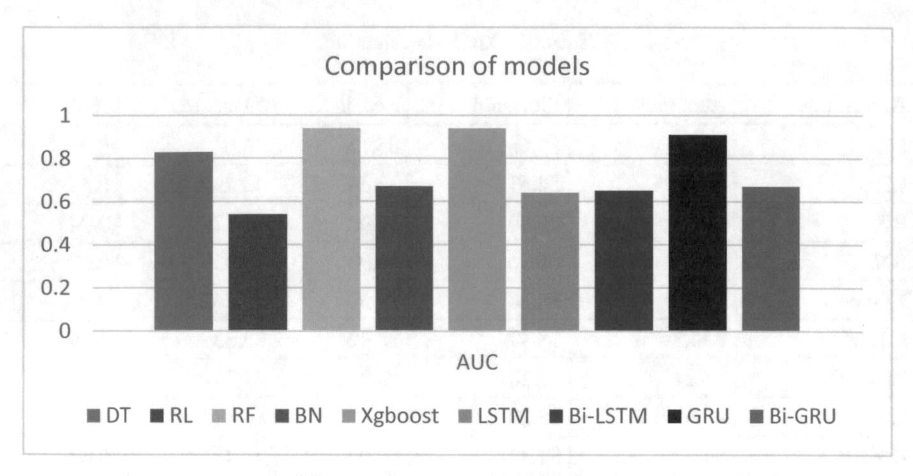

Fig. 3. Comparison of models using the ROC curve.

5 Conclusions

Highway safety has remained a challenge for researchers and practitioners. Therefore, accident risk prediction helps to investigate and resolve safety issues, and helps to improve operations and safety management.

In this article, we went through a deep study and analysis of the articles, comparisons of different approaches, and finalized by an experiment to evaluate the performance of the most used techniques to predict accident risks.

From the comparative study of the articles, we came out with many conclusions about the problematic (prediction of accident risks), data, algorithms and technique. In the experimentation, we used a UK Government database with 146,342 accidents to predict accident severity. This study investigated the efficiency of machine learning and deep learning algorithms to build classifiers that are precise and reliable. It shows that based on the area under the ROC curve and confusion matrix the RF and XGboost algorithms seem to perform better than the other models 90.46% accuracy and 0.94 AUC.

For future work, we plan to exploit multi-source datasets to analyze spatial and temporal dependencies to help provide a different comparison for predicting accident risk. On the other hand, the prediction of accident risk may not only be associated with traffic data, but human behavior could also contribute to an accident. Our objective is also to study, analyze and classify the behavior of drivers through the detection of facial emotions.

References

1. Mfenjou, M.L.: Methodology and trends for an intelligent transport system in developing countries. Sustain. Comput. Inf. Syst. **19**, 96–111 (2018). https://doi.org/10.1016/j.suscom.2018.08.002

2. Wang, J., Kong, Y., Fu, T.: Expressway crash risk prediction using back propagation neural network: a brief investigation on safety resilience. Accid. Anal. Prev. **124**, 180–192 (2019). https://doi.org/10.1016/j.aap.2019.01.007
3. Wu, M., Shan, D., Wang, Z., Sun, X., Liu, J., Sun, M.: A Bayesian network model for real-time crash prediction based on selected variables by random forest. In: ICTIS 2019 - 5th International Conference on Transportation Information and Safety, pp. 670–677 (2019). https://doi.org/10.1109/ICTIS.2019.8883694
4. Zhao, H., Yu, H., Li, D., Mao, T., Zhu, H.: Vehicle Accident risk prediction based on AdaBoost-SO in VANETs. IEEE Access. **7**, 14549–14557 (2019). https://doi.org/10.1109/ACCESS.2019.2894176
5. Peng, Y., Li, C., Wang, K., Gao, Z., Yu, R.: Examining imbalanced classification algorithms in predicting real-time traffic crash risk. Accid. Anal. Prev. **144**, 105610 (2020). https://doi.org/10.1016/j.aap.2020.105610
6. Zhai, B., Lu, J., Wang, Y., Wu, B.: Real-time prediction of crash risk on freeways under fog conditions. Int. J. Transp. Sci. Technol. **9**, 287–298 (2020). https://doi.org/10.1016/j.ijtst.2020.02.001
7. Li, P., Abdel-Aty, M., Yuan, J.: Real-time crash risk prediction on arterials based on LSTM-CNN. Accid. Anal. Prev. **135**, 105371 (2020). https://doi.org/10.1016/j.aap.2019.105371
8. Huang, T., Wang, S., Sharma, A.: Highway crash detection and risk estimation using deep learning. Accid. Anal. Prev. **135**, 105392 (2020). https://doi.org/10.1016/j.aap.2019.105392
9. Bao, J., Liu, P., Ukkusuri, S.V.: A spatiotemporal deep learning approach for citywide short-term crash risk prediction with multi-source data. Accid. Anal. Prev. **122**, 239–254 (2019). https://doi.org/10.1016/j.aap.2018.10.015
10. Yan, Y., Zhang, Y., Yang, X., Hu, J., Tang, J., Guo, Z.: Crash prediction based on random effect negative binomial model considering data heterogeneity. Phys. A. Stat. Mech. Appl. **547**, 123858 (2020). https://doi.org/10.1016/j.physa.2019.123858
11. Zhou, Z., Chen, L., Zhu, C., Wang, P.: Stack ResNet for short-term accident risk prediction leveraging cross-domain data. In: Proceedings - 2019 Chinese Automation Congress, CAC 2019, pp. 782–787 (2019). https://doi.org/10.1109/CAC48633.2019.8996483
12. Ren, H., Song, Y., Wang, J., Hu, Y., Lei, J.: A Deep Learning Approach to the Citywide Traffic Accident Risk Prediction. In: IEEE Conference on Intelligent Transportation Systems, Proceedings, ITSC. 2018-Novem, pp. 3346–3351 (2018). https://doi.org/10.1109/ITSC.2018.8569437
13. Li, M.K., Yu, J.J., Ma, L., Zhang, W.: Modeling and mitigating fatigue-related accident risk of taxi drivers. Accid. Anal. Prev. **123**, 79–87 (2019). https://doi.org/10.1016/j.aap.2018.11.001
14. Bohan, H., Yun, B.: Traffic flow prediction based on BRNN. In: ICEIEC 2019 - Proceedings of 2019 IEEE 9th International Conference on Electronics Information and Emergency Communication (2019)
15. Almamlook, R.E., Kwayu, K.M., Alkasisbeh, M.R., Frefer, A.A.: Comparison of machine learning algorithms for predicting traffic accident severity. In: 2019 IEEE Jordan International Joint Conference on Electrical Engineering and Information Technology, JEEIT 2019 - Proceedings. 272–276 (2019). https://doi.org/10.1109/JEEIT.2019.8717393
16. AlKheder, S., AlRukaibi, F., Aiash, A.: Risk analysis of traffic accidents' severities: an application of three data mining models. ISA Trans. **106**, 213–220 (2020). https://doi.org/10.1016/j.isatra.2020.06.018
17. Chawla, N.V., Bowyer, K.W., Hall, L.O., Kegelmeyer, W.P.: SMOTE: Synthetic Minority Over-Sampling Technique. J. Artif. Intell. Res. (2002). https://doi.org/10.1613/jair.953
18. James, G., Witten, D., Hastie, T., Tibshirani, R.: An introduction to statistical learning with application in R (2013)

19. Ma, D., Sheng, B., Jin, S., Ma, X., Gao, P.: Short-term traffic flow forecasting by selecting appropriate predictions based on pattern matching. IEEE Access. **6**, 75629–75638 (2018). https://doi.org/10.1109/ACCESS.2018.2879055
20. Liao, S., Chen, J., Hou, J., Xiong, Q., Wen, J.: Deep convolutional neural networks with random subspace learning for short-term traffic flow prediction with incomplete data. In: Proceedings of the International Joint Conference on Neural Networks, 1–6 July 2018. https://doi.org/10.1109/IJCNN.2018.8489536
21. Zang, D., Fang, Y., Wei, Z., Tang, K., Cheng, J.: Traffic flow data prediction using residual deconvolution based deep generative network. IEEE Access. **7**, 71311–71322 (2019). https://doi.org/10.1109/ACCESS.2019.2919996
22. Wang, J., Hu, F., Xu, X., Wang, D., Li, L.: A deep prediction model of traffic flow considering precipitation impact. In: Proceedings of the International Joint Conference on Neural Networks, July 2018. https://doi.org/10.1109/IJCNN.2018.8489033
23. Formosa, N., Quddus, M., Ison, S., Abdel-Aty, M., Yuan, J.: Predicting real-time traffic conflicts using deep learning. Accid. Anal. Prev. **136**, 105429 (2020). https://doi.org/10.1016/j.aap.2019.105429
24. Jiang, F., Yuen, K.K.R., Lee, E.W.M.: A long short-term memory-based framework for crash detection on freeways with traffic data of different temporal resolutions. Accid. Anal. Prev. **141**, 105520 (2020). https://doi.org/10.1016/j.aap.2020.105520
25. Al Mamlook, R.E., Ali, A., Hasan, R.A., Mohamed Kazim, H.A.: Machine learning to predict the freeway traffic accidents-based driving simulation. In: Proceedings of the IEEE National Aerospace Electronics Conference, NAECON, July 2019, pp. 630–634 (2019). https://doi.org/10.1109/NAECON46414.2019.9058268
26. Qi, W., Wang, Z., Tang, R., Wang, L.: Driving risk detection model of deceleration zone in expressway based on generalized regression neural network. J. Adv. Transp. (2018). https://doi.org/10.1155/2018/8014385

A Systematic Literature Review of Machine Learning Applications in Software Engineering

Houda Mezouar(✉) and Abdellatif El Afia

ENSIAS, Mohammed V University of Rabat, Rabat, Morocco
houda.mezouar@gmail.com, a.elafia@um5s.net.ma

Abstract. Machine Learning (ML) has been a concern in Software Engineering (SE) over the past years. However, how to use ML and what it can offer for SE is still subject to debate among researchers. This paper investigates the application of ML in SE. The goal is to identify the used algorithms, the addressed topics and the main findings. It performs a Systematic Literature Review (SLR) of peer-reviewed studies published between 1995 and 2020. Data extracted from the studies show that ML algorithms are of great practical value in the different activities of software development process, especially "Software specification" and "Software validation" since "Software bug prediction" and "Software quality improvement" are the most recurring research topics.

Keywords: Machine Learning · Software Engineering · Software development process · Systematic Literature Review

1 Introduction

Nowadays we hear more and more about the terms artificial intelligence and ML. Artificial intelligence requires finding a solution to a hard problem that even a specialist in a given domain cannot express mathematically. So, it is considered as the development of computer systems to solve problems associated with human intelligence [1]. While ML is a form of artificial intelligence, which can be used to automate decision-making and make predictions [2]. More explicitly, ML is making computers able to improve a performance measure through example data or experience data. Various parameters defines a model (that can be predictive, descriptive or both), and the learning is optimizing those parameters by a program that uses the training data or experience data. ML algorithms have showed to be of a high value in different fields [3]. On the other hand, SE needs to apply the engineering principles, computer science, design skills, mathematical formalism and good management practice. It is the duty of SE to tie together these independent fields of expertise and to adapt them to requirements, design, implementation, verification and maintenance of software projects [4]. In fact, SE is a continuously growing research field derived from computer science since the 1960s. Correspondently, the importance of software engineering has been widely recognized by more and more scholars all over the world in the past five decades [5]. Not surprisingly, SE proved being a field where various tasks of software development and maintenance could be

M. Lazaar et al. (Eds.): BDIoT 2021, LNNS 489, pp. 317–331, 2022.
https://doi.org/10.1007/978-3-031-07969-6_24

approached by ML algorithms [6]. Thanks to ML, setting up software applications that could never have been written using a classic algorithm, is possible. Therefore, it is considered that the next software revolution will be geared towards ML. Moreover, a SRL, as defined in [7], is a comprehensive review of all published selected papers to address a specific question using a systematic method of identifying relevant studies in order to minimize biases and error. Among the multiples benefits of conduction a SRL, we can cite: the evaluation of the present state of research on a subject, the recognition of the experts on a specific subject, the description of main questions about a subject that need more research, and the determination of past studies used approaches for the same or similar research subjects. To take advantage of these benefits in the field of ML applications in SE, in this paper, the SLR protocol is respected, and it provides a mapping of the most relevant research works related to this topic, published between 1995 and February 2020, and selected according to described criteria.

The reminder of this paper is organized as follows. Section 2 presents the research methodology, by giving an overview of the selected studies, after explaining the criteria of their selections. While Sect. 3, answers each research question, and analyzes the results of the review. Followed by conclusion in Sect. 4.

2 Research Methodology

A SLR is a way allowing us to evaluate and to interpret researches related to a specific research question, or to a research subject. It focuses on giving an objective evaluation of a research subject trough a credible methodology [8]. Once the research questions are specified, a protocol is established, this covered definitions of "Inclusion and exclusion criteria", "Search strategy", "Study selection", "Data extraction" and "Selected studies overview".

2.1 Research Questions

The structure of the SRL protocol starts with the research questions (RQ) choice, which will lead to the review process. The essential aim of this exploratory work is to give a mapping of researches related to ML in SE, by answering those consecutive RQs:

RQ1: What SE Research Topics have been addressed in the selected studies?

RQ2: What ML algorithms have been used in the selected studies?

RQ3: What are the main outcomes of the selected studies?

2.2 Inclusion and Exclusion Criteria

The used inclusion and exclusion criteria are "Only papers published in SpringerLink, ACM digital library, ScienceDirect, and IEEE Xplore, were included", "Only papers presenting research on ML applications in SE were included", "Only papers published between January 1995 and February 2020 were included.", "If the same paper contains in more than one source, only one of them is included." In addition, "Only papers writing in English were included".

2.3 Search Strategy

A process, joining manual and automated searches, was handled to determine peer-reviewed papers published (or available online) between January 1995 and February 2020. It used the four aforementioned indexing systems. Manual Search was performed on the following journals: "Applied Discrete Mathematics and Heuristic Algorithms journal", "International Journal of Computer Applications", "International Journal of Computer Science and Mobile Computing" and "Software Quality Journal". Search terms used were: ("Machine Learning") AND ("software project" OR "software engineering" OR "software development").

2.4 Study Selection and Data Extraction

Figure 1 illustrates the study selection process. Results from the automatic search (n = 1120) were evaluated by looking at the title and excluding the studies that were clearly not relevant. The resulting articles (n = 203) were merged with four potentially articles obtained from the manual search, and the duplicates one were removed. Finally, we applied the inclusion and exclusion criteria on the set of potentially relevant studies (n = 134), resulting in 76 articles selected for the review.

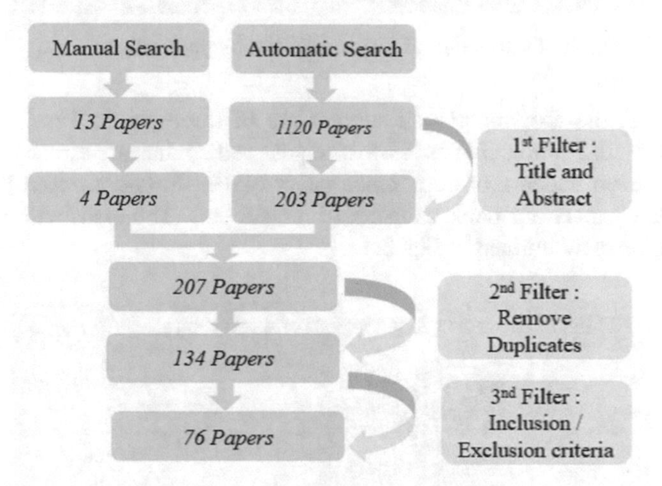

Fig. 1. Stages of the study selection process

A database of publications is managed using Mendeley and relevant information are recorded in a Microsoft Excel spreadsheet. To each paper was assigned a unique identifier (9–84). For each paper, those information were elicited: the source, the title, the authors, the year of publication, the publication type, the source name, the keywords, the research topic addressed, and the RQ.

2.5 Selected Studies Overview

Figure 2 shows the selected studies distribution through the various sources. Out of seventy-six studies, twenty-five (33%) came from IEEE Xplore. Eighteen studies (24%)

are from Elsevier ScienceDirect. Fifteen studies (20%) are from ACM Digital Library. Fourteen studies (18%) are from SpringerLink, and four (5%) from other sources (due to a manual and direct search in the following journals: "Software Quality Journal", "International Journal of Computer Science and Mobile Computing", "International Journal of Computer Applications" and "Applied Discrete Mathematics and Heuristic Algorithms").

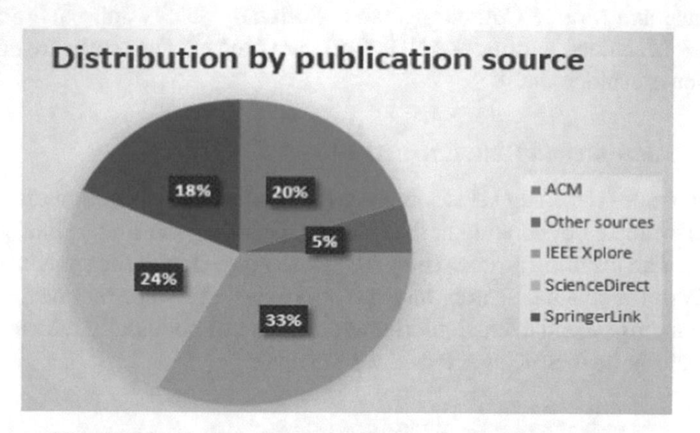

Fig. 2. Distribution of selected studies by publication source

Papers from the aforementioned sources are of three types: "conference paper", "journal paper" and "book chapter". Most of selected studies are either journal papers (49%, thirty-seven papers), or conference paper (47%, thirty-six papers) and just 4% of the selected papers are book chapters (three papers). The distribution through the publication type is mentioned in Fig. 3.

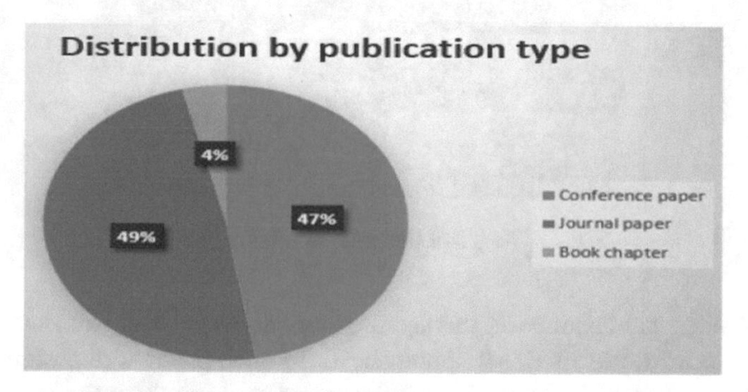

Fig. 3. Distribution of selected studies by publication type

The distribution of selected studies according to the published year can be seen in Fig. 4. Most papers are published from 2017, the decrease mentioned in the figure for the year 2020, is not credible since the selection is made just for the first two months of the year. This timeline shows that the subject of our SLR is clearly a highly research issue of major importance nowadays.

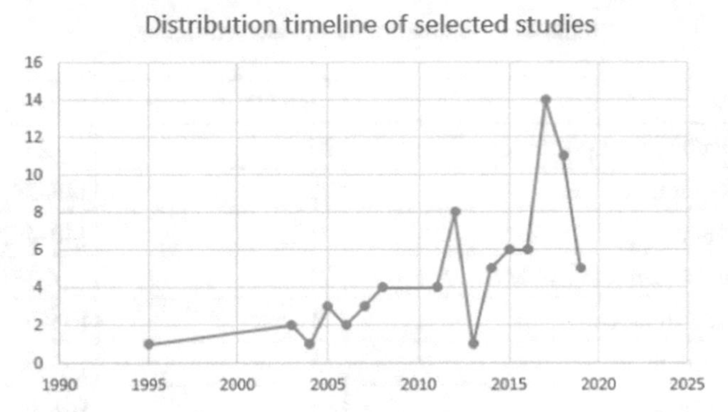

Fig. 4. Distribution timeline of selected studies

3 Results

This section presents the results related to each research question.

3.1 RQ1: What SE Research Topics have been Addressed in the Selected Studies?

A variety of topics was addressed in the selected studies, the following table (Table 1) gives all these topics, mentioning the number of papers by topic and their references.

The studies focused in nineteen different topics, the most addressed research topic is "the prediction of software defect, bug or fault". Out of 76 papers, 20 ones approached this topic. Reference [9], presents a proposed tool that uses ML to automatically detect coding defects as programmers write code, as for [10], they proposed an approach that uses Markov logic to localize software bugs; while [11] introduced a new bug prediction technique that works at the granularity of an individual file level change. Moreover, [12] and [17] gave a comparison of ML algorithms for bug prediction in open source software; and a survey of software metrics used for predicting software fault by using ML algorithm is examined by [13]. In addition, [14] performed a comparison between news networks and support vector machine when applied to the problem of classifying modules as faulty or error-free, and [15], reviewed all the studies between the period of 1991 and 2013 to give a SRL of ML techniques for Software Fault Prediction. On the other hand, to predict software defect [16] explained a methodology that combined naïve bayes, decision tree and support vector machine, while [18] proposed a cognitive approach for software fault classification using a sequential forward search with extreme learning machine. Within the same framework, [19] suggested the use of Gene Expression Programming and Support Vector Regression as a solid regression algorithms enabling to predict error through entropy of changes. Reference [20], considered 20 metrics kinds for developing a model for fault prediction. Reference [21], analyzed how ML algorithms can be used to achieve the classification of requests done by testers or developers as bug or improvement request. In the same context, [22] explained a bug prediction method that can, for an immense software, select then test a percentage of modules, and then build a bug prediction model for the rest of the modules. Reference [23], proposed a defect prediction framework that considers the feature extraction and

Table 1. Research topics of selected studies.

Research topic	Number of papers	Papers references
Software defect/bug/fault prediction	20	[9–28]
Software quality improvement	11	[29–39, 85]
Data classification	7	[40–46]
Introduction to fundamentals of ML	7	[47–53]
Software effort estimation	4	[54–57]
Build recommendations	3	[58–60]
Code changes detection	3	[61–63]
Design pattern detection	3	[64–66]
Software process evaluation	3	[67–69]
Software reliability prediction	3	[70–72]
Software testing improvement	3	[73–75]
Software verification	2	[76, 77]
Customizing software	1	[78]
Grading system automation	1	[79]
Software learning	1	[80]
Software maturity assessment	1	[81]
Software requirement specifications analysis	1	[82]
Software risk identification	1	[83]
Software run-time prediction	1	[84]

class imbalance issues, moreover [24] proposed a framework for defect prediction using ML techniques with android software. Similarly, [25] presented a model of software defect prediction based on support vector machine and local tangent space alignment. While [26] demonstrated the positive effects of combining feature selection and ensemble learning on the performance of defect classification. Reference [27], proposed a software defect prediction that incorporates a reject option in the classifier to improve the reliability in the decision making process. Moreover, [28] proposed an approach for feature selection, which is used as a classification technique for the software fault prediction.

Eleven papers focused on the second topic, which is "Software quality improvement". Reference [29], explained how to choose a ML model for software effort estimation. Similarly, [30] proposed a method to select the web service configurations, which reduce energy consumption. While [31] examined the difficulties that arise in applying statistical ML as a tool for software development. Moreover, [32] used Case-Based reasoning to improve software quality through early prediction of error patterns. Reference [33], compared seven missing data techniques and showed how a combination of those techniques and decision tree algorithm leads to a significant improvement in prediction performance. In the same context, [34] compared the result of using deep learning technique and traditional ML algorithm to solve software insecurity issues. While [35], were rather interested in improving the effectiveness of software behavior learning using k-nearest neighbor. In addition, [36] showed that the integration of models with high quantitative performance and intelligible models is desirable to reach a compromise between performance and intelligibility. Within the same framework, [37] focused on improving the performance of anti-spam system using naive bayes. Reference [38], dealt the issue of how to choose the appropriate software process model, and cast it into a classification problem. Finally yet importantly, [39] examined the application of ML algorithms towards predicting the software quality characteristics. Reference [85] proposed a solution to obtain a self-adaptive process based on Q-Learning.

The third topic is "Data classification", seven papers tackled this one. Having as goal the automatic classification of program execution data as coming from passing or failing program executions, [40] presented and studied a technique for remote analysis and monitoring of software systems based on random forests ML algorithm. Reference [41], identified software artifacts types produced by 383 open source projects. Then, they proposed an automatic approach based on ML to classify those types. As for [42], the authors were interested on classification rules for documents, the ML algorithm (C4.5) was found to produce concise models for this issue. While [43] applied classification ML algorithms used in data mining on data obtained from individuals during the vocational guidance process, to determine the most appropriate algorithm. Reference [44], proposed a model based on neural network to classify software requirements without labor-intensive feature engineering. Within the same framework, [45] developed three types of generalized rough set models, which are useful in the study of knowledge acquisition in incomplete information systems and incomplete decision tables. Reference [46], extended the rough set theory in incomplete and complete information system. The authors also presented the method and technology dealing with information in generalized information system.

The fourth topic is "Introduction to fundamentals of ML", out of 76 studies, 7 interested in this one. Reference [47] gave a technical briefing of ML algorithms, their assumptions and guarantees. In the same context [48] is a book chapter that provide the fundamentals of ML. As for [49] and [50], they gave an overview of ML technologies, with concrete case studies from code analysis. Reference [51], provided a SLR that gave a ML algorithms analogy within the community of web spam detection. Reference [52], provided the characteristics and applicability of some utilized ML algorithms in the field of SE. The last but not the least, [53] gave a SLR on Supervised ML Algorithms and Boosting Process.

Four studies were interested in the fifth topic, which is "Software effort estimation". Reference [54], used Random forest algorithm to propose a ML approach for prediction of teamwork effectiveness in SE education. As for [55], the authors proposed an approach based on neural networks, support vector machines, and generalized linear models to estimate software project duration and effort. While [56] performed a SLR of studies on ML for software effort estimation published between 1991 and 2010. Moreover, [57] focused on methods to calculate agile projects effort and especially the story point approach (SPA) and analyzed its prediction precision.

The sixth topic is "Building recommendations", 3 papers focused on this one. Reference [58], proposed a system that aimed to help users to mine information on the Web. As for [59], presented issues related to the application of ML in recommendation systems. Moreover, [60] aimed to optimize the task parallelism of big data applications.

The seventh topic is "Code changes detection", three papers treated this topic. Reference [61], presented prediction models to determine either a source file was touched by a kind of source code changes or not. As for [62], the authors proposed a classification model that applies ML to the judgments of every user concerning the code clones. Moreover, [63] described a study performed on two industrial projects, where the authors used problems obtained from various versions of the projects to predict changes in code.

The eighth topic is "Design pattern detection", 3 papers focused on it. Reference [64], applied ML algorithms to predict parallel patterns and compared these algorithms in terms of precision and speed. As for [65], it presented a method to permit the use of ML algorithms for a design patterns modeling able to represent pattern instances composed of many classes. Moreover, [66] proposed a framework to discuss some issues that may help beginner developers to choose the right design pattern for a given design issue.

The ninth topic is "Software process evaluation", three papers discussed this one. Reference [67], presented a method to evaluate software process, they formulated the research issue as a series of classification task, which is solved using a proposed method. As for [68], they proposed an approach to categorize software projects without any source code using a small number of API calls as attributes. Moreover, [69] presented a novel semi-automated approach to software process evaluation using the classification ML algorithms.

The tenth topic is "Software reliability prediction", three papers focused on it. Based on past failures of a software, [70], applied ML algorithms to predict its reliability. The authors demonstrated that support vector machines outperformed the model predicted using the other algorithms. As for [71], the study proposed the use of several ML algorithms for software reliability prediction and evaluated them using some performance keys, on different datasets collected from industrial software. While [72] utilized a deep learning model to predict the software faults number and asses its reliability.

The eleventh topic is "Software testing improvement", three papers tackled it. Reference [73], presented an approach, that starts by organizing test data into identical functional clusters, then generating classifiers that had various applications. As for [74], the study provided a short overview of researches done related to software testing, and announced on a set of new applications the authors' works covered. Moreover, [75] is a book chapter that presented an overview of ML algorithms for various software testing issues.

The twelfth topic is "Software verification", two papers have treated this topic. Reference [76], presented a ML approach to predict software verification tools rankings. Moreover, [77] proposed a deductive verification method of embedded assembly program for verifying real-time safety.

The thirteenth topic is "Customizing software", one paper tackled this topic. To learn a user's preferences, [78] employed a decision tree algorithm in a multi agent framework, and they conducted experiments on a mobile environment.

The fourteenth topic is "Grading system automation", one paper tackled this topic. Reference [79], proposed a system that, once taught by a field expert, can learn to do the basics word's difficulty grading, and can return the difficulty level of a word.

The fifteenth topic is "Software learning", one paper tackled it. Reference [80], proposed a smart agent-based system of single machine scheduling that can select the perfect dispatching guideline for various system objectives, and it can operate well for all criteria.

The sixteenth topic is "Software maturity assessment", one paper tackled this topic. Reference [81], proposed an evaluation model of open source project maturity established by the decision trees and support vector machine which has a high accuracy.

The seventeenth topic is "Software requirement specifications analysis", one paper tackled it. Reference [82], focused on the extraction of semantic information from software requirement specifications. It proposed a method in which, a semantic labeling approach and ML were combined.

The eighteenth topic is "Software risk identification", one paper tackled it. Reference [83], developed a trained neural network algorithm to generate risk prompts, which is based on software project characteristics and other factors.

The last topic is "Software run-time prediction", two papers tackled this topic. Reference [84] proposed a method that trains on various runs of a chosen program, and then it predicts the next phase run-time, for the same program and different inputs.

3.2 RQ2: What ML Algorithms have been used in the Selected Studies?

Panoply of fifteen ML algorithms were used in the selected studies, as shown in Fig. 5. The most used algorithms are decision tree (23% of studies) and support vector machine (23% of studies), followed with random forest algorithm (12% of studies). While 10% of the selected studies used naive Bayes algorithm, and 8% used artificial neural networks. K-nearest neighbor was used by 7% of the selected studies, and extreme learning machine by 5% of the selected studies. Logistic regression was used by 3% of the selected studies, as for multilayer perceptron that was also used by 3% of the selected studies. Each of

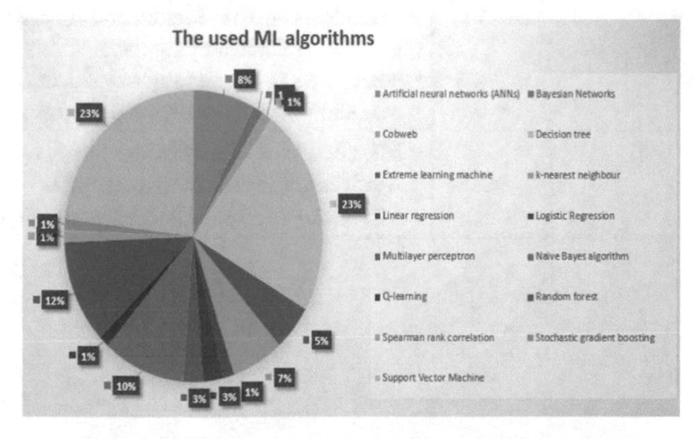

Fig. 5. The ML algorithms used in selected studies

Bayesian networks, cobweb, linear regression, Q-learning, spearman rank correlation and stochastic gradient boosting was used by 1% of the selected studies.

3.3 RQ3: What are the Main Outcomes of the Selected Studies?

The selected studies applied different ML algorithms to deal with various software developments process issues. This SLR confirmed the usefulness of the ML techniques in many steps of all the activities of software development process: requirements, design, implementation, verification and maintenance. Table 2 groups the outcomes released from the selected studies.

Table 2. Main outcomes in the different software development process activities.

Software development process activity	Main outcomes
Requirements	• ML allows extraction of semantic information from software requirement specifications document • ML allows making the judicious choice of software process model
Design	• ML gives a correct estimation of software project effort • ML allows the prediction of teamwork effectiveness • ML allows design patterns classification • ML allows design patterns detection
Implementation	• ML allows the prediction of coding rules violations • ML allows to measure the consumption of energy for web service configuration • ML facilitates the choice of tools by classifying the different software artifacts • ML allows to offer customized software applications • ML allows offering recommendations to developers during the implementation
Verification	• ML allows software fault and bug prediction • ML allows software bug localization • ML allows software fault classification • ML allows software defect detection
Maintenance	• ML allows the classification of software applications into domain categories • ML allows to exploit the software execution data

4 Conclusion

This study investigates the current state of art on the use of ML in SE by means of a SLR. During the SLR, seventy-six studies were selected from different online data sources, in order to answer three research questions. The adopted search strategy showed that this SLR topic has been a focus for researchers since 1995 and that numbers of research works (journal papers, conference papers and book chapters) devoted to this subject has begun to increase since 2017. The main finding of RQ1 showed that the use of ML in SE has touched a variety of current SE issue; in fact, nineteen research topics were treated. The recurrent ones were the software defect/bug/fault prediction and the software quality improvement. As for RQ2, it showed that the use of ML in SE does not consume a given type of ML algorithms; in fact, fifteen different ML algorithms were used. The most used ones are decision trees and support vector machine. Moreover, RQ3 explained how the use of those algorithms in SE improved all activities of the software process development.

References

1. Rothman, D.: Artificial Intelligence by Example, 1st edn. Packt Publishing Ltd. Livery Place 35 Livery Street Birmingham B3 2PB, UK (2018)
2. Tack, C.: Artificial intelligence and machine learning | applications in musculoskeletal physiotherapy. Musculoskeletal Sci. Pract. **39**, 164–169 (2019)
3. Alpaydın, E.: Introduction to Machine Learning, 2nd edn. The MIT Press Cambridge, Massachusetts (2010)
4. Partridge, D.: Artificial Intelligence and Software Engineering Understanding the Promise of the Future. AMACOM American Management Association, 1601 Broadway New York, New York 10019, USA (1998)
5. Dimitra, K., Yihao, L., Elvira, M.A., Misirlis, N., Wong, W.E.: A bibliometric assessment of software engineering scholars and institutions (2010–2017). J. Syst. Softw. **147**, 246–261 (2019)
6. Zhang, D., Tsai, J.P.: Machine Learning Applications in Software Engineering (Series on Software Engineering and Knowledge Engineering). World Scientific Publishing Company, Singapore (2005)
7. Jesson, J., Matheson, L., Lacey, F.M.: Doing Your Literature Review: Traditional and Systematic Techniques, 1st edn. SAGE Publications Ltd, Newbury Park (2011)
8. Kitchenham, B., Charters, S.: Guidelines for performing systematic literature reviews in software engineering. Technical Report EBSE-2007-01, School of Computer Science and Mathematics, Keele University (2007)
9. Hamou-Lhadj, W., Nayrolles, M.: A project on software defect prevention at commit-time: a success story of university-industry research collaboration. In: The 5th International Workshop on Software Engineering Research and Industrial Practice on Proceedings, Gothenburg, Sweden, pp. 24–25 (2018)
10. Sai, Z., Congle, Z.: Software bug localization with Markov logic. In: The 36th International Conference on Software Engineering, Hyderabad, India, pp. 424–427 (2014)
11. Sunghun, K., Whitehead, E.J., Yi, Z.: Classifying software changes: clean or buggy? IEEE Trans. Softw. Eng. **34**(2), 181–196 (2008)
12. Ruchika, M., Laavanye, B., Sushant, S., Pragati, P.: Empirical comparison of machine learning algorithms for bug prediction in open source software. In: The International Conference on Big Data Analytics and Computational Intelligence (ICBDAC), Chirala, India, pp. 40–45 (2017)
13. Meiliana, Syaeful, K., Harco, L.H.S.W., Ford, L.G., Edi, A.: Software metrics for fault prediction using machine learning approaches: a literature review with PROMISE repository

dataset. In: The International Conference on Cybernetics and Computational Intelligence (CyberneticsCom), Phuket, Thailand, pp. 19–23 (2017)
14. Iker, G.: Applying machine learning to software fault-proneness prediction. J. Syst. Softw. **81**, 186–195 (2008)
15. Ruchika, M.: A systematic review of machine learning techniques for software fault prediction. Appl. Soft Comput. **27**, 504–518 (2015)
16. Jyoti, D.: A review of improving software quality using machine learning algorithms, Int. J. Comput. Sci. Mob. Comput. **6**(3), 148–153 (2017)
17. Hanchate, D.B., Bichkar, R.S.: The machine learning in software project management: a journey. Part II. Appl. Discrete Math. Heuristic Alg. **1**(4), 29–58 (2015)
18. Pandey, A.K., Gupta, M.: Software fault classification using extreme learning machine: a cognitive approach. Evol. Intel. **1**, 1–8 (2018)
19. Kaur, A., Kaur, K., Chopra, D.: An empirical study of software entropy based bug prediction using machine learning. Int. J. Syst. Assur. Eng. Manage **8**(2), 599–616 (2017)
20. Kumar, L., Tirkey, A., Rath, S.-K.: An effective fault prediction model developed using an extreme learning machine with various kernel methods. Front. Inform. Technol. Electron. Eng. **19**(7), 864–888 (2018)
21. Nitish, P., Debarshi, K., Hudait, A., Amitava, S.: Automated classification of software issue reports using machine learning techniques: an empirical study. Innov. Syst. Softw. Eng. **13**(4), 279–297 (2017)
22. Li, M., Zhang, H., Rongxin, W., Zhou, Z.-H.: Sample-based software defect prediction with active and semi-supervised learning. Autom. Softw. Eng. **19**(2), 201–230 (2012)
23. Zhou, X., et al.: Software defect prediction based on kernel PCA and weighted extreme learning machine. Inf. Softw. Technol. **106**, 182–200 (2019)
24. Malhotra, R.: An empirical framework for defect prediction using machine learning techniques with Android software. Appl. Soft Comput. **49**, 1034–1050 (2016)
25. Hua, W., et al.: Establishing a software defect prediction model via effective reduction. Inf. Sci. **477**, 399–409 (2019)
26. Laradji, I.H., et al.: Software defect prediction using ensemble learning on selected features. Inf. Softw. Technol. **58**, 388–402 (2015)
27. Diego, P.P.M., et al.: Classification with reject option for software defect prediction. Appl. Soft Comput. **49**, 1085–1093 (2016)
28. Turabiedh, H., Mafarja, M., Li, X.: Iterated feature selection algorithms with layered recurrent neural network for software fault prediction. Expert Syst. Appl. **122**, 27–42 (2019)
29. Minku, L.L., Yao, X.: A principled evaluation of ensembles of learning machines for software effort estimation. In: The 7th International Conference on Predictive Models in Software Engineering, Banff, Alberta, Canada (2011)
30. Murwantara, I.M., Bordbar, B., Minku, L.L.: Measuring energy consumption for web service product configuration. In: The 16th International Conference on Information Integration and Web-based Applications & Services, Hanoi, Viet Nam, pp. 224–228 (2014)
31. Patel, K., Fogarty, J., Landay, J.A., Harrison, B.: Investigating statistical machine learning as a tool for software development. In: The SIGCHI Conference on Human Factors in Computing Systems, Florence, Italy, pp. 667–676 (2008)
32. Rashid, E., Patnayak, S., Bhattacherjee, V.: A survey in the area of machine learning and its application for software quality prediction. ACM SIGSOFT Softw. Eng. Notes **37**(5), 1–7 (2012)
33. Twala, B., Cartwright, M., Shepperd, M.: Ensemble of missing data techniques to improve software prediction accuracy. In: The 28th International Conference on Software Engineering, Shanghai, China, pp. 909–912 (2006)

34. Cesar, J.C., Fehmi, J., Yasir, M.: Is predicting software security bugs using deep learning better than the traditional machine learning algorithms? In: The International Conference on Software Quality, Reliability and Security (QRS), Lisbon, Portugal, pp. 95–102 (2018)
35. Feng, Y., Chen, Z.: Multi-label software behavior learning. In: The 34th International Conference on Software Engineering, Zurich, Switzerland, pp. 1305–1308 (2012)
36. Lounis, H., Gayed, T.F., Boukadoum, M.: Machine-learning models for software quality: a compromise between performance and intelligibility. In: The 23rd International Conference on Tools with Artificial Intelligence, Boca Raton, FL, USA, pp. 919–921 (2011)
37. Zhang, P., Su, Y., Wang, C.: Statistical machine learning used in integrated anti-spam system. In: The 6th International Conference on Machine Learning and Cybernetics on Proceedings, Hong Kong, pp. 4055–4058 (2007)
38. Qinbao, S., et al.: A machine learning based software process model recommendation method. J. Syst. Softw. **118**, 85–100 (2016)
39. Mehta, P., Srividya, A., Verma, A.K.: Application of machine learning paradigms for predicting quality in upstream software development life cycle. OPSEARCH 42(4), 332–339 (2005)
40. Haran, M., et al.: Applying classification techniques to remotely collected program execution data. In: the 10th European software engineering conference, Lisbon, Portugal, pp. 146–155 (2005)
41. Yazhan, M., et al.: Automatic classification of software artifacts in open-source applications. In: The 15th International Conference on Mining Software Repositories on Proceedings, New York, USA, pp. 414–425 (2018)
42. Cunningham, S.J., Summers, B.: Applying machine learning to subject classification and subject description for information retrieval. In: The 2nd New Zealand International Two-Stream Conference on Artificial Neural Networks and Expert Systems on Proceedings, Dunedin, New Zealand, pp. 243–246 (1995)
43. Bulbul, H.I., Unsal, O.: Comparison of classification techniques used in machine learning as applied on vocational guidance data. In: The 10th International Conference on Machine Learning and Applications and Workshops, Honolulu, HI, USA, pp. 298–301 (2011)
44. Navarro-Almanza, R., Juarez-Ramırez, R., Licea, G.: Towards supporting software engineering using deep learning: a case of software requirements classification. In: The 5th International Conference in Software Engineering Research and Innovation, Mérida, Mexico, pp. 116–120 (2017)
45. Wei-Zhi, W., Xiao-Ping, Y.: Information granules and approximations in incomplete information systems. In: The 6th International Conference on Machine Learning and Cybernetics, Hong Kong, pp. 3740–3745 (2007)
46. Zhou, J., Zhang, Q., Tong, S.: Decision rules based on rough set theory in generalized information systems. In: The 5th International Conference on Machine Learning and Cybernetics, Dalian, China, pp. 1477–1482 (2009)
47. Meinke, K., Bennaceur, A.: Machine learning for software engineering models, methods, and applications. In: The 40th International Conference on Software Engineering, Gothenburg, Sweden, pp. 548–549 (2018)
48. Myeongsu, K., Noel, J.J.: Prognostics and Health Management of Electronics: Fundamentals, Machine Learning, and the Internet of Things, 1st edn. John Wiley and Sons Ltd, USA (2018)
49. Louridas, P., Ebert, C.: Machine Learning. IEEE Softw. **33**(5), 110–115 (2016)
50. Meinke, K., Bennaceur, A.: Machine learning for software engineering. In: The 40th International Conference on Software Engineering, Sweden, pp. 548–549 (2018)
51. Kwang, L.G., Ashutosh, K.S.: Comprehensive literature review on machine learning structures for web spam classification. Procedia Comput. Sci. **70**, 434–441 (2015)
52. Zhang, D., Tsai, J.J.P.: Machine learning and software engineering. Softw. Qual. J. **11**, 87–119 (2003)

53. Praveena, M., Jaiganesh, V.: A literature review on supervised machine learning algorithms and boosting process. Int. J. Comp. Appl. **169**(8), 32–35 (2017)
54. Dragutin, P., et al.: Work in progress: a machine learning approach for assessment and prediction of teamwork effectiveness in software engineering education. In: The Frontiers in Education Conference, Seattle, WA, USA, pp. 1–3 (2012)
55. Pospieszny, P., Czarnacka-Chrobot, B., Kobylinski, A.: An effective approach for software project effort and duration estimation with machine learning algorithms. J. Syst. Softw. **137**, 184–196 (2018)
56. Wen, J., Li, S., Lin, Z., Hu, Y., Huang, C.: Systematic literature review of machine learning based software development effort estimation models. Inf. Softw. Technol. **54**, 41–59 (2012)
57. Satapathy, S.M., Rath, S.K.: Empirical assessment of machine learning models for agile software development effort estimation using story points. Innov. Syst. Softw. Eng. **13**(2–3), 191–200 (2017)
58. Tao, L., Li, Y.: A synthetic intelligent system for web information mining. In: The International Conference on Machine Learning and Cybernetics, Shanghai, China, pp. 1357–1360 (2005)
59. Nawrocka, A., Kot, A., Nawrocki, M.: Application of machine learning in recommendation systems. In: The 19th International Carpathian Control Conference, Szilvasvarad, Hungary, pp. 328–331 (2018)
60. Hernandez, A.B., Perez, M.S., Gupta, S.: Using machine learning to optimize parallelism in big data applications. Futur. Gener. Comput. Syst. **86**, 1076–1092 (2018)
61. Giger, E., Pinzger, M., Gall, H.C.: Can we predict types of code changes? An empirical analysis. In: The 9th IEEE Working Conference on Mining Software Repositories, Zurich, Switzerland, pp. 217–226 (2012)
62. Yang, J., Hotta, K., Higo, Y., Igaki, H., Kusumoto, S.: Classification model for code clones based on machine learning. Empir. Softw. Eng. **20**(4), 1095–1125 (2014)
63. Tollin, I., Fontana, F.A., Zanoni, M.: Change prediction through coding rules violations. In: The 21st International Conference on Evaluation and Assessment in Software Engineering, New York, USA, pp. 61–64 (2017)
64. Deniz, E., Sen, A.: Using machine learning techniques to detect parallel patterns of multi-threaded applications. Int. J. Parallel Program. **44**(4), 867–900 (2016)
65. Zanoni, M., Fontana, F.A., Stella, F.: On applying machine learning techniques for design pattern detection. J. Syst. Softw. **103**, 102–117 (2015)
66. Hussain, S., et al.: Automated framework for classification and selection of software design patterns. Appl. Soft Comput. **75**, 1–20 (2019)
67. Chen, N., Hoi, S.C.H., Xiao, X.: Software process evaluation: a machine learning approach. In: The 26th International Conference on Automated Software Engineering, Lawrence, KS, USA, pp. 333–342 (2011)
68. Linares-Vásquez, M., McMillan, C., Poshyvanyk, D., Grechanik, M.: On using machine learning to automatically classify software applications into domain categories. Empir. Softw. Eng. **19**(3), 582–618 (2014)
69. Chen, N., Hoi, S.C.H., Xiao, X.: Software process evaluation: a machine learning framework with application to defect management process. Emp. Softw. Eng. 19(6), 1531–1564 (2014)
70. Kumar, P., Singh, Y.: An empirical study of software reliability prediction using machine learning techniques. Int. J. Syst. Assur. Eng. Manage. **3**(3), 194–208 (2012)
71. Jaiswal, A., Malhotra, R.: Software reliability prediction using machine learning techniques. Int. J. Syst. Assur. Eng. Manage. **9**(1), 230–244 (2016)
72. Jinyong, W., Ce, Z.: Software reliability prediction using a deep learning model based on the RNN encoder decoder. Reliab. Eng. Syst. Saf. **170**, 73–82 (2018)
73. Lenz, A.R., Pozo, A., Vergilio, S.R.: Linking software testing results with a machine learning approach. Eng. Appl. Artif. Intell. **26**, 1631–1640 (2013)

74. Briand, L.C.: Novel applications of machine learning in software testing. In: The 8th International Conference on Quality Software, Oxford, UK, pp. 3–10 (2008)
75. Gove, R., Faytong, J.: Machine learning and event-based software testing: classifiers for identifying infeasible GUI event sequences. Adv. Comput. **86**, 109–135 (2012)
76. Czech, M., Hullermeier, E., Jakobs, M., Wehrheim, H.: Predicting rankings of software verification tools. In: The 3rd ACM SIGSOFT International Workshop on Software Analytics, New York, USA, pp. 23–26 (2017)
77. Satoshi, Y.: Deductively verifying embedded software in the era of artificial intelligence = Machine Learning + Software Science. In: The 6th Global Conference on Consumer Electronics, Nagoya, Japan, pp. 1–4 (2017)
78. Lee, W.-P., Cheng-Che, L.: Customising WAP-based information services on mobile networks. Personal Ubiquitous Comput. **7**(6), 321–330 (2003)
79. Chang, C., Liu, H., Lin, J.: Constructing grading information system for words' difficulty using a supervised learning method. In: The International Conference on Machine Learning and Cybernetics, Hong Kong, China, pp. 3991–3996 (2007)
80. Kong, L., Wu, J.:Dynamic single machine scheduling using Q-learning agent. In: The International Conference on Machine Learning and Cybernetics, Guangzhou, China, pp. 3237–3241 (2005)
81. Liu, Q., Li, X., Zhu, H., Fan, H.: Acquisition of open source software project maturity based on time series machine learning. In: The 10th International Symposium on Computational Intelligence and Design, Hangzhou, China, pp. 296–299 (2017)
82. Wang, Y.: Automatic semantic analysis of software requirements through machine learning and ontology approach. J. Shanghai Jiaotong Univ. (Sci.) **21**(6), 692–701 (2016)
83. Harry, R.J.: Poster: software development risk management: using machine learning for generating risk prompts. In: The 37th IEEE International Conference on Software Engineering, Florence, Italy, pp. 833–834 (2015)
84. Chiu, M., Moss, E.: Run-time program-specific phase prediction for python programs. In: The 15th International Conference on Managed Languages & Runtimes, Linz, Austria (2018)
85. Mezouar, H., El Afia, A.: A 4-level reference for self-adaptive processes based on SCOR and integrating Q-Learning. In: The 4th International Conference on Big Data and Internet of Things (BDIoT 2019), 23 and 24 October 2019, Tangier-Tetuan, Morocco. ACM (2019). https://doi.org/10.1145/3372938.3372953

Machine Learning for Used Cars Price Prediction: Moroccan Use Case

Faouzia Benabbou(✉), Nawal Sael, and Imad Herchy

Laboratory of Information Technology and Modeling, Faculty of Sciences Ben M'Sik, Hassan II University, Casablanca, Morocco
faouzia.benabbou@univh2c.ma

Abstract. Nowadays, due to increasing competition and to have an advantage over their competitors, companies are increasingly opting for solutions based on artificial intelligence techniques for classification, forecasting, decision support, or prediction. Estimating the price of a product in a given context is one of the challenged tasks for researchers. In this paper, we present a state of the art of solutions based on supervised learning algorithms for the task of predicting the price of used cars. Then we propose a prediction system for second-hand cars in the Moroccan context. For this purpose, we collected data using a web crawler, and then applied feature selection, listwise deletion and other pre-processing techniques. As a result, XGBoost gives the best performance with an R2 of 90.7%. The model was deployed using a flask application making it easier to use for common users.

Keywords: Price prediction · Machine learning · Used cars · Regression · Ensemble methods

1 Introduction

Given the demand for private cars, most buyers focus on the used car market because of the reasonableness and affordability of their prices. Many factors can influence the price of a used car, such as the age, state, color, mileage, etc., which makes the used car market non-persistent. Thus, car price prediction models are required to help buyers and vendors in the estimation process. Machine learning techniques can be used to perform different tasks. For instance, in the case of data lacking labels machine learning can be used for unsupervised tasks such as clustering to study similarities between data samples or even pattern recognition for instance association rules learning. In the case of labeled data which is our case we speak of supervised learning. Supervised learning techniques can be also used for classification purpose, for example a car price can be classified into different classes (Expensive, Normal, and Bargain), or it can be a regression task in which the output will be a continues numerical value which's in this case the price. Many studies were conducted to predict a price of products as fossil fuel, natural gas, stock, bitcoin, natural gas, houses, mobile, etc.

M. Lazaar et al. (Eds.): BDIoT 2021, LNNS 489, pp. 332–346, 2022.
https://doi.org/10.1007/978-3-031-07969-6_25

In this paper we conducted a comparative study of several machine learning techniques such as Multiple Linear Regression (MLR), bagging algorithms, boosting algorithms, and stacking for used car price prediction purpose. Each of these algorithms was trained on data scraped from automobile ads in www.avito.ma website. The main objective for this study is to find the best performing model for used cars price prediction in Moroccan context.

The structure for this research paper is the following. In Sect. 2, we reviewed similar works that have been done to accomplish the prediction price task and we presented a comparative study. In Sect. 3 we described some background of the algorithms used in the study and we presented our methodology. Section 4 is dedicated to the results of the experiment and the deployment of our car prediction system. Conclusion and perspectives are discussed in Sect. 5.

2 Literature Review and Comparison

An accurate used car price evaluation is a catalyst for the healthy development of used car market. Data mining and Machine Learning have been applied to predict used product prices in several studies. Moreover, the task of price prediction can be tackled in different approaches mainly forecasting and predicting. Forecasting a price or a market is to predict the changes in the prices for a certain good such as energy sources price changes [1], crude oil prices [2], natural gas [3], Bitcoin [3, 4] or stock [5]. When the task doesn't consider the changes in a specific market with time forecasting turns to plain predicting. Three different studies focused on the task of predicting house prices. using an Extracted dataset from Virginia, Park et al. [6] achieved an average error of 0.2488 using Repeated Incremental Pruning To Produce Error Reduction (RIPPER). Phan et al. [7] used Melbourne Housing dataset resulting in a Mean Squared Error (MSE) of 0.0561 using Support Vector Machine (SVM), and using Housing Price in Beijing dataset. Truong et al. [8] opted for Random Forest (RForest) which achieved a RMSE score of 0.1298. Asim and Khan [9], and Balakumar et al. [10] focused on the task of price prediction for mobile phones, using J48, and K-Nearest Neighbors (KNN) which achieved the highest accuracy of 92%.

Our study is focused on used car price prediction and its correlation with the Moroccan context. The discriminant features for used car prices can change from one country to another. In this context, Voß and Lessmann [11] contributed by using sales data from a German car manufacturer. the results of this study suggested that the methods most widely used in resale price modeling are least effective. In particular, Linear Regression (LR) methods predict significantly less accurately than advanced methods such as RForest and Ensemble Selection (ES), the best performance was accomplished by ES scoring a Mean Absolute Error (MAE) of 3.97. Another interesting approach based on images for predicting car prices was proposed by Yang et al. [12]. To our knowledge it's the only study in which visual features were used to estimate the price for a vehicle. Bicycles and cars labeled datasets were used and LR, multiclass SVM, Transfer learning, and PriceNet, which is a NN architecture from scratch, were explored. Strong results were achieved on both datasets, and deep CNNs significantly outperformed the other models in a variety of metrics scoring specially R2 with 0.98. Meng et al. [13] proposed a system for price prediction of Taiwan used car auctions. Data sample included 504 successful transactions of trading in Taiwan used car auction marketing with each record containing 12 features. This study was simply statistic. The Log-Linear Multiple Regression (LLMR) model was applied to evaluate this data taking in consideration multiple features, mainly Engine, Years, and Mileage. According to the results, polynomial LLMR models was more suitable for this research and produced the best result of 0.758 for R2. In China context, Chen et al. [14] did a comparative analysis of used car price prediction models on dataset collected from several used car markets in Shanghai and a used car auction website that operates nationwide. More than 100,000 records were collected. Each record within the dataset had 19 attributes of which 13 features were manually selected based on business intuition. Two algorithms were used: LR and RForest which had the upper hand scoring 0.024 for the Normalized Mean Squared Error (NMSE).

Lessmann and Voß [15] showcased the power of ensemble selection using Germen cars dataset with features such as Model year, Engine type, and Age, scoring 5.18 MAE outperforming models like stepwise regression, MLR, and SVR in predicting used cars prices. Thai et al. [16] proposed a system for used car price prediction using two datasets with 370 000 records and 33 605 records respectively. RForest, XGBoost and LightGBM were used to model the data based on features like Brand, Seller, Gearbox, and model year, out of which RForest gave the best performance for the first dataset with 0.8634 for R2 score while LightGBM had the edge for the second dataset with 0.8270 for R2 score. In the Turkey context [17], Çelik and Osmanoğlu worked on a dataset of 5041 cars with features such as Brand, Model, Model Year, Fuel Type, Horsepower, Kilometer. Multivariable Linear Regression model was used and performed 89.1% for R2 score. In Bosnia and Herzegovina context, Gegic et al. [18] used scraped data collected from the web portal and applied Artificial NN (ANN), RForest and SVM models to classify (they used multiple classes for the price.) the data based on features like brand, model, and fuel type, and the best result was performed by SVM with an accuracy of 90.48%. Another case study showing the value in using RForest was proposed by Pal et al. [19]. The authors used the "CARS" database available within Kaggle that contains over 370,000 used cars with the cleaned data contained features like The Brand, Model, and Gearbox, etc. LR was used to compare to the performance of RForest which outperformed LR with determination score R2 of 83.63%. From China, Sun et al. took on the challenge of using Back-Propagation Neural Networks (BPNN) model to predict a secondhand car prices [20]. In this experiment, an amount of data of 18055 records with 10 attributes (City, Mileage, and Productive Year etc.) out of which one was a nominal attribute was used (condition). BPNN performed an Absolute Error (AE) of 0.113 and a Relative Error of 0.78.

From Romania, Caciandone and Chiru [21] generated prediction based on automobile ads, mining the content of the autovit.ro website, and extracted from it over 16 000 ads using features like fuel type, mileage, and age. MLR and RForest were used with the obvious result of RForest scoring the better accuracy of 90%. From Thailand, Monburinon et al. [22] used that Kaggle cars dataset discussed in earlier work to benchmark different models, the data size was 371 528 containing 12 features. Gradient boosting, RForest and MLR were used as models and gradient boosting scored the higher MSE of 0.28 based on features like Seller, Brand, Fuel type, and Model, which puts another stamp proving that ensemble methods outperform single models when dealing with regression problems. Another study was conducted in Mauritius by Pudaruth [23] using historical data collected from daily newspapers. After data cleaning only 97 records remained with the following features: brand, cylinder capacity, year, and mileage. MLR, K-Nearest Neighbors (KNN), Naive Bayes (NB) and J48 were compared and the best performance of accuracy score was 70%.

From Pakistan, Noor et al. [24] showed the power of MLR model when it's used on a good quality dataset of 2000 records of used cars gathered from Pakwheels containing 16 variables. After using feature selection only three variables made the final cut: model year, model, and engine type. The model scored 98% for precision.

In the comparative study below, we focused on systems dedicated for price car prediction and used a set of criteria shown in Table 1 such as: the context, dataset used, feature selection method if used, machine learning techniques and performance. For every paper, we try to exhibit the top 6 features based on for prediction and show if any implementation is proposed.

Table 1. Comparative study

Paper	Country case study	Dataset	Feature selection method	Top 6 features	Model	Performance	Deployment
[11]	Germany	Extracted	–	–	ES	MAE = 3.97	–
[12]	USA	CARS (Kaggle)	–	224 × 224 pixels	CNN	R2 = 0.98	–
[13]	Taiwan	Extracted	–	Confiscated, engine, years, mileage, body, interior	PMLR	R2 = 0.758	–
[14]	China	Extracted	Manual	Age, mileage, condition, region, ownership, manufacturer	RForest	NMSE = 0.024	–
[15]	Germany	Extracted	Manual	Age, mileage, customer, EngineType, PrevUsage, lacquer	ES	MAE = 5.18	–
[16]	Vietnam	Ebay car sales data (Kaggle)	Manual	Brand, name, actuator, gearbox, age, fuel	Rforest	R2 = 0.8634	–
		Extracted			LightGBM	R2 = 0.827	–
[17]	Turkey	Extracted	Correlation	Brand, model, model year, fuel type, horsepower, kilometer	MLR	R2 = 0.891	IBM SPSS

(*continued*)

Table 1. (*continued*)

Paper	Country case study	Dataset	Feature selection method	Top 6 features	Model	Performance	Deployment
[18]	Bosnia and Herzegovina	Extracted	Manual	Brand, model, fuel, power, year, miles	SVM	Accuracy = 90.5%	Java
[19]	India	Ebay car sales data (Kaggle)	Manual	Kilometers, brand, and vehicle type	RForest	Accuracy = 83.6%	–
[20]	China	Extracted	–	Vehicle number, year, city, mileage, condition, sales year	BPNN	Absolute error 0.113	Matlab
[21]	Romania	Extracted	Manual/Wrapper	Year, mileage, horsepower, engine, make, model	RForest	Accuracy = 90%	Python
[22]	Thailand	Ebay car sales data (Kaggle)	Manual/Correlation	Brand, power, kilometer, model, fuel, gearbox	Gboost	MSE = 0.28	–
[23]	Mauritius	Extracted	Manual	Make, year, cylinder vol, mileage	MLR, KNN, NBayes J48	R2 = 0.662	WEKA
[24]	Pakistan	Extracted	–	Model year, model, engine	MLR	Precision = 98%	Minitab

What we can derive from all these papers working on price prediction especially the studies taking a regression approach we see a clear dominance of ensemble methods over simple models especially when dealing with high dimension data. Most of the papers discussed in this section were extracted either using web crawlers or actually being manually extracted. These latter when approaching the problematic as a regression task use the coefficient of determination (R2) because of how easy it can be interpreted, the best performance recorded to our knowledge on an extracted dataset is 89.1 for R2 using MLR model [17]. The features selection is based on business intuition approach or a correlation method. For car price prediction case, the discriminant characteristics differ from one country to another but some feature as mileage, brand are usually used as shown in Fig. 1.

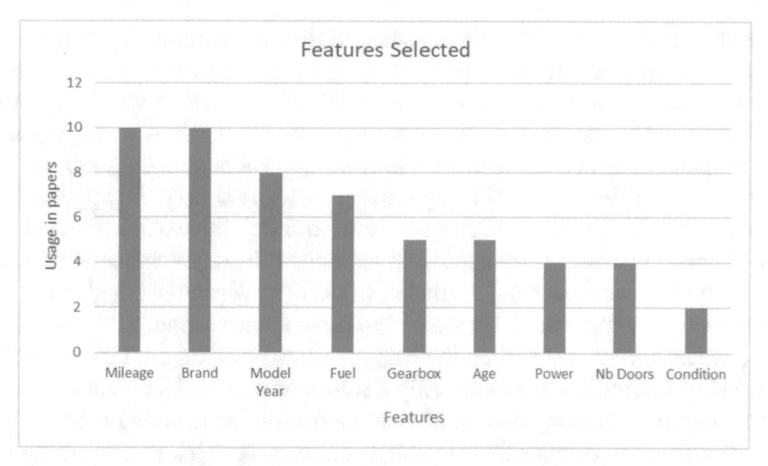

Fig. 1. Most selected features for used cars price prediction.

3 Materials and Methods

3.1 Background Theory

The task our study is aiming to accomplish a real value prediction for used cars based on a spectrum of variables in Moroccan context. To tackle this problem multiple methods were explored as addressed in the state of art, and the majors were Multiple Linear Regression, and ensemble approaches.

Multiple Linear Regression (MLR)
Multiple linear regression attempts to model the relationship between two or more explanatory variables and a target variable by fitting a linear equation to observed data. Mathematically linear regression calculates the prediction using each feature as a variable and its importance as a coefficient, the coefficients can be positive or negative values depending on how that specific variable affects the outcome of the prediction, and the formula is the following:

$$y = \beta_0 + \beta_1 X_1 + \cdots + \beta_n X_n + \varepsilon \tag{1}$$

y: The predicted value of the dependent variable (The price).
β_0: The y-intercept (value of y when all other parameters are set to 0).
$\beta_i X_i$: The regression coefficient multiplied by the variable (feature).
ε: Model error (a.k.a. how much variation there is in our estimate of y).

Ensemble Methods
Ensemble learning helps improve machine learning results by combining several models. This approach allows the production of better predictive performance compared to a single model. During this study we explore three different ensemble approaches: stacking, bagging and boosting.

Stacking is an ensemble learning technique that combines multiple classification or regression models via a meta-classifier or a meta-regressor, the base level models are trained based on a complete training set, and then the meta-model is trained on the outputs of the base level model as features. Bagging stands for bootstrap aggregation, it works by bootstrapping multiple samples of the dataset, training multiple individual models also called weak learners, and the output is calculated by averaging the weak learners' outputs. In our experiment we use two different bagging algorithms, Random Forest and Bagging Regressor. The bagging regressor is simply the application of the bagging approach for a regression problem, after training the weak learners it averages all the predictions provided to give the final prediction, those weak learners can be trees or any other type of weak learners available, In thc case where the bagging uses trees as weak learners most people think the algorithm is the same as random forest but it is not, although random forest uses as well the bagging approach, Random forests differ from bagged trees by forcing the tree to use only a subset of its available predictors to split on in the growing phase creating greater diversity amongst the resulting models. Boosting refers to a family of algorithms that are able to convert weak learners to strong learners. The main principle of boosting is to fit a sequence of weak learners, models that are only slightly better than random guessing, such as small decision trees to weighted versions of the data. More weight is given to examples that were misclassified by earlier rounds. The predictions are then combined through a weighted sum to produce the final prediction. During this study we focus on three boosting algorithms: XGBoost, LightGBM, and CatBoost.

Feature Selection Methods

Feature selection is intended to reduce the number of inputs to those that are believed to be most useful for the model in order to predict the target variable. In our study we opt to use two feature selection techniques RFE and FI as described below.

Recursive Feature Elimination (RFE)

RFE is a wrapper-type feature selection algorithm. This means that a different machine learning algorithm is given and used in the core of the method, is wrapped by RFE, and used to help select features. Different than correlation which only studies the relationship between the features within the dataset to select features, selected features within RFE can change depending on the model trained for establishing the prediction task. The choice of this FS algorithm is motivated by the good results achieved by similar studies using wrapper methods, also RFE is agnostic to the input datatype which makes it even more reliable since our dataset contains multiple encoded features.

Feature Importance (FI)

FI is a widely used feature selection techniques for regression problems, and a very beneficial approach for assigning scores to input features to a predictive model that indicates the relative importance of each feature when making a prediction. The idea of how feature importance work is: after evaluating the performance of a model, we permute the values of a feature of interest and reevaluate model performance.

Correlation Heat-Map

In order to study the correlation between features we used correlation heat-map method. This is a visual representation of correlation between features, features with high correlation tend to be irrelevant for training models, and the heat-map makes it easier to locate those correlated features for deletion afterwards.

3.2 Methodology

To fulfill the task at hand the system proposed in Fig. 2 follows a specific process made-up of different blocks.

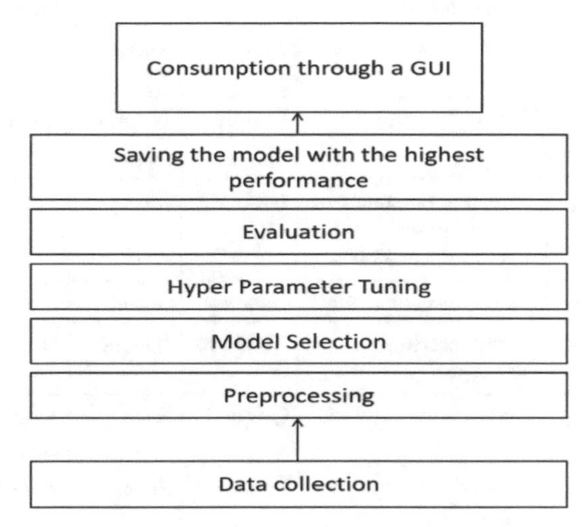

Fig. 2. Used cars price prediction system process.

Data Collection
To retrieve the data from the automobile ads website www.avito.ma we used a web crawler implemented in Python. The raw data contained 51134 unique records; each record contained 18 different attributes. The table 2 below shows each attribute and datatype with a description.

Data Preprocessing
To clean the data the following measures were taken. The first step into our preprocessing strategy is intuitive feature selection, this process is purely motivated by business logic, and the features dismissed during this step are: Name, Type, Location, Sector, Phone, and URL. We opted for listwise deletion as a second step to exclude records with missing values. After this step 18000 records remained to pass for the next stage of our preprocessing system. The business preprocessing came third to help dismissing irrelevant data like car mileage surpassing 500 000 km because of the bad condition of a car. Electric, LPG, and Hybrid cars were dismissed also because they represent outliers to Moroccan market. Old models dismissed based on prospection study. Another step in our business preprocessing stage is creating the "Age" feature. This attribute replaces Post Time in our raw data, the logic behind this decision is the fact age gives more value compared to Post Time because of the fact that age incorporates the model year as well (Age = Post Time − Model Year). Lastly, we revised the wrong entries using an excel file containing the price range for a specific car, this way in case a record has an incorrect price we can assign a random price from its price range. After business preprocessing

Table 2. Raw data attributes.

Attribute	Type	Description	Attribute	Type	Description
Name	String	A concatenation of the model and year	Sector	String	The seller's sector (e.g. Hay Hassani)
Price	Integer	The price for the car	Type	String	The ads type (e.g. offer)
Location	String	The seller's city	Post time	Datetime	The publication time
Model year	Integer	The year	Phone	String	The seller's phone
Mileage	String	The driven distance in a class format (e.g. 4999–14999)	URL	String	The ad's url
Fuel type	String	Diesel, electrical, etc	Gearbox	String	Manual, automatic
Brand	String	Manufacturing company	Nb Doors	Integer	Number of doors
Model	String	The car's model	Origin	String	Morocco or other country
Fiscal power	String	The car's fiscal power (e.g. 8 CV)	Firsthand	String	Yes or no

multiple features were subject of data transformation to normalize and standardize the data to make it easier to use with the models.

To discriminate features, we applied three methods: 1) correlation-based feature selection to detect the features with high correlation; 2) Recursive Feature Elimination (RFE) which is a wrapper method for feature selection, and 3) Feature importance which assigns a scores to the input features to indicate the relative importance of each feature when making a prediction.

Model Selection

From the state of art, we saw that MLR and ensemble methods were the most used for price prediction. That's why in this paper we used MLR which is the most common form of linear regression analysis, and different families of ensemble approaches. Stacking or Stacked generalization is an ensemble machine learning algorithm. It involves training a new learning algorithm to combine the predictions of several base learners. In our case we stack the different algorithms we use that provided the best performance, LightGBM, XGBoost, RF, and CatBoost. For the bagging approach we use both Random forest and bagging regressor. For boosting we opted for XGBoost, LightGBM, and CatBoost which gave the highest performance in the literature review.

4 Experiments and Results

This section goes into the details of the experiment conducted in this work. To implement this study we used python programming language because of how common it is between data scientists and also how easy it is, we use jupyter notebook for its fast processing and documenting aspect, and also we use flask which is a python framework used for creating web pages to consume the model. These experiments were carried out on a Dell machine with an i7-4800MQ Intel CPU with (8 CPUS) and 12 GB of ram.

4.1 Hyper Parameter Tuning

Tuning parameters for a model to get the best results is a very important step to get the best performance possible for a model, there are several ways to execute this process from manual to automatic, and in this case, we use an exhaustive approach called Grid Search. Which exhaustively generates candidates from a grid of parameter values specified with the param_grid parameter. The Table 3 presents the hyper parameters values used for the machine learning algorithms used.

Table 3. Hyper parameter values.

Model	Hyper parameter	Value
Bagging regressor	base_estimator	None
	bootstrap	True
	max_samples	1.0
	n_estimators	1000
XGBoost	colsample_bytree	0.8
	gamma	0.0
	max_depth	3
	min_child_weight	2.0
	n_estimators	1000
	subsample	1.0
Light GBM	colsample_bytree	0.8
	learning_rate	0.1
	max_depth	5
	min_child_weight	0.1
	n_estimators	1000
	subsample	0.6
CatBoost	boosting_type	Plain
	depth	5
	learning_rate	0.2
	n_estimators	1000
	sampling_frequency	PerTreeLevel

(*continued*)

Table 3. (*continued*)

Model	Hyper parameter	Value
	subsample	1.0
Random Forest	bootstrap	True
	max_depth	None
	max_features	auto
	min_samples_split	5
	n_estimators	200

4.2 Evaluation Metric

To evaluate the performance of prediction we used the R2 score, this latter indicates how fitting the model is to the data, a high R2 corresponds to a good model while a low R2 means the model cannot predict the data. A secondary performance measurement we opt for is the mean absolute error (MAE), which simply refers to the result of measuring the difference between the real price and the prediction averaged for all the observations in the test data.

4.3 Results

For linear regression the results are the following in Table 4. The results achieved for LR can be explained by how nonlinear the data is in the Moroccan market. As far as ensemble methods are concerned, we can see that using RFE feature selection with XGBoost outperform the rest scoring with 90.7% for R2 in the best experiment possible. Also, our approach result is better than LightGBM result presented in [16].

Table 4. Linear regression results.

Experiment	R2	MAE
No FS	−22%	0.491
Correlation heat-map	−21.9%	0.491
RFE FS	−21.9%	0.5
Feature importance	−21.9%	0.5

For ensemble methods the results are the following (Table 5).

Table 5. Ensemble methods results.

Algorithm	Experiment	R2	MAE
Stacking Regressor	No FS	90%	0.169
	Correlation heat-map	90.1%	0.168
Bagging Regressor	No FS	85.8%	0.183
	Correlation heat-map	85.9%	0.183
Random Forest	No FS	85.8%	0.182
	Correlation heat-map	85.8%	0.183
	RFE FS	86.8%	0.182
	Feature importance	86.5%	0.182
XGBoost	No FS	89.6%	0.174
	Correlation heat-map	89.7%	0.173
	RFE FS	**90.7%**	**0.161**
	Feature importance	89.8%	0.172
LightGBM	No FS	87.8%	0.181
	Correlation heat-map	88%	0.181
	RFE FS	88.2%	0.18
	Feature Importance	88.3%	0.178
CatBoost	No FS	89%	0.173
	Correlation heat-map	89.1%	0.173
	RFE FS	89.1%	0.173
	Feature importance	89.2%	0.172

4.4 Model Deployment

After model training in order to avoid retraining the model every time we save the model in a pickle format, "Pickling" is the process whereby a Python object hierarchy is converted into a byte stream, this latter can be moved and loaded to be used without the need of retraining. Once the trained model is added to the app, a GUI is added as an interaction layer with the user to input the features needed to predict the car's price. A request is sent to the backend layer of the Flask app where the inputs are treated to be fit for the model to treat them (Encoding, Scaling, Normalization, etc.). Once the features are ready, we feed them to our model and get the prediction that get sent in a reply to the GUI to be displayed to the user. The following figure shows an example of usage. As shown in Fig. 3 and Fig. 4, we firstly fill in the characteristics of the car whose price we want to predict on the Moroccan market, such as the manufacturer, model, age, etc.

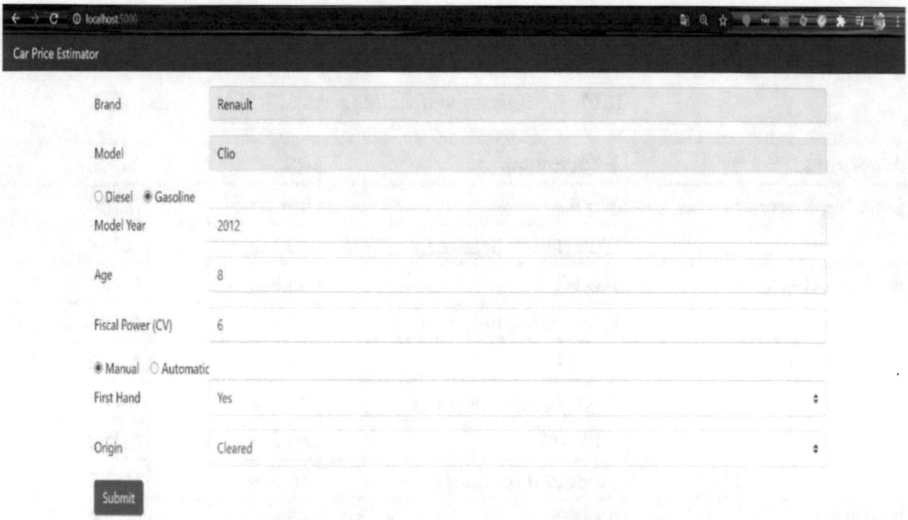

Fig. 3. Request for car price prediction

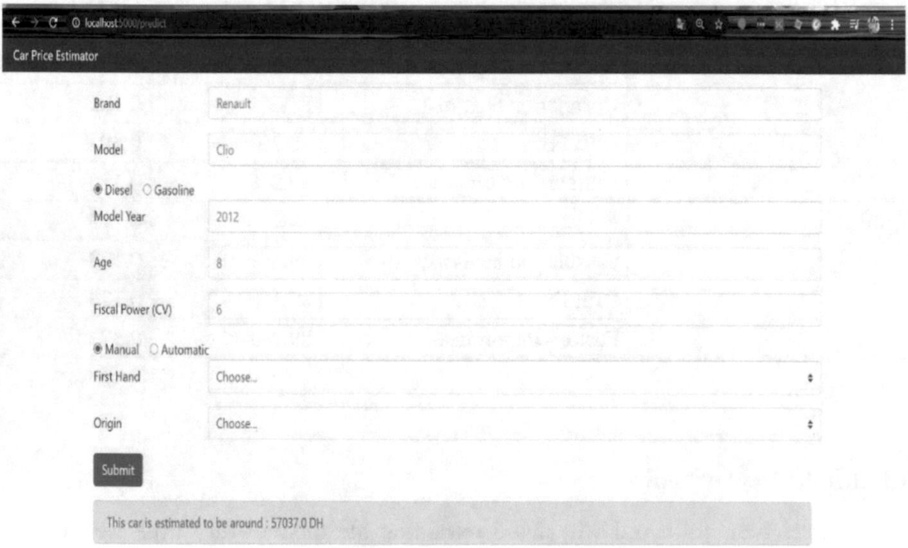

Fig. 4. Car price prediction response.

The predicted price in this case is 57034 MAD. From the different examples we used, the model on the prediction is so close to real life prices in the Moroccan market, which further proves the effectiveness of our proposition.

5 Conclusion

In this paper we conducted a comparative performances study of several regression approaches for used cars price prediction in the Moroccan market. Our study shows the value of a used car is determined by multiple variables, and also sheds light upon the importance of data reliability and how data preprocessing depends generally on the raw data to handle. From an academic stand point this study contributes both by consolidating similar studies by proving the power of using ensemble methods for price prediction, and by being a stepping stone for future research work, also focusing on the studies conducted to answer the task of used cars price prediction based on extracted data our model surpassed the work of [17] with R2 score of at 90.7. Keeping the current model as baseline, we intend to explore the possibility of using a deep learning approach possibly building a company specific neural network architecture that takes other variables that may impact the prices into account such as client reviews, and social media data and consuming it through a graphical user interface making it easier for other employees to use the predictive tool.

References

1. Bristone, M., Prasad, R., Abubakar, A.A.: CPPCNDL: crude oil price prediction using complex network and deep learning algorithms. Petroleum (2019)
2. Čeperić, E., Žiković, S., Čeperić, V.: Short-term forecasting of natural gas prices using machine learning and feature selection algorithms. Energy **140**, 893–900 (2017)
3. McNally, S., Roche, J., Caton, S.: Predicting the price of bitcoin using machine learning, pp. 339–343 (2018)
4. Velankar, S., Valecha, S., Maji, S.: Bitcoin price prediction using machine learning, pp. 144–147 (2018)
5. Selvin, S., Vinayakumar, R., Gopalakrishnan, E., Menon, V.K., Soman, K.: Stock price prediction using LSTM, RNN and CNN-sliding window model, pp. 1643–1647 (2017)
6. Park, B., Bae, J.K.: Using machine learning algorithms for housing price prediction: the case of Fairfax County, Virginia housing data. Expert Syst. Appl. **42**(6) (2015). Art. no. 6
7. Phan, T.D.: Housing price prediction using machine learning algorithms: the case of Melbourne city, Australia, pp. 35–42 (2018)
8. Truong, Q., Nguyen, M., Dang, H., Mei, B.: Housing price prediction via improved machine learning techniques. Procedia Comput. Sci. **174**, 433–442 (2020)
9. Asim, M., Khan, Z.: Mobile price class prediction using machine learning techniques. Int. J. Comput. Appl. **975**, 8887 (2018)
10. Balakumar, B., Raviraj, P., Gowsalya, V.: Mobile Price prediction using Machine Learning Techniques (n.d.)
11. Voß, S., Lessmann, S.: Resale price prediction in the used car market. Int. J. Forecast. (2017)
12. Yang, R.R., Chen, S., Chou, E.: AI blue book: vehicle price prediction using visual features, *arXiv preprint* arXiv:1803.11227 (2018)
13. Meng, S.-M., Liu, L.-J., Kuritsyn, M., Pechnikov, V.: Price determinants on used car auction in Taiwan. Int. J. Asian Soc. Sci. **9**(1), 48–58 (2019)
14. Chen, C., Hao, L., Xu, C.: Comparative analysis of used car price evaluation models, **1839**(1), 020165 (2017)
15. Lessmann, S., Voß, S.: Car resale price forecasting: the impact of regression method, private information, and heterogeneity on forecast accuracy. Int. J. Forecast. **33**(4), 864–877 (2017)

16. Van Thai, D., Son, L.N., Tien, P.V., Anh, N.N., Anh, N.T.N.: Prediction car prices using quantify qualitative data and knowledge-based system, pp. 1–5 (2019)
17. Çelik, Ö., Osmanoğlu, U.Ö.: Prediction of the prices of second-hand cars. Avrupa Bilim ve Teknoloji Dergisi **16**, 77–83 (2019)
18. Gegic, E., Isakovic, B., Keco, D., Masetic, Z., Kevric, J.: Car price prediction using machine learning techniques. TEM J. **8**(1), 113 (2019)
19. Pal, N., Arora, P., Kohli, P., Sundararaman, D., Palakurthy, S.: How much is my car worth? a methodology for predicting used cars' prices using random forest. In: Arai, Kohei, Kapoor, Supriya, Bhatia, Rahul (eds.) FICC 2018. AISC, vol. 886, pp. 413–422. Springer, Cham (2019). https://doi.org/10.1007/978-3-030-03402-3_28
20. Sun, N., Bai, H., Geng, Y., Shi, H.: Price evaluation model in second-hand car system based on BP neural network theory, pp. 431–436 (2017)
21. Caciandone, S., Chiru, Costin-Gabriel.: Using machine learning to generate predictions based on the information extracted from automobile ads. In: Dichev, Christo, Agre, Gennady (eds.) AIMSA 2016. LNCS (LNAI), vol. 9883, pp. 36–45. Springer, Cham (2016). https://doi.org/10.1007/978-3-319-44748-3_4
22. Monburinon, N., Chertchom, P., Kaewkiriya, T., Rungpheung, S., Buya, S., Boonpou, P.: Prediction of prices for used car by using regression models, pp. 115–119 (2018)
23. Pudaruth, S.: Predicting the price of used cars using machine learning techniques. Int. J. Inf. Comput. Technol **4**(7), 753–764 (2014)
24. Noor, K., Jan, S.: Vehicle price prediction system using machine learning techniques. Int. J. Comput. Appl. **167**(9), 27–31 (2017)

Improving Team Performance by Using Clustering and Features Reduction Techniques

Zbakh Mourad[1](✉), Aknin Noura[1], Chrayah Mohamed[2], and Elkadiri Kamal Eddine[1]

[1] FS, Abdelmalek Essaadi University, Tetouan, Morocco
mourad.zbakh@etu.uae.ac.ma, {noura.aknin,kelkadiri}@uae.ac.ma
[2] ENSATE, Abdelmalek Essaadi University, Tetouan, Morocco

Abstract. In a global context where competition is increasing, companies are constantly looking for sustainable competitive advantages that will allow them to improve their market shares and profit margins, improving team performance is one of the key factors in this process which allows these organizations to increase their productivity, their competitiveness, and their profitability.

Evaluating this team performance is one of the major challenges of human resources management, which has experienced in recent years a profound digital transformation of data and their management, current IT tools are no longer able to use the mass of data resulting from several sources and which does not stop multiplying from one day to another, or to find correlations between them to draw new knowledge and to anticipate future events.

The purpose of our research is to establish a team classification model according to several performance factors using Machine Learning algorithms, in particular for dimensionality reduction and clustering, The result of this work represents a decision support model for companies to develop a tailor-made team about the overall strategy of the company, to set up an action plan adapted to each team cluster and to anticipate future events, namely departures.

Keywords: Human resources management · Key performance indicators · K-means · Cluster · Principal component analysis

1 Introduction

For decades, several studies have tried to demonstrate the effect of Human resources management practices, improving the performance of teams on the overall results of companies.

Barney's article published in 1991 [1] constitutes an essential element in the emergence of the theory of management by the resource-based view (RBV), which affirms that the way in which companies mobilize their teams is a key element in the development of overall performance, in fact, according to this theory, the development of skills such as Valuable, Rare, Inimitable and Non-substitutable (VRIN) contributes to overall performance by doing things differently. According to the same study, companies must put considerable effort into identifying, understanding, and classifying essential skills.

M. Lazaar et al. (Eds.): BDIoT 2021, LNNS 489, pp. 347–361, 2022.
https://doi.org/10.1007/978-3-031-07969-6_26

To achieve this, and with the aim of developing the skills of their teams in general, companies are constantly looking for new ways to use both computer technologies and the mass of data currently available through internal sources. Like human resource information systems (HRIS) or external like social networks.

In this regard, the human resources department is one of the most relevant application sectors for Big Data and machine learning to transform the way companies manage their teams for optimal management. According to a study conducted by Towers Watson [2], companies spend a large part of their budget dedicated to Human Resources technologies in Big Data software.

Certainly, many empirical studies on the factors affecting team performance have been published, but these studies focused on one or at most two factors, namely the effectiveness of training (Arthur, Bennett, Edens, & Bell, 2003 [3], its general impact on overall performance (Aguinis & Kraiger, 2009) [4], and the impact of the chosen training strategy on financial results ([AASanchez, MIBAragon & RSValle, 2003) [5].

In the light of the above, the objective of our study is to propose an innovative model to evaluate the performance and subsequently to classify the collaborators of a team in relation to the different factors that affect their performance, our model is based on the Principal Component Analysis algorithm for dimensionality reduction and the K-Means algorithm for clustering, we chose a database published by IBM which collects information on age, income, seniority and certain data concerning the profile of 1470 employees.

In sum, The end result of this work represents a decision support tool that will be at the service of companies in order to develop a tailor-made team in relation to both the overall strategy of the company and the skills gaps, to sets up an action plan to adapt to each team cluster and to anticipate future events, namely departures and recruitment, in fact, this tool will allow managers to adopt an appropriate management style in relation to each cluster, to plan targeted training in relation to specific needs and manage critical skills in terms of gender and number.

2 Big Data, Machine Learning and Human Resources Management

2.1 Big Data

In the literature, the term "Big Data" refers to the technologies, processes, and techniques that allow an organization to create, manipulate and manage data at scale and extract new knowledge from it to create economic value. this concept of Big Data is described by the theory of 5 V (Volume, Variety, Velocity of data, veracity, and added value). Volume is the size of data streams that continuously come in at an exponential size. In particular, in terms of human resources, these flows are varied between data internal to the company (HRIS, SAP, etc.), external data (social networks, telephone applications, etc.) structured data (documents, images, etc.), or unstructured (publications, GPS data, etc.) that conventional computer tools cannot process. the third characteristic is the velocity which corresponds to the speed of production of the data, the fourth characteristic is the veracity of the data, finally, the last characteristic is the added value of these data and their applications [6]. The real richness of a Big Data project is to correlate this data in real-time and to derive new knowledge from it.

2.2 Machine Learning

The application of "Big Data" has emerged as one of the major issues related to the development of new technologies within the organization. It is considered to be the engine of innovation, customer satisfaction, and the achievement of greater profit margins (Barton & Court, 2012) [7]. It allows better productivity when decisions are made based on analyzes and data cross-referencing. A study by McAfee and Brynjolfsson (2012) [8] found that companies that adopted advanced data analysis techniques achieve higher rates of productivity and profitability than their competitors. It also allows better management of information in terms of use and classification of information by priority. According to Kaufman (1973) [9], managers do not need to acquire more information in order to make better decisions, but the rather better organization and better use of the information at their disposal.

Traditional analytical tools are not performing well enough to fully exploit the value of big data. The volume of data is too large for comprehensive analyzes, and the correlations and relationships between these data are too large for analysts to test all hypotheses to derive value from the data.

Basic analytical methods are used by business intelligence and reporting tools for reporting amounts, for doing accounts, and for performing SQL queries. Online analytical processing is a systemized extension of these basic analytical tools that require human intervention to specify what needs to be calculated.

2.3 Human Resources Management

According to the literature, human resources management is defined as the set of practices implemented to administer, mobilize and develop the human resources involved in the activity of an organization.

Of course, companies have a lot of data on their employees. However, for a long time, this data is not fully exploited to build new knowledge but rather is used only for descriptive reporting purposes, as in the social report for example.

More recently, a new trend has emerged called "HR analytics" (Marler and Boudreau, 2016) [10]. It is presented as a more sophisticated way of mobilizing data, notably by using more complex statistical methods, but above all by aiming at a different objective. It is no longer a question of providing only descriptive reporting, but of mobilizing data to better understand a phenomenon, to improve decision-making.

In terms of HR, very little research works on big data, in particular, its application to improve team performance [11]. Thus, the development of a decision support model to develop both individual and collective results represents an urgent and important issue. Which is the objective of our research.

3 Related Works

Human Resource Analytics has become a new trend, it is a process that collects and analyzes Human Resource data from traditional sources or by using another source now available in order to improve the improvement of overall performance by improving that of the teams [12].

In performance management, first of all, the quantification of individual performance in an objective way by referring to the factors that best reflect this performance, but also the quantification of the risks of individuals leaving are two big challenges in hr analytics.

Much research has focused on the factors that influence this performance, training is one of these factors, in fact, the article published in 2019 by researchers Joshua S. Bendickson and Timothy D. Chandler [13] was the study that we were inspired to conduct this research.

The researchers referred to data from 2003 to 2011, including 30 Major League Baseball organizations (as well as their affiliates), which were analyzed using regression models to examine the impact of human capital development programs (HCDP) on financial performance through operational performance. The results support the assumption that better human capital development programs lead to operational performance, which leads to increased revenue and sales.

To test the theoretical model, a regression model including control variables is used to determine if the development ranking has a significant impact on team wins. Next, mediation models are used to determine the impact of development rankings on earnings and average attendance across team wins.

This study proves that teams win more games two years after the increase in the development ranking (the number of the development ranking decreases), this model represents an inspiration to find out how a skill ranking (that varies from observed to observed) can be used to increase overall performance.

However, in an industrial context, this task is even more complex and complicated, with traditional HR tools many questions remain unanswered, namely: how to measure an individual's contribution to collective success? why does an employee seek to leave a given company? that he is the right candidate to recruit?

In this section, we will try to give a summary of the works that have tried to establish models that make performance predictions based on data collected from individuals.

Iwamoto et al. [14] have proposed an approach based on a multiple regression model which assesses individual performance according to the purely financial results of employees who influence a performance indicator of the organization, this represents a first step towards the objective assessment of individual performance but our aim is to establish the link with other factors as well, years later, Abdullah et al. [15] established a model that demonstrates the importance of knowledge versus skills through a case study in Malaysia, The analytical hierarchy process is used to integrate the multi-faceted preferences of the five criteria of human capital to determine the importance of the four indicators identified.

Chen and Chen [16] applied a data mining algorithm, based on decision trees and association rules, to employee characteristics and performance, which represents a starting point for our research. which aims to establish first a resource classification model to assess their performance in an objective manner and to predict starting risks.

4 Improving Team Performance By Using Clustering and Features Reduction Techniques

4.1 Presentation of the Data

IBM has collected information on age, income, seniority, and some data concerning the profile of 1470 of their employees, the variables are summarized in a data set is made of 1470 rows and 35 columns (Table 1).

Table 1. Description of the variables

Name	Description
AGE	Numerical value
ATTRITION	Employee leaving the company (0 = no, 1 = yes)
BUSINESS TRAVEL	(1 = no travel, 2 = travel frequently, 3 = travel rarely)
DAILY RATE	Numerical value - salary level
DEPARTMENT	(1 = HR, 2 = R&D, 3 = Sales)
DISTANCE FROM HOME	Numerical value - the distance from work to home
EDUCATION	Numerical value
EDUCATION FIELD	(1 = HR, 2 = Life Sciences, 3 = Marketing, 4 = Medical Sciences, 5 = Others, 6 = Technical)
EMPLOYEE NUMBER	Numerical value - employee id
ENVIROMENT SATISFACTION	Numerical value - satisfaction with the environment
GENDER	(1 = Female, 2 = Male)
HOURLY RATE	Numerical Value - Hourly Salary
JOB INVOLVEMENT	Numerical Value - Job Involvement
JOB LEVEL	Numerical Value - Level of Job
JOB ROLE	(1 = HC Rep, 2 = HR, 3 = Lab Technician, 4 = Manager, 5 = Managing Director, 6 = Research Director, 7 = Research Scientist, 8 = Sales Executive, 9 = Sales Representative)
JOB SATISFACTION	Numerical value - satisfaction with the job
MARITAL STATUS	(1 = divorced, 2 = married, 3 = single)
MONTHLY INCOME	Numerical value - monthly salary
MONTHLY RATE	Numerical value - monthly rate
NUM COMPANIES WORKED	Numerical value - no. Of companies worked at
OVER 18	(1 = yes, 2 = no)
OVERTIME	(1 = no, 2 = yes)
PERCENT SALARY HIKE	Numerical value - percentage increase in salary
PERFORMANCE RATING	Numerical value - Performance rating
RELATIONS SATISFACTION	Numerical value - relations satisfaction

(*continued*)

Table 1. (*continued*)

Name	Description
STANDARD HOURS	Numerical value - standard hours
STOCK OPTIONS LEVEL	Numerical value - stock options
TOTAL WORKING YEARS	Numerical value - total years worked
TRAINING TIMES LAST YEAR	Numerical value - hours spent training
WORK-LIFE BALANCE	Numerical value - time spent between work and outside
YEARS AT COMPANY	Numerical value - total number of years at the company
YEARS IN CURRENT ROLE	Numerical value -years in current role
YEARS SINCE LAST PROMOTION	Numerical value - last promotion
YEARS WITH CURRENT MANAGER	Numerical value - years spent with current manager

4.2 Model Building

We will propose a new approach to the classification of teams according to their profiles according to the following steps illustrated in (Fig. 1). This method is based on the analysis of the main components, which we choose for its ability to compress and synthesize a large data set in comparison with other compression technics after we will classify the team members using the K-means clustering algorithm, favorable for its ease, speed and for its ability to be implemented on a large dataset in comparison with other clustering algorithms.

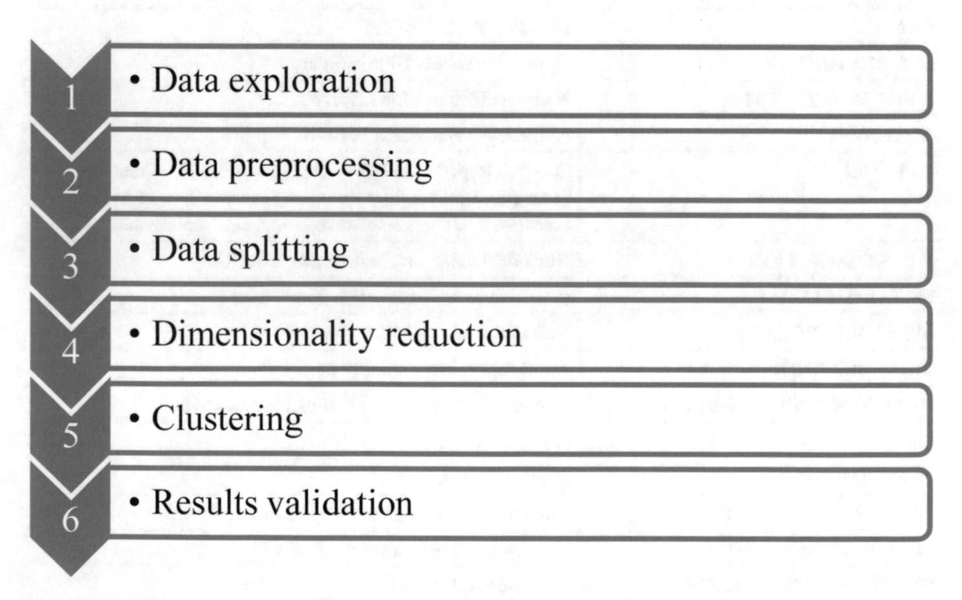

Fig. 1. Steps of model construction

Data Preprocessing

After checking that there is no missing data, we standardize the active variables to make them comparable and factor the additional variables, we chose the variables that best reflect the importance of the profile, these variables are shown as follows (Table 2 and Table 3).

Table 2. Active variables

Name	Description
AGE	Age
JOB LEVEL	Level of job
MONTHLY INCOME	Monthly salary
NUM COMPANIES WORKED	Number of companies worked at
TOTAL WORKING YEARS	Total years worked
YEARS AT COMPANY	Total number of years at the company
YEARS IN CURRENT ROLE	Years in current role

Table 3. Supplementary variables.

Name	Description
DEPARTMENT	Department
EDUCATION	Education
PERFORMANCE RATING	Performance rating

Data Splitting

We split the data into a training sample and a test sample - using a random split technique. The distribution used is 80% for learning, 20% as testing data.

Dimensionality Reduction: Principal Component Analysis

Typically, unsupervised algorithms make inferences from data sets using only input vectors without referring to known or labeled results.

The Principal Component Analysis is part of dimensionality reduction algorithms, it consists of transforming interrelated variables (called "correlated" in statistics) into new variables decorrelated from each other. These new variables are called "principal components", or principal axes [17].

To apply the Principal Component Analysis algorithm to the training set, we used the "FactoMineR" package on the R software, the results are as follows.

Eigenvalues and Eigenvectors/Variance

From the graph below (Fig. 2), we might want to stop at the second principal component. 74.9% of the information (variances) contained in the data is kept by the first two principal components.

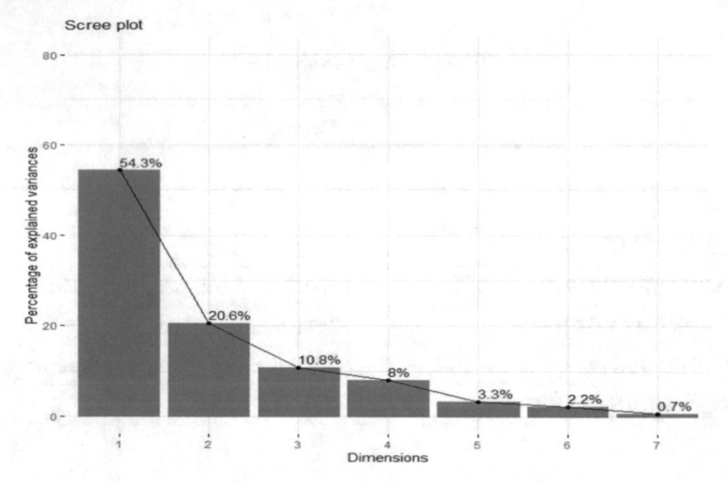

Fig. 2. Eigenvalues and Eigenvectors

Variable Correlation

The graph below shows the correlation between the variables (Fig. 3), the contribution of each variable on the first two factors is detailed in the graph below.

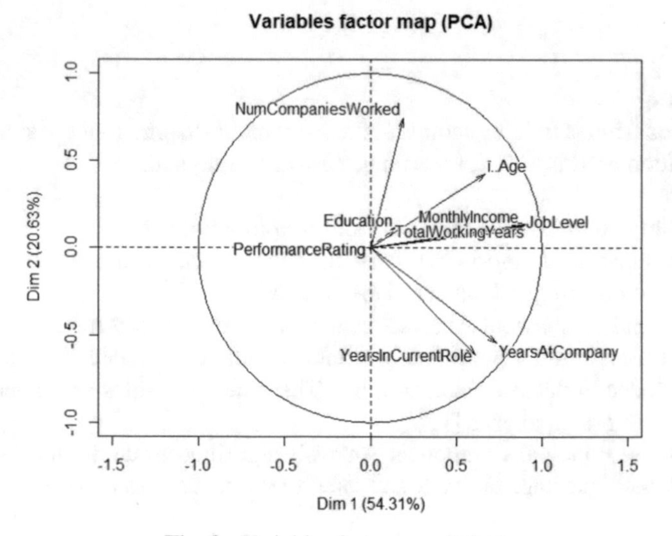

Fig. 3. Variables factor map (PCA)

Graph of Individuals

The graph below (Fig. 4) shows the distribution of individuals from each department with concerning the first two factors of the Principal Component Analysis of the data set that we have chosen.

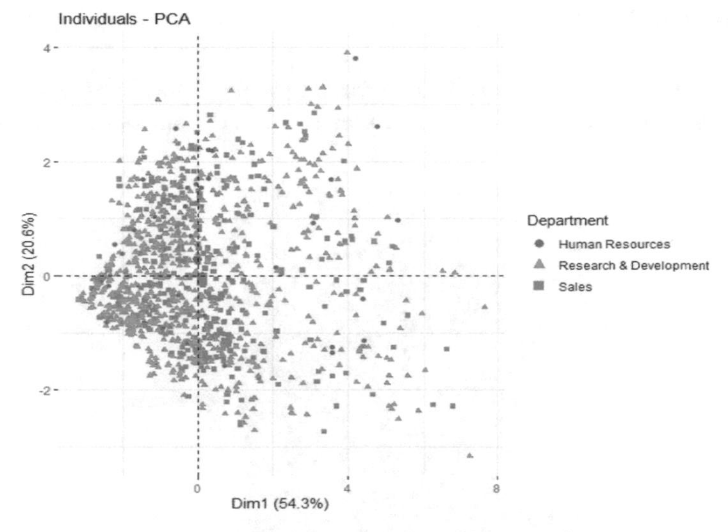

Fig. 4. Individuals (PCA)

The distribution of variables in each factor makes it possible to name factor 1, experience factor (53.4% of the variance), factor 2, dynamic factor (20.6%) (Table 4). Factor 1, the experience factor, is the most important in terms of variance.

It includes the variables that assess the age, the level of the position, the monthly income, the total number of years of experience, and the length of service in the company. As for factor 2, dynamic factor, it comprises the variables which assess the number of antecedent companies and the length of service in the company.

Table 4. The distribution of variables

Variables	Factor 1 (expertise)	Factor 2 (risk of departure)	Factor 3
AGE	0,6,650,504		
JOB LEVEL	0,896,769		
MONTHLY INCOME	0,8,808,675		
NUM COMPANIES WORKED		0,7,451,965	0,5,383,931
TOTAL WORKING YEARS	0,915,279		
YEARS AT COMPANY	0,7,347,542	-0,5,407,864	

- Values greater than 0.50 have been retained as significant, values less than 0.50 have been deleted for the clarity of the table.
- A high level of Factor 1 can be explained by a high starting risk, and subsequently, on a population, we can consider that the stability within the team is risky.
- A high level of Factor 2 can be explained by a risk of having a critical skill or having an overqualified skill.

Below is a report by the department according to the factors (Fig. 5 and Fig. 6) to guide Managers to target their actions according to the structure of their teams.

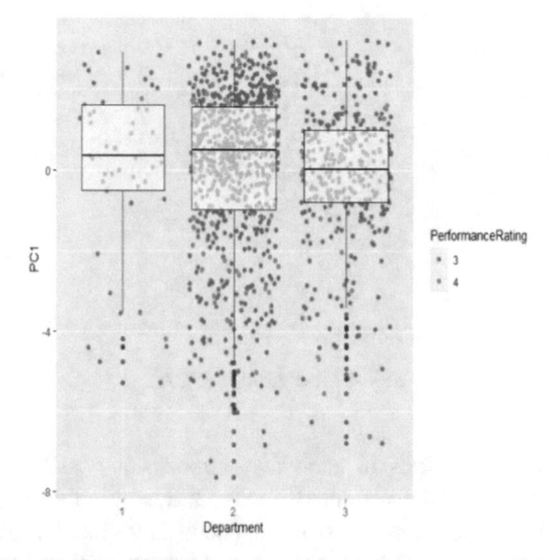

Fig. 5. Distribution of individuals by each department according to PCA1

Compared to hierarchical algorithms, the k-Means algorithm is the most suitable to use, this algorithm is one of the most widespread, especially for its simplicity, efficiency, flexibility, and ease of interpretation of its results.

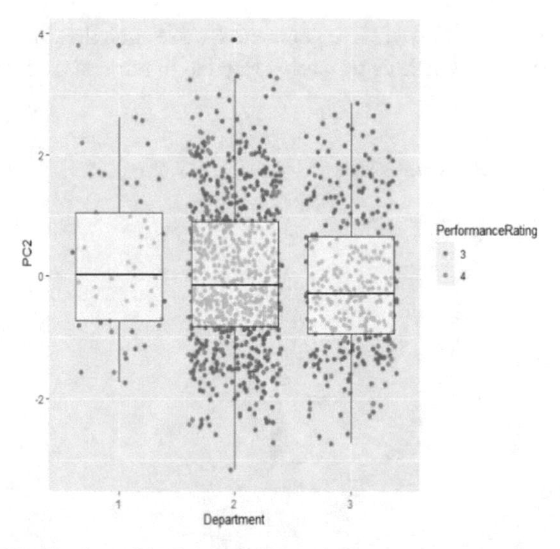

Fig. 6. Distribution of individuals by each department according to PCA2

The K-means algorithm allows to group observations of similar data and discovers the underlying models. To achieve this goal, K-means searches for a fixed number (k) of clusters in a dataset [18] according to the following steps (Fig. 7).

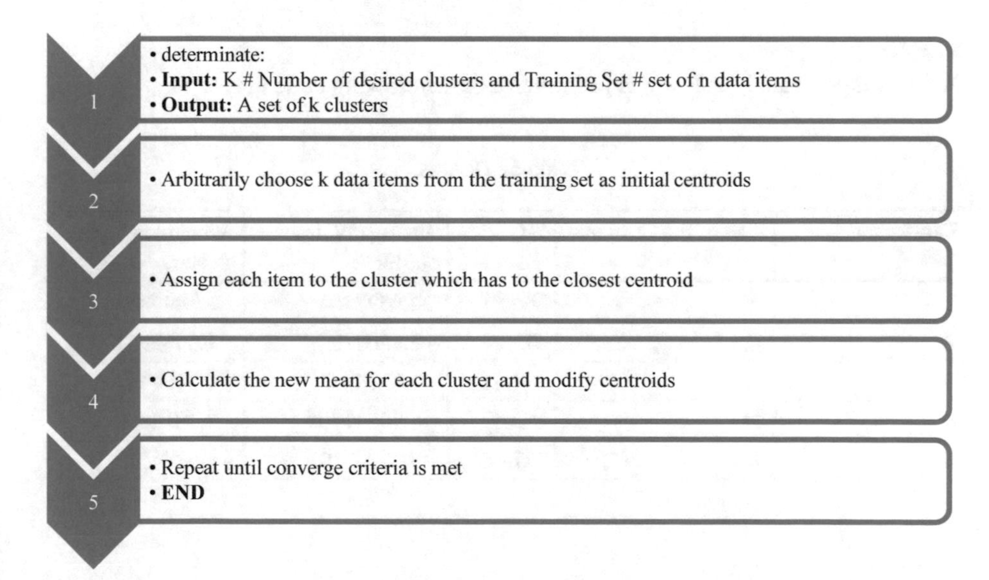

Fig. 7. Schematic algorithm of K-means method

To apply the K-means algorithm on our database we used the "factoextra" package on the R software,

To define the number of clusters k to generate we use such that the total within-cluster sum of square (WSS) is as small as possible (Fig. 8), in the rest of the study we decided to take k = 5.

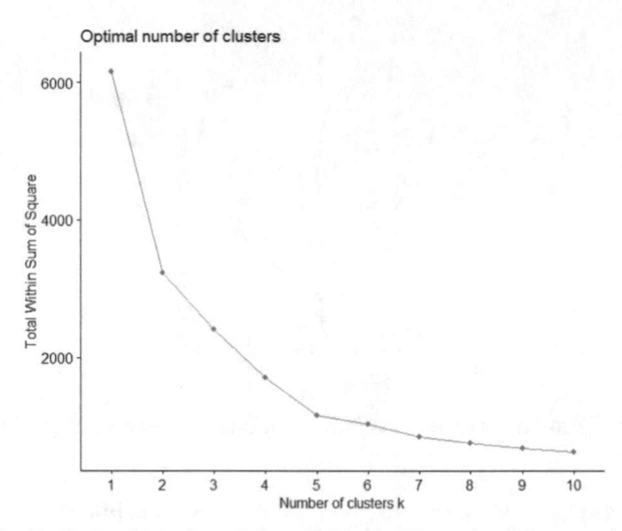

Fig. 8. Optimal number of cluster identification using Elbow method

Results

The result generated for k = 5 is presented in the following table (Table 5).

Table 5. Results

Cluster number	Size of the Cluster	WCSS	Average PC1 per cluster	Average PC2 per cluster
1	303	260,81	0.4527564	1.1897820
2	111	173,94	−2.7204782	1.6069972
3	368	243,85	1.8368283	−0.3454493
4	84	193,47	−4.3964170	−1.0816469
5	310	287,66	−0.4576319	−1.0351514

with:

$$\text{BSS}/\text{TSS} = 81.2\% \quad (1)$$

The result of the classification of individuals by our model is represented by the graph below (Fig. 9).

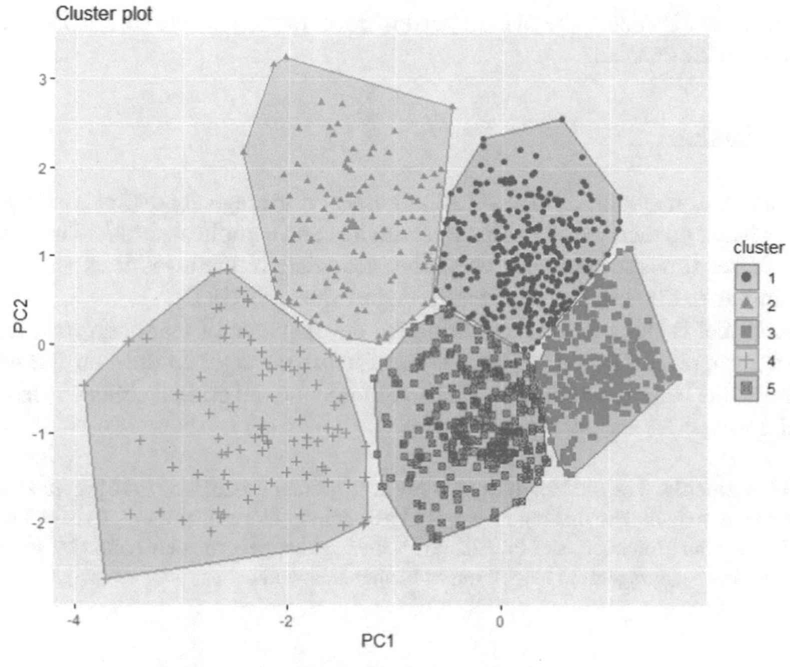

Fig. 9. Cluster graph

Discussion

According to the results previously obtained, we can categorize the clusters as summarized in the following table (Table 6).

Table 6. Interpretation of clusters

Cluster	Interpretation
1	Unstable Expert
2	Unstable senior
3	Stable expert
4	Junior, very early to evaluate
5	Senior stable

Also, for a new recruit, this model makes it possible to assess his performance in relation to his profile and to assign him to a cluster to adapt a suitable career plan to him, also this model makes it possible to assess his departure risk.

Perspectives

For the rest of our study, first, we will compare our results with those obtained on the model proposed by CF Chen, LF Chen [16], secondly, we will measure the individual factors that most influence the overall performance. to propose the correlation that best reflects this performance.

5 Conclusion

With traditional methods of classification based on average scores, classifying team members based on their performance is a complex and complicated task. The proposed method makes it possible to have clustering according to the level of expertise (PC1) and according to the stability which reflects the initial risk (PC2).

This Model is suitable for monitoring the development of team performance. The result of its application is decision support for the manager to develop the level of expertise of his team (according to PC1) and to define the critical skills concerned by the initial risk (PC2) and set up an action plan.

Acknowledgements. I would like to express my deep gratitude to all who have provided me with the opportunity to complete this report. I would particularly like to thank my advisors Pr. Aknin Noura, Pr Chrayah Mohamed, and Pr. Elkadiri Kamal Eddine whose contribution by stimulating suggestions and encouragement helped me to finalize this work.

References

1. Barney, J.B.: Firm resources and sustained competitive advantage. J. Manag. **17**(1), 99–120 (1991)
2. Elinor, F., Andrew, H., Klayton, S.: Insurance Big Data Insurance Big Data Can Improve Business. Towers Watson and Willis (2006)
3. Arthur, W., Bennett, W., Edens, P.S., Bell, S.T.: Effectiveness of training in organizations: a meta-analysis of design and evaluation features. J Appl Psychol. **88**(2), 234–45 (2003)
4. Aguinis, H., Kraiger, K.: Benefits of training and development for individuals and teams, organizations, and society. Annu. Rev. Psychol. **60**, 451–474 (2009)
5. Aragón Sánchez, A., Barba Aragón, M.I., Sanz Valle, R.: Effect of training on business results. Int. J. Human Res. Manag. **14**(6), 956–980 (2003)
6. Sagiroglu, S., Sinanc, D.: Big data: a review, collaboration technologies and systems (CTS). In: International Conference on Digital Object Identifier, pp. 42–47 (2013)
7. Barton, D., Court, D.: Marketing advanced analytics work for you. Harvard Business Rev. **90**(10), 78–83 (2012)
8. McAfee, A., Brynjolfsson, E.: Big data: the management revolution. Harvard Business Rev. **90**(10), 60–68 (2012)
9. Kaufmann, W.: Without guilt and justice: From decidophobia to autonomy. P.H. Wyden, NewYork (1973)
10. Marler, J.H., Boudreau, J.W.: An evidence-based review of HR analytics. Int. J. Human Res. Manag. **28**(1), 3–26 (2016)
11. Arthur, J.B.: Effects of human resource systems on manufacturing performance and turnover. Academy of Manag. J. **37**(3), 670–687 (1994)
12. Mishra, S.N., Lama, D.R., Pal, Y.: Human resource predictive analytics (HRPA) for HR management in organizations. Int. J. Sci. Technol. Res. **5**(5), 33–35 (2016)

13. Bendickson, J.S., Chandlere, T.D.: Operational performance: the mediator between human capital developmental programs and financial performance. J. Business Res. **94**, 162–171 (2019)
14. Iwamoto, H., Takahashi, M.: A quantitative approach to human capital management. Proc.-Soc. Behav. Sci. **172**, 112–119 (2015)
15. Abdullah, L., Jaafar, S., Taib, I.: Ranking of human capital indicators using analytic hierarchy process. Proc.-Soc. Behav. Sci. **107**, 22–28 (2013)
16. Chen, C.F., Chen, L.F.: Data mining to improve personnel selection and enhance human capital: a case study in high-technology industry. Expert Syst. Appl. **34**(1), 280–290 (2008)
17. Philippeau, G.: Comment Interpréter les Résultants d'une Analyse en Composantes Principales. Cited 61 times. Paris: Institut Techniques des Céréales et Fourrages (1986)
18. Fahim, A.M., Salem, A.M., Torkey, F.A., Ramadan, M.A.: An efficient enhanced k-means clustering algorithm. J. Zhejiang Univ. Sci. A. **7**(10), 1626–1633 (2006)

Home Automation and Machine Learning Models for Health Monitoring

Lamiae Eloutouate[1(✉)], Fatiha Elouaai[1], Hicham Gibet Tani[2], and Mohammed Bouhorma[1]

[1] Laboratory of Informatics Systems and Telecommunications (LIST), Department of Computer Engineering, Faculty of Sciences and Technologies, Abdelmalek Essaadi University, Route Ziaten, B.P. 416, Tangier, Morocco
lamiae.elo@gmail.com

[2] Laboratory of Advanced Sciences and Technologies (STA), Department of Computer Science, Multidisciplinary Faculty of Larache, Abdelmalek Essaadi University, Tetouan, Morocco

Abstract. Medical surveillance has been constantly linked to hospitals and infirmaries. However, the recent increase in demand for health assistance, especially with the current covid-19 pandemic, has made it clear that relying on placing patients on hospitals for surveillance is deprecated. In the same context, this paper strives to illuminate the significance of using trending paradigms such as home automation and artificial intelligence to better advance and modernize healthcare systems. Accordingly, the main contribution of this study is a demonstration of a novel smart home architecture and an evaluation of machine-learning algorithms aimed at predicting a health condition severity based on the patient data gathered from several sensors and wearables. In respect to the need of providing a real time alerting system, several classification algorithms are highlighted with their advantages in mitigating the remote health monitoring problematic. The results assessment of the machine learning algorithms emphasizes the convenience of using artificial intelligence for health monitoring regardless of time and place constraints.

Keywords: Machine Learning · Medical surveillance · Smart home

1 Introduction

According to the World Health Organization (WHO) [1] life expectancy has increased by over 8% globally between 2000 and 2016. Nevertheless, the WHO has stressed repeatedly the severity of life changing diseases like heart diseases, diabetes, cancer… and the actual coronavirus pandemic (COVID19) and their ravaging impact on healthcare systems and the world population. Consequently, several studies [2–4] have exposed the escalating bed shortage in several countries across the world and their forecasts on the ascending level of this shortage. Putting all of this into perspective, we can clearly establish that maintaining the classical model of healthcare is not the appropriate path to improve individuals' health nor lifestyle. One approach to solve this issue involves the use

M. Lazaar et al. (Eds.): BDIoT 2021, LNNS 489, pp. 362–372, 2022.
https://doi.org/10.1007/978-3-031-07969-6_27

of innovative technologies from whom we can mention: Home Automation (Smart Home or domotics), Cloud Computing, Internet of Things (IoT), wearables... Despite the fact that these technologies are being implemented as solutions in various fields, they have rarely been studied directly as a solution for remote medical surveillance. Among the studies that tackles these technologies, the investigation carried by the authors of paper [5] that proposed cloud computing as a solution to create a network for patients, doctors and healthcare institutions that is mainly used to share health applications and centralizes the data storage and management on the cloud. Shinde S. P. and Phalle V. N [6] explored IoT applications for healthcare, whereas the authors presented the different advantage of IoT required to build the paradigm of smart hospitals. Mike K. in [7] discussed the technical implementation of wearable technologies in hospitals. The author of this paper examined the advantages of IoT for hospitals and the major challenges related to the acceptance and the integration feasibility in healthcare critical facilities. The paper [8] is a presentation of using wireless technologies to track patient vital parameters and transmit these information's to doctors or medical authorities. Malasinghe, L.P. et al. [9] reviewed the most recent developments in remote healthcare with a discussion of their issues. The authors investigated several case studies of remote health monitoring namely heart and blood systems, brain and nervous systems, diabetes. Malasinghe, L.P. et al. research also analyzed monitoring contactless camera-based methods. The authors of paper [10] proposed two real time health-monitoring systems that employs machine-learning techniques to classify patients in intensive care units, which have considerably enhanced the existing systems by decreasing false alarms.

In general, several solutions have been proposed to renovate healthcare system issues; however, a sustainable solution would combine several technologies and methods to create a comprehensive and a patient centric model for medical surveillance that take into account the patient health condition and lifestyle. This paper introduces a new model for smart home for medical surveillance with the key technologies and aspects required for health monitoring. The proposed model utilizes machine-learning algorithms to inspect a patient health condition and predicts the required or possible intervention. Moreover, a case study of heart disease monitoring using machine learning (ML) algorithms is explored to test and validate the advantages of the evaluated algorithms. The rest of this paper will be organized as follow, the smart home architecture is exposed on Sect. 2 followed by a presentation of machine learning algorithms and their application field in Sect. 3. Section 4 presents the evaluation, testing and validation of machine learning algorithms usability for medical surveillance. This paper is then concluded in Sect. 5.

2 Home Automation and Health Monitoring

Home automation [11] is a composition of different techniques that make it possible to control, program and automate a home. This innovative paradigm uses the fields of electronics, Information Technology (IT), telecommunications and robotics. Home automation operates in a very vast technical and IT field. It allows you to program and remotely control most appliances and electrical devices in the house, from lighting and heating to audiovisual equipment, including opening/closing windows and doors. It also facilitates home control by managing alarm systems, fire prevention, and room temperature.

Amid home automation benefits is that everything is conceivable and possible through a wide range of applications [12]:

- **Household electrical devices control:** control and management of any electrical device within the home. Hours, programs, features, start-up, shutdown, etc. Everything is possible remotely from inside (using a local or wireless network) or outside the house (using the internet) and with simple gestures (using smart phone or tablet).
- **Energy management:** managing the energy sources inside the house. Heating, air conditioning, lighting, opening and closing of shades, water temperature.
- **Home security:** manages the house security, its occupants and properties by controlling alarms, access authorizations by voice recognition or magnetic card, tokens, biometrics, movement detectors, fire and flood protection devices.
- **Daily life assistance:** making everyday life easier for people who are dependent (elderly or with a health condition) and/or whose mobility is reduced or compromised, temporarily or permanently.

2.1 Home Automation Technologies for Health Monitoring

Home automation is constantly evolving, on the one hand, with the advancement of new technologies and, on the other hand, with changes in thinking and society [13]. When home automation is used at the scale of a house, it mainly uses:

- **Microcomputers**: Small sized computers that are wired or wirelessly connected to various devices in order to automate or control a house functionality [14]. The most known microcomputers technologies are Arduino and Raspberry Pi.
- **Wireless technologies**: it represents the most used communication and networking protocols suitable for home automation [14] such as 6LoWPAN, Bluetooth, DASH7, Zigbee, Wi-Fi…
- **Mobile Applications**: Smart phone (or tablet) mobile application are at the center of home automation technologies as they are used as controlling and managing endpoints. There is a variety of platforms [15] for building mobile applications such as Windows Mobile, Symbian, iOS and Android.
- **Sensors**: They are specialized equipment that take information on the behavior of the operative part and transform it into information that can be used by the control part. A collection of sensors can be used in the home automation system [16]: Temperature Sensors, Motion Detectors, Sensors for lighting control, Sensors for access control, Smoke detectors and Gas detectors.
- **Wearables**: They are a category of sensor devices that can detect parameters such as blood pressure and oxygen levels in blood, these devices can measure sleep and asthma, related to anxiety and sleep disorders, as well as general eldercare [17].
- **Machine Learning**: is an artificial intelligence technology that enables computers to learn and make predictions or decisions. Machine learning [18] is vastly used in home automation applications such as behavioral analysis, voice and face recognition and data analysis.

2.2 The Proposed Home Automation Architecture for Health Monitoring

Need. Home automation can make our life easier and automated [19], however, a health risk of aging, accident or disease can make daily life challenging and require special attention. According to this, home automation should respect the following:

- Hold and maintain the patient (user) schedule for medication and physical activities.
- Collect and manage the patient (user) health data.
- Monitor the patient (user) movement inside the house.
- Predict a patient (user) health emergency and call for help.
- Admit emergency help personal and guide them to the patient (user).

Proposed Architecture. Figure 1 represents the proposed architecture and design for a health monitoring oriented smart home:

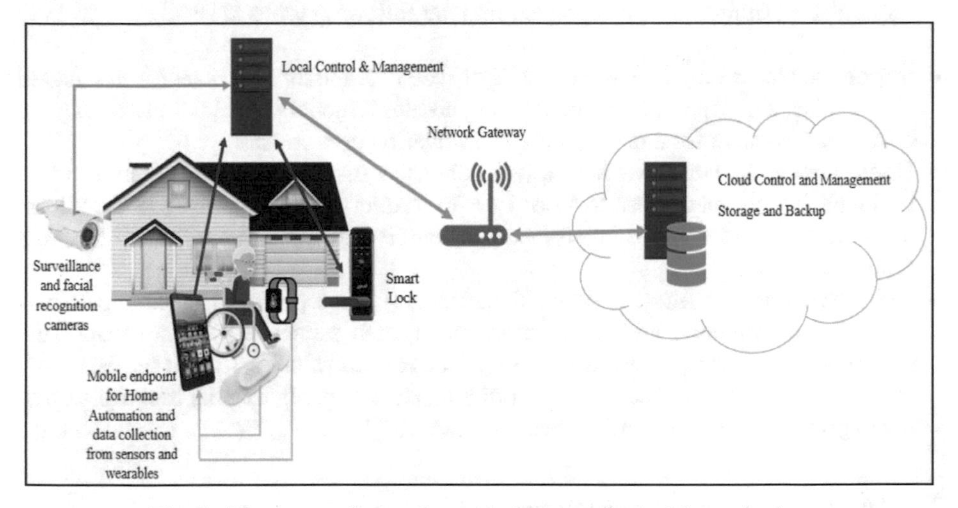

Fig. 1. The proposed smart home architecture for health monitoring

Objectives. The proposed home automation for health monitoring considers the following:

- Local control and management server with a centralized platform for managing the house automation aspects including the patient health condition and the required machine learning algorithms for monitoring the patient medical state. Additionally, the centralized platform controls the house unlocking in case of an emergency.
- Sensors and wearable items attached to the patient collect and centralize data in the local control and management platform installed inside the patient house.
- Surveillance cameras monitor the patient mobility inside the house and possible incidents or falling.

- The patient smart phone enables him to control the house elements including locking and unlocking the house doors/windows.
- The patient smart phone is constantly updated from the control and management platform (Medical schedule: medication, exercise, doctor appointments…).
- The cloud is used to back up the home automation system and data. Moreover, the cloud become the house control and management platform when the patient is outside the house.

3 Machine Learning for Health Monitoring

Machine Learning (ML) [20] is a recent discipline for discovering recurrences (patterns) in one or more data streams and making predictions based on mathematical statistics. Actually, Machine Learning is based on data mining, which supports pattern recognition and provide predictive analysis.

Machine learning can be categorized into the following types [21]:

- **Supervised learning**: we have a set of labeled objects and for each object an associated target value; this type of machine learning creates a model capable of predicting the correct target value for a new object (Classification or regression).
- **Unsupervised learning**: we have a set of objects without any associated target value (unlabeled); this model has to learn how to extract the regularities present within the objects in order to better visualize or understand the structure of all the data (Clustering).
- **Reinforcement learning**: A set of decision sequences (political or strategic) in a dynamic environment, and for each action of each sequence a reward value (the reward value of the sequence is the sum of the reward values actions it implements). In this category, the output model becomes capable of predicting the best decision to make given a state of the environment (Q Learning).

3.1 Machine Learning Applications

Machine learning techniques are used in various fields. Following are several application domains that allow us to illustrate the usefulness and effects of this technology in our daily life [22, 23]:

- Customer recommendations.
- Medical diagnostic.
- Image processing and recognition.
- Facial recognition.
- Voice recognition.
- Spam detection.
- Fraud detection.
- Autonomous vehicles.
- Robotics…

3.2 Machine Learning for Health Monitoring

Health monitoring is the fact of placing a patient under observation using specialized sensors to gather data regarding the patient vital signs (Heart rate, blood pressure, body temperature, blood oxygen levels (SpO2), respiration rate). Afterwards, a health practitioner assesses the collected data to make specific decision regarding medications, hospitalization, surgery, intensive care, checkups, nutritional interventions, environmental alterations…

In light of this information, machine learning can help create intelligent models capable of learning a specific health condition patterns (by means of training the model on data congregated from previous cases) and predicts the correct interventions. As long as health data is labeled, supervised machine learning algorithms can be employed to train models on every health condition possible.

Supervised machine learning can be divided in two types [24]:

- **Classification**: It is the process of finding or discovering a model (function) which allows data to be separated into several categorical classes.
- **Regression**: It is the process of finding a pattern or function that distinguishes data in continuous real values instead of using classes.

There are numerous supervised learning algorithms, which have a vast application field [25]:

- Logistic regression: This linear classification model is the counterpart of linear regression. In general, the model is used when Y must take only two possible values (0 or 1, true or false…).
- K nearest Neighbor (KNN): The principle of this model consists on choosing the "K" data closest to the point studied in order to predict its value.
- Decision Tree: A decision tree functioning starts with a several characteristics, the decision begins with one of these characteristics; if this is not enough, we use another one, and so on.
- Random Forest: This machine-learning algorithm makes it possible to make predictions based on the aggregation of several decision trees.
- Multi-Layer Perceptron (Artificial Neural Network): This is model simulates the brain learning mechanism to acquire the ability to make predictions.
- Naïve Bayes: Naive Bayes classifiers are a family of algorithms that is based on the common principle that a feature specific value is independent of the value of any other feature. They allow us to predict the likelihood of an event occurring based on conditions we know for the events in question.
- Support Vector Machine: this model builds a real-value classifier that breakdown any problem into two sub-problems: Non-linear transformation of inputs and choosing the optimal linear separation.
- Linear Discriminant Analysis: this classification model can predict an individual's membership to a predefined class (group) from their characteristics measured using predictor variables.

4 Case Study: Heart Disease Monitoring Using Machine Learning

A case study of heart disease monitoring is proposed in pursuance of highlighting the significance of machine learning adoption for health monitoring. The case study was founded on data provided by the Center for Machine Learning and Intelligent Systems at University of California, Irvine (UCI) - School of Information and Computer Sciences (ICS).

The data was collected from the four following locations:

1. Cleveland Clinic Foundation.
2. Hungarian Institute of Cardiology, Budapest (Andras Janosi, M.D.)
3. V.A. Medical Center, Long Beach, CA (Robert Detrano, M.D., Ph.D)
4. University Hospital, Zurich, Switzerland (William Steinbrunn, M.D, Matthias Pfisterer, M.D.)

The data consists on the attributes of:

1. Age (years),
2. Sex: (Binary, 1 for male and 0 for female),
3. Chest-pain type (takes four possible values: 1 = typical angina, 2 = atypical angina, 3 = non-anginal pain, 4 = asymptotic),
4. Resting Blood Pressure (in mmHg),
5. Serum Cholesterol (in mg/dl),
6. Fasting Blood Sugar (Binary, 1 = fasting blood sugar > 120 mg/dl, 0 = normal),
7. Resting ECG (takes three possible values: 0 = normal, 1 = having ST-T wave abnormality, 2 = left ventricular hypertrophy),
8. Max heart rate achieved (numerical),
9. Exercise induced angina (Binary: 1 = yes, 0 = no),
10. ST depression induced by exercise relative to rest (numerical),
11. Peak exercise ST segment (1 = upsloping, 2 = flat, 3 = downsloping),
12. Number of major vessels (numerical),
13. Thalassemia (3 = normal, 6 = fixed defect, 7 = reversible defect)

The last column contains the heart disease diagnosis according to the the New York Heart Association (NYHA) Functional Classification (Table 1):

By means of z-score method, the input data was normalized to provide the ML algorithms with comprehensive values regardless of their metrics:

$$zscore = \frac{value - \mu}{\sigma}$$

Whereas μ represents the mean value of the parameter on a specific column and σ is the standard deviation of the parameter on that same column.

The data was also divided in two sets: Training set (90% of data) and Testing set (10% of data).

The implementation of the ML classification algorithms has been done using Python programming language and the scikit-learn library that provides a massive access to simple and efficient tools for predictive data analysis.

Table 1. Heart disease diagnosis

Diagnosis description	Value
Absence of heart disease	0
Patients with cardiac disease but without resulting in limitation of physical activity. (Mild)	1
Patients with cardiac disease resulting in slight limitation of physical activity. (Mild)	2
Patients with cardiac disease resulting in marked limitation of physical activity. (Moderate)	3
Patients with cardiac disease resulting in the inability to carry on any physical activity without discomfort. (Severe)	4

4.1 Machine Learning Application Results

Table 2 illustrates the results of the evaluated ML algorithms (Sect. 3.2) in regards of the data presented on the precedent section:

Table 2. ML classification algorithms results

ML algorithm	Accuracy score (%)	Time to finish training (seconds)	Configuration
Logistic regression	78.33	0.0365	–
K nearest neighbor	74.04	0.0025	Metric = Minkowski with 3 neighbors
Decision tree	73.33	0.0045	Maximum depth of 3
Random forest	73.33	0.1876	–
Multi-layer perceptron	75.16	1.3527	2 hidden layers with 13 and 26 nodes respectively (1000 iteration)
Naïve Bayes	70	0.0028	–
Support vector machine	83.33	0.0070	–
Linear discriminant analysis	71.66	0.0085	–

Figure 2 demonstrates a comparison of the ML classification accuracy score by percentage:

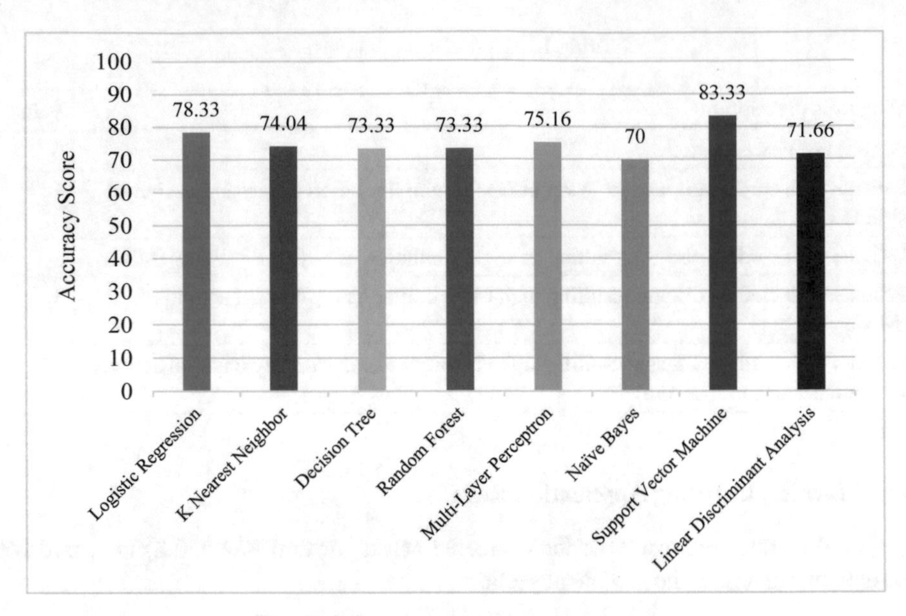

Fig. 2. ML accuracy score representation

The following figure (Fig. 3) portrays the time necessary to finish the training step before the model can make a prediction:

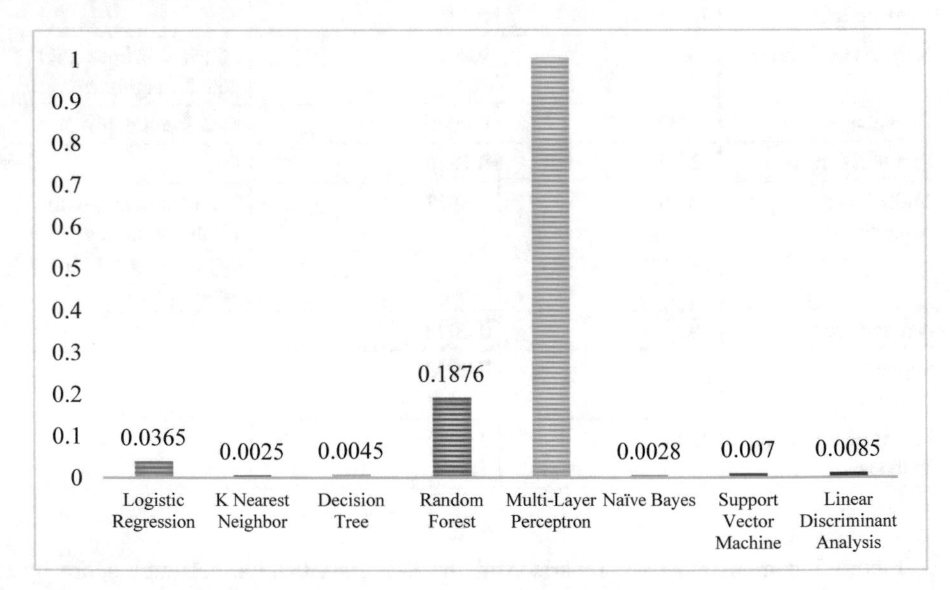

Fig. 3. Time to finish the ML classification algorithms training

4.2 Results Discussion

Through the analysis of the results obtained from Fig. 2, we can agree that almost all the ML classification algorithms showed interesting accuracy scores, which nominate them as possible model for health monitoring. From these analyzed models, Support Vector machine (SVM) and Logistic Regression can be considered as the best models for the studied issue with accuracy scores of 83.33% and 78.33% respectively.

Figure 3 showcases the time required to train the ML classification models, where we can see the slowest algorithms on the training step are: Multi-Layer Perceptron, Random Forest and Logistic regression in this order. The rest of the ML classification algorithms demonstrated a significant fast training period. By congregating the results of both Figs. 2 and 3 we can distinguish Support Vector Machine as the best ML model for health monitoring. KNN and Decision Tree manifested motivating results and should be assessed furthermore as possible solutions for real time health monitoring.

5 Conclusion

The findings of this study can be considered as a promising aspect for the application of machine learning in health monitoring and home automation. By using home automation technologies (wearables, Internet of Things, sensors, surveillance cameras), remote health monitoring can be the ultimate solution for healthcare systems that aims to solve hospitals beds shortage. Accordingly, a sustainable intelligent home should consider the employment of Machine learning capabilities for patient health monitoring. The present study results showed outstanding results in case of a patient with a heart condition.

We believe that apart from looking for the best machine-learning algorithm for health monitoring, future research should look for an artificial intelligence ecosystem capable of tackling every aspect of a smart home dedicated for health monitoring or medical surveillance. This would provide a good starting point for evaluating several health monitoring case studies using machine learning, engaging facial and voice recognition for home control and ultimately a home personalization system based on habits and health aspects which would be suitable for the elderly, people with reduced mobility or health problems.

References

1. WHO: World health statistics 2020: monitoring health for the SDGs, sustainable development goals (2020). ISBN 978–92–4–000510–5
2. Author, F., Author, S.: A study on shortage of hospital beds in the Philippines using system dynamics. In: 2018 5th International Conference on Industrial Engineering and Applications (ICIEA), IEEE Xplore, 18 June 2018
3. Bucci, S., et al.: Emergency department crowding and hospital bed shortage: Is lean a smart answer? A systematic review. Eur. Rev. Med. Pharmacol. Sci. **20**(Wp), 4209–4219, November 2016
4. Francis, A.D.: The impact of hospital bed and beddings on patients: the Ghanaian healthcare consumer perspectives. Int. J. Innovative Res. Adv. Stud. (IJIRAS) **6**(1), 138–145 (2019)

5. Priyanga, P., MuthuKumar, V.P.: Cloud computing for healthcare organization. Int. J. Multi. Res. Dev. **2**(4) (2015)
6. Shinde, S.P., Phalle, V.N.: A survey paper on internet of things based healthcare system. Int. Adv. Res. J. Sci. Eng. Technol. **4**(4) (2017)
7. Mike, K.: Wearable technology in health care – acceptance and technical requirements for medical information systems. In: 2020 6th International Conference on Information Management (ICIM), IEEE Xplore, 30 April 2020
8. Priyanka, D., et al.: A review paper on patient monitoring system. J. Appl. Fundam. Sci. **1** (2015)
9. Malasinghe, L.P., Ramzan, N., Dahal, K.: Remote patient monitoring a comprehensive study. J. Ambient Intell. Hum. Comput. **10**, 57–76 (2019)
10. Ben Rejab, F., Nouira, K., Trabelsi, A.: Health monitoring systems using machine learning techniques. In: Chen, L., Kapoor, S., Bhatia, R. (eds.) Intelligent Systems for Science and Information. SCI, vol. 542, pp. 423–440. Springer, Cham (2014). https://doi.org/10.1007/978-3-319-04702-7_24
11. Nikita, B., Prem, K.: A review paper on home automation. Int. J. Eng. Tech. **4**(1) (2018)
12. Mussab, A., et al. : A review of smart home applications based on Internet of Things. J. Netw. Comput. Appl. (2017). https://doi.org/10.1016/j.jnca.2017.08.017
13. Kuppusamy, P.: Smart home automation using sensors and Internet of Things. Asian J. Res. Soc. Sci. Hum. **6**(8), 2642–2649 (2016)
14. Gabriele, L., et al.: A review of systems and technologies for smart homes and smart grids. Energies **9**, 348 (2016). https://doi.org/10.3390/en9050348
15. Tanish, S., Shubham, M.: Home automation using IOT and mobile App. Int. Res. J. Eng. Technol. **04**(02) (2017)
16. Karishma, Y., Rajat, J.: Sensors for home automation. Int. J. Sci. Dev. Res. **1**(4) (2016)
17. Thanos, G.S., et al.: IoT wearable sensors and devices in elderly care: a literature review. Sensors 20, 2826 (2020). https://doi.org/10.3390/s20102826
18. Garima, T., et al.: Home automation system using artificial intelligence. Int. J. Res. Appl. Sci. Eng. Technol. **5**(8) (2017)
19. Saha, J., et al.: Advanced IOT based combined remote health monitoring, home automation and alarm system. In: 2018 IEEE 8th Annual Computing and Communication Workshop and Conference (CCWC), Las Vegas, pp. 602–606 (2018). https://doi.org/10.1109/CCWC.2018.8301659
20. Shinde, P.P., Shah, S.: A review of machine learning and deep learning applications. In: 2018 Fourth International Conference on Computing Communication Control and Automation (ICCUBEA), Pune, India, pp. 1–6 (2018). https://doi.org/10.1109/ICCUBEA.2018.8697857
21. Ayon, D.: Machine learning algorithms: a review. Int. J. Comput. Sci. Inf. Technol. **7**(3), 1174–1179 (2016)
22. Angra, S., Ahuja, S.: Machine learning and its applications: a review. In: 2017 International Conference on Big Data Analytics and Computational Intelligence (ICBDAC), Chirala, pp. 57–60 (2017). https://doi.org/10.1109/ICBDACI.2017.8070809
23. Francesco, M., et al.: An overview on application of machine learning techniques in optical networks. https://arxiv.org/abs/1808.07647, 1 December 2018
24. Vladimir, N.: An overview of the supervised machine learning methods. In: St Kliment Ohridski University – Bitola Repository (2017). https://doi.org/10.20544/HORIZONS.B.04.1.17.P05
25. Priyadharsan, D.M.J., et al.: Patient health monitoring using IoT with machine learning. Int. Res. J. Eng. Technol. **06**(03) (2019)

A Study of Machine Learning Based Approach for Hotels' Matching

El Mostafa Jaouhar[1], Said El Kafhali[1(✉)], and Youssef Saadi[2]

[1] Faculty of Sciences and Techniques, Computer, Networks, Mobility and Modeling Laboratory (IR2M), Hassan First University of Settat, B.P. 577, 26000 Settat, Morocco
{m.jaouhar,said.elkafhali}@uhp.ac.ma

[2] Information Processing and Decision Support Laboratory, Faculty of Sciences and Technologies, Sultan Moulay Slimane University, 23000 Beni Mellal, Morocco
y.saadi@usms.ma

Abstract. Hotels' matching is a big challenge faced by the actors of the business of selling hotels such as online travel agencies (OTAs), tour operators (TO) and Business to Business (B2B) companies. Nowadays, machine learning classification is used for numerous tasks and it can help to classify a pair of hotels as identical or different. In this paper, we described different approaches and techniques applied in hotel data matching and we proposed a solution which combines machine learning classification and text similarity algorithms. The machine learning algorithms used in this study are logistic regression (LR), naive Bayes (NB), support vector machine (SVM), k-nearest neighbours (KNN), decision tree (DT) and random forest (RF). The obtained results show the performance of our model in terms of accuracy, precision, recall and F1 score. The best results were obtained with the help of random forest and decision tree classifiers where all performance metrics values are greater than 99%.

Keywords: Hotel matching · Entity resolution · Machine learning · Text similarity

1 Introduction

The tourism industry has been completely transformed by new Internet-related technologies. Over the last decade, OTAs have replaced classical travel agencies by offering clients many online booking services related to their trips [1]. In order to offer the best prices to their partners and customers, the companies in the hotel sales industry are working with different suppliers. Every supplier has its own hotel data. Therefore, these companies are confronted to the challenging task of matching their hotel data because the same hotel can be provided with different information by their suppliers.

M. Lazaar et al. (Eds.): BDIoT 2021, LNNS 489, pp. 373–384, 2022.
https://doi.org/10.1007/978-3-031-07969-6_28

Entity resolution, also known as data matching and record linkage [2] is the task of identifying records that refer to same entity. It can be defined as [3]: Given two datasets A and B from two semantically related data sources S_A and S_B, the entity matching problem consists in finding all matches between data records in $A \times B$ that refer to the same entity in the real world. Research on data matching has been present for decades, as the problem appears in many applications for different public sector related to transportation, health, finance, and so forth [4].

Duplicate detection or deduplication is a similar problem in which the matching is performed over records from the same dataset [5]. It is an approach which identifies and eliminates duplicate representations of identical real world entities in a dataset in order to avoid data redundancy. It is a pervasive problem and of high importance for data integration and data quality, for example to identify duplicate clients in corporate databases or to match article offers for cost comparison portals. Deduplication technique typically compare couple of entities by calculating multiple similarity measures to take efficient match decisions [6].

The task of hotel data matching is very important and must be accurate for many reasons. If there are duplicates of hotels, a same hotel may be returned many times to the customer with different prices. Matching errors cause real problems for the customer support and hurt the company's reputation toward its partners. The most accurate solution is the manual matching by professional content managers but it is expensive and time-consuming [7]. Therefore, the objective of this article is to provide a new method of hotel data matching, based on machine learning involving hotel data coming from different providers and evaluate it with suitable performance metrics.

The rest of this paper is organized as follows. Section 2 provides some related studies. Section 3 presents our dataset and describes its content in details. Section 4 describes the proposed solution. Results and discussion are presented in Sect. 5. Finally, Sect. 6 presents our conclusions and some insights for future work.

2 Related Work

The data matching has earned a lot of attraction, and there are many published papers regarding the hotels' matching. In this section, we present some related works to the field of entity resolution and especially with hotel data. For instance, the authors of [8], have proposed an approach based on unsupervised feature extraction that focuses on two tasks of the product integration pipeline, namely product matching and categorization. Additionally, they used a convolutional neural network (CNN) to produce embedded images from product images, further improving outcomes on both tasks. Kozhevnikov *et al.* [7] worked for the Russian market in their article and covered more than 2.8 million hotels. They first divided the hotels into buckets, built pairs of hotels to compare and then used images data with the basic features of the hotel to find similar hotels. They found that the best result of classification using machine learning was obtained with a random forest classifier of 30 trees. When they used hotel image features in the machine learning section to identify similar records, they got better results. Aksoy *et al.* [9] conducted a study where they searched solutions for the problems of hotel data coming from different providers with different names and addresses. In this study, the Map-Reduce process was performed by using the Soundex algorithm to match the hotel data of a tourism agency with the hotel data provided by their partners. The time required for data matching was significantly reduced. When comparing some String similarity algorithms in terms of the data they correctly matched and the process time, they observed that Dice coefficient algorithm yielded a better result.

Mohammed *et al.* [10] proposed a framework adopted by a full-service hotel in Hong Kong to define the hotel's corporate identity, analyze the market for potential competitors, and match hotels with similar corporate identities. The important objective aimed to provide additional information that can be used by industry practitioners to identify potential competitors.

Perez [11] made a study where he tried to find records which represent the same hotel using machine learning method. They found better results with AdaBoost [12] and XGboost [13] algorithms. Using data from a company in the tourism sector, Bayrak *et al.* [14] tried to detect duplicate customer records with similarity algorithms and machine learning methods. They obtained more successful results in the classification model where they used the Supporting Vector Machine (SVM) algorithm. Koumarelas *et al.* [15] utilized character-based and vector-based similarity methods in their studies. They determined duplicate records of hotels in the United States where address and location information were available. They used geocoding and reverse-geocoding methods to reduce the error rate in the enrichment process. Zheng *et al.* [16] made a study using classification algorithms to find similar records representing the same record on a dataset containing geographic locations and fed from various sources. They have achieved more successful results than the one-to-one matching and rule-based approach.

In contrast to prior work reported in the literature, in our proposed solution, we use only the most common hotel attributes that all providers possess. Other studies compare other hotel fields like phone number, website, email and images data which are not always available and could be totally different. For example, hotel images could be taken from different places and at different times.

3 Dataset

In this study, we used a dataset composed of 259851 hotel pairs for which we know in advance if the hotels are identical or not. We used the most common attributes provided by the majority of providers as shown in Fig. 1.

Our data is gathered from two different sources, so it is strongly probable to find errors, missing values and outliers which can skew matching results. Therefore, a phase of data pre-processing was required before using this data as input for our matching process. For example, in Table 1, we have in the second

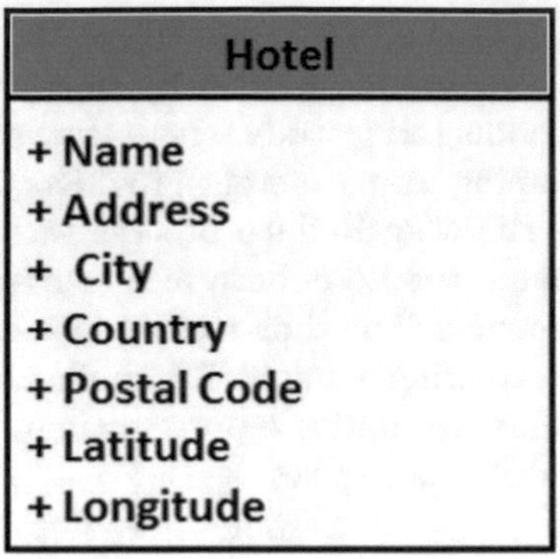

Hotel

+ Name
+ Address
+ City
+ Country
+ Postal Code
+ Latitude
+ Longitude

Fig. 1. Used hotel attributes

row an erroneous value of Longitude attribute while we have in the third row the value of the Country Name instead of the Country Code used in the other records.

Table 1. Sample of hotel records

Name	Address	Postal code	City	Country	Latitude	Longitude
Days Inn by Wyndham Greensboro Airport	501 S Regional Rd	27409	Greensboro	US	36.08244	−79.95644
Agriturismo Poderi Minori	Loc. Marena-Podere Archiano	52011	Bibbiena	IT	43.71867	**1181919**
La Mamounia	Avenue Bab-Jdid	40040	Marrakesh	**Morocco**	31.62207	−7.99760

4 Proposed Solution

As illustrated in the flow diagram of Fig. 2, the first step is data pre-processing where data is cleaned, corrected and standardized. Then we measure similarity between hotels of every pair of the dataset and build new features using similarity algorithms. After the extraction of the features, machine learning model trained on training data and tested using test data. Finally we evaluate our model using performance metrics.

4.1 Data Pre-processing

The dataset requires an initial processing before it is given as input to the machine learning model, called pre-processing. This is a crucial step in building

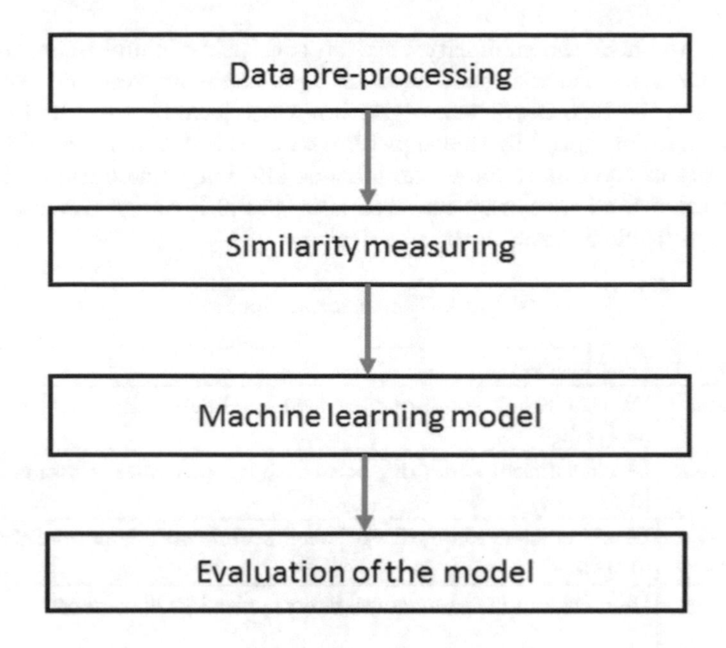

Fig. 2. Flow diagram of the proposed solution

machine learning models so that meaningful information is drawn from the data. Different pre-processing methods are data normalization, categorical features handling, missing data handling, label noise handling, outlier removal, etc. The hotel data used in our study is issued from two different sources, thus data has to be pre-processed in order to make the machine learning model more accurate and efficient. Pre-processing was applied on all hotel attributes to improve the quality of the dataset. Coordinate data was validated and corrected. Country code was used as a standardized value of the Country attribute. To pre-process textual attributes, the following steps were performed:

- Converting all punctuation marks to spaces.
- Normalizing multiple white spaces by replacing any sequence of white space symbols with one white space.
- Replacing any letter with accent with its equivalent without accent.
- Lowercasing all letters.
- Removing the most repetitive words like hotel, hostel, suites, inn and resort.

4.2 Similarity Measuring

Text similarity measures play an important role in text related research and applications [17]. They are involved in tasks such as information retrieval, text classification, document clustering, topic detection, machine translation, text summarization and others.

In order to check the similarity between each pair of hotel pairs, text similarity measurement methods were used for hotel name, address and postal code data. We used the Dice Coefficient algorithm which gave best results in [11], the similarity values produced by this algorithm will be used as features. To measure the similarity of coordinate data, the latitude and longitude data of every pair were subtracted from each other and normalized to [0,1] range to create two new features. The Table 2 describe the new features.

Table 2. Similarity attributes

Attribute	Description
Hotel name similarity	Dice coefficient similarity between hotel names is normalized to [0, 1] range
Address name similarity	Dice coefficient similarity between hotel addresses is normalized to [0, 1] range
Postal code similarity	Dice coefficient similarity between postal codes is normalized to [0,1] range
Latitude similarity	Difference in latitude values is normalized to [0,1] range
Longitude similarity	Difference in longitude values is normalized to [0,1] range
Same city	1 if the city information of the pair of hotels is identical 0 if not
Same country	1 if the country information of the pair of hotels is identical 0 if not

The Dice similarity coefficient, also known as the Sørensen-Dice index, is a statistical method which allows to measure the similarity between two sets of data. The equation for this similarity method is:

$$Dice\ coefficient = 2 * \frac{|X \cap Y|}{|X| + |Y|} \tag{1}$$

where $|X|$ and $|Y|$ are the cardinalities of the two sets. The Sørensen-Dice index is equal to twice the number of elements of the the two sets intersection divided by the sum of the number of elements in each set.

In this work, we used the Min-Max normalization technique. It maps the values of a given attribute A into range of [0, 1]. Equation (2) shows the computation of Min-Max normalization technique.

$$v_n = \frac{v - min_A}{max_A - min_A} \tag{2}$$

where

- v is a value of the attribute A,
- v_n is the new value of v in the range [0,1],
- min_A and max_A are the minimum and the maximum values of the attribute A.

4.3 Machine Learning Model

In machine learning, classification refers to a predictive modeling problem where a class label is predicted for a given example of input data. Binary classification refers to those classification tasks that have two class labels. Hence, predicting if a pair of hotels are similar or not can be considered as a binary classification, with two classes Identical and Different. We implemented the most frequently used machine learning algorithms for this kind of tasks [18], and compare their results.

- **Logistic Regression**: It is a machine learning algorithm borrowed from the field of statistics used for classification tasks. This algorithm uses a logistic function (also called sigmoid function) to model a binary dependent variable based on a given set of independent variables.
- **K-Nearest Neighbors**: K-Nearest Neighbors (or kNN for short) is one of the simplest machine learning algorithms that can give highly accurate results. It can deal with both classification and regression tasks but mostly it is used for classification problems. It is one of the most popular distance-based algorithms. This algorithm classifies data points based on the points that are most similar to them.
- **Decision Trees**: Decision trees are a supervised learning method capable of performing both regression and classification but mainly used for classification. It is one of the most widely used classifiers for classification problems. It generates a set of decision sequences that permit predicting class labels.
- **Random Forest**: Random Forest is one of the popular and robust machine learning algorithms that can be used for regression and classification tasks. A random forest model is a collection of individual decision trees, called estimators. The random forest model combines the predictions of its estimators to produce a more accurate prediction. Random Forest models are less interpretable than a single decision tree but their predictive performance and accuracy are better.
- **Support Vector Machine**: Support vector machine (or in short SVM) is a supervised machine learning algorithm that can be used for both classification and regression tasks. This algorithm performs classification by finding a hyperplane that linearly separates data points belonging to different classes.
- **Naive Bayes**: Naive Bayes is a simple and effective classification algorithm used for binary and multi-class classification problems. It is based on Bayes theorem which assumes the independence between features. Naive Bayes classifier is a probabilistic classification algorithm which is easy to implement and particularly very useful for large datasets.

4.4 Evaluation of the Model

To evaluate the performance of our models in terms of accuracy of each classification algorithm, we used the confusion matrix described in Table 3 and we measured the metrics accuracy, precision, recall and F1 score.

A confusion matrix is a table that allows to visualize the performance of a classification model, the rows represent the predicted class and the columns represent the actual class.

Table 3. Confusion matrix for binary classification

	Actual positive class	Actual negative class
Predicted positive class	True positive (TP)	False Negative (FN)
Predicted negative class	False positive (FP)	True Negative (TN)

With:

- True Positive (TP) refers to the collection of hotel pairs that are classified correctly as identical.
- True Negative (TN) refers to the collection of hotel pairs that are classified correctly as different.
- False Positive (FP) refers to the collection of hotel pairs incorrectly classified to as different.
- False Negative (FN) refers to the collection of hotel pairs incorrectly classified as identical.

Further details explaining the used performance metrics [19] are described below.

- **Accuracy**: In general, the accuracy metric measures the ratio of correct predictions over the total number of instances evaluated.

$$Accuracy = \frac{TP + TN}{TP + TN + FP + FN} \tag{3}$$

- **Precision**: It allows to measure the positive patterns that are correctly predicted from the total predicted patterns in a positive class.

$$Precision = \frac{TP}{TP + FP} \tag{4}$$

- **Recall**: Recall is used to measure the fraction of positive patterns that are correctly classified.

$$Recall = \frac{TP}{TP + TN} \tag{5}$$

- **F1 score**: This metric represents the harmonic mean between recall and precision values.

$$F1\ score = \frac{2 * precision * recall}{precision + recall} \tag{6}$$

5 Results and Discussion

We used scikit-learn library to implement machine learning algorithms. We divided our preprocessed dataset into training and testing subsets using stratification in a ratio of 7.5:2.5 in order to avoid over fitting. The training dataset was used to fit the model and the evaluation was made on the testing dataset. Table 4 and Figs. 3, 4, 5, 6, 7 and 8 summarize the implementation results for each algorithm used in our study. All used classifiers gave great results with the contribution of all features. The best results were obtained with the help of random forest and decision trees which outperformed other classifiers in terms of accuracy, precision, recall and f-score. All performance metrics values were concluded as greater than 99% for random forest and decision trees algorithms.

Table 4. Different algorithms results

	Accuracy	F1 score	Recall	Precision
Logistic regression	0.95129535	0.95935147	0.93424246	0.98584744
Naive Bayes	0.97236889	0.97695616	0.95696788	0.99779724
SVM	0.98115851	0.98448512	0.97218676	0.99709862
KNN	0.98523775	0.98523775	0.98518648	0.99064286
Decision tree	0.99812201	0.99847210	0.99794718	0.99899757
Random forest	0.99918415	0.99933349	0.99929579	0.99937120

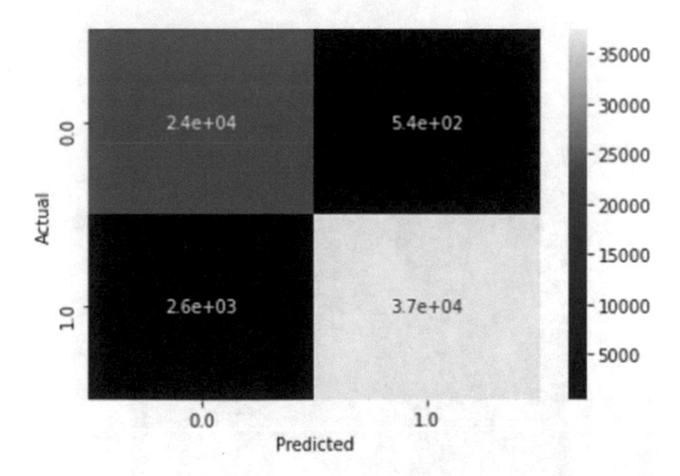

Fig. 3. Logistic regression confusion matrix

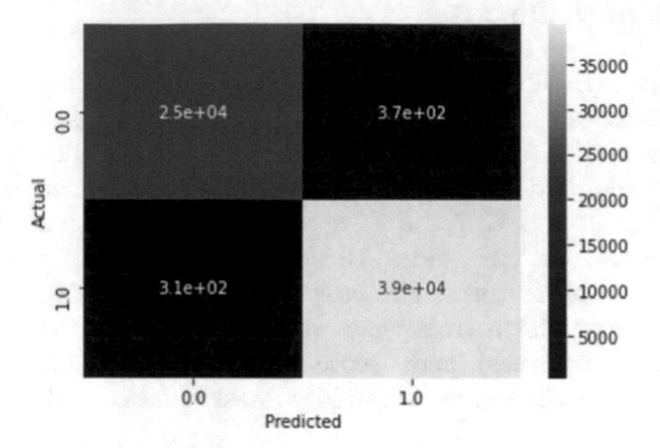

Fig. 4. KNN confusion matrix

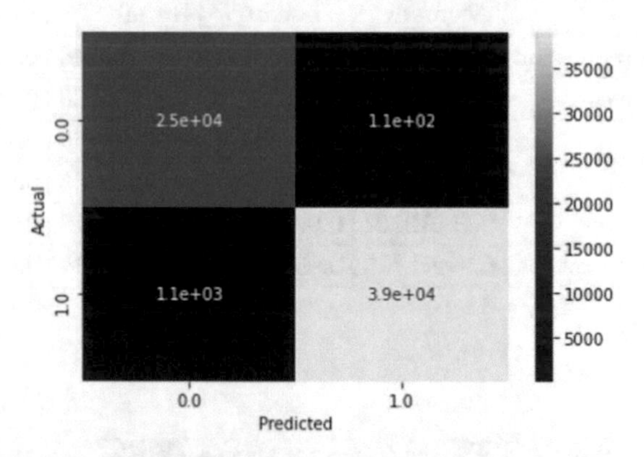

Fig. 5. SVM confusion matrix

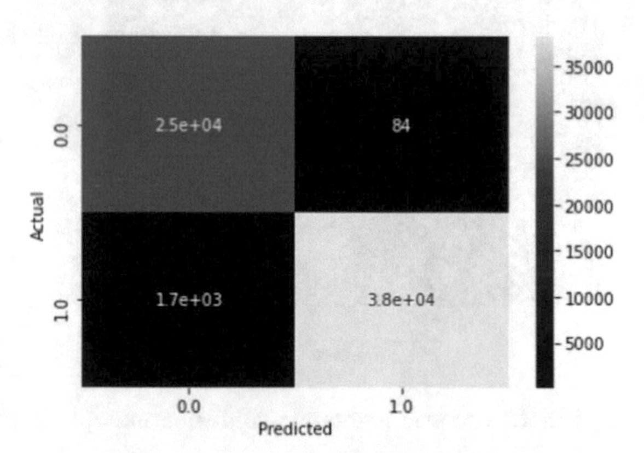

Fig. 6. Naive Bayes confusion matrix

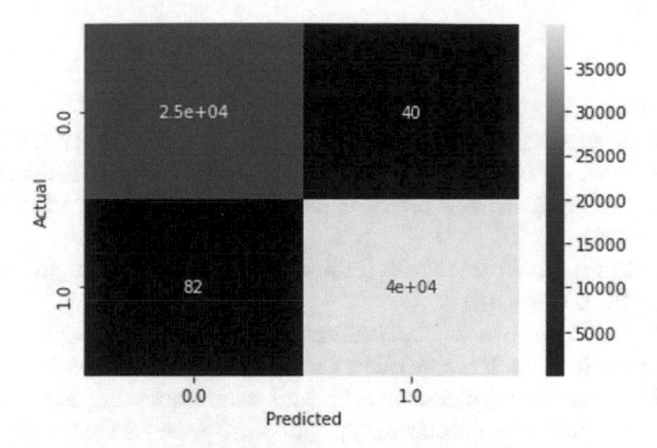

Fig. 7. Decision tree confusion matrix

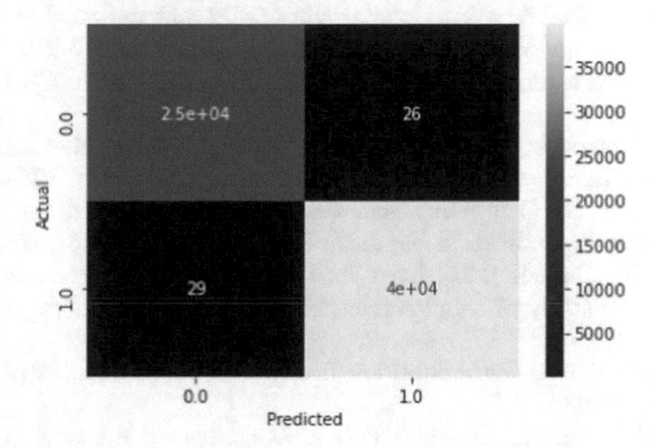

Fig. 8. Random forest confusion matrix

6 Conclusion

In this paper, we proposed a method based on machine learning solving the entity matching problem for hotel data coming from different providers. Our contributions are focused on the three main tasks: data pre-processing, classification model and evaluation of the model. We performed the data pre-processing task which consists to clean, correct and complete hotel data. Our machine learning model was trained and tested using a dataset composed of pairs of identical and different hotels, to which we added new features that we created by measuring the similarity of the categorical attributes of every pair. Finally, the model was evaluated with the appropriate performance metrics. According to the evaluation task, the proposed solution is accurate and it showed great results. As a future work, we will try to use more data from different sources in order to study and test the effectiveness of our solution.

References

1. Zhang, J., Zhou, Y., Tang, W., Gu, H., Yan, J., Wang, H.: An agent-mediated tendering mechanism for intelligent hotel reservation. In: 2017 IEEE 14th International Conference on e-Business Engineering (ICEBE), pp. 307–311. IEEE (2017)
2. Fellegi, I.P., Sunter, A.B.: A theory for record linkage. J. Am. Stat. Assoc. **64**(328), 1183–1210 (1969)
3. Kopcke, H., Rahm, E.: Frameworks for entity matching: a comparison. Data Knowl. Eng. **69**(2), 197–210 (2010)
4. Koumarelas, L., Papenbrock, T., Naumann, F.: MDedup: duplicate detection with matching dependencies. Proc. VLDB Endow. **13**(5), 712–725 (2020)
5. Christen, P.: Data matching: concepts and techniques for record linkage, entity resolution, and duplicate detection, pp. 23–35. Springer-Verlag (2012)
6. Kolb, L., Rahm, E.: Parallel entity resolution with dedoop. Datenbank-Spektrum **13**(1), 23–32 (2013). https://doi.org/10.1007/s13222-012-0110-x
7. Kozhevnikov, I., Gorovoy, V.: Comparison of different approaches for hotels deduplication. In: Ngonga Ngomo, A.-C., Křemen, P. (eds.) KESW 2016. CCIS, vol. 649, pp. 230–240. Springer, Cham (2016). https://doi.org/10.1007/978-3-319-45880-9_18
8. Ristoski, P., Petrovski, P., Mika, P., Paulheim, H.: A machine learning approach for product matching and categorization. Semant. Web **9**(5), 707–728 (2018)
9. Aksoy, B., Uğuz, S., Oral, O.: Comparison of the data matching performances of string similarity algorithms in big data. J. Eng. Sci. Des. **7**(3), 608–618 (2019)
10. Mohammed, I., Guillet, B.D., Law, R.: Competitor set identification in the hotel industry: a case study of a full-service hotel in Hong Kong. Int. J. Hosp. Manag. **39**, 29–40 (2014)
11. Pérez Sena, F.D.: Accommodations deduplication (2018). https://addi.ehu.es/handle/10810/28984
12. Ying, C., Qi-Guang, M., Jia-Chen, L., Lin, G.: Advance and prospects of AdaBoost algorithm. Acta Automatica Sin. **39**(6), 745–758 (2013)
13. Mitchell, R., Frank, E.: Accelerating the XGBoost algorithm using GPU computing. PeerJ Comput. Sci. **3**, e127 (2017)
14. Bayrak, A.T., Yılmaz, A.İ., Yılmaz, K.B., Düzağaç, R., Yıldız, O.T.: Near duplicate detection in relational databases. In: Proceedings of the 26th Signal Processing and Communications Applications Conference (SIU), pp. 1–4. IEEE (2018)
15. Koumarelas, I., Kroschk, A., Mosley, C., Naumann, F.: Experience: enhancing address matching with geocoding and similarity measure selection. J. Data Inf. Qual. **10**(2), 1–16 (2018)
16. Zheng, Y., Fen, X., Xie, X., Peng, S., Fu, J.: Detecting nearly duplicated records in location datasets. In: Proceedings of the 18th SIGSPATIAL International Conference on Advances in Geographic Information Systems, pp. 137–143, ACM (2010)
17. Gomaa, W.H., Fahmy, A.A.: A survey of text similarity approaches. Int. J. Comput. Appl. **68**(13), 13–18 (2013)
18. Sen, P.C., Hajra, M., Ghosh, M.: Supervised classification algorithms in machine learning: a survey and review. In: Mandal, J.K., Bhattacharya, D. (eds.) Emerging Technology in Modelling and Graphics. AISC, vol. 937, pp. 99–111. Springer, Singapore (2020). https://doi.org/10.1007/978-981-13-7403-6_11
19. Hossin, M., Sulaiman, M.N.: A review on evaluation metrics for data classification evaluations. Int. J. Data Min. Knowl. Manag. Process **5**(2), 1–11 (2015)

Natural Processing Language

Empirical Study: What is the Best N-Gram Graphical Indexing Technique

Latifa Rassam(✉), Ahmed Zellou, and Taoufiq Rachad

SPM Team, ENSIAS, Mohammed V University in Rabat, Rabat, Morocco
rassamlatifa@gmail.com

Abstract. Recent research shows that indexing techniques, such as n-gram models, are useful at a wide variety of software engineering tasks, e.g., data extraction, document indexing, etc. However, these models require some numerous parameters. Moreover, the different ways one can extract data essentially yield different models. In this paper, we focus on n-gram models and evaluate how the use different set of methods and n values impact the predicting ability of these models. Thus, we compare the use of multiple techniques and sets of different parameters (n values character gram and word gram) with the aim of identifying the most appropriate indexing technique.

Keywords: Indexing · n-gram · Extraction

1 Introduction

Indexing is considered as one of the main phases of the information retrieval cycle, carrying out a search, with the aim of knowing how doing it, while generally seeking to directly reach the most basic information satisfying the user's need. The information retrieval process must therefore be particularly fast and efficient, hence the interest of treating the technical document as a molecular construction. While decomposing it to give birth to new usable units and this for carrying out specific readings. This will allow, on the one hand, a fine representation of its content, and on the other hand, a more localized access to information facilitating the task of consultation for the user [1].

Two types of indexing are generally distinguished: statistical indexing and graphical indexing. Statistical approaches include indexing methodologies such as: methodologies based on linguistic, lexical, cognitive, thematic and frequency analysis TF-IDF (Term Frequency – Inverse Document Frequency). However, for graphical indexing, many methodologies exist, such as: indexing of images based on graphs and hypergraphs, GraphGrep, gIndex, tree-B, tree-B +, tree-kd, and the most used and recent technique n-gram graphs... [2].

The N-gram graph approach is essentially based on several techniques, namely: word attraction rank, position rank and the simple structural character/word N- gram… [2].

This leads us to elaborate the following problematic: what is the most widespread technique in terms of textual indexing based on n-gram graphs?

M. Lazaar et al. (Eds.): BDIoT 2021, LNNS 489, pp. 387–398, 2022.
https://doi.org/10.1007/978-3-031-07969-6_29

So, this article aims to specify the most convincing indexing N-gram technique in term of data precision. In this article we will first tackle the different indexing based N-gram techniques, and then carry out with a simulation state of each one of them and finally discuss them in order to identify the most appropriate indexing technique.

2 Indexing Techniques Based on n-gram Graphs

In the domain of natural language processing, there have been a number of methods using n-grams. A n-gram is a, possibly ordered, set of words or characters, containing n elements. N-grams have been used in summarization and summary evaluation [3].

2.1 History

In the n-gram graph method, no term extraction is required and the graph is based directly on the text, without further background, such as a corpus for the calculation of TF-IDF or any other weighting factor. We can just apply a graph matching method, that offers a graded similarity indication between two document graphs. Moreover, in order to compare a whole set of documents we can represent the set of documents with a single graph [4].

Indexing, according to AFNOR (The French Association for Standardization), is the process intended to represent by elements of a documentary or natural language data results of the analysis of a document or part of this document. The objective of this process is to allow or facilitate the search for documents by their content [5].

At the indexing of graphics (structural approaches) each object is represented by a structure. It is a question of defining a structural model which describes at the same time the various components of the object in question and the relations which link them. This representation is characterized by a more expressive representative capacity than that via characteristic vectors (Statistical indexing) [4]. Among the data structures which are widely used in the recognition of structural forms, there are particularly trees and graphs. From an algorithmic point of view, graphs generalize a large set of data structures [6].

2.2 N-gram Graph Concepts

In the domain of natural language processing, there have been a number of methods using n-grams. A n-gram is a, possibly ordered, set of words or characters, containing n elements. N-grams have been used in summarization and summary evaluation [3].

Here are some examples of n-grams from the sentence: “This is a sentence”.

- Word unigrams: this, is, a, sentence
- Word bigrams: this is, is a, a sentence
- Character bigrams: th, hi, is, s, a…
- Character 4-g: this, his, is…

Since there is no indication about the appropriate value of N, we investigated the options between 2 and 10. In case of a long text, a large N can depict in a more accurate way the sequence of words but the graph and the complexity becomes higher. On the other hand, if N is small then the graph is smaller but the accuracy of the method is negatively affected. In short text, a small N can be more efficient than the use of large N [7].

2.3 N-gram Graph Process

The n-gram graph is a graph G = {VG, EG, L, W}, where:

- VG is the set of vertices
- EG is the set of edges
- L is a function assigning a label to each vertex and to each edge
- W is a function assigning a weight to every edge [8].

The graph has n-grams as its vertices vG ∈ VG and the edges eG ∈ EG connecting the n-grams indicate the proximity of the corresponding vertex n-grams (the superscript G will be omitted where easily assumed).

The edges can be weighted by the distance between the two neighboring n-grams in the original text, or the number of co-occurrences within a given window. We note that the meaning of distance and window size changes by whether we use character or word n-grams [9].

The labeling function L for edges assigns to each edge the concatenation of the labels of its corresponding vertex labels in a predefined order. For direct graphs the order is the order of the edge direction while in undirected graphs the order can be the lexicographic order of the vertex labels. To ensure that no duplicate vertices exist, we require that the labelling function is a one-to-one function [10].

More formally, here is a mathematical definition:

If S = {S1, S2,…}, Sk different from Sl for k different from l, k, l belong to N is the set of distinct n-grams extracted from a text Tl, and Si is the i-th extracted n- gram, then G = {Vg, Eg, L, W} is a graph where Vg = S is the set of vertices v, Eg is the set of edges e of the form e = {v1,v2}, L:Vg → M is a function assigning a label l(v) from the set of possible labels M to each vertex v and W:Eg → R is a function assigning a weight w(e) to every edge [11].

In our example, the edges E are assigned weights of ci,j where ci,j is the number of times a given pair Si , Sj of n-grams happen to be neighbors in a string within some distance Dwin of each other. We use the weight to determine which Sj is useful. The vertices vi , vj corresponding to n-grams Si , Sj that are located within this parameter distance Dwin are connected by a corresponding edge e ≡ {vi , vj} [12].

A vertex represents a text's N-Gram and an edge joins adjacent N-grams. The frequency of adjacencies can be denoted as weights on the graph edges [7].

The edges can be weighted or unweighted. Unweighted edges simply represent that some N-Grams are close in a text. Weighted edges can represent the frequency of the order of appearance between subsequent N-Grams. Despite the intuitive belief that weighted NGG can have better accuracy, the experiments as described in the Evaluation section proved in most of the cases the opposite. This is explained by the fact that many common N-Grams coexist very often in various texts from irrelevant categories. These common neighbor N-Grams lead to the failure to classify the texts based on the real important and informative word sequences but based on common and unimportant N-Gram sequences [13].

2.4 K-core Graph Process

For a graph G let C (k; G) be the maximal subgraph of G with the minimal degree at least k. It is not hard to see that C (k; G) is well defined and can be obtained from G as a result of the process of removing from a graph vertices of degree less than k.

Let G (n, p) be a graph with n labelled vertices in which each edge is present independently with probability p = p(n) and let C (k; n, p) be the maximal subgraph of G (n,p) with the minimal degree at least k = k(n). We estimate the size of C (k; n, p) and consider the probability that C(k; n, p) is k-connected when n tends to infinity [14].

We call C (k; G) the k-core of G. If no subgraph H of G has the property Lambda(H) >= k we say that the k-core of G is empty(Lambda is the minimal degree). Now let G(n, p) be a random graph with n labelled vertices in which each from possible edges is present independently with probability p [14].

We shall study the size and k-connectivity of the k-core of G(n, p), denoted by C(k; n, P) = C(k; G(n, P)). We shall assume that both k and p may depend on n and consider only the asymptotical property when n is infinite. Instead of p we shall use also average degree of G(n, p) defined as c = c(n) = (n − 1)p(n) as another parameter characterizing density of G(n, p) [14].

2.5 N-gram Graph Indexing Algorithm

If we choose to extract the n-grams (S n) of a text T, the (elementary) algorithm is indicated as Algorithm 1. The algorithm's complexity is linear to the size |T| of the input text T.

```
Input: text T // T is the text we analyze
Output: n-gram set SSn

1-  SS n ← Ø;
2-   for all i in [1,length(T)-n+1]
        do SSn ← SSn
3-   U Si,i+n−1
4- 4 end
```

The algorithm applies no preprocessing (such as extraction of blanks, punctuation or lemmatization). Furthermore, it obviously extracts overlapping parts of text, as the sliding window of size n is shifted by one position and not by n positions at a time. This technique is used to avoid the problem of segmenting the text. The redundancy apparent in this approach proves to be useful similarly to a convolution function: summing similarities over a scrolling window may prove useful if we do not know the exact center of the match between two subparts of a string. In the case of summary evaluation against a model summary, for example, the extracted n-grams are certain to include n-grams of the model summary, if such a n-gram exists, whereas a method where the text would be segmented in equally sized n-grams might not identify similar n-grams [15].

3 Empirical Study

3.1 Data Set

Here is an example of the algorithm application on textual document, extracted from the Guardian's British Journal.

Considering the pandemic situation of the whole world, we had recourse to the choice of this important subject: covid-19, by choosing an article extracted from one of the largest British newspapers; The Guardian (Fig.1).

2d ago
00:01

UK to relax travel quarantine from July 10

The quarantine policy for passengers arriving in England from "lower risk countries" such as Spain, France, Italy and Germany will be lifted, the Department for Transport (DfT) has announced.

A full list of countries from which arrivals will not need to self-isolate for 14 days will be published later on Friday. The new measures come into force from July 10.

All passengers except for those in certain categories will continue to be required to provide contact information on arrival.

The Foreign and Commonwealth Office will set out a number of destinations which will be exempt from its policy of advising against all non-essential overseas travel.

That change will come into effect on Saturday, allowing people to take holidays overseas with regular travel insurance policies.

The DfT said the devolved administrations "will set out their own approach", which means passengers arriving in Scotland, Wales and Northern Ireland "should ensure they follow the laws and guidance which applies there".

Fig. 1. An extract from the Guardian's British Journal [16].

3.2 Character N-gram Graph

A requested n-gram character size of 4 in the first paragraph would return 168 n-gram words:

{'Theq ', 'hequ', 'equa', 'quar', 'uara', 'aran', 'rant', 'anti', 'ntin' ,'inep' ,..., 'nced'}

In this case the first paragraph of our example above will be represented by almost 10 6-gram character graphs. Each of these graphs will contain 14 vertexes as shown in the representation below, while in the case of bigram word graph, the paragraph will be represented by 2 graphs of 14 vertexes (Fig. 2).

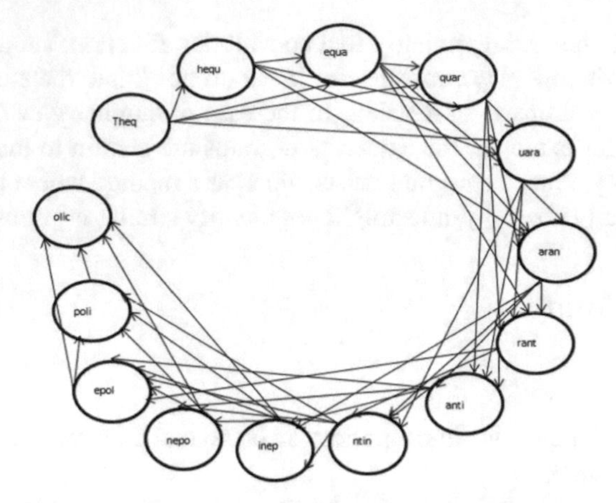

Fig. 2. 4-Gram graph representation of the first part's paragraph.

A requested n-gram character size of 5 in the first paragraph would return 152 n-gram words:

{'Thequ', 'hequa', 'equar', 'quara', 'uaran', 'arant', 'ranti', 'antin', 'ntine', 'inepo', 'nepol', 'epoli' , 'polic', 'olicy', 'licyf', 'icyfo', 'cyfor',….. 'unced'}

In this case the first paragraph of our example above will be represented by almost 10 5-gram character graphs and each of these graphs will contain 14 vertexes as shown in the representation below. While in the case of bigram word graph, the paragraph will be represented by 2 graphs of 14 vertexes (Fig. 3).

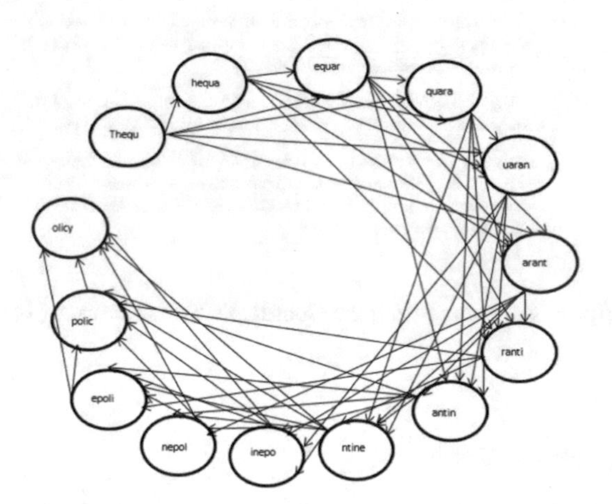

Fig. 3. 5-Gram graph representation of the first part's paragraph.

A requested n-gram character size of 6 in the first paragraph would return 148 n-gram words:

{'Thequa ', 'hequar','equara','quaran',, 'uarant','aranti', 'rantin', 'antine','ntinep' ,'inepol','nepoli' ,'epolic' ,'policy' ,'olicyf' ,'licyfo' ,'icyfor' ,'cyforp','yforpa' ,'forpas'

,'orpass' ,'rpasse' ,'passen' ,'asseng' ,'ssenge' ,'senger' ,'engers','ngersa' ,'ersarr' ,'rsarri' ,..... 'ounced'}

The first paragraph of our example above will be represented by almost 10 6-gram character graphs and each of these graphs will contain 14 vertexes as shown in the representation below, while in the case of bigram word graph, the paragraph will be represented by 2 graphs of 14 vertexes as shown in the figure below (Fig. 4).

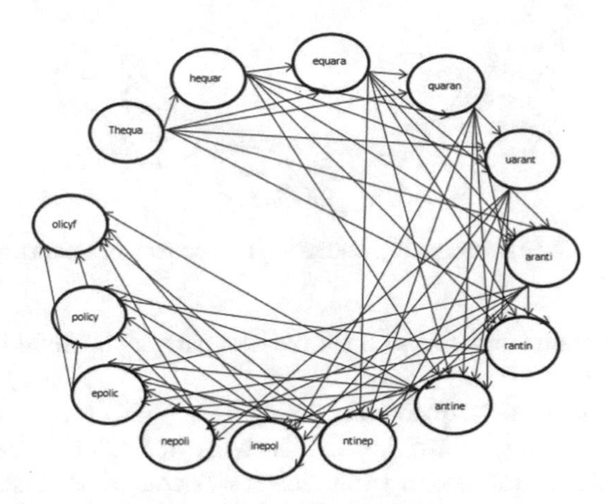

Fig. 4. 6-Gram graph representation of the first part's paragraph.

The table below defines the set of perfect indexes of each graph based on the number N. The weight of an n-gram is the main factor that allows the identification of the index (Table 1).

Table 1. The set of generated indexes depending on the N value for a character N-gram.

	N = 4	N = 5	N = 6
Set of indexes	{aran, rant, anti}	{arant, ranti, antin}	{aranti, rantin, antine}
Number of N-gram	268	252	248

3.3 Word N-gram Graph

A requested n-gram word size of 2 in the first paragraph would return 28 n- gram words. The punctuation marks and spaces would be omitted:

{'The quarantine', 'quarantine policy', 'policy for', 'for passengers', 'passengers arriving', 'arriving in', 'in England', 'England from', 'from lower', 'lower risk', 'risk countries', 'countries such', 'such as', 'as Spain', 'Spain France', 'France Italy', 'Italy and', 'and Germany', 'Germany will', 'will be', 'be lifted', 'lifted the', 'the department', 'department for', 'for transport', 'transport DFT', 'DFT has', 'has announced' } (Fig. 5).

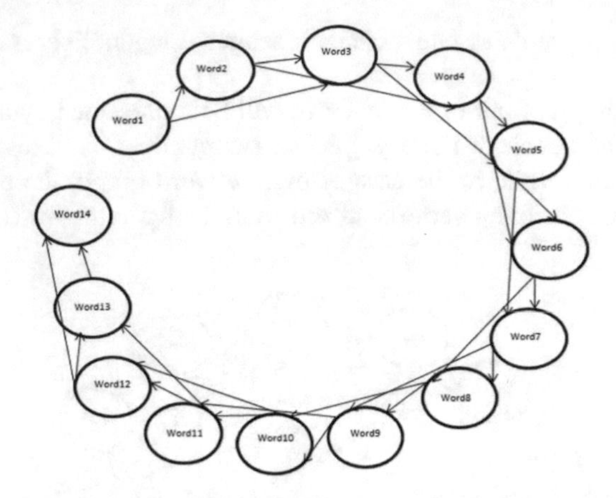

Fig. 5. 2-Gram graph representation of the first part's paragraph.

A requested n-gram word size of 3 for the first paragraph would return 27 n- gram words:

{'The quarantine policy', 'quarantine policy for', 'policy for passengers', 'for passengers arriving', 'passengers arriving in', 'arriving in England', 'in England from', 'England from lower', 'from lower risk', 'lower risk countries', 'risk countries such', 'countries such as', 'such as Spain', 'as Spain France', 'Spain France Italy', 'France Italy and', 'Italy and Germany', 'and Germany will', 'Germany will be', 'will be lifted', 'be lifted the', 'lifted the department', 'the department for', 'department for transport', 'for transport DFT', 'transport DFT has', 'DFT has announced'} (Fig. 6)

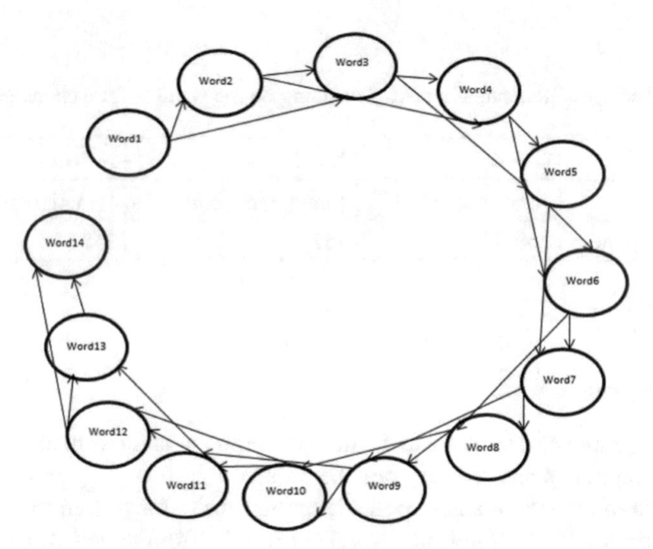

Fig. 6. 3-Gram graph representation of the first part's paragraph.

A requested n-gram word size of 4 for the first paragraph would return 26 n-gram words:

{'The quarantine policy for', 'quarantine policy for passengers', 'policy for passengers arriving, 'for passengers arriving in', 'passengers arriving in England', 'arriving in England from', 'in England from lower', 'England from lower risk', 'from lower risk countries', 'lower risk countries such', 'risk countries such as', 'countries such as Spain', 'such as Spain France', 'as Spain France Italy', 'Spain France Italy and', 'France Italy and Germany', 'Italy and Germany will', 'and Germany will be', 'Germany will be lifted', 'will be lifted the', 'be lifted the department', 'lifted the department for', 'the department for transport', 'department for transport DFT', 'for transport DFT has', 'transport DFT has announced'}

In the three cases, our test's paragraph will be represented by 2 graphs and each one of them will contain 14 vertexes for n having two and three and four as test's values (Fig. 7 and Table 2).

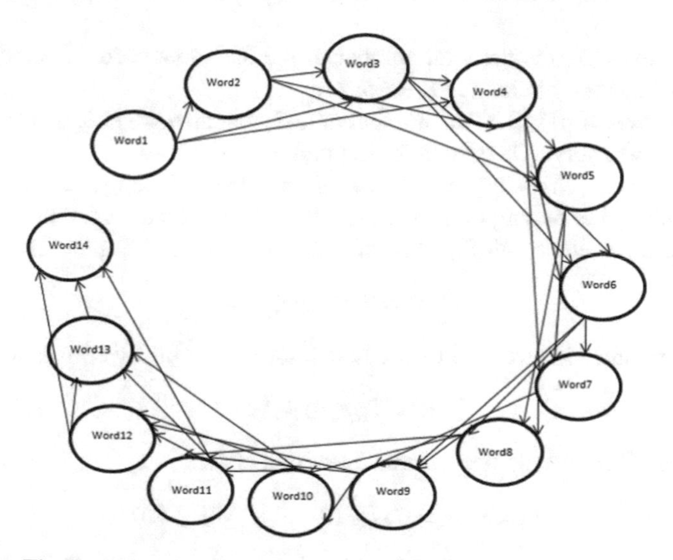

Fig. 7. 4-Gram graph representation of the first part's paragraph.

Table 2. The set of generated indexes depending on the N value for a word N-gram.

	N = 2	N = 3	N = 4
Set of indexes	{'arriving in', 'in England', 'England from'}	{'arriving in England', 'in England from', 'England from lower'}	{'arriving in England from', 'in England from lower', 'England from lower risk'}
Number of N-gram	28	27	26

4 Discussion

To properly assess the quality of the indexing performed and that of the indexes generated, it was essential to define another quality factor other than the number of n-grams generated and the indexes representing, this is the precision of indexing, the latter which strongly depends on the number N. For this we will discover the nature of the relationship between these two factors.

As known, the N value varies from two to ten. The table below describes the values of a graph precision depending on the N value; this table was extracted from an experimental study [17].

Macro Precision, Recall and precision express the quality of a classification. The macro precision for one class denotes the fraction of retrieved instances that are relevant and the recall denotes the fraction of extracted instances that are retrieved. The term Category is used for a category of texts as formed in the baseline dataset. The term Class A is used for the corresponding prediction of category A as estimated by a classification method [18].

True Positives (Tp) is the total number of the texts that correctly predicted in the Class A and they really belong in Category A.

False Positives (Fp) is the total number of texts that do not belong in Category A but they predicted wrongly to belong in the Class A.

False Negatives (Fn) is the total number of texts that do belong in the Category A but were not predicted to belong in the Class A. We now continue with defining the macro precision Equation, the recall Equation and the precision Equation metrics [19]

$$\mathrm{MP} = \mathrm{Tp}/(\mathrm{Tp} + \mathrm{Fp})$$

The macro precision MP is a metric that describes the purity of a class.

$$\mathrm{R} = \mathrm{Tp}/(\mathrm{Tp} + \mathrm{Fn})$$

The Recall R shows the text completeness of a class.

$$\mathrm{Precision} = 2\mathrm{x}((\mathrm{MP} - \mathrm{R})/(\mathrm{MP} + \mathrm{R}))$$

The precision P describes the purity and the text completeness of a class.

We used a scale of 10 to the power of 5 for the precision values: 1/10000 → 1 (Table 3)

Table 3. The values of a graph precision depending on the N value

N value	2	3	4	5	6	7	8	9	10
Precision by Character	8114	8756	9033	9181	9208	9226	9197	9168	9110
Precision by Word	8296	8774	8892	8974	8965	8959	8928	8877	8808

The graph below describes decently the proportional relation between the N value of a gram, the gram type and the graph's precision value (Fig. 8).

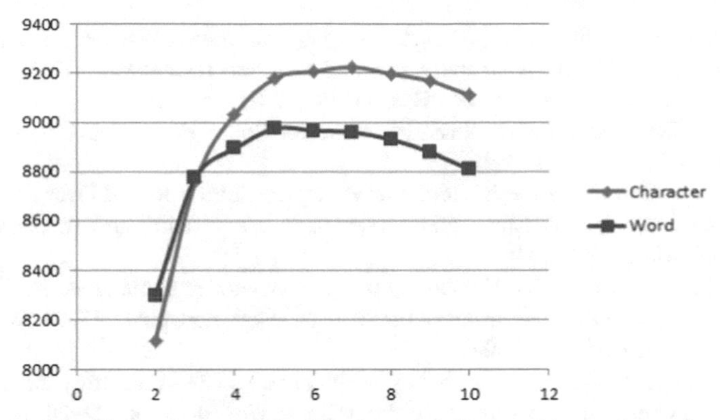

Fig. 8. The graph's relation between the N values and the graph's precision for character and word grams.

5 Conclusion and Perspectives

The objective of this paper consisted of the study of choice of the perfect indexing technique based on one of the most characterizing indexing methodology. We can conclude to a relevant technique for the based N-graph data indexing. To achieve this goal, we started with the definition data indexing, its history, its concepts, its technical process and algorithms. Then, we have led an empirical study about N- gram graphs indexing, its types by defining each one of them based on some examples of textual documents to make clear the N value effect on the indexes choice, ending with a discussion about the importance of a graph precision in such an indexing technique depending on the N value.

As a result of the empirical study, we conclude to an indexing technique which brings together several criteria such as the N value, and the precision value so that the classical indexing technique will be the most relevant. This gives rise to a new problem: could we improve find more graph based indexing techniques for all data types including images, relational databases and others?

References

1. La Segmentation des documents techniques en amont de l'indexation : définition d'un modèle OUERFELLI Tarek Institut Supérieur de Documentation Campus Universitaire de la Manouba - BP 600 Tunisie (2010)
2. Chin-Yew, L., Eduard, H.: Automatic evaluation of summaries using n-gram co- occurrence statistics. In: NAACL '03: Proceedings of the 2003 Conference of the North American Chapter of the Association for Computational Linguistics on Human Language Technology, pp. 71–78, Morristown, NJ, USA, Association for Computational Linguistics. (2003)
3. Baldwin, B., Donaway, R., Hovy, E., et al.: An Evaluation Roadmap for Summarization Research. Technical report (2000)
4. N-gram graphs: representing documents and document sets in summary system evaluation, George giannakopoulos university of Trento, Italy and Vangelis Karkaletsis ncsr demokritos, Greece (2009)

5. Une ontologie pour éditer des schémas de description audiovisuels, extension pour l'inférence sur les descriptions Thomas Dechilly, Bruno Bachimont Institut national de l'audiovisuel, Direction de la Recherche 4, av. de l'Europe - 94366 Bry-sur-Marne
6. Riesen, K.: Classification and Clustering of Vector Space Embedded Graphs. PhD thesis, University of Bern, Bern, (2009)
7. Text Classification Using the N-Gram Graph Representation Model Over High Frequency Data Streams John Violos, Konstantinos Tserpes, Iraklis Varlamis and Theodora Varvarigou Front. Appl. Math. Stat., (2018)
8. Copeck, T., Szpakowicz, S.: Vocabulary usage in newswire summaries. In: Text Summarization Branches Out: Proceedings of the ACL-04 Workshop, pages 19–21. Association for Computational Linguistics (2004)
9. Copeck, T., Szpakowicz, S.: Vocabulary usage in newswire summaries. In: Text Summarization Branches Out: Proceedings of the ACL-04 Workshop, pp. 22–24. Association for Computational Linguistics (2004)
10. Copeck, T., Szpakowicz, S.: Vocabulary usage in newswire summaries. In: Text Summarization Branches Out: Proceedings of the ACL-04 Workshop, pp. 24–26. Association for Computational Linguistics (2004)
11. George, G.: Automatic Summarization from Multiple Documents. PhD thesis, Department of Information and Communication Systems Engineering, University of the Aegean, Samos, Greece, (2009)
12. Giannakopoulos, G., Karkaletsis, V., Vouros, G., Stamatopoulos, P.: Summarization system evaluation revisited: N-gram graphs. ACM Trans. Speech Lang. Process. **5**(3), 1–39 (2008)
13. Wang, L., Shen, H., Tian, H.: Weighted ensemble classification of multi-label data streams. In: Advances in Knowledge Discovery and Data Mining, Lecture Notes in Computer Science. Presented at the Pacific-Asia Conference on Knowledge Discovery and Data Mining, p. 551–62 Springer , Cham (2017)
14. Tuczak, T.: Size and connectivity of the k-core of a random graph, Discrete Mathematics **91**, 61–63 (1991)
15. Michele, B., Lucy, V.: Using n-grams to understand the nature of summaries. In: Daniel, M.S.D., Salim, R., editors, HLT-NAACL 2004: Short Papers, pages 1–2, Boston, Massachusetts, USA. Association for Computational Linguistics (2004)
16. https://www.theguardian.com/world/live/2020/jul/02/coronavirus-live-updates-latest-news-new-zealand-health-minister-quits-who-middle-east-us-trump-brazil
17. Tuczak, T.: Size and connectivity of the k-core of a random graph. Discrete Mathematics **91**, 64–68 (1991)
18. Xing, Y., Shen, F., Luo, C., Zhao, J.: L3-SVM: a lifelong learning method for SVM. In: 2015 International Joint Conference on Neural Networks (IJCNN). Presented at the 2015 International Joint Conference on Neural Networks (IJCNN). Killarney, pp. 1–8 (2015)
19. Bunke, H.: Error-tolerant graph matching: a formal framework and algorithms. Adv. Pattern Recognition, LNCS **1451**, 1–14 (1998)

Transformer Model and Convolutional Neural Networks (CNNs) for Arabic to English Machine Translation

Nouhaila Bensalah[1(✉)], Habib Ayad[1], Abdellah Adib[1], and Abdelhamid Ibn El Farouk[2]

[1] Team Networks, Telecoms and Multimedia, University of Hassan II Casablanca, Casablanca 20000, Morocco
nouhaila.bensalah@etu.fstm.ac.ma, abdellah.adib@fstm.ac.ma

[2] Teaching, Languages and Cultures Laboratory Mohammedia, Casablanca, Morocco

Abstract. Arabic remains one of the richest languages in the world in terms of vocabulary with a large set of morphological features and relatively few resources compared to English. Given the challenge posed by the morphological richness of this language, Arabic Natural Language Processing (NLP) tasks like Named Entity Recognition (NER), Sentiment Analysis (SA), Question Answering (QA) and Machine Translation (MT) have proved to be very challenging to handle. Recently, the transformers based models, have proved to be very effective in terms of language understanding, and have obtained state-of-the-art results for many NLP tasks and in particular MT. In this paper, we proposed a novel Deep Learning architecture based on the Convolutional Neural Networks (CNNs) and the transformer model to further improve the results obtained on the Arabic-English Neural Machine Translation task. Moreover, a special preprocessing of the Arabic sentences based mainly on Farasa and AraBERT is carried out. Experiments on the UN Arabic-English datasets show that our approach outperforms the state-of-the-art Arabic MT systems.

Keywords: Arabic machine translation · Arabic preprocessing · Arabert · Farasa · Transfomers · CNNs

1 Introduction

Neural Machine Translation (NMT) models have advanced the previous state-of-the-art by learning mappings between natural languages using neural networks and attention mechanisms [8,29]. Early MT systems read and generate word sequences using a series of Recurrent Neural Networks (RNNs), Long Short-Term Memory (LSTM) [19] or Gated Recurrent Unit (GRU) [13] units. However, given the sequential nature of these units, it has been proved that it is difficult to fully take advantage of modern computing devices such as Graphics Processing Units (GPUs) or Tensor Processing Units (TPUs) which rely on parallel processing.

M. Lazaar et al. (Eds.): BDIoT 2021, LNNS 489, pp. 399–410, 2022.
https://doi.org/10.1007/978-3-031-07969-6_30

Since then, numerous attempts have continued to push the boundaries of recurrent language models and encoder-decoder architectures [20,22]. Recently, the transformer model has achieved state-of-the-art translation performance and was conceived to solve these problems [31]. Briefly, it is an encoder-decoder model which is relying entirely on self-attention to compute representations of the input and the output sequences without using RNNs.

Arabic is among the most commonly spoken languages worldwide with more than 300 million native speakers. However, the complex morphology of this language which is substantially different from that of Indo-European languages (such as English and French) continues to present challenges to Arabic Word Sense Disambiguation researchers [3]. Morphologically, it is characterized by the following features:

1. There are three types of sentences in Arabic: Nominative, genitive as well as accusative.
2. Almost 85% of Arabic words originate from tri-lateral roots.
3. Pronouns, gender and conjugation may be used in combination in a single word. For example, ولقطتها (waluqutatiha) is one word. However, the prefix و (wa) refers to and, the letter ل (li) refers to the word for, قطة (qita) stands for cat, and the suffix ها (ha) corresponds to the gender pronoun her.
4. The Arabic language could be viewed as a pro-drop language in which the subject of a verb might be indirectly embedded into its morphology. For example, the following sentence "she participated in the conference" can be translated into Arabic as "شاركت في المؤتمر". The subject "she" as well as the verb "participated" are translated into Arabic with the following verbal form – > "شاركت". Finally, "in the conference" is translated as – > "في المؤتمر".

Considering all these intricacies, Arabic MT has turned out to be very tricky to handle. Fortunately, substantial improvement has been achieved as a result of the recent development in the preprocessing and the data-driven translation paradigms [17].
In this paper, we present "The Transformer-CNN", a Deep Learning (DL) model based on the CNNs [21] and the transformer model for Arabic to English MT in the pursuit of achieving the same success that the transformers did for the other European languages. Moreover, different Arabic preprocessing settings are carried out which significantly affect the final translation quality [25]. The rest of this paper is organized as follows. Section 2 provides an overview of some Arabic MT works. Section 3 describes our proposed approach. Section 4 illustrates the experimental settings and the results. Finally, we conclude the paper and mention the future work in Sect. 5.

2 Related Works

The majority of the research that has been carried out on Arabic MT falls within the Statistical MT (SMT) and NMT paradigms. Among these works, Habash and Sadat [15] and Sadat and Habash [27] employed several Arabic morphological pre-processing schemes in order to assess their relevance to the task of Arabic SMT. They observed that the separation of the conjunction and particles (which they named scheme D2) is more efficient when a large amount of training data is utilized. But, with a limited amount of training data, a sophisticated morphological analysis is required to obtain enhanced results. So as to tackle the phenomena of reordering in phrase-based SMT, Chen et al. [12] presented a reordering mechanism for the Arabic-to-English MT task. This method intends to automatically retrieve the reordering rules at the syntactic level within a parallel corpus. This method resulted in very slight enhancement when using the 2004 and 2005 IWSLT Arabic-English evaluation test sets. Hadj Ameur et al. [5] suggested a post-processing scheme for n-best list re-scoring list within the context of the Arabic-English NMT. A set of characteristics covering the syntactic, lexical as well as semantic features of the translation candidates were studied. So as to obtain optimal weights of the embedded features, the Quantum Particle Swarm Optimization (QPSO) algorithm was adopted. They discovered that their algorithm leads to substantial gains of more than 1.5 BLUE points compared to their basic NMT schemes over a variety of Arabic-English in-domain and out-of-domain test sets. In [28], a morphological word incorporation model was introduced by extending the standard Word2vec model by permitting it to accommodate morphological lemmas derived from a language-specific morphological analyzer (MADAMIRA). On the basis of the experimental results, they achieved an augmentation of 0.4 BLUE points when compared to the original Word2vec model. More recently, Oudah et al. [25] reviewed the effectiveness of several segmentation schemes on Neural as well as statistical Arabic-English MT models. The overarching conclusion was that the Penn Arabic Treebank (ATB) schema was found to be advantageous for both NMT and SMT models. Meanwhile, combining ATB with the byte pair coding (BPE) technique yielded the best performance for SMT models but did not lead to an increase compared to ATB segmentation for NMT models.

3 Our Proposed Approach

The Transformer-CNN architecture is based on the transformer and the CNNs. Specifically, we used the encoder part of the transformer which is composed of two blocks A and B; see Fig. 1. The block A consists of a positional encoding to capture the notion of a token position in the input sequence. Without this block, the output of the block B for the sentences "أنا غير سعيدة لأنها مريضة" (I'm not happy because she is sick) and "أنا سعيدة لأنها غير مريضة" (I'm happy because she is not sick) would be the same. The block B is composed of a stack

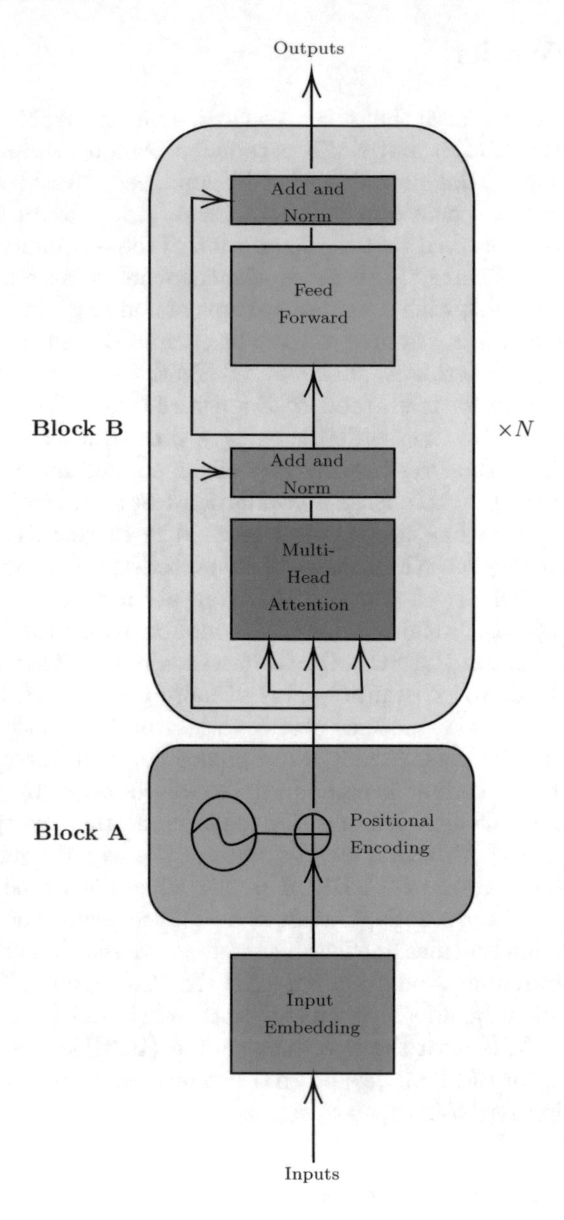

Fig. 1. The transformer-the encoder part

of $N = 20$ identical layers. Each layer has two sub-layers which are: a multi-head self-attention mechanism and a position-wise fully connected feed-forward network. Rather than estimating a single attention function, several attention blocks over the source sentence are measured.

The obtained output values are then combined and projected linearly, leading to the final values. In addition to the multi-head self-attention mechanism, a fully connected feed-forward network is employed. This consists of two linear transformations with a Rectified Linear Unit (ReLU) activation function in between. Moreover, a residual connection [18] around each of the two sub-layers is employed, followed by a normalization layer [7].

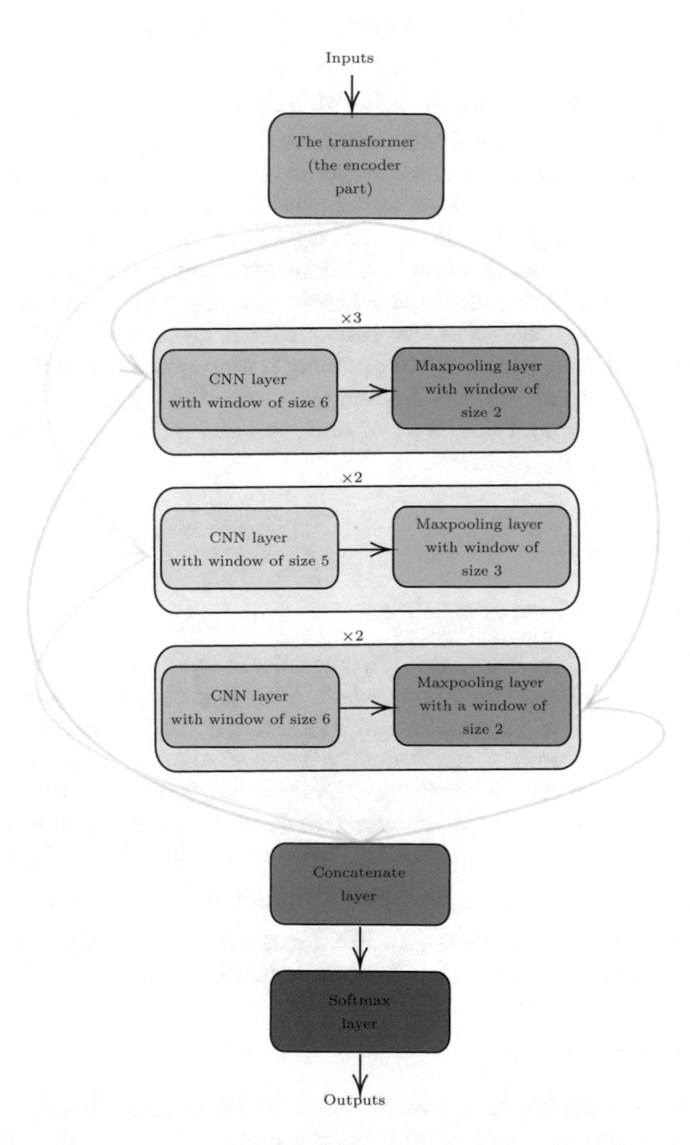

Fig. 2. Schematic illustration of the Transformer-CNN architecture

The one dimensional CNNs are now ubiquitous in various applications such as MT [10,14,23], Question Answering (QA) [9,33] and Sentiment Analysis (SA) [11,24]. Generally, the Convolution1D (Conv1D) process is achieved by sliding a kernel $k(n) \in [1, M]$ along the input data. In this way, the ith output $y_i(n)$ of the feature maps generated by the Conv1D layer can be defined as follows:

$$y_i(n) = \sum_{l=1}^{M} k_i(l)x(n-l), i = 1, .., L \tag{1}$$

where L is the number of convolutional filters $k_i(n)$.

In order to capture several features from the same input sentence, the outputs of the transformer model are used as input of three multi-layer stacked CNN networks. The first one consists of three convolutional layers with a window of size six. Then, a Max pooling layer with a pooling length of two is added after each convolution. The second one is built by stacking two convolutional layers with a kernel size of five followed by a Max pooling with a pooling length of three. The third one consists of two convolutional layers with a window of size six. As the other CNN networks, Max pooling is employed after each convolution with a pooling length of two. Finally, the outputs of the three CNN networks are concatenated. The overall flowchart of our proposed algorithm is schematically illustrated in Fig. 2.

4 Experiments

4.1 Dataset and Preprocessing

Arabic MT research has been fueled by the release of numerous datasets such as Tanzil [30], QED [2] and Arab-Acquis [16]. We evaluate our proposed approach on UN dataset [30], which consists of over 73,000 Arabic- English sentences. In the preprocessing step, digits, punctuation and symbols were removed from the dataset. For the Arabic sentences, we normalize some letters such as the Alif Maqsura (ى) and the ta marbouta (ة) which are replaced with Ya (ي) and Ha (ه), respectively. Then, we segment the words using Farasa [1] into stems, prefixes and suffixes. For example (المدرسة- Almadrasa), which means the school, becomes (ال+مدرس+ة- Al+madras+a). Finally, the AraBERT tokenizer [6] is used to decompose each sentence into a list of tokens (Table 1).

4.2 Hyperparameters

Hyperparameters are important for DL algorithms as they directly control the behaviors of training algorithms and have a major impact on the performance of DL models. In this study, the proposed model parameters were optimized by RMSprop. Then, we used the LearningRateScheduler callback to adjust the learning rate dynamically. In this work, we found that the optimal learning

Table 1. Cleaning, normalization and segmentation applied to an example

Setting	Example
Original	وإذ تشيد بالتطورات الإيجابية الجارية في شبه الجزيرة الكورية في أعقاب اجتماع قمة الكوريتين
Normalization	واذ تشيد بالتطورات الإيجابيه الجاريه في شبه الجزيره الكوريه في اعقاب اجتماع قمه الكوريتين
Segmentation	و+اذ تشيد ب+ال+تطور+ات ال+إيجابي+ه ال+جاري+ه في شبه ال+جزير+ه ال+كوري+ه في أعقاب اجتماع قم+ه ال+كوريتين

rate using this technique is approximately 10^{-4}. Moreover, the early stopping technique [32] based on the validation error was used to avoid overtraining. The details of the proposed model hyperparameters are illustrated in Table 2.

Table 2. The details of the proposed model hyperparameters

Hyperparameter	Values
Validation size/ Test size	0.2/ 0.1
Embedding size	200
Hidden units (for the Transformer)	512
Encoder depth (for the Transformer)	20
The maximum sentence length	30

4.3 Experimental Details and Results

In order to assess the effectiveness of our proposed approach, investigations were performed on a subset of the online available UN dataset. Hence, we tried the following combinations:

1. Gated Recurrent Unit (GRU) (instead of the Transformer model discussed in 3) + The Arabic preprocessing (discussed in 4.1).
2. Long Short Term Memory (LSTM) (instead of the Transformer model discussed in 3) + The Arabic preprocessing (discussed in 4.1).
3. Bidirectional GRU (instead of the Transformer model discussed in 3) + The Arabic preprocessing (discussed in 4.1).

4. Bidirectional LSTM (instead of the Transformer model discussed in 3) + The Arabic preprocessing (discussed in 4.1).
5. Our approach without the Arabic preprocessing (discussed in 4.1).

As an evaluation metric, we used the BLEU score [26] which is the most commonly used in MT task. Table 3 illustrates the experimental results of applying our approach on Arabic-English MT, compared to different combinations of approaches and the sate-of-the-art-works.

Table 3. Performance of the proposed approach on Arabic MT compared to different combinations of approaches and the sate-of-the-art-works

	BLEU score (%)
GRU + The Arabic preprocessing	57.11
LSTM + The Arabic preprocessing	57.58
Bidirectional GRU + The Arabic preprocessing	58.50
Bidirectional LSTM + The Arabic preprocessing	60.40
Our approach without the Arabic preprocessing	67.40
[25]	42.38
[4]	41.14
[10]	60.40
Our approach	**68.60**

As expected, adding features computed by the Transformer model improves the results over the RNN (LSTM/GRU/Bidirectionnal LSTM/Bidirectional GRU) performance. The best performance was achieved when we used both the Transformer-CNN and the Arabic preprocessing based on AraBERT and Farasa. In comparing the obtained BLEU scores for Arabic MT on UN database with that of literature works, it was found that our model has the advantage over the other works. It improves the BLEU score by approximately 25%, 27% and 8% compared to the BLEU scores obtained by Oudah et al. [25], Alrajeh [4] and Bensalah et al. [10]; respectively. In order to further evaluate the effect of using our model on the Arabic-English translation task, four Arabic sentences were translated using the above DL models. From the results illustrated in Table 4, we can clearly remark that the four models translate the first three Arabic sentences fluently. For the fourth translation sample, we can observe that the two first models did not capture the semantic meaning of the source sentence. Moreover, the third model misunderstand the word الى (to), which leads to an inaccurate translation result. Instead, our model was able to generate an appropriate translation result. These findings may be explained by the fact that the Arabic preprocessing can deal with the out-of-vocabulary problem. Furthermore, the Transformer-CNN model uses a self-attention mechanism that immediately extracts the relationships between all the words in a given input sentence, regardless of their respective position. The proposed model was able to preserve long-range dependencies in a sentence, encode its regional (n-gram) information and hence achieve state-of-the-art Arabic translation performance.

Table 4. A few examples of translations generated by our model and the sate-of-the-art approaches

Source	إن الجمعية العامة
Oudah et al. [25]	The General Assembly
Alrajeh [4]	The General Assembly
Bensalah et al. [10]	The General Assembly
Our model	The General Assembly
Truth	The General Assembly
Source	لاحظ مع بالغ القلق
Oudah et al. [25]	notice with great concern
Alrajeh [4]	notice with great concern
Bensalah et al. [10]	notice with great concern
Our model	notice with great concern
Truth	notice with great concern
Source	تقرير المحكمة الجنائية الدولية
Oudah et al. [25]	report of international criminal court
Alrajeh [4]	report of international criminal court
Bensalah et al. [10]	report of the international criminal court
Our model	report of the international criminal court
Truth	report of the international criminal court
Source	تقديم المساعدة إلى اللاجئين والعائدين والمشردين في أفريقيا
Oudah et al. [25]	providing aid to persons in Africa
Alrajeh [4]	providing assistance to and persons in Africa
Bensalah et al. [10]	Providing assistance in Africa as returnees returnees and displaced persons
Our model	providing assistance for refugees returnees and displaced people in Africa
Truth	providing assistance to refugees returnees and displaced persons in Africa

5 Conclusion

In this paper, we proposed a new neural network architecture, called the Transformer-CNN, which is able to preserve long-range dependencies in a given input sentence, capture its regional information and solve the issue of data sparsity. We evaluated the proposed model on the task of Arabic to English MT using UN dataset. Quantitatively and qualitatively, we were able to show that the new model achieves state-of-the-art Arabic-English translation performance. As future work, we plan to use transfer learning to further enhance the translation quality for the low-resource and morphologically rich languages, especially the Moroccan language.

References

1. Abdelali, A., Darwish, K., Durrani, N., Mubarak, H.: Farasa: a fast and furious segmenter for arabic. In: Proceedings of the 2016 Conference of the North American Chapter of the Association for Computational Linguistics: Demonstrations, pp. 11–16 (2016)
2. Abdelali, A., Guzmán, F., Sajjad, H., Vogel, S.: The AMARA corpus: building parallel language resources for the educational domain. In: Proceedings of the Ninth International Conference on Language Resources and Evaluation, LREC, pp. 1856–1862 (2014)
3. Alqudsi, A., Omar, N., Shaker, K.: Arabic machine translation: a survey. Artifi. Intell. Rev. **42**(4), 549–572 (2012). https://doi.org/10.1007/s10462-012-9351-1
4. Alrajeh, A.: A recipe for Arabic-English neural machine translation. CoRR abs/1808.06116 (2018)
5. Hadj Ameur, M.S., Guessoum, A., Meziane, F.: Improving Arabic neural machine translation via n-best list re-ranking. Mach. Translation **33**(4), 279–314 (2019). https://doi.org/10.1007/s10590-019-09237-6
6. Antoun, W., Baly, F., Hajj, H.M.: Arabert: transformer-based model for arabic language understanding. CoRR abs/2003.00104 (2020)
7. Ba, L.J., Kiros, J.R., Hinton, G.E.: Layer normalization. CoRR abs/1607.06450 (2016)
8. Bahdanau, D., Cho, K., Bengio, Y.: Neural machine translation by jointly learning to align and translate. In: 3rd International Conference on Learning Representations, ICLR 2015, San Diego, CA, USA, May 7–9, 2015, Conference Track Proceedings (2015)
9. Bensalah, N., Ayad, H., Adib, A., Ibn el farouk, A.: Combining word and character embeddings for Arabic chatbots. In: Kacprzyk, J., Balas, V.E., Ezziyyani, M. (eds.) AI2SD 2020. AISC, vol. 1417, pp. 571–578. Springer, Cham (2022). https://doi.org/10.1007/978-3-030-90633-7_48
10. Bensalah, N., Ayad, H., Adib, A., El Farouk, A.I.: Arabic machine translation based on the combination of word embedding techniques. In: Gherabi, N., Kacprzyk, J. (eds.) Intelligent Systems in Big Data, Semantic Web and Machine Learning. AISC, vol. 1344, pp. 247–260. Springer, Cham (2021). https://doi.org/10.1007/978-3-030-72588-4_17
11. Bensalah, N., Ayad, H., Adib, A., Farouk, A.I.E.: Arabic sentiment analysis based on 1-D convolutional neural network. In: International Conference on Smart City Applications, SCA20, Safranbolu, Turkey(2020)
12. Chen, B., Cettolo, M., Federico, M.R.: Rules for phrase-based statistical machine translation. In: International workshop on spoken language translation. IWSLT 2006, pp. 182–189 (2006)
13. Chung, J., Gülçehre, Ç., Cho, K., Bengio, Y.: Empirical evaluation of gated recurrent neural networks on sequence modeling. CoRR abs/1412.3555 (2014)

14. Gehring, J., Auli, M., Grangier, D., Dauphin, Y.N.: A convolutional encoder model for neural machine translation. In: Barzilay, R., Kan, M. (eds.) Proceedings of the 55th Annual Meeting of the Association for Computational Linguistics, ACL, pp. 123–135 (2017)
15. Habash, N., Sadat, F.: Arabic preprocessing schemes for statistical machine translation. In: Human Language Technology Conference of the North American Chapter of the Association of Computational Linguistics (2006)
16. Habash, N., Zalmout, N., Taji, D., Hoang, H., Alzate, M.: A parallel corpus for evaluating machine translation between Arabic and European languages. In: Proceedings of the 15th Conference of the European Chapter of the Association for Computational Linguistics, Vol. 2, Short Papers, pp. 235–241 (2017)
17. Hadj Ameur, M., Meziane, F., Guessoum, A.: Arabic machine translation: a survey of the latest trends and challenges. Comput. Sci. Rev. **38**, 22 (2020)
18. He, K., Zhang, X., Ren, S., Sun, J.: Deep residual learning for image recognition. In: 2016 IEEE Conference on Computer Vision and Pattern Recognition, CVPR 2016, Las Vegas, NV, USA, June 27-30, 2016. IEEE Computer Society, pp. 770–778 (2016)
19. Hochreiter, S., Schmidhuber, J.: Long short-term memory. Neural Comput. **9**(8), 1735–1780 (1997)
20. Józefowicz, R., Vinyals, O., Schuster, M., Shazeer, N., Wu, Y.: Exploring the limits of language modeling. CoRR abs/1602.02410 (2016)
21. LeCun, Y., Bottou, L., Bengio, Y., Haffner, P.: Gradient-based learning applied to document recognition. In: Proceedings of the IEEE (1998)
22. Luong, M., Pham, H., Manning, C.D.: Effective approaches to attention-based neural machine translation. CoRR abs/1508.04025 (2015)
23. Meng, F., Lu, Z., Wang, M., Li, H., Jiang, W., Liu, Q.: Encoding source language with convolutional neural network for machine translation. In: Proceedings of the 53rd Annual Meeting of the Association for Computational Linguistics and the 7th International Joint Conference on Natural Language Processing of the Asian Federation of Natural Language Processing, ACL, pp. 20–30 (2015)
24. Ombabi, A.H., Ouarda, W., Alimi, A.M.: Deep learning CNN–LSTM framework for Arabic sentiment analysis using textual information shared in social networks. Soc. Netw. Anal. Min. **10**(1), 1–13 (2020). https://doi.org/10.1007/s13278-020-00668-1
25. Oudah, M., Almahairi, A., Habash, N.: The impact of preprocessing on Arabic-English statistical and neural machine translation. In: Proceedings of Machine Translation Summit XVII vol. 1: Research Track, MTSummit, pp. 214–221 (2019)
26. Papineni, K., Roukos, S., Ward, T., Zhu, W.-J.: Bleu: a method for automatic evaluation of machine translation. In: Proceedings of the 40th Annual Meeting on Association for Computational Linguistics, Association for Computational Linguistics, pp. 311–318 (2002)
27. Sadat, F., Habash, N.: Combination of Arabic preprocessing schemes for statistical machine translation. In: ACL 2006, 21st International Conference on Computational Linguistics (2006)
28. Shapiro, P., Duh, K.: Morphological word embeddings for Arabic neural machine translation in low-resource settings. In: Proceedings of the Second Workshop on Subword/Character LEvel Models, pp. 1–11 (2018)
29. Sutskever, I., Vinyals, O., Le, Q.V.: Sequence to sequence learning with neural networks. In: Advances in Neural Information Processing Systems 27: Annual Conference on Neural Information Processing Systems 2014, Montreal, Quebec, Canada, 8–13 December 2014, pp. 3104–3112 (2014)

30. Tiedemann, J.: Parallel data, tools and interfaces in OPUS. In: Proceedings of the Eighth International Conference on Language Resources and Evaluation, LREC, pp. 2214–2218 (2012)
31. Vaswani, A., Shazeer, N., Parmar, N.: Attention is all you need. In: NIPS (2017)
32. Yao, Y., Rosasco, L., Caponnetto, A.: On early stopping in gradient descent learning. Constructive Approximation **26**, 289–315 (2007)
33. Zhang, G., Fan, X., Jin, C., Wu, M.: Open-domain document-based automatic QA models based on CNN and attention mechanism. In: 2019 IEEE International Conference on Big Knowledge, ICBK, pp. 326–332 (2019)

An Overview of Word Embedding Models Evaluation for Arabic Sentiment Analysis

Youssra Zahidi[1(✉)], Yacine El Younoussi[1], and Yassine Al-Amrani[2]

[1] Information System and Software Engineering Laboratory, Abdelmalek Essaadi University, Tetuan, Morocco
{youssra.zahidi-etu,yacine.elyounoussi}@uae.ac.ma

[2] Information Technology and Modeling Systems (TIMS) Research Team, Abdelmalek Essaadi University, Tetuan, Morocco
yassine.alamrani@uae.ac.ma

Abstract. Sentiment Analysis (SA) or opinion mining has become one of the essential research fields whose application is clearly visible in a large variety of domains. Due to Arabic structure complexities at the level of morphology, orthography, and dialects, manual feature extraction is a time-consuming and challenging mission, Besides, Arabic Sentiment analysis (ASA) is very difficult and is considered a more difficult task compared to other languages.

In this paper, we discuss the issue of word embedding models that is very important in the field of Arabic Sentiment Analysis also provide a comparative analysis of the most famous word embedding models namely: Word2Vec, Glove, FastText, Elmo, and Bert. We found through this deep evaluation, that FaxtText, Elmo, and Bert embedding models outperform other Models.

Keywords: Arabic Sentiment Analysis · Word Embeddings · Word2Vec · Glove · Elmo · Bert · FastText

1 Introduction

A large variety of users express their feelings about multiple issues in Social media platforms and a very high percentage of the world's data is not unstructured in a predefined way. For this reason, there is a very big need to classify these comments. SA Systems is a great manner to resolve this topic [1], where the goal is to identify the feeling conveyed in some text.

As part of our ASA task, we aim to present a comparative study of various existing Word Embeddings techniques to reach a general view on the most valuable ones in ASA field.

Word embeddings are an alternative for hand-crafted features in ASA because Manual feature extraction is a time-consuming and very difficult task, especially in the Arabic language. Due to its complexities, Arabic is considered as the most difficult languages compared to the English language for example. This structure together with the various Arabic dialects and the scarcity of its resources make a large variety of challenges in

M. Lazaar et al. (Eds.): BDIoT 2021, LNNS 489, pp. 411–427, 2022.
https://doi.org/10.1007/978-3-031-07969-6_31

the ASA research field. Selecting the most suitable set of Word Embeddings techniques that satisfies our needs is very challenging work and requires a deep study. To resolve this issue, we base on a large group of good aspects in this deep evaluation.

This deep study of the techniques is very important in that it would enable Arabic Sentiment Analysis researchers to choose the suitable set of techniques in their projects to make the right decisions.

The rest of this work is divided as follows: Sect. 2 gives the importance of Sentiment Analysis and various Sentiment Analysis Levels together with a deep description of the major feature types adopted in this field of research. Section 3 outlines word embeddings in Arabic Sentiment Analysis with a comparative analysis of these Word Embeddings models. In Sect. 4, the comparison between Word Embeddings techniques is discussed in detail and the paper is concluded with final thoughts in Sect. 5.

2 Arabic Sentiment Analysis

Arabic Sentiment analysis [2] faces various challenges problems such as:

- Big differences in informal written Arabic: Most Arabic speakers don't care much about grammar when writing Arabic informally online (online service reviews, forum posts, tweets) while Modern Standard Arabic has clear grammatical rules. Users commonly make spelling mistakes. utilizing various versions of the same character. Moreover, some may use diacritics, others may not…
- Lack of Resources: Arabic Language, suffer from a scarcity of resources, research, and support while there is a growing number of tools available for sentiment analysis in other languages like English for example. Besides, Large Arabic datasets are few. Furthermore, many of the available software used for sentiment analysis either provide limited support of Arabic letters or do not support them at all.
- Large variety of Dialectal Arabic: Most Arabic speakers in their online posts may use various dialects which use various words that have the same meaning and also, they may use many slang words rather than the use of Modern Standardized Arabic in informal settings.

2.1 Arabic Sentiment Analysis Levels

Arabic Sentiment Analysis [3] can be explored fundamentally at three granularity levels: sentence level, document level and aspect level.

- Document Level Sentiment Analysis: [4, 5] is to assign an overall polarity or sentiment orientation to an opinion document. It relies on the whole document as the major unit of information and supposes that the document is defined to be opinionated. In this type of document level classification, each document defines an opinion on a single structure about a single topic and has only single opinion holder. However, in the case of blogs or forums, customers may compare one product with the other that has the same characteristics. for this reason, document level analysis inappropriate for blogs and forums.

- Sentence Level Sentiment Analysis: aims to identify positive, neutral or negative sentiment orientation form multiple sentences in a document. This level of sentiment analysis is associated with subjectivity classification.

  Here, the sentiment orientation of each sentence is concluded and calculated and then same document level classification methods are used for the sentence level classification problem. Then the objective and subjective sentences must be discovered.

  The subjective sentences must include opinion words which allow to find the sentiment about entity. After that the sentiment classification or orientation is done into three-class negative, positive, and neutral classes.
- Aspect Level Sentiment Analysis: [6] performs finer-grained analysis. Its purpose is to find out the sentiment on entities. The two key tasks in this level of sentiment analysis are aspect sentiment classification and aspect extraction. In the first one, the attributes are identified. In the second, the sentiment orientation of large variety of aspects are selected. A positive sentiment on an object can be positive on just an object attribute, but not on the object as a whole.

2.2 Arabic Sentiment Analysis Feature Types

Various Arabic Sentiment Analysis works in general, based on one of two feature types: 1. hand-crafted features that are generated with the morphological analyzers assistance, lexical resources and Natural Language Processing techniques or 2. text embedding features which are learned automatically from the text itself through supervised/unsupervised learning strategies, such that the usage of words besides the words' semantic, synonymous or syntactic relations can be incorporated within the produced features: The following tables: Table 1 express various works that adopt the hand-crafted features and Table 2 shows different papers that adopt the Text embedding feature-based models.

3 Word Embeddings in Arabic Sentiment Analysis

Word Embedding emerged from the domain of NLP which is an intersection of Artificial Intelligence, Computer Science, computational linguistics and Machine Learning. Word embedding is a powerful text mining technique of setting up connection between words in textual data (Corpus). The semantic and syntactic meanings of words are realized from the context in which they are used. The distributional hypothesis concept proposes that words occurring in the same context are semantically similar. Word embedding relies on two broad approaches prediction-based embeddings and Count based embeddings.

The use of word embedding models has become one of the most significant advancements in the ASA field because it is considered as the major feature source for opinion mining in Arabic text like consumer reviews, news articles, and tweets. Manual feature extraction is a time consuming and very difficult task, especially in the Arabic language. Most systems of Arabic sentiment analysis still base on costly hand-crafted features, where features representation requires manual pre-processing for getting the desired accuracy. Word embedding models are an alternative for such hand-crafted features in Arabic sentiment analysis. where Arabic words or phrases from the vocabulary

Table 1. Hand-crafted features-based models.

	ASA Works	Description
Hand-crafted features-based models	[7–10]	The most common features employed in hand-crafted feature-based models are bag-of-words and n-gram schemes in addition to a large variety of stylistic/linguistic features usually extracted by morphological analyzers
	[7]	A comparison between supervised and lexicon-based SA models was performed by this work. Both models were applied to MSA/ Jordanian tweets. Light stemming and Root stemming were proceeded to increase the lexicon size in the lexicon-based model and to enrich the bag-of-words features in the supervised model and. The models were evaluated using the ArTwitter dataset. The experimental evaluation indicated the outperformance of the supervised model as it gained an accuracy of 87.2% using SVM, compared to 59.6% scored by the lexicon-based model
	[8]	Numerous combinations of n-gram features were employed in this work to execute SA on MSA/Tunisian tweets. The produced features were reduced utilizing the information gain (IG) method which can elicit the most important features with respect to the class attribute. Six various classifiers like Naïve Bayes (NB), K-nearest neighbor (KNN), support vector machines (SVMs), decision trees (DTs), and random fields (RF) were trained with the selected features. Two classification types were adopted: binary classification for positive/negative tweets and multiclass classification with neutral tweets considered. The best performance for binary classification was achieved by SVM with an F-measure of 63%.and an accuracy of 71.1%
	[9]	High-level n-grams were employed to classify the sentiment of Jordanian/MSA tweets. With NB and SVM used as classifiers, the authors investigated the impact of stemming/ light stemming on the performance of the classification. It was noted that for bigram training features, which were optimized and reduced by TF-IDF and stemming, SVM was the best performing classifier, as it outperformed NB with an accuracy of 88.7% compared to 81.4%
	[10]	This work aims to rely on the most informative n-gram features, weighting schemes like term frequency/term frequency-inverse document frequency (TF/TF-IDF) are usually employed. This was examined in this work, where TF/TF-IDF was used along with combinations of n-gram features to train NB and SVM supervised classifiers. Features reduction was conducted using the score 'Between Group to within Group Sum of Squares' (BSS/WSS) strategy. For evaluating the suggested model, it was applied to an MSA/Moroccan dataset of Facebook comments. The best performance gained when unigrams and bigrams were both adopted with TF-IDF weighting scheme, gaining an accuracy of 78%
	[11]	Authors in this work produced linguistic features to mine the sentiment of Tunisian customers' reviews within a lexicon-based SA framework. To create the linguistic features, the author built a Tunisian Arabic morphological analyzer along with a transliteration machine to handle Arabizi. Besides, a Tunisian version of the SentiWordNet lexicon was employed. The model was applied on a manually collected corpus (TAC) which combined neutral, positive, and negative tweets. The binary classification accuracy of the employed Tunisian lexicon was 72.1% considering negative/positive tweets

Table 2. Text embedding feature-based model.

ASA Works	Embedding model	Datasets	Algorithm	Results
[1]	CBOW	ASTD [12], ArTwitter [7] and QCRI [13]; LABR [14] and MSA news articles [15]	LinearSVC,Rnd.Forest GaussianNB,NuSVC Log.Reg,SGDClassifier	The results display that the CBOW model achieves considerable performance and the best accuracy than the top hand-crafted methods The model's performance was a little superior to [13] in subjectivity classification, while for the polarity classification of the Twitter datasets, the top metric values were scored employing Nu-SVM classifier with an F-measure of 79.6% and an accuracy of 80.2%
[16]	CBOW and Skip-Gram	LABR, ASTD [12] and ArTwitter [7]	CNN, Linear SVM	The results appeared that the presented model performance generally outperformed all the state-of-the-art systems where for the ArTwitter dataset, the accuracy was 85.0%
[17]	Glove, Skip-Gram and CBOW	ASTD [12] and SemEval-2017 [18] datasets	CNN and SVM	The results revealed that employing pre-trained word embeddings led to preferable evaluation measures compared to the baseline systems. In the ASTD dataset, for example, the superior F-measure scored by CNN-ASAWR was 72.14% compared to 62.60% achieved by [12], while for the SemEval dataset, an F-measure of 63% is achieved against 61% scored by the system of [19]
[20]	doc2vec	OCA [21], LABR [14] and a manually annotated Tunisian Sentiment Analysis Corpus (TSAC) obtained from Facebook comments about popular TV shows	SVM, Bernoulli NB (BNB) and multilayer perceptron (MLP) classifiers	The prominent results were scored by MLP classifier when TSAC corpus was solely employed as a training set where it gained an F-measure equals 78% and an accuracy of 78%

(continued)

Table 2. (*continued*)

ASA Works	Embedding model	Datasets	Algorithm	Results
[22]	word2vec techniques: CBOW and skip-gram	ASTD [12] and ArTwitter by [7]	CNN, LSTM	The results showed that updating the word embeddings during learning gained prominent results in most model configurations. Besides, while LSTM outperformed generally CNN, combined LSTMs architecture was the best-performing model gaining an accuracy of 87.2% for the ArTwitter dataset
[23]	fastText and AraVec-Web	The corpus consisted of 5K airline service-related tweets in Arabic	support vector machine classifier for both, aspect detection, and sentiment polarity classification	The results revealed that fastText word embedding model performed slightly better than AraVec-Web
[24]	Word2vec, Fasttext and Glove	Twitter datasets for Sentiment Analysis. large Arabic corpus (OSAC)	GaussianNB, LinearSVC, NuSVC, LogisticRegression, SGD and RandomForest	The results present the efficiency of Fasttext compared to other models
[25]	AraVec and fastText	Arabic Gold Standard Twitter Data for Sentiment Analysis ASTD: Arabic Sentiment Tweets Dataset Twitter Data set for Arabic Sentiment Analysis Twitter corpus of movies in Arabic Language	Gaussian NB, Linear SVC, NuSVC, Logistic regression, SGD, Random forest	The combination of the two models, gains the best results in terms of accuracy, principally with the NuSVC classifier which is a type of SVM
[26]	word2vec, BERT	TSAC dataset and TUNIZI dataset	Convolutional Neural Networks and Bidirectional Long Short-Term Memory	The results offer the outperformance of the M-BERT embeddings over those generated by word2vec in the TUNIZI Dataset the BERT embedding combined with the CNN leads to the remarkable performance with a 93,2% accuracy compared to 92,6% scored by Bi-LSTM in the TSAC dataset

(*continued*)

Table 2. (*continued*)

ASA Works	Embedding model	Datasets	Algorithm	Results
[27]	Glove,FastText and ELMo	The Restaurants dataset	SVM, LSTM, Deep Memory Network (MemNet), Interactive Attention Networks (IAN), Recurrent Attention on Memory (RAM), Transformation Networks (TNet), Content Attention Based Aspect based Sentiment Classification (CABASC), Attention-over-Attention (AOA) Neural Networks, Granular Computing-based Multi-level Interactive Attention networks (GC-MIA), Co-attention based networks, MGAN	Word embedding techniques FastText and ELMo can make better results
[28]	Glove and Elmo	The dataset consists of short 3-turn dialogues between two speakers	SVM, NN, Long Short-term Memory (LSTM), Gated Recurrent Units (GRU)	The best result on the test dataset was gained with ELMo

are mapped to vectors of numbers. Conceptually it involves a mathematical embedding from a space with several dimensions per word to a continuous vector space with a much lower dimension. This allows many researchers to represent sentiment features as dense vectors instead of the conventional sparse representations. According to various sentiments classification experiments, the use of the embeddings in several feature representation for training multiple binary classifiers to identify subjectivity and sentiment in both Dialectal Arabic and Modern Standard Arabic achieve a very high performance compared to existing techniques. Therefore, Arabic sentiment analysis systems that base on word embedding have been very successful in a large number of Natural Language Processing modern tasks, especially in the field of Arabic sentiment Analysis where this technique allows many researchers to improve the classifiers performance and gain a significant result in the field of sentiments classification. The most famous word embedding models in our work, can be mainly divided into five models namely:

3.1 Word2Vec

The Word2Vec model is a prediction-based algorithm. It is a set of related models that are used to build word embeddings. These models are shallow, two-layer neural networks that are trained to reproduce words linguistic contexts. The basic principle the word2vec model is that words existing in the same context are related. That are mean Word vectors are positioned in the vector space such that words that have the same contexts in the corpus are located close to one another in the vector space.

Word2vec is an unsupervised learning algorithm used to generate low-dimensional vector representations of words. It accepts as its input a big corpus of text and builds a vector space, typically of a large variety of dimensions, with each unique word in the corpus being assigned a corresponding vector in the space.

Word2vec has two model architectures in order to build a distributed representation of words: continuous skip-gram and continuous bag-of-words (CBOW). The first model architecture relies on the current word to predict the window of surrounding context words. The skip-gram architecture weighs nearby context words more heavily than more distant context words.

The second model architecture predicts the current word from the surrounding window of context words. The order of context words does not impact prediction (bag-of-words assumption). Continuous bag-of-words (CBOW) is very faste compared to skip-gram that is slower but does a better job for infrequent words.

3.2 Global Vectors (GloVe)

Global Vectors (GloVe) is also an unsupervised learning algorithm for getting low-dimensional distributed representations for words. This is done by mapping words into a meaningful space where the distance between words is relied on semantic similarity. Training is performed on aggregated global word-word co-occurrence statistics from a corpus, and the resulting representations exhibit good linear substructures of the word vector space. It is created as an open-source project at Stanford. As log-bilinear regression model for unsupervised learning of word representations, it joins the features of two model families, namely the local context window and global matrix factorization methods.

Glove lets to identify connection between words such as company product relations, synonyms, zip codes, and cities etc. It is also used by the spaCy model to create semantic word embeddings vectors while computing the top list words that match with distance measures like Cosine Similarity and Euclidean distance approach.

3.3 FastText

FastText is an open-source library, it is another word embedding method that lets users to learn text representations and text classifiers. It works on standard, generic hardware. Models can later be reduced in size to even fit on mobile devices.

FastText is an extension of the word2vec model. It represents each word as an n-gram of characters instead of learning vectors for words directly. Therefore, take the following word with n = 3, "الاصطناعية", the fastText representation of this word is < اصط, ناع, ية ال, >, where the beginning and end of the word are indicated by using angular brackets.

This allows to identify the shorter words meaning and helps the embeddings to understand prefixes and suffixes. Once the word has been expressed by character n-grams, a skip-gram model is trained to learn the embeddings. This model is considered as a bag of words model with a sliding window over a word because no internal texture of the word is taken into consideration. As long as the characters are within this window, the order of the n-grams doesn't matter.

FastText works very well with rare words. Therefore, even if a word wasn't seen during training step, the word can be broken down into n-grams to get its embeddings.

Word2vec and GloVe both fail to give any vector representation for words that don't exist the model dictionary. This is an important advantage of this method.

3.4 Elmo

ELMo (Embeddings from Language Models): [29] aims to model both: complex characteristics of word use (syntax and semantics), and how these uses vary across linguistic contexts (to model polysemy).

In order to be able to produce contextual word embedding. Elmo uses a bi-directional LSTM trained on a specific topic, ELMo is a deep contextualized word representation that improves language understanding and language modelling. The ELMo LSTM, after being trained on a very big dataset, can then be used as a component in other NLP models that are for language modelling. ELMo aims to predict the next word in a sentence, which is, essentially, what Language Models do. When trained on a massive dataset, the model also starts to pick up on language patterns.

Such models allow you to guess the next word in the example, if you see the sentence like "اكتب ب => I write with a", the word "القلم => the pen" seems to be a more reasonable next word than "المسطرة => the ruler". Elmo uses bi-directional LSTM in training, so that its language model not only understands the next word, but also the previous word in the sentence. It contains a 2-layer bidirectional LSTM backbone.

3.5 Bert

BERT: Bidirectional Encoder Representations from Transformers released in late 2018 [30], it is a simple and powerful method of pretraining language representations that was used to create models that ASA practitioners can then download and use for free. Unlike recent language representation models, BERT is built to pretrain deep bidirectional representations from unlabeled text by jointly conditioning on both right and left context in all layers. Therefore, BERT model can be finetuned with just one additional output layer to create state-of-the-art models for a large variety of tasks, like Sentiment Analysis, without substantial task specific architecture modifications.

BERT is used to extract features (word and sentence embedding vectors) from text data. these word and sentence embedding vectors are used as high-quality feature inputs to downstream models. There are natural language models like CNNs or LSTMs need inputs in the form of numerical vectors, and this means translating features like the vocabulary into numerical representations. previously, words have been expressed either as uniquely indexed values (one-hot encoding), or more usefully as neural word embeddings where vocabulary words are matched against the fixed-length feature embeddings that result from FastText or Word2Vec models for example. BERT provides a great usefulness over models such as Word2Vec, because while each word has a fixed representation under Word2Vec regardless of the context within which the word appears, BERT produces word representations that are dynamically informed by the words around them. Such as, given two poetic verses from Alkhalia poem: "تلاعب في أعطافه التيه والخال" => Vanity and arrogance: الخال.

"وإن لام عمي الطيب الأصل والخال" =>The uncle: الخال

Word2Vec would create the same word embedding for the word "الخال"in both poetic verses, while under BERT the word embedding for "الخال"would be different for each poetic verse. Aside from identifying clear differences such as polysemy, the context-informed word embeddings capture other forms of information that lead to more accurate representations of features, which in turn leads to great model performance.

The following tables: Table 3, Table 4, Table 5, Table 6, and Table 7 highlights the numerous criteria in order to evaluate these Word Embeddings Models. We can deduce throw these tables that each Word Embedding Model has its own characteristics.

Table 3. The characteristics of WORD2Vec embedding model.

Word embeddings model		Created by	Techniques	Advantages	Disadvantages
Word2Vec	Skip Gram	A team of researchers led by Tomas Mikolov at Google. 2013	Skip-Gram and CBOW	• Skip-gram embedding model can catch two semantics for an individual word. For example, it will have two vector representations of "Apple". One for the fruit and the other for the company. • Skip-gram embedding model with negative sub-sampling outperforms generally every other method	• Softmax function is computationally costly. • The time desired for training this algorithm is high
	CBOW			• Being probabilistic in nature, it is assumed to proceed generally superior to deterministic methods. • It does not require to hold large RAM requirements like that of the co-occurrence matrix where it requires to store three massive matrices. Therefore, it is low on memory	• CBOW adopts the context average of a word. For instance, the word "Apple" can be both a company and a fruit but CBOW adopts an average of both the contexts and puts it in between a cluster for companies and fruits. • Training a CBOW from scratch can take forever if not properly optimized

4 Results and Discussion

Arabic Sentiment Analysis, in general, employs one of two feature types: hand-crafted features or text embedding features. In this evaluation, we adopt the second point because

Table 4. The characteristics of Glove Embedding Model.

Word Embeddings Model	Created by	Techniques	Advantages	Disadvantages
Glove	It is developed as an open source project at Stanford University 2014	Co-occurrence	• The goal of the Glove model is very straightforward, which means, to enforce the vectors of a word to catch sub-linear relationships in the vector space. Therefore, in the word analogy tasks, it demonstrates to perform better than the Word2vec model • By considering the relationships between word pair and word pair rather than word and word. Glove embedding model adds some more practical meaning into word vectors. • Glove awards lower weight for highly frequent word pairs so as to block the meaningless stop words such as "an", "the" will not control the training progress	The model is trained on the co-occurrence matrix of words, which takes a lot of memory for storage. Essentially, if you change the hyper-parameters related to the co-occurrence matrix, you have to rebuild the matrix again, which is very time-consumption

Table 5. The characteristics of FastText Embedding Model.

Word Embeddings Model	Created by	Techniques	Advantages	Disadvantages
FastText	Facebook's AI Research 2015	BOW + Subword Information: based on the skip-gram model, each word is represented as a bag of character n-grams	• Sentence Vectors can be facilely computed. • FastText works better on small datasets in comparison to gensim. • FastText fairs equally well in the case of semantic performance and performs superior to gensim in terms of syntactic performance	• This is not a standalone library for NLP since it will demand another library for the steps of pre-processing. • Although, this library has a python implementation. It is not officially supported

Table 6. The characteristics of Elmo Embedding Model.

Word Embeddings Model	Created by	Techniques	Advantages	Disadvantages
Elmo	2018 by AllenNLP	It uses a bi-directional Long Short-Term Memory	• Deep context-based and character-level word representations adapted for more complex tasks. • Shown to perform very better than simple embeddings on many tasks. • Pre-trained version facilely accessible	• Struggles with long-term context dependencies (vs transformers-based models for long sentences). • The complex Bi-LSMT structure makes it very slow to train and produce embeddings

Table 7. The characteristics of Bert Embedding Model.

Word Embeddings Model	Created by	Techniques	Advantages	Disadvantages
Bert	Google's BERT 2018	It uses a bidirectional Transformer	• BERT considers t he completion of each word by observing the previous and subsequent ones, to comprehend the intention of the search queries • BERT is the first unsupervised system for NLP, which is significant because a large variety of plain text is publicly available on the web in many languages	• The major downside of employing BERT and other big neural language models is the computational resources needed to train, fine-tune, and make inferences. However, more modern research has proposed various methods to overcome this issue

of their great results and major advantages in the field of modern Arabic Sentiment Analysis.

We aimed to rely on our review of the literature on the most famous and useful Word Embedding Models, namely: Word2Vec, Glove, FastText, Elmo, and Bert. We can utilize each of them in different ways and scenarios. We attempted to give you a global view of them, and we trust it can help you create the right choice for your issue.

1. FastText, Glove and Word2Vec. The major difference is Glove deals with each word in the corpus such as an atomic entity and creates a vector for each word. For this reason, Glove is very similar to word2vec because both deals with words as the smallest unit to train on.

   Fasttext word embedding model which is basically an extension of the word2vec model. It beats the restrictions of word2vec in consideration of the internal form of the words. This model deals with each word as consisted of character n-grams. Therefore, the vector for a word is made of the sum of this character n-grams. This variance enables Fasttext to:

- Produce better word embeddings for rare words (Their character n-grams are still shared with other words even if words are rare, hence the embeddings can still be perfect).

- Out of vocabulary words (even if the word does not seem in the training corpus, they can create the vector for a word from its character n-grams). Both word2vec and Glove models cannot.

2. Glove and Word2vec word embeddings are context independent, these models output just one embedding (vector) for each word, joining all the various senses of the word into one vector.

   - That is the one numeric representation of a word (vector/embedding) regardless of the different meanings, the words may have and regardless of where they occur in a sentence. For example, after we train Glove/word2vec on a corpus (unsupervised training - no labels needed) we obtain as output one vector representation such as the word "cell". Therefore, even if we had a sentence as "The Doctor went to the jail cell with his cell phone to extract blood cell samples from prisoners", where the word "cell" has various meanings relied on the context of the sentence, these models just collapse them all in their output into one vector for "cell".

3. ELMo and BERT models can produce various word embeddings for a word that captures the word context, that is its placement in a sentence.

   - For example, for the same example above "The Doctor went to the jail cell with his cell phone to extract blood cell samples from prisoners», both BERT and Elmo models would produce various vectors for the three vectors for the word "cell". The first cell (the jail cell case), for example, would be closer to words such as imprisonment, crime penitentiary… whereas the second "cell" (The phone case) would be nearer to words as android, iPhone, galaxy…

The major variance above is a result of the fact Glove and Word2vec models do not take into consideration in their training the word order. However, BERT and ELMo models take into account word order (BERT utilized Transformer and ELMo adopts LSTMS).

A functional implication of this variance is that we can employ Glove and word2vec vectors trained on a very large corpus directly for downstream tasks. So, we need the vectors for the words and there is no need for the model itself that was employed to train these vectors.

However, since the BERT and ELMo models are context-dependent, we need the model that was employed to train the vectors even after training, since the models produce the vectors for a word based on the context. We can just utilize the context-independent vectors for a word if we choose too (just obtain the raw trained vector from the trained model) but would overcome the very purpose/advantage of these present models.

There is a key variance between the manner BERT produces its embeddings and all the other models - Glove, Word2vec, FastText and ELMo:

- Word2vec and Glove are word-based models that means the models adopt as input words and output word embeddings.

- The FastText embedding model exploits subword information to build word embeddings. Representations are learned of character n-grams, and words represented as the n-gram vectors sum. This expands the word2vec type models with subword information. This aids the embeddings understand prefixes and suffixes. A skip-gram model is trained to learn the embeddings once a word is represented using character n-grams.
- Elmo embedding model in contrast is a character-based model utilizing character convolutions and can handle out of vocabulary words for this purpose. The learned representations are words in this case.
- BERT embedding model shows input as subwords and learns embeddings for subwords. Therefore, it owns a vocabulary that is about 30,000 for a model trained a corpus with a very big number of unique words (almost millions), which is very smaller in contrast to a Word2vec, Glove, or Elmo model trained on the same corpus. Showing input as sub words as opposed to words has become the newest trend because it gains a balance between word-based and character-based representations - the most prominent advantage being the avoidance of out of vocabulary (OOV) cases which the other two embedding models (Word2vec and Glove) suffer from.

As a conclusion of this part, These Word Embedding Models are all very useful in the field of Arabic Sentiment Analysis. However, according to our deep evaluation, the literature, and a large variety of powerful and significant works in the domain of Arabic Sentiment Analysis, we conclude that the FastText, Elmo, and Bert embedding models have numerous benefits and great results in the domain of Arabic Sentiment Analysis compared to other embedding models.

5 Conclusion and Future Work

Machine learning and Deep Learning algorithms expect the input data to be in a numerical representation. The Texts can be encoded with a large variety of methods and tools, from simple ones such as the bag of words approach (BOW) and great results can often be accomplished with the TF-IDF method.

More amelioration in recent years was the development of GloVe and Word2Vec embedding models, both based on distributional hypothesis or observation that exist in the same contexts often have comparable senses. FastText embedding model soon followed which allowed producing great vector representations for words that were either rare or did not seem in the training corpus.

ELMo embedding model successfully treated the weakness and shortcoming of the previous models that could not appropriately believe that the meaning of several words depends on the context.

BERT embedding model gives good interest over models like Word2Vec because while each word has a stable representation under Word2Vec regardless of the context during which the word seems, BERT creates word representations that are dynamically informed by the words around them.

In this work, we conducted a deep evaluation and we described a variety of Word Embeddings models which are considered as the most powerful and useful. However, there are other Hand-crafted feature-based models that could be helpful. Besides, we have tried to evaluate the various models using several aspects and multiple criteria.

It can be deduced that each Word Embedding Model has its own benefits and characteristics for the ASA task. However, according to our deep comparative study, the literature, and many great works, we found that FaxtText, Elmo, and Bert outperform other Models.

References

1. Altowayan, A.A., Tao, L.: Word embeddings for Arabic sentiment analysis. In: Proceedings - 2016 IEEE International Conference on Big Data, Big Data 2016, pp. 3820–3825 (2016)
2. Zahidi, Y., El Younoussi, Y., Al-Amrani, Y.: Different valuable tools for Arabic sentiment analysis: a comparative evaluation. Int. J. Electr. Comput. Eng. **11**(1), 753–762 (2021)
3. Zahidi, Y., El Younoussi, Y., Al-Amrani, Y.: A powerful comparison of deep learning frameworks for Arabic sentiment analysis. Int. J. Electr. Comput. Eng. **11**(1), 745–752 (2021)
4. Al Amrani, Y., Lazaar, M., El Kadiri, K.E.: Recovery of the opinions through the specificities of documents text. In: 2019 International Conference on Wireless Technologies, Embedded and Intelligent Systems, WITS 2019 (2019)
5. Al Amrani, Y., Lazaar, M., El Kadiri, K.E.: A novel hybrid classification approach for sentiment analysis of text document. Int. J. Electr. Comput. Eng. **8**(6), 4554-4567 (2018)
6. Ashi, M.M., Siddiqui, M.A., Nadeem, F.: Pre-trained word embeddings for Arabic aspect-based sentiment analysis of airline tweets. In: Hassanien, A.E., Tolba, M.F., Shaalan, K., Azar, A.T. (eds.) AISI 2018. AISC, vol. 845, pp. 241–251. Springer, Cham (2019). https://doi.org/10.1007/978-3-319-99010-1_22
7. Abdulla, N.A., Ahmed, N.A., Shehab, M.A., Al-Ayyoub, M.: Arabic sentiment analysis: Lexicon-based and corpus-based. In: 2013 IEEE Jordan Conference on Applied Electrical Engineering and Computing Technologies, AEECT 2013 (2013)
8. Sayadi, K., Liwicki, M., Ingold, R., Bui, M.: Tunisian dialect and modern standard arabic dataset for sentiment analysis: Tunisian election context. Proc. 17th Int. Conf. Intell. Text Process. Arab. Comput. Linguist., no. March 2014 (2016)
9. Alomari, K.M., ElSherif, H.M., Shaalan, K.: Arabic tweets sentimental analysis using machine learning. In: Benferhat, S., Tabia, K., Ali, M. (eds.) IEA/AIE 2017. LNCS (LNAI), vol. 10350, pp. 602–610. Springer, Cham (2017). https://doi.org/10.1007/978-3-319-60042-0_66
10. Elouardighi, A., Maghfour, M., Hammia, H., Aazi, F.Z.: A machine Learning approach for sentiment analysis in the standard or dialectal Arabic Facebook comments. In: Proceedings of 2017 International Conference of Cloud Computing Technologies and Applications, CloudTech 2017, vol. 2018-Janua, pp. 1–8 (2018)
11. Nadia, K.: Tunisian Arabic Customer's Reviews Processing and Analysis for an Internet Supervision System. PhD dissertation, Sfax University, Tunisia (2017)
12. Nabil, M., Aly, M., Atiya, A.F.: ASTD: Arabic sentiment tweets dataset. In: Conference Proceedings - EMNLP 2015: Conference on Empirical Methods in Natural Language Processing, pp. 2515–2519 (2015)
13. Mourad, A., Darwish, K.: Subjectivity and sentiment analysis of modern standard Arabic and Arabic Microblogs. In: Proceedings of the 4th Workshop on Computational Approaches to Subjectivity, Sentiment and Social Media Analysis, pp. 55–64 (2013)
14. Aly, M., Atiya, A.: LABR: A Large Scale Arabic Book Reviews Dataset. (2013)
15. Banea, C., Mihalcea, R.: Multilingual Subjectivity : Are More Languages Better ? no. August, pp. 28–36 (2010)
16. Dahou, A., Xiong, S., Zhou, J., Haddoud, M.H., Duan, P.: Word Embeddings and Convolutional Neural Network for Arabic Sentiment Classification (2016)

17. Gridach, M., Haddad, H., Mulki, H.: Empirical Evaluation of Word Representations on Arabic Sentiment Analysis (2017)
18. Rosenthal, S., Farra, N., Nakov, P.: SemEval-2017 Task 4: Sentiment analysis in Twitter. In: Proceedings of the 11th International Workshop on Semantic Evaluation (SemEval-2017), pp. 502–518 (2017)
19. El-Beltagy, S.R., El kalamawy, M., Soliman, A.B.: NileTMRG at SemEval-2017 Task 4: Arabic Sentiment Analysis. In: Proceedings of the 11th International Workshop on Semantic Evaluation (SemEval-2017), pp. 790–795 (2017)
20. Medhaffar, S., Bougares, F., Estève, Y., Hadrich-Belguith, L.: Sentiment analysis of tunisian dialects: linguistic ressources and experiments. In: Proceedings of the Third Arabic Natural Language Processing Workshop, pp. 55–61 (2017)
21. Rushdi-Saleh, M., Martín-Valdivia, M.T., Ureña-López, L.A., Perea-Ortega, J.M.: OCA: Opinion corpus for Arabic. Journal of the American Society for Information Science and Technology, John Wiley & Sons, Ltd, pp. 2045–2054 (2011)
22. Al-Azani, S., El-Alfy, E.S.M.: Hybrid deep learning for sentiment polarity determination of Arabic Microblogs. In: Lecture Notes in Computer Science (including subseries Lecture Notes in Artificial Intelligence and Lecture Notes in Bioinformatics), vol. 10635 LNCS, pp. 491–500 (2017)
23. Ashi, M.M., Siddiqui, M.A., Nadeem, F.: Pre-trained Word Embeddings for Arabic Aspect-Based Sentiment Analysis of Airline Tweets Pre-trained Word Embeddings for Arabic Aspect-Based Sentiment Analysis of Airline Tweets (2018)
24. Kaibi, I., Nfaoui, E.H.: A comparative evaluation of word embeddings techniques for twitter sentiment analysis. In: 2019 International Conference Wireless Technololgy Embedded Intelligent System, pp. 1–4 (2019)
25. Kaibi, I., Nfaoui, E.H., Satori, H.: Sentiment analysis approach based on combination of word embedding techniques. In: Bhateja, V., Satapathy, S.C., Satori, H. (eds.) Embedded Systems and Artificial Intelligence. AISC, vol. 1076, pp. 805–813. Springer, Singapore (2020). https://doi.org/10.1007/978-981-15-0947-6_76
26. Messaoudi, A., Haddad, H., Hajhmida, B., Fourati, C., Ben Hamida, A.: Learning Word Representations for Tunisian Sentiment Analysis (2020)
27. Talafha, B., Al-Ayyoub, M., Abuammar, A., Jararweh, Y.: Outperforming state-of-the-art systems for aspect-based sentiment analysis. In: Proceedings of IEEE/ACS International Conference on Computer Systems and Applications, AICCSA, vol. 2019-Novem, pp. 1–5 (2019)
28. Mohammadi, E., Amini, H., Kosseim, L.: Neural feature extraction for contextual emotion detection. In: Proceedings of the International Conference on Recent Advances in Natural Language Processing, RANLP, pp. 785–794 (2019)
29. Barhoumi, A., Camelin, N., Aloulou, C., Estève, Y., Hadrich Belguith, L.: An Empirical Evaluation of Arabic-Specific Embeddings for Sentiment Analysis, pp. 34–48 (2019)
30. Zhang, C., Abdul-Mageed, M.: BERT-Based Arabic Social Media Author Profiling (2019)

A Hybrid Learning Approach for Text Classification Using Natural Language Processing

Iman El Mir[1], Said El Kafhali[2(✉)], and Abdelkrim Haqiq[2]

[1] Hassan First University of Settat, Institute of Sports Sciences, Computer, Networks, Modeling, and Mobility Laboratory (IR2M), B.P. 539, 26000 Settat, Morocco
iman.elmir@uhp.ac.ma

[2] Hassan First University of Settat, Faculty of Sciences and Techniques, Computer, Networks, Modeling, and Mobility Laboratory (IR2M), B.P. 577, 26000 Settat, Morocco
{said.elkafhali,abdelkrim.haqiq}@uhp.ac.ma

Abstract. Text classification and categorization is a hot topic that involves assigning tags or categories to a text based on its content. It is one of the important tasks of automatic natural language processing (NLP) in many applications such as topic tagging, sentiment analysis, intent detection, spam filtering, and email routing. Machine learning text classification can support businesses to automatically analyze and structure their textual documents promptly and inexpensively, to automate processes and improve data-driven decisions. In this article, we propose a new algorithm to classify textual documents using a hybrid approach that combines a set of given algorithms, using the best for each class. These documents can be classified into a set of possible class labels given a priori. Two machine learning algorithms are used to evaluate our proposed approach: Naive Bayesian (NB) and Logistic Regression (LR). The obtained results showed that the proposed hybrid algorithm is more efficient than NB and LR algorithms with an accuracy of 91.86%.

Keywords: Text classification · Machine learning · Natural language processing · Naive Bayesian · Logistic Regression

1 Introduction

The classification and categorization of textual documents is the activity of the classic problem in NLP, which consists in automatically classifying documentary resources, generally, from a corpus [1]. This classification can take an infinite number of forms. We can cite the classification by genre, by theme, or by opinion. The task of classification is performed with specific algorithms, implemented by information processing systems. It is a task of automating a classification process, which most often involves numerical methods. The document classification

M. Lazaar et al. (Eds.): BDIoT 2021, LNNS 489, pp. 428–439, 2022.
https://doi.org/10.1007/978-3-031-07969-6_32

activity is essential in many economic fields: it makes it possible to organize documentary corpus, sort them, and help to exploit them in different sectors such as administration, aeronautics, research on the internet, science, etc. [2]. The text classification algorithms have already been implemented for commercial and academic purposes. They can be mainly arranged in the following phases, namely, feature extraction, dimension reductions, classifier selection, and evaluation. Among the techniques of text classification, we find term weighting methods that conceive appropriate weights to the explicit terms to improve the performance of text classification.

The application and use of NLP help in rapid recognition, text analysis, language translation, natural language understanding, natural language generation, as well as other functions [3]. Usually, it includes two methods, such as statistical NLP and semantic NLP. Machine language is the basis for statistical NLP (including deep neural networks), which has increased the level of accuracy in text classification. Text classification is one of the most typical tasks of supervised machine learning. Assigning categories of text documents, which can be a library book, web page, gallery, media articles, etc. have many applications like sentiment analysis, intent detection, spam filtering, email routing, topic tagging, etc. In the context of text classification, machine learning is of particular importance that estimates an unknown dependence between data and output of the considered system based on available samples for proper classification. A system classification learns to classify new features into predefined discrete issue classes.

One of the most important applications of text classification is sentiment analysis that analyzes people's opinions and feelings, attitudes, and emotions [4]. It is one of the most active research areas in NLP and it is also widely studied in data mining. This research has extended outside of science to management science and social science. The growing importance of sentiment analysis coincides with the growth of social media such as Twitter and social media. For the first time in human history, we have a considerable volume of opinions in digital form for analysis [5]. Sentiment analysis systems are applied in almost all areas of business and society because opinions are central to almost all human activities and are key influencers of our behavior. Our beliefs and perceptions of reality, and the choices we make, are largely conditioned by how others see and value the world. For this reason, when we have to make a decision, we often seek the opinions of others. This is true not only for individuals but also for organizations. Sentiment analysis can determine tone, emotional behaviors, and patterns in documents to assess whether the opinion expressed on a topic is positive or negative using machine learning techniques [6].

There is a generous number of machine learning techniques that are used in the literature for text classification. However, all of them do not have identical accuracy, that is, one may have low accuracy, while others may have higher accuracy than the other. In this paper, the text classification is done using two classical machine learning algorithms that have been implemented on the AG's new topic classification dataset. The two popular classification machine learning used are logistic regression classifier and naive Bayesian classifier. The two used algorithms are playing a significant role in text classification and obtained the best results in different existing research works [7]. In addition to the two algorithms, we have proposed a hybrid algorithm for text classification and compare its performance with the above-mentioned algorithms. The main task here is to assign unlabeled new text documents to predefined classes. The main objective is to present an approach to determine how textual documents can be classified using a hybrid algorithm, based on a simple concept; choose the best algorithm for each class. This approach can improve the accuracy of the classification results.

The remainder of this paper is organized as follows. Section 2 presents some related works in the field of text classification. The text classification process is described in Sect. 3. The classifier models and the proposed hybrid algorithm are presented in Sect. 4. Section 5 discusses the evaluation of the used methods as well as our proposed algorithm for text classification. Finally, Sect. 6 concludes the paper.

2 Related Work

There have been important research papers on the advantages of machine learning algorithm development for text classification. In this section, we present some related works in the field of text classification. Numerous techniques, datasets, and evaluation metrics have been introduced in the literature and provided in multiple contributions. Following the text and models used for feature extraction and classification, the authors [8] proposed a taxonomy for text classification. Indeed, they compared the different existing techniques and developed a detailed study of the technical developments and datasets that enable prediction tests. The authors [9] examine the text classification techniques and deal with different term weighting to compare the different classification techniques using machine learning algorithms. The authors [10] presented the structure and technical implementations of text classification systems in terms of the pipeline. It is mandatory to choose the best classification algorithm for document categorization. The pipeline is divided into two sections predicting the test set and evaluating the model.

Some researchers apply feature selection models for text classification including filter, wrapper, embedded, and hybrid. The authors [11] presented the primary feature selection techniques for text classification. They introduced the Nearest Neighbor method, Naive Bayes, Support Vector Machine, Decision Tree, and Neural Networks as the most common text classifiers to discuss document representation schemes, and similarity between documents. To enhance the classification accuracy, several studies have been done using optimization algorithms. For example, the authors [12] proposed a novel firefly algorithm-based feature selection for Arabic text classification. Consequently, they suggested a new Arabic Text Categorizer system; to validate the proposed feature selection method for enhancing Arabic Text Classification accuracy; they carried out some experiments using real datasets and compare the state-of-the-art techniques with the proposed method. Due to the problem of small samples and high dimensionality for text classification, the evaluation of feature selection methods is still complicated because it involves multiple criteria. To develop a better evaluation method, several criteria must be taken into account. The authors [13] investigated multiple criteria decision-making-based methods to evaluate feature selection methods for text classification with small sample datasets according to classification performance, stability and efficiency.

In other words, the authors [7] presented a BBC news text classification model based on machine learning algorithms. Indeed, they proposed logistic regression, random forest, and K-nearest neighbor algorithms that describe each aspect of the model in detail providing the evaluation metrics. The accuracy remains the most important parameter that must be taken into consideration when machine learning algorithms are implemented on a particular data set. The obtained experimental results prove the efficiency of the BBC news text classification model according to the algorithms tested on the data set. The authors [1] studied some deep learning models for Arabic news classification. To avoid the need for the pre-processing phase, they applied deep learning hence to produce a highly accurate and robust classifier for Arabic news articles. Authors in [14], have proposed to classify text documents within a certain hierarchy. However, deep neural models remain difficult to apply and define the levels of documents hierarchically because they need a large amount of training data. The authors [15] proposed a novel Text Classification Framework called SS3, whose aim is to provide support for incremental classification, early classification for simple and effective early depression detection.

Long Short-Term Memory (LSTM) is the most common architecture, which is designed to catch long term dependencies. The LSTM network is applied in each language to model the documents [16]. The classification is performed by using a hierarchical attention mechanism, where the sentence-level attention model determines which sentences of a document are more significant for giving the overall sentiment. While the word-level attention model elaborates which words in each sentence are decisive. Comparing Bidirectional Encoder Representations with Transformer (BERT) Bidirectional Encoder Representations with directional models like RNN and LSTM, it is noted that these latter sequentially consider each input while BERT is nondirectional and read the whole sentence as the input instead of sequential processing. BERT remains more efficient in terms of results and learning speed [17]. Once pretrained, in an unsupervised manner, it has its linguistic representation. It is then possible, based on this initial representation, to customize it for a particular task. It can be trained in an incremental mode in a supervised fashion to specialize the model quickly and with little data. As, it can work in a multimodel way, taking as input data of different types like images, text. Finally, the authors [18] used the deep learning technique to deal with the sentiment classification. Indeed, they proposed convolutional neural networks for sentiment classification. They tackle the sentiment analysis that classifies a relatively longer text into one of the sentiment categories.

3 Text Classification Process

The objective of the so-called classical Machine Learning is to give a machine the ability to learn to solve a problem without having to explicitly program each rule. The idea of machine learning is therefore to solve problems by modeling behavior through data-driven learning. However, before being able to model a problem through a machine-learning algorithm, it is often necessary to perform several transformations on the data. These transformations, which are done manually, are dictated by the business problem we are trying to solve, and by the choice of the algorithm used. This data processing, called feature engineering, is often very time consuming and may require business expertise to be relevant. We can consider these data transformations as the construction of a representation of the problem that can be easily understood and interpreted by the machine learning algorithm.

In text classification process, at the start, text documents are read from the collection, then preprocessing like tokenization, stop word elimination, stemming. After that, text representation in a form so that learning algorithms can be applied. Finally, important features are selected from the feature vector to remove the irrelevant features and minimize the dimensionality of feature space. To do that, we should follow the steps as illustrated in the flow diagram of Fig. 1.

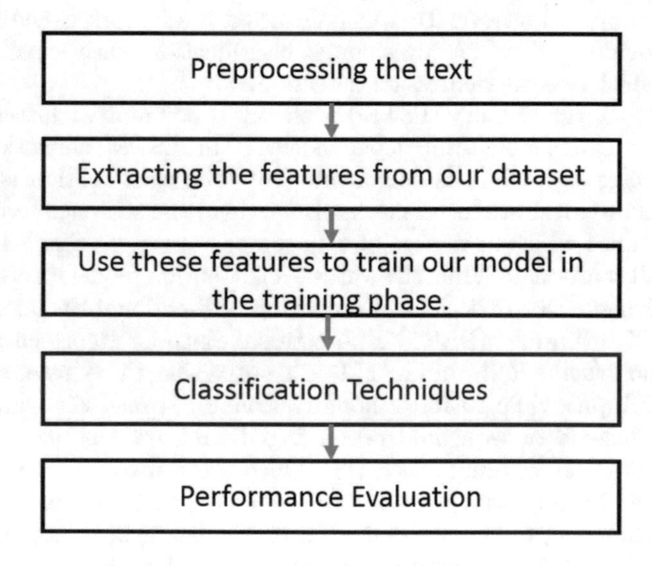

Fig. 1. Flow diagram of the proposed approach

3.1 Preprocessing

Tokenization. Tokenization is the first pre-processing stride of natural language processing and corpus generation. It is the process of replacing the meaningful sentence into individual words with space as the delimiter and it retains all valuable information. Each word is known as tokens. These tokens are the key elements of NLP.

Stop Words Elimination. Stop words are a part of the natural language that does not have so much meaning in a retrieval system. The reason that stop words should be removed from a text is that they make the text look heavier and less important for analysts. Removing stop words reduces the dimensionality of the term space. The most common words are in text documents are prepositions, articles, and pronouns, etc. that do not provide the meaning of the documents. These words are treated as stop words.

Stemming. The stemming technique is used to determine the root/stem of a word. It converts the words into their stems, which incorporates much of the language-dependent linguistic knowledge.

Feature Selection. Is an essential technique in dimensionality reduction to extract the most important features. It is an operation of converting text into a collection of features as a real number next to a vector which will be used as the input of classifier [19]. The operation of extracting a feature from the text, use the bag of words that changes all text in a dictionary consists of all words that appear in all texts. It creates a collection of entities as real numbers inside a vector for each text where the value of each entity inside the vector will be founded on the rate of occurrence of each word enumerated in the text.

After doing the preprocessing, we should go to the feature selection step. In this step, the primary task is to minimize the feature space dimensionality. The subset entities available in the dataset are the selected keywords. The selected features receive the highest scores based on a function that measures the importance of the feature to the text classification task. The used functions to measure the importance of the feature are most significant. The simple and effective function is the term frequency of a term, which is only those terms that appear in the highest numbers in a document are kept.

3.2 Text Representation

There are many techniques for text representation, including the Term Frequency, Inverse Document Frequency (TF-IDF) technique and others of its modifications by applying dimensionality reduction techniques such as Latent Semantic Analysis (LSA) and linear discriminant analysis (LDA).

In this article, we have used TF-IDF [20] that is a numerical statistic whether reveals that a word is how important to a text document in a collection. TF-IDF is frequently used as a weighting factor in information retrieval and text mining. The value of TF-IDF increases proportionally to the number of times a word appears in the text document but is counteracted by the frequency of the word in the corpus. This can help to supervision the fact that a number of words are mostly more common than others. TF-IDF can be successfully used for stop-words filtering in various subject fields including text classification and summarization. TF-IDF is the product of two statistics that are term frequency and inverse document frequency. To more distinguish them, the number of times each term occurs in each text document is counted and sums them together.

Term Frequency (TF) is defined as the number of times a term occurs in a document:

$$TF(word, document) = 0.5 + \frac{(0.5 \times f(word, document))}{\text{Maximum occurrences of words}} \tag{1}$$

An Inverse Document Frequency (IDF) is a statistical weight used for measuring the significance of a term in a text document collection. IDF feature is incorporated which minimizes the weight of terms that occur very frequently in the document set and increases the weight of terms that occur rarely:

$$IDF(word, document) = \beta \times \log(\frac{\text{Total number of documents}}{\text{Number of documents in which word appears}}) \tag{2}$$

with β is the number of times word appears in document.

Then the TF-IDF is calculated as:

$$TF - IDF = TF(word, document) \times IDF(word, document) \tag{3}$$

4 Classification Techniques

Several techniques are utilized to classify the text documents like NB, support vector machine (SVM), k-nearest neighbor (KNN), and LR. In this paper, we have used NB and LR algorithms, which are presented in the next sections.

4.1 Naive Bayesian Algorithm

NB is a classification machine learning technique based on an assumption of independence between predictors or what is called Bayes' formula. An NB classifier assumes that the presence of a particular characteristic in a class is not related to the presence of another characteristic [9]. The BN classifier works by taking a text document to classify and calculate the probability that the text document falls into each of the categories with which the system has trained. It generates a unique training category with the highest probability of containing the document to be classified and produces a probability for each possible class to identify several categories to which a document can belong. To categorize and classify text documents by NB, we use the following equation:

$$P(class|document) = \frac{P(class) \times P(document|class)}{P(document)} \tag{4}$$

The probabilities used in the previous equation are defined as follows:

- $P(class|document)$: is the probability that a given text document belongs to a given class.
- $P(document)$: is the probability of a document. It is a constant, we can ignore it.
- $P(class)$: is the probability of a class that is calculated from the number of text documents in the category divided by the text documents number in all categories. Let D_c represent the number of documents in a given class and D_t represent the number of documents we have total. We can calculate the probability of a class by:

$$P(class) = \frac{P(D_c)}{P(D_t)} \tag{5}$$

- $P(document|class)$: is the probability of a text document given class.

Text documents can be represented by a sets of words, hence we have the following equations:

$$P(document|class) = \prod_i P(word_i|class) \tag{6}$$

$$P(document|class) = P(class) \times \prod_i P(word_i|class) \tag{7}$$

With $P(word_i|class)$ is the probability that a given word occurs in all text documents of a given class.

4.2 Logistic Regression Algorithm

LR comes under the supervised learning classification algorithm. In recent years, this algorithm has achieved importance and its use has increased extensively in different field. LR is a classification machine learning algorithm that estimates discrete values (binary values such as 0/1, yes/no, true/false) based on a given set of independent variables. It predicts the probability of an event occurring. RL is a discriminant method that consists in calculating $P(y|x)$ by discriminating between the different possible values of the class y based on the given input class x that is as shown below:

$$P(c|x) = \sum_{i=1}^{N} w_i f_i(x) \tag{8}$$

We cannot calculate the value $P(y|x)$ directly by using the Eq. (8) because it will result in a value from $-\infty$ to $+\infty$ which means it will not result from an output between value 0 and 1.

To get a value of output that between the value 0 and 1, the following equation is used

$$P(c|x) = \frac{exp(\sum_{i=1}^{N} w_i f_i(x))}{\sum_c exp(\sum_{i=1}^{N} w_i f_i(x))} \tag{9}$$

N specify the number of features.

It is usual in language processing to employ binary-valued features. The features are not only a property of the observation x, but also are a property for both the observation x and the candidate output class c [21]. So, instead of $f_i(x)$, we use the function $f_i(c, x)$ that consists to whereby feature i of class c is assigned to the given entry of x. Therefore, the equation to calculate the probability of y being of class c given x becomes

$$P(c|x) = \frac{exp(\sum_{i=1}^{N} w_i f_i(c, x))}{\sum_{j \in c} exp(\sum_{i=1}^{N} w_i f_i(j, x))} \tag{10}$$

4.3 The Proposed Hybrid Approach

This approach combines the two algorithms above. In this approach, we will choose the best algorithms for each class, which means that every class is predicted by one and only one algorithm. The algorithms must be trained apriori over the dataset. Our proposed algorithm can be presented as in Algorithm 1.

Require: Decision_table: tab_decision
Ensure: Accuracy_table: tab_accuracy
1. i=0
2. **while** algo ∈ algorithmes **do**
3.   tab_accuracy←accuracy(algo)
4. **end while**
5. **for** row ∈ test **do**
6.   Initialise tab_decision by 0 for each class
7.   **for** algo ∈ algorithms **do**
8.   class_predicted←predict algo(row)
9.   tab_decision(class_predicted)+=tab_accuracy(algo)
10.   **end for**
11.   classe_predicted=key(max((tab_decision))
12.   **if** classe_predicted is True **then**
13.   i+=1
14.   **end if**
15. **end for**
16. Accuracy= i/size(test)
17. **Return** Accuracy

Algorithm 1: Hybrid algorithm

5 Performance Evaluation

5.1 Classification Dataset

We have used the AG's news topic classification dataset [22] that is a set of over one million news articles (war, sport, business, science, and technology). News articles have been gathered from over 2000 news sources in over a year of activity. The AG's news topic classification dataset is provided by the university community for research goals in data mining, data streaming, data compression, and every other noncommercial activity.

The AG's news topic classification+ dataset is constructed by choosing the four largest classes from the original corpus in September 2015, and it is used as a text classification benchmark by Zhang *et al.* [23]. Each class contains 30,000 training samples and 1,900 testing samples. There are 120,000 articles for the training set and 7,600 articles for the test.

5.2 Results and Discussion

We use the accuracy parameter to evaluate the efficiency of our proposed technique. Accuracy can be defined as the ratio between the number of correctly categorized texts and the entire text. Table 1 presents a confusion matrix is utilized to calculate the accuracy metric.

Table 1. Confusion matrix

		Predicted class	
		Categorized positive	Categorized negative
Actual class	Categorized positive	True positive	False negative
	Categorized negative	False positive	True negative

The accuracy is defined by Eq. (11) that measures the ratio of correct predictions over the total number of instances evaluated.

$$Accuracy = \frac{TP + TN}{TP + TN + FP + FN} \tag{11}$$

With:

- True Positive (TP) refers to the collection of text documents that are assigned correctly to the given category.
- True Negative (TN) refers to the collection of text documents that are not assigned correctly to the given category.
- False Positive (FP) refers to the collection of text documents incorrectly assigned to the given category.
- False Negative (FN) refers to the collection of text documents incorrectly not assigned to the given category.

Table 2 shows the accuracies for each algorithm and the hybrid algorithm. We found that the Hybrid classifier outperformed NB and LR.

Table 2. Accuracies for each algorithm

Algorithm	Accuracy
Naive Bayesian algorithm (NB)	90.27%
Logistic Regression algorithm (LR)	91.40%
Hybrid algorithm	91.86%

6 Conclusion

We have presented a hybrid approach for text classification that combines a set of given algorithms, using the best for each class. Feature reduction and selection methods are used in combination with NB and LR learning algorithms to increase the accuracy of the classifier. The proposed approach has proved its efficiency by outperforming the given algorithms in our case. The future direction of this work is to increase the volume of the dataset and use other algorithms of machine learning and deep learning, with different feature selection methods like doc2vec. We also try to experiment with more hybrid algorithms to obtain high benefits from machine learning techniques and to reach better classification results.

References

1. Elnagar, A., Al-Debsi, R., Einea, O.: Arabic text classification using deep learning models. Inf. Process. Manag. **57**(1), 102121 (2020)
2. Hartmann, J., Huppertz, J., Schamp, C., Heitmann, M.: Comparing automated text classification methods. Int. J. Res. Mark. **36**(1), 20–38 (2019)
3. Liu, H., Yin, Q., Wang, W.Y.: Towards explainable NLP: a generative explanation framework for text classification. In: Proceedings of the 57th Annual Meeting of the Association for Computational Linguistics, Association for Computational Linguistics, pp. 5570–5581 (2019)
4. Yadav, A., Vishwakarma, D.K.: Sentiment analysis using deep learning architectures: a review. Artif. Intell. Rev. **53**(6), 4335–4385 (2019). https://doi.org/10.1007/s10462-019-09794-5
5. Medhat, W., Hassan, A., Korashy, H.: Sentiment analysis algorithms and applications: a survey. Ain Shams Eng. J. **5**(4), 1093–1113 (2014)
6. Jain, A.P., Dandannavar, P.: Application of machine learning techniques to sentiment analysis. In: 2016 2nd International Conference on Applied and Theoretical Computing and Communication Technology (iCATccT), pp. 628–632. IEEE (2016)
7. Shah, K., Patel, H., Sanghvi, D., Shah, M.: A comparative analysis of logistic regression, random forest and KNN models for the text classification. Augmented Hum. Res. **5**(1), 1–16 (2020). https://doi.org/10.1007/s41133-020-00032-0

8. Li, Q., et al.: A survey on text classification: from shallow to deep learning. ACM Comput. Surv. **37**(4), 1–35 (2020)
9. Kadhim, A.I.: Survey on supervised machine learning techniques for automatic text classification. Artif. Intell. Rev. **52**(1), 273–292 (2019). https://doi.org/10.1007/s10462-018-09677-1
10. Kowsari, K., Jafari Meimandi, K., Heidarysafa, M., Mendu, S., Barnes, L., Brown, D.: Text classification algorithms: a survey. Information **10**(4), 150 (2019)
11. Deng, X., Li, Y., Weng, J., Zhang, J.: Feature selection for text classification: a review. Multimedia Tools Appl. **78**(3), 3797–3816 (2018). https://doi.org/10.1007/s11042-018-6083-5
12. Marie-Sainte, S.L., Alalyani, N.: Firefly algorithm based feature selection for Arabic text classification. J. King Saud Univ.-Comput. Inf. Sci. **32**(3), 320–328 (2020)
13. Kou, G., Yang, P., Peng, Y., Xiao, F., Chen, Y., Alsaadi, F.E.: Evaluation of feature selection methods for text classification with small datasets using multiple criteria decision-making methods. Appl. Soft Comput. **86**, 105836 (2020)
14. Meng, Y., Shen, J., Zhang, C., Han, J.: Weakly-supervised hierarchical text classification. In: Proceedings of the AAAI Conference on Artificial Intelligence, vol. 33, pp. 6826-6833, AAAI Press, Palo Alto, California USA (2019)
15. Burdisso, S.G., Errecalde, M., Montes-y-Gómez, M.: A text classification framework for simple and effective early depression detection over social media streams. Expert Syst. Appl. **133**, 182–197 (2019)
16. Sachan, D. S., Zaheer, M., Salakhutdinov, R.: Revisiting lstm networks for semi-supervised text classification via mixed objective function. In: Proceedings of the AAAI Conference on Artificial Intelligence, vol. 33, pp. 6940–6948, AAAI Press, Palo Alto, California USA (2019)
17. Sun, C., Qiu, X., Xu, Y., Huang, X.: How to fine-tune BERT for text classification? In: Sun, M., Huang, X., Ji, H., Liu, Z., Liu, Y. (eds.) CCL 2019. LNCS (LNAI), vol. 11856, pp. 194–206. Springer, Cham (2019). https://doi.org/10.1007/978-3-030-32381-3_16
18. Kim, H., Jeong, Y.S.: Sentiment classification using convolutional neural networks. Appl. Sci. **9**(11), 2347 (2019)
19. Dzisevič, R., Šešok, D.: Text classification using different feature extraction approaches. In: 2019 Open Conference of Electrical, Electronic and Information Sciences (eStream), pp. 1–4. IEEE (2019)
20. Christian, H., Agus, M.P., Suhartono, D.: Single document automatic text summarization using term frequency-inverse document frequency (TF-IDF). ComTech Comput. Math. Eng. Appl. **7**(4), 285–294 (2016)
21. Indra, S. T., Wikarsa, L., Turang, R.: Using logistic regression method to classify tweets into the selected topics. In: 2016 International Conference on Advanced Computer Science and Information Systems (ICACSIS), pp. 385–390. IEEE (2016)
22. http://www.di.unipi.it/~gulli/AG_corpus_of_news_articles.html
23. Zhang, X., Zhao, J., LeCun, Y.: Character-level convolutional networks for text classification. In: Proceedings of the 28th International Conference on Neural Information Processing Systems, pp. 649–657 (2015)

Automatic Key-Phrase Extraction: Empirical Study of Graph-Based Methods

Lahbib Ajallouda[1(✉)], Fatima Zahra Fagroud[2], Ahmed Zellou[1], and El Habib Benlahmar[2]

[1] ENSIAS Mohammed V University in Rabat, Rabat, Morocco
lahbiblahbib@hotmail.fr, ahmed.zellou@um5.ac.ma
[2] LTIM–FSBM Hassan II University in Casablanca, Casablanca, Morocco

Abstract. Key-phrases in a document are phrases that provide a high-level description of its content without reading it completely. In some research articles, authors specify key-phrases in the articles they have written. However, the vast majority of books, articles, and web pages published every day, lack key-phrases. The manual extraction of these phrases is a tedious task and takes a long time. For this reason, automatic key-phrase extraction (AKE), which is an area of Text Mining, remains the best solution to overcome these difficulties. Because they are used in many Natural Language Processing (NLP) applications, such as text summarization and text classification. This article presents a comparison of some methods of extracting key-phrases from documents. Especially the graph-based approaches. These approaches are evaluated by their abilities to extract key-phrases. Our work focuses on the study of the performance of these methods in extracting key-phrases, whether from short or long texts, with the aim of providing information that contributes to improving their efficiency.

Keywords: Automatic key-phrase extraction · Text mining · Natural language processing · Graph-based

1 Introduction

Key-phrases extraction is a task that consists of identifying phrases in a document that characterize its content. However, every day a large number of documents are published, for example, 16,866 publications related to COVID-19 published in the first half of 2020 [1]. There are two main approaches to extracting key-phrases [2]. The first approach is manual analysis, that is, manually process the data and start recording the information that seems useful. The other approach is automatic which would use computer technologies to perform text analysis. AKE is widely used in many fields, such as information search [3], digital content management [4], natural language processing [5], contextual advertising [6] and recommendation systems [7].

In recent years, a wide variety of AKE methods have been introduced, which use different machine learning techniques (supervised and unsupervised). The objective of this article is to compare unsupervised approaches to extracting key phrases that are

M. Lazaar et al. (Eds.): BDIoT 2021, LNNS 489, pp. 440–456, 2022.
https://doi.org/10.1007/978-3-031-07969-6_33

based on graphs. For each approach, we examine the evaluation parameters, the data sets used, their accuracy and discuss the results. In order to provide researchers with information on the proposed solutions and their limitations.

This paper is structured as follows. Section 2 presents the AKE process. The approaches of AKE are presented in Sect. 3. Our comparison is detailed in Sect. 4. Section 5 presents the results of this study as well as future directions.

2 AKE Process

This section is devoted to presenting the automatic key-phrase extraction process. According to [8], the AKE process consists of four main steps: preprocessing; identification of a list of candidate phrases; selection of key-phrases; and finally, the evaluation of the results, See Fig. 1.

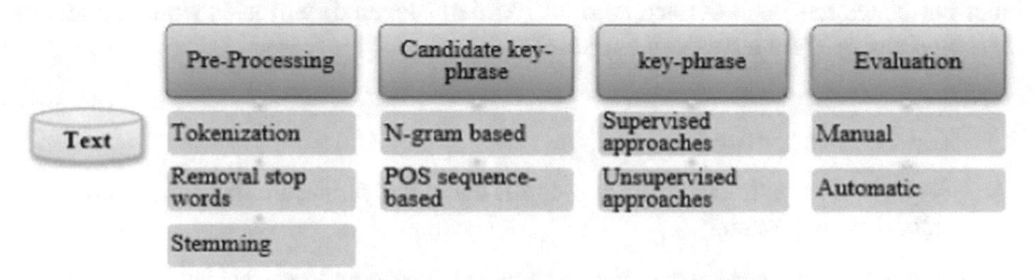

Fig. 1. Automatic key-phrases extraction process

2.1 Preprocessing

The first step of the key-phrases extraction process is the preprocessing, this phase involves several operations such as tokenization, removal stop word, and stemming [9].

Tokenization. The preprocessing is started by the decomposition of the analyzed text in units called tokens. This step is called, tokenization and tokens can be words, phrases, or even whole sentences [10].

Removal Stop Word. A stop word is a non-significant word in a text. The second step in the preprocessing process is the removal of stop words. This step is very important for natural language processing applications. Several algorithms are proposed for the deletion of stop words as Z-Methods and TBRS [11].

Stemming. Stemming is a process of determining the racing of a word through the removal of derivative and inflectional affixes. The stemmer is mainly used in NLP domains. Several methods developed for stemming as Snowball stemmer and Lancaster stemmer [12].

2.2 Candidate Key-Phrases

Candidate key-phrases are potentially key-phrase, that is, terms that do not contain punctuation or stop words and have noun-adjective morphosyntactic structures (for example, "Big data", "Computer engineering" etc.) [13]. The selection of candidate key-phrases is a preliminary step in the extraction of key phrases, it helps to remove unnecessary textual units. In order to eliminate noise in the basic text and decrease the processing time required for key-phrase extraction. Several techniques help to select candidate key-phrases [14], the most used are:

N-grams. This method divides the text into sequences of n adjacent words. It provides a set of candidate terms, this set includes candidates with useful clues, but also irrelevant candidates. To overcome this problem, an anti-dictionary is used to filter the candidates.

POS (Part Of Speech). This method captures the idea of key terms having a certain syntactic property, but it is based on empirical evidence in the training data. A set of POS label templates has been defined, and all (partially labeled) words or word sequences matching any of them have been extracted.

2.3 Key-Phrases Extraction

The key-phrases are extracted from among the candidate phrases, either via a supervised or unsupervised approach.

Supervised Approaches. Are methods capable of learning to extract key phrases from a document. Learning takes place through a corpus whose documents are annotated in key phrases. Supervised approaches mainly learn to classify candidate terms into "key-phrases" or "non-key-phrases" using a binary classification function f, denoted by:

$$f : \mathbb{R}^m \rightarrow \{1, -1\} \tag{1}$$

f is a function which takes as argument a vector represents a candidate key-phrase $p = (f_1, f_2 \ldots f_m)$.

f returns a value of 1 if candidate p is a key-phrase in document D, and a value of –1 if it is not. The key-phrases of a D document can be represented by:

$$M = \bigcup_{p_i \epsilon D} \{p_i : f(p_i) = 1\} \tag{2}$$

Unsupervised Approaches. Consider the key-phrase extraction task as a ranking problem. After extracting a set of candidates key-phrases from a document, they are then ordered according to one or more weighting coefficients. The subset of candidate key-phrases associated with a weight value that exceeds a given threshold is the set of key-phrases.

We denote by $W = \{p_1, p_2 \ldots, p_n\}$ a set of n candidate key-phrases for a document D and $S(p_i)$ the weighting of the candidate p_i according to a given coefficient. Unsupervised extraction generally corresponds to the following function: (confusion entre word et phrase).

$$W_k = \bigcup_{p \in W} \{p_i : S(p_i) > min\} \tag{3}$$

W_k is the set of key-phrases in document D.
$S(p_i)$: score of pi. It is calculated according to the approach used.
Min: A numeric value, candidate key-phrases having a score greater than this value are considered to be key-phrases.

2.4 The Evaluation

The evaluation process aims to know the performance of the approach used to extract the key phrases. This assessment is based on a set of data available in the literature (scientific articles, notices and current affairs news), and can be performed either manually or automatically.

Manual Evaluation. The results of the approach used are evaluated on the basis of human judgments. This evaluation has been used by several authors [15]. Despite this type gives correct judgments it is very expensive and time-consuming. For this, many authors prefer automatic evaluation.

Automatic Evaluation. To automatically evaluate the output of a key-phrases mining system, metrics such as precision (P), recall (R), and F1-score (F) are used by researchers.

Precision: The P precision of an Automatic Key Phrase Extraction System is the ratio of the number of correctly extracted key phrases to the number of all extracted key phrases. It is defined as follows:

$$P = \frac{KP_{Correct}}{KP_{Extracted}} \quad (4)$$

where, $KP_{Correct}$ is the number of correctly extracted key-phrases and $KP_{Exrtacted}$ is the number of extracted key-phrases.

Recall: Recall R is the ratio between the number of correctly extracted key phrases and the number of key phrases annotated by the author. The R of an AKE System is defined as follows:

$$R = \frac{KP_{Correct}}{KP_{Author}} \quad (5)$$

where, $KP_{Correct}$ is the number of correctly extracted key phrases and KP_{Author} is the number of key phrases annotated by the author.

F1-Score: F-Score F, is the harmonic mean of the precision P and the recall R. It is defined as follows:

$$F = \frac{2 \times P \times R}{P + R} \quad (6)$$

Dataset. In order to evaluate and develop their approaches. The authors need textual sources. That is to say scientific publications, news documents, and abstracts of articles. Eirini in [16] groups the best-known textual sources into three types. Table 1 presents these sources.

Table 1. Popular evaluation datasets grouped by their type of source.

Type	Dataset	Created By	Docs	Language	Annotation
Full-text Papers	NUS	Nguyen and Kan (2007)	211	English	Authors/ Readers
	Krapivin	Krapivin et al.(2008)	2304	English	Authors
	Semeval	Kim et al.(2010b)	248	English	Authors/ Readers
	Citeulike	Medelyan et al. (2009)	180	English	Readers
Paper Abstracts	Inspec	Hulth (2003)	2000	English	Authors/ Readers
	KDD	Gollapalli and Caragea(2014)	755	English	Authors
	KP20k	Meng et al.(2017)	567830	English	Authors
	WWW	Gollapalli and Caragea(2014)	1330	English	Authors
News	DUC-2001	Wan and Xiao(2008)	308	English	Readers
	500N-KPCorwd	Marujo et al.(2012)	500	English	Readers
	Wikinews	Bougouin et al.(2013)	100	French	Readers

3 Graph-Based Key-Phrases Extraction

The use of graphs allows a simple presentation of the content of a document. In this section we present, based-graph key phrases extraction approaches, which we will compare later in Sect. 4.

3.1 TextRank Approach

TextRank [18] was the first to use a graph to extract key terms from a document.

Graph. Formally, let G = (V, E) be an undirected graph. The vertices V are the words of the document and E are the edges that connect them. Two vertices V_i and V_j are linked by an edge if they are co-occured in a window of 2 to 10 words.

The most important words are considered keywords. A word is more important in the graph if it's connected with a large number of words and if the words with which it's connected are also important.

Importance Score. Once the graph is built, the score associated with each vertex Vi is fixed at an initial value of 1. Then, the score of a word is calculated using formula (7) which uses the principle of the algorithm PageRank [19]. It is executed several times until $WS^j(Vi) - WS^{j+1}(Vi)$, converges to a threshold of 0.0001

$$WS(C_i) = (1 - \lambda) + \lambda * \sum_{C_j \in In(C_i)} \frac{w_{ij}}{\sum_{C_k \in Out(C_j)} w_{jk}} WS(C_j) \quad (7)$$

In (C_i): the set of vertices associated with C_i.
Out (C_j) the set of vertices associated with C_j.
λ: damping factor. 0.85 is used as the default for λ.
W_{ij}: the weight of an edge between two vertices C_i and C_j.

Key-Phrases Extraction. The vertices are sorted in descending order of their final score and the N well ranked vertices are considered as key-phrases. Usually N is between 5 and 20. Adjacent key-phrases sequences are grouped into a key-phrases consisting of several words.

3.2 ExpandRank Approach

ExpandRank [20] uses the same principle of TextRank. Except that, ExpandRank increases the precision of order by using an enlarged co-occurrence window and documents similar to the document analyzed according to the measure of similarity of the cosine vector.

Graph. ExpandRank uses documents close to the analyzed document d_0.

Let $D = \{d_0, d_1, d_2, \ldots d_k\}$ The document set. Formally, given the developed set of documents D, let G = (V, E) be an undirected graph. V is the set of vertices and each vertex is a candidate word in the set of documents D. Two vertices are connected by an edge if the corresponding words occur at least once in a window of w words, where w can be defined between 2 and 20 words.

E is the set of edges, each edge e_{ij} in E is associated with an affinity weight $aff(v_i, v_j)$ between the words v_i and v_j.

The affinity weight aff (v_i, v_j) is defined as follows:

$$aff(v_i, v_j) = \sum_{d_p \in D} sim_{doc}(d_0, d_p) \times count_{d_p}(v_i, v_j) \quad (8)$$

Sim_{doc} (d_0, d_p) is the cos similarity between d_0 and d_p. It is calculated using (9).

$Count_{dp}$ (v_i, v_j) is the number of co-occurrences between the words v_i and v_j in the document d_p

$$sim_{doc}(d_i, d_j) = \frac{\vec{d_i} \cdot \vec{d_j}}{\| \vec{d_i} \| x \| \vec{d_j} \|} \quad (9)$$

$\vec{d_i}$ and $\vec{d_j}$ are vectors tf-idf, represent the documents d_i and d_j.

To facilitate the manipulation of the graph, the set of weights of the edges are represented in the form of a matrix M. The weight of an edge M_{ij}, is defined as follows:

$$M_{i,j} = \begin{cases} aff(v_i, v_j), & \text{if } v_i \text{ connected with } v_j \\ 0, & \text{else} \end{cases} \tag{10}$$

M is normalized to $\tilde{M}$

$$\breve{M}_{ij} = \begin{cases} M_{i,j} \Big/ \sum_{j=1}^{|V|} M_{i,j}\ , & \text{si } \sum_{j=1}^{|V|} M_{i,j} \neq 0 \\ 0, \text{ Sinon} & \end{cases} \tag{11}$$

Importance Score. Once the graph is built, the score associated with each vertex V_i is fixed at an initial value of 1. Then, the score of a word is calculated using formula (12) which uses the principle of the algorithm PageRank. It is executed several times until $WS^j(V_i)–WS^{j+1}(V_i)$, converges to a threshold of 0.0001.

$$WS(V_i) = \lambda. \sum_{allj \neq i} WS(V_j).\ \tilde{M}_{j,i} + \frac{(1-\lambda)}{|V|} \tag{12}$$

λ is the damping factor generally set at 0.85.
|V|: the total number of words in D.
The score of a phrase p_i is the sum of the scores of the words contained in p_i.

$$\text{PhraseScore}(p_i) = \sum_{v_j \in p_i} \text{WordScore}(v_j) \tag{13}$$

Key-Phrases Extraction. The candidate phrases of the analyzed document are sorted in descending order according to their scores, and the N well ranked vertices are considered as key-phrases. Key phrases made up of several words are no longer generated, but ordered from the sum of the importance scores of the words that compose them.

3.3 TopicRank approach

TopicRank [21] is a graph-based method that aims to extract key phrases from the most important topics from a document. The subjects are defined as clusters of key candidates.

Graph. Before building a graph, TopicRank groups the candidates into Topics, using the HAC algorithm [22]. The analyzed document is represented by a complete graph where the nodes are the subjects. Each subject is connected to the others by a weighted edge. The closer the candidate keyword terms of two subjects are in the document, the higher the edge weighting between the two subjects.

Edge Weight. $\omega_{i,j}$ is the edge weight between topics t_i and t_j, where:

$$\omega_{i,j} = \sum_{c_i \in t_i} \sum_{c_j \in t_j} dist(c_i, c_j) \tag{14}$$

dist (c_i, c_j) represents the semantic force between the candidate c_i and c_j, it calculated from their respective positions, pos (c_i) and pos (c_j), in the document.

$$dist(c_i, c_j) = \sum_{p_i \epsilon pos(c_i)} \sum_{p_j \epsilon pos(c_j)} \frac{1}{|p_i - p_j|} \tag{15}$$

Topic Score. Once the graph is built. The score for each topic is calculated using formula (16). The score associated with each vertex t_i is fixed at an initial value of 1. It is executed several times until S^j (V_i)–S^{j+1} (V_i), converges to a threshold of 0.0001.

$$S(t_i) = (1 - \lambda) + \lambda \times \sum_{t_j \in V_i} \frac{\omega_{i,j} \times S(t_j)}{\sum_{t_k \in V_j} \omega_{j,k}} \tag{16}$$

where V_i are the subjects voting for t_i and λ is a damping factor generally set at 0.85.

Key-Phrases Extraction. To extract the N key-phrases from the document, TopicRank selects the N topics that obtained a high score. In each topic group, TopicRank chooses only one term as the key-phrase. The selection of these phrases is based on three strategies.

First, the key-phrase chosen for a topic is that appears first in the analyzed document.

Second, the key-phrase chosen for a topic is the most commonly used in the analyzed document.

Third, the key-phrase chosen for a topic is the most similar to the other phrase in the topic.

3.4 K-Core Approach

Unlike previous approaches that use the PageRank algorithm to calculate the importance scores of vertices. K-Core [23] uses the Batagelj and Zaveršnik algorithm [24].

Graph. Formally, let G = (V, E) be an undirected graph. The vertices V are the terms of the analyzed document and E are the edges which connect them. Two vertices V_i and V_j are linked by an edge if they are cooccurred in a window of N terms. k-core or core of order k is a subgraph $H_k = (V_0 \subseteq V, E_0 \subseteq E)$, where $\forall$ v $\in$ V_0, deg (v) $\geq$ k.

Vertex Core. The k-core of a graph is a maximal connected subgraph whose vertices are at least of degree k. The core of a vertex v is the highest order of the subgraph that contains this vertex. The maximum order core is called the main core and the set of all k-cores in a graph between 0-core and main core. The calculation of the core of a vertex is done via the Batagelj and Zaveršnik algorithm.

Key-Phrases Extraction. The vertices are ordered in descending order of their core values. The number of key phrases is not decided by the reader. Key phrases are phrases that have the main K-core.

3.5 WordAttractionRank Approach

WordAttractionRank [25] represents a document as a weighted undirected graph, where vertices represent candidate key terms and edges are co-occurrence relationships between terms in a phrase. WordAttractionRank also uses the PageRank algorithm to rank candidate key terms.

Graph. Formally, let G = (V, E) be an undirected graph. The vertices V are the terms of the analyzed document and E are the edges which connect them. Edge weights are defined and calculated using the values provided by the embedded words and local statistical information.

Edge Weight. The weight of an edge is calculated using two parameters, the attraction force and the dice coefficient.

Attraction Force. Two words cannot be considered important for a document if they both have very low frequencies, even though they may have a very strong semantic relationship. Conversely, two words can be said to be important if one of them has a high frequency while another has a very low frequency and they have a very strong semantic relationship. Formally, the attraction force of word w_i and w_j in a document D is computed as:

$$f(w_i, w_j) = \frac{freq(w_i) \times freq(w_j)}{d^2} \tag{17}$$

where freq (w) is the frequency of occurrence of word w in D and d is the Euclidean distance between w_i and w_j.

Dice Coefficient. The dice coefficient is used to measure the probability of two words co-occurring in a pattern, or by chance. Formally, the dice coefficient is calculated by:

$$dice(w_i, w_j) = \frac{2 \times freq(w_i, w_j)}{freq(w_i) + freq(w_j)} \tag{18}$$

where freq (w_i, w_j) is the frequency of co-occurrence of words w_i and w_j, and freq (w) is the frequency of occurrence of w in D.

The weight of the edge is calculated by the product of the dice coefficient and the attraction force.

$$attr(w_i, w_j) = dice(w_i, w_j) \times f(w_i, w_j) \tag{19}$$

Importance Score. The score associated with each vertex V_i is fixed at an initial value of 1. Then, the score of a word is calculated using formula (20) which uses the principle of the algorithm PageRank.

$$S(v_i) = (1 - \delta) + \delta \times \sum_{v_j \in C(v_i)} \frac{attr(v_i, v_j)}{\sum_{v_k \in C(v_j)} attr(v_j, v_k)} S(v_j) \tag{20}$$

$C(v_i)$ the set of vertices Linked to v_i.

$S(v_i)$ is calculated several times until $S^j(V_i) - S^{j+1}(V_i)$, converges to a threshold of 0.0001.

The score of a phrase p_i is the sum of the scores of the words contained in p_i.

$$PS(p) = \sum_{w_i \in p} S(w_i) \tag{21}$$

w_j is a word contained in p.

Key-Phrases Extraction. The candidate phrases of the analyzed document are sorted in descending order according to their scores, and the N well ranked vertices are considered as key-phrases.

3.6 PositionRank Approach

PositionRank [26], is a graph-based key phrase extraction approach. The importance score for each candidate phrase is calculated via PageRank which simultaneously integrates the position of words and their frequency in a document.

Graph. PositionRank constructs an undirected word graph G = (V, E). Two vertices v_i and v_j are connected by an edge $(v_i, v_j) \in E$ if the words corresponding to these nodes co-appear in a window of N words. The weight of an edge $(v_i, v_j) \in E$ is calculated on the basis of the co-occurrence count of the two words v_i and v_j. In order to simplify the manipulation of the weights, a matrix M is created representing the weights of the edges. The weight of an edge M_{ij}, is defined as follows:

$$M_{i,j} = \begin{cases} \text{weight}(v_i, v_j), & \text{if } v_i \text{ connected with } v_j \\ 0, & \text{else} \end{cases} \tag{22}$$

M is normalized to $\tilde{M}$.

Importance Score. The idea of PositionRank is to favor the most frequent weights of the words are found at the beginning of the analyzed document. The position weight of each candidate word is the inverted position in the document. If the same word appears more than once in the document, all of its position weights are added. For example, if a word is in the following positions: 2, 5, and 10, then its weight there is equal to: $1/2 + 1/5 + 1/10 = 0.8$. the position weights of all the candidate words are represented by a normalized vector P as follows:

$$\breve{P} = \left[\frac{p_1}{p_1 + \ldots + p_{|V|}}, \frac{p_2}{p_1 + \ldots + p_{|V|}}, \ldots, \frac{p_{|V|}}{p_1 + \ldots + p_{|V|}} \right] \tag{23}$$

p_i: is the position weight of the word w_i.
|V|: Number of candidate words.

For all $v_i \in V$. The score of a vertex v_i, is calculated recursively by the following equation:

$$S(v_i) = (1 - \lambda).\breve{p}_i + \lambda. \sum_{v_j \in Adj(v_i)} \frac{w_{ji}}{O(v_j)} S(v_j) \tag{24}$$

$$O(v_j) = \sum_{v_k \in Adj(v_j)} w_{jk} \tag{25}$$

λ is the damping factor generally set at 0.85.

$\breve{p}_i$ is the position weight for the vertex v_i.

The initial values of S (v_i) are set to 1/|V|. Where |V| is the number of vertices.

Word scores are calculated recursively until 100 iterations or the difference between two consecutive iterations is less than 0.001.

Key-Phrases Extraction. The candidate phrases of the analyzed document are sorted in descending order according to their scores, and the N well ranked vertices are considered as key-phrases. The score of a phrase composed of several candidates is the sum of the scores of the words which compose it.

3.7 Multipartite Graph Approach

Multipartite graph [27] based on the TopicRink principle [21]. It uses a graph structure, to represent documents as closely related sets of candidates related to topics. This representation allows the ranking algorithm to fully exploit the mutually reinforcing relationship between topics and candidates and allows for the incorporation of intra-topic keyword selection preferences.

Graphic. a complete directed graph represents the analyzed document is created by multipartite graph. The vertices are candidate words that are only connected if they belong to different subjects. The edge weighting is calculated based on the distance between two candidates in the document. Edges are oriented, which gives more control over the impact of individual candidates on the overall ranking

Formally, the weight w_{ij} of an edge which connects the vertices ci and c_j is calculated by the sum of the inverse distances between the occurrences of the candidates c_i and c_j:

$$\omega_{i,j} = \sum_{p_i \epsilon pos(c_i)} \sum_{p_j \epsilon pos(c_j)} \frac{1}{|p_i - p_j|} \tag{26}$$

Pos (c_i) is the set of positions of candidate c_i.

Weight Adjustment. The position of the candidate within the document is the most reliable method for selecting a topic keyword [21]. So, the candidates who appear at the beginning of the document are promoted according to the other candidates belonging to the same topic. In order to exploit this, Multipartite graph adjusts the weights of the edges corresponding to the first appearing candidate of each subject. The weights of these edges are modified by the following equation:

$$\omega_{ij} = \omega_{ij} + \alpha.e^{\frac{1}{p_i}} \sum_{c_k \epsilon T(c_j) \backslash \{c_j\}} \omega_{ki} \tag{27}$$

w_{ij} is the weight of the edge between the vertices c_i and c_j.

T (c_j) is the set of candidates belonging to the same topic as c_j.

p_i is the position of the first occurrence of candidate c_i.

α is a hyperparameter that controls the strength of the weight adjustment. Its default value is $\alpha = 1.1$

Importance Score. The score of each candidate word is calculated recursively by the following formula:

$$S(c_i) = (1 - \lambda) + \lambda * \sum_{c_j \epsilon IN(c_i)} \frac{w_{ij} * S(c_j)}{\sum_{c_k \epsilon Out(c_j)} w_{jk}} \tag{28}$$

IN (c_i) is the set of predecessors of c_i.

Out (c_j) is the set of successors of c_j.

λ is a damping factor set at 0.85.

For any candidate word. The initial values of S are set to 1. Word scores are calculated recursively until 100 iterations or the difference between two consecutive iterations is less than 0.001.

Key-Phrase Extraction. The candidate phrases of the analyzed document are sorted in descending order according to their scores, and the N well ranked vertices are considered as key-phrases. The score of a phrase composed of several candidates is the sum of the scores of the words which compose it.

3.8 RaKUn Approach

RaKUn [28], is a key phrase extraction method based on the centrality of a vertex in a graph.

Graph. The analyzed document composed of tokens $\{t_1, \ldots, t_n\}$. This set is ordered according to the appearance of the tokens in the document. The graph G = (V, E) created by *RaKUn* is a directed graph. *RaKUn* forms a directed edge $(t_i, t_{i+1}) \in E$ between each tokens t_i, and its successor t_{i+1}., These edges are weighted according to the number of times they appear in the document.

Load Centrality. Centrality is a measure used to estimate the importance of vertices in a graph. This measure is defined according to the number of shortest paths passing through a given vertex w. it's calculated by the following formula

$$c(w) = \sum_{u,v \in V} \frac{\sigma(u, v|w)}{\sigma(u, v)}; u, v \neq w \tag{29}$$

σ (u, v|w) represents the number of shortest paths that go from vertex u to vertex v via w.

σ (u, v) the number of all the shortest paths between u and v.

The vertices of the graph with the highest centrality are considered by RinkUp as key-phrases.

N-grams. The centralities of phrases composed of two words or most, equal the average of the centralities of words which compose.

$$C(p) = \frac{\sum_{w_i \in p} c(w_i)}{N} \tag{30}$$

N: number of words that make up the phrase p.

Key-Phrase Extraction. The candidate key-phrases of the analyzed document are sorted in decreasing order according to their centrality, and the N well-classified vertices are considered as key-phrases.

4 Evaluation

In this section, we will compare via an empirical study the key-phrase extraction method previously presented in the previous section, based on the results of evaluation metrics such as Precision (P), Recall (R), and F1-Score (F1).

4.1 Datasets

To evaluate the performance of the key phrase extraction approach, a dataset must be adopted. In our study, we used two datasets Inspec [29] and SemEval [30] which provided us with a set of English data.

Inspec Dataset. is composed of 2000 summaries of scientific articles (short texts). Each article has two lists of key phrases:

Controlled: Are the key phrases assigned by the article authors.
Uncontrolled: Are sentences freely assigned by the readers of articles.

Semeval 2010 Dataset. includes 284 articles collected from ACM Digital Library. 100 items used as a test set, 144 used as a training set and a size 40 test set. Each item has two lists of controlled and uncontrolled key phrases.

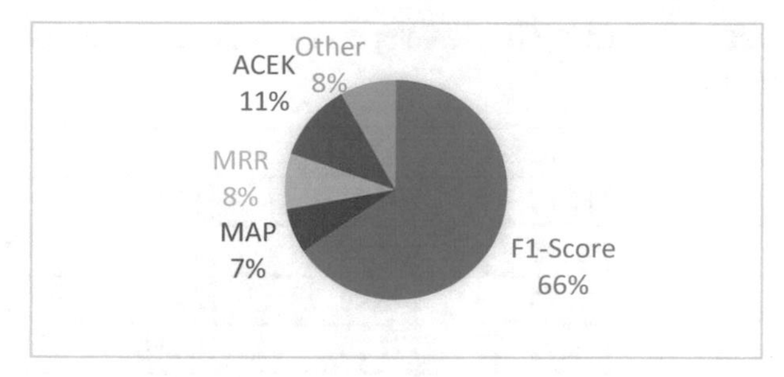

Fig. 2. The percentage of use of evaluation metrics in related work to key-phrases extraction.

4.2 Evaluation Metric

There are several metrics that allow us to evaluate the methods for extracting key-phrases as Mean Reciprocal Rank (MRR), Mean Average Precision (MAP), and Average of Correctly Extracted Key-phrases - (ACEK) [16]. In our study, to assess and compare the precision and effectiveness of the approaches presented in the previous section, we used F1-score, which is one of the most commonly used methods, calculated on the basis of precision and recall measures. In Fig. 2, we show the percentage of use of evaluation metrics in related work to key-phrases extraction.

The values for these metrics vary depending on the number of key-phrases extracted. In this article, the evaluation of each approach is done by extracting 5 and 10 key-phrases. For F1-Score we use as F1-Score@5 and F1-Score@10.

4.3 Results

The results of the evaluation were determined using the two datasets Semeval, which contains fairly long texts (articles of 6 pages or more), and Inspec, which contains abstracts of scientific articles, which can be considered as short texts. These results will allow us to compare the precision and efficiency of the approaches presented in our work, whether applied to long or short texts. Table 2 shows the results of the F-Score@5 and F-Score@10 for each method according to the Semeval and Inspec dataset.

4.4 Discussion

F1-score results showed that using graphics-based methods to extract key phrases from short texts (Inspec) is better than using them from long texts (Semeval). Figure 3 presents the average score of these approaches by dataset. The ineffectiveness of these methods for long texts is due to the large amount of information that these texts contain, which makes it difficult to find key phrases from the graph.

On the other hand, the results obtained showed the superiority of the ExpandRank method in short texts compared to other methods. This gives us the impression that exploiting a corpus of texts similar to the text to be analyzed, can give encouraging results in the process of extracting key-phrases from short texts. In addition, Multipartite

Table 2. Results of F1@5 and F@10 according to the Semeval and Inspect dataset for the compared approaches.

Approach	Semeval (%)		Inspec (%)	
	F1@5	F1@10	F1@5	F1@10
TextRank	2.50	3.55	27.10	34.10
ExpandRank	2.75	3.90	29.90	35.70
TopicRank	9.70	12.30	25.30	29.35
K-Core	3.00	6.00	16.30	23.10
WordAttractionRank	6.10	11.50	29.20	36.40
PositionRank	10.65	12.25	23.50	30.50
Multipartite	12.20	14.55	25.90	30.65
RaKUn	2.50	5.95	5.40	10.60

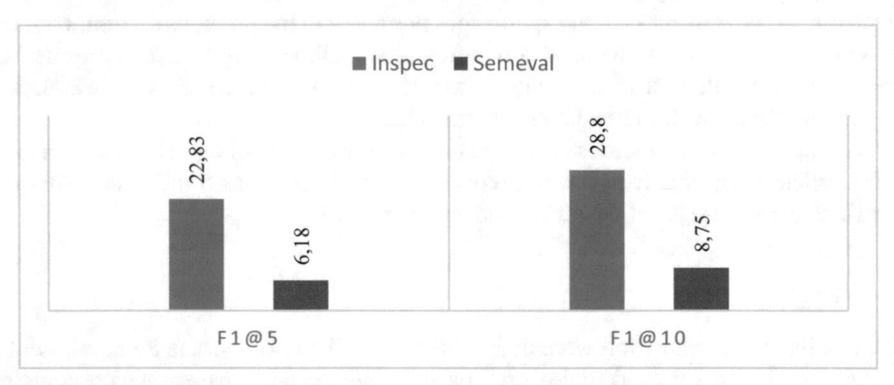

Fig. 3. F1-score average for extracting key-phrases for graph-based methods using the Inspect and Semeval dataset.

has shown superiority over other methods in long texts, which makes us say that graph-based approaches can benefit from the integration of different types of information, such as topical information, position and statistics to improve their precision in extracting key-phrases from long texts.

5 Conclusion

Key-phrases in a document are phrases that describe its content without having to read or analyze it. Our paper reviewed eight graph-based key-phrase extraction methods. We also compared the precision of the key term extraction for these approaches using the Inspect and Semeval2010 dataset through the F1-Score calculation. The results obtained have shown that the efficiency of graph-based methods is better when applied to short texts rather than long texts. But methods that integrated other information, such as topical, positional, and statistical information, gave us indications that if the graph-based methods exploit other information we might get better results in long texts, and this is what we will work on in the future. Word preprocessing and the size of the word window are among the factors affecting the key-phrase extraction processes of a document. their improvement will be helped to increase the efficiency of the key-phrase extraction methods.

References

1. Aristovnik, A., Ravšelj, D., Umek, L.. A bibliometric analysis of COVID-19 across science and social science research landscape (2020)
2. Nasar, Z., Jaffry, S.W., Malik, M.K.: Textual keyword extraction and summarization: State-of-the-art. Inf. Process. Manage. **56**(6), 102088 (2019)
3. Zhai, C., Lafferty, J.: A study of smoothing methods for language models applied to ad hoc information retrieval. ACM SIGIR Forum, 268–276 (2017)
4. Vallez, M., Pedraza-Jiménez, R., Codina, L., Blanco, S., Rovira, C.: A semi-automatic indexing system based on embedded information in html documents. Libr. Hi Tech **33**(2), 195–210 (2015)
5. Reilly, R.G., Sharkey, N.: Connectionist Approaches to Natural Language Processing. Routledge, Abingdon (2016)
6. Sterckx, L., Demeester, T., Deleu, J., Develder, C.: Creation and evaluation of large key-phrase extraction collections with multiple opinions. Lang. Resour. Eval. **52**(2), 503–532 (2018)
7. Idrissi, N., Zellou, A.: A systematic literature review of sparsity issues in recommender systems. Soc. Netw. Anal. Min. **10**(1), 15 (2020)
8. Merrouni, Z.A., Frikh, B., Ouhbi, B.: Automatic key-phrase extraction: An overview of the state of the art. In: Proceedings of the 2016 4th IEEE International Colloquium on Information Science and Technology (CiSt), pp. 306–313. IEEE (2016)
9. Vijayarani, S., Ilamathi, M.J., Nithya, M.: Preprocessing techniques for text mining-an overview. Int. J. Comput. Sci. Commun. Netw. **5**(1), 7–16 (2015)
10. Vijayarani, S., Janani, R., et al.: Text mining: Open source tokenization tools-an analysis. Adv. Comput. Intell. Int. J. (ACII) **3**(1), 37–47 (2016)
11. Kaur, J., Buttar, P.K.: A systematic review on stopword removal algorithms. Int. J. Futur. Revolut. Comput. Sci. Commun. Eng. **4**(4), 207–210 (2018)
12. Jivani, A.G.: A comparative study of stemming algorithms. Int. J. Comp. Tech. Appl. **2**(6), 1930–1938 (2011)
13. Sarkar, K.: A hybrid approach to extract key-phrases from medical documents (2013). arXiv preprint arXiv:1303.1441
14. Rafiei-Asl, J., Nickabadi, A.: Tsake: A topical and structural automatic key-phrase extractor. Appl. Soft Comput. **58**, 620–630 (2017)
15. Matsuo, Y., Ishizuka, M.: Keyword extraction from a single document using word co-occurrence statistical information. Int. J. Artif. Intell. Tools **13**(01), 157–169 (2004)
16. Papagiannopoulou, E., Tsoumakas, G.: A review of key-phrase extraction. Wiley Interdisc. Rev. Data Min. Knowl. Discovery **10**(2), e1339 (2020)
17. Meng, R., Zhao, S., Han, S., He, D.: Deep key-phrase generation. In: Proceedings of the 55th Annual Meeting of the Association for Computational Linguistics, vol. 1, Long Papers. Association for Computational Linguistics, Vancouver, pp. 582–592 (2017)
18. Mihalcea, R., Tarau, P.: Textrank: Bringing order into text. In: Proceedings of the 2004 conference on empirical methods in natural language, pp. 404–411 (2004)
19. Brin, S., Page, L.: The anatomy of a large-scale hypertextual web search engine. Comput. Netw. ISDN Syst. **30**(1), 107–117 (1998)

20. Wan, X., Xiao, J.: Single document keyphrase extraction using neighborhood knowledge. AAAI **8**, 855–860 (2008)
21. Bougouin, A., Boudin, F., Daille, B.: Topicrank: Graph-based topic ranking for key-phrase extraction. Int. Joint Conf. Nat. Lang. Process. (IJCNLP), 543–551 (2013)
22. Shinde, P., Govilkar, S.: A systematic study of text mining techniques. Int. J. Nat. Lang. Comput. **4**(4), 54–62 (2015)
23. Rousseau, F., Vazirgiannis, M.: Main core retention on graph-of-words for single document keyword extraction: Advances in Information Retrieval, 382–393 (2015)
24. Batagelj, V., Zaveršnik, M.: Fast algorithms for determining (generalized) core groups in social networks. Adv. Data Anal. Classif. **5**(2), 129–145 (2011)
25. Wang, R., Liu, W., McDonald, C.: Corpus-independent generic key-phrase extraction using word embedding vectors. In: Proceedings of the Software Engineering Research Conference, pp. 1–8 (2014)
26. Florescu, C., Caragea, C.: Positionrank: An unsupervised approach to key-phrase extraction from scholarly documents. In: Proceedings of the 55th Annual Meeting of the Association for Computational Linguistics, vol. 1, Long Papers, pp. 1105–1115 (2017)
27. Boudin, F.: Unsupervised key-phrase extraction with multipartite graphs (2018). arXiv preprint arXiv:1803.08721
28. Škrlj, B., Repar, A., Pollak, S.: RaKUn: Rank-based keyword extraction via unsupervised learning and meta vertex aggregation. In: Proceedings of the International Conference on Statistical Language and Speech Processing, pp. 311–323. Springer, Cham (2019)
29. Hulth, A.: Improved automatic keyword extraction given more linguistic knowledge. In: Proceedings of the 2003 Conference on Empirical Methods in Natural Language Processing, pp. 216–223 (2003)
30. Kim, S.N., Medelyan, O., Kan, M.Y., Baldwin, T., Pingar, L.P.: Semeval-2010 task 5: Automatic key-phrase extraction from scientific articles. In: Proceedings of the 5th International Workshop on Semantic Evaluation, pp. 21–26 (2010)

Sentiment Analysis of Moroccan Dialect Using Deep Learning

Boutaina Hdioud[1](✉) and Mohammed El Haj Tirari[2]

[1] High National School for Computer Science and Systems Analysis - ENSIAS Cordially, Rabat, Morocco
hdioud.boutaina@hotmail.fr

[2] National Institute of Statistics and Applied Economics (INSEA), Rabat, Morocco

Abstract. In recent years, with the proliferation of the internet and the social media, there are massive numbers of users who share their contents over wide range of social networks. Thus, a huge volume of electronic data is available on the Internet containing the users' thoughts, attitudes, views and opinions towards certain products, events, news or any interesting topics. Therefore, Sentiment Analysis (SA) became one of the most active research areas in Natural Language Processing (NLP). Most research efforts in the area of opinion mining deal with English text and little work is done with Arabic text. In this paper we presents a tool for sentiment analysis comments written in Moroccan dialect. This tool is based on an approach combining the use of the Convolutional Neural Network (CNN) model and Word2Vect. We evaluated our approach using a dataset of moroccan dialect. Our result obtained achieves a score of 60%.

Keywords: Sentiment analysis · CNN · Word2Vec

1 Introduction

The Over the last few years, sentiment analysis has attracted much interest from researchers, Sentiment analysis was and still one of the main fields in natural language processing, its considered as the study of people's opinions, emotions and sentiments towards products, events services, organizations, individuals, issue, etc.

With the growth of data that have been produced with social media in the world like Facebook, Youtube and Twitter, sentiment analysis considered as the efficient way to analysis this data and generate important information for other purpose, it has spread from computer science to management sciences and social sciences such as marketing, finance, political science, communications, health science, and even history, to business and society. This is due to its importance and the fact that opinions are the key influence of others people choices that they make based on what others think of the same choice. However, finding and analyzing opinions on the Web and social media and the information contain in them is a huge task. Existing research has produced several techniques for sentiment analysis with machine learning methods, that have been done a great job to improve the accuracy of SA, but since years deep learning considered as a powerful

M. Lazaar et al. (Eds.): BDIoT 2021, LNNS 489, pp. 457–466, 2022.
https://doi.org/10.1007/978-3-031-07969-6_34

machine learning technique and produced state of-the-art results in many application domains such as computer vision, speech recognition to NLP.

Applying deep learning to sentiment analysis has also become very popular recently in order to benefit from two fields, the main contribution in this paper is to propose an approach for Moroccan sentiment analysis using a CNN approach.

Moroccan dialect is a chellenging language it contains several problems but what we notice is that shares a lot of difficulties with the leading languages (English) like misspelling words and same words mean different context we choose to collect several comments from youtube videos divided into two classes positive and negative, we build a word2vec simulation from this data as a begining as a pretrained words representation and applied the CNN algorithm to this data and compare results, the results was realand efficient.

The remainder of the paper is organized as follows; Sect. 2 we present sentiment analysis techniques, Sect. 3 presents some related work. In Sect. 4 we describe our proposed approach. Results have been presented in Sect. 5. Finally, we end with a conclusion of this research.

2 Sentiment Analysis

This section gives an overview on the sentiment analysis techniques. Since the paper focuses on sentiment analysis of moroccan dialect, this section starts with some basic characteristics of Moroccan dialect.

2.1 Moroccan Dialect

In recent years in moroccan social network we observe that users try to write differents kind of their opinion by dialect moroccan with arabic letters this choice of written it became widespread in youtube videos, facebook post, etc. what make it helpful in NLP task is that the problem of written a single word with different forms that we had before with moroccan dialect with foreign letters Slightly faded or become less common. Moroccan data is a challening language to analysis from severals factors, Among them written same words with differents meaning (ambiguity) this issues Not limited to a specific language we can face the same problem with english, the second problem that constitutes an obstacle in sentiment analysis with moroccan dialect is that when we remove or add a caractere it can change the word meaning in a phrases, for the algorithm each one of them mean an unique word and have an unique indice in the vocabulary, this can be the same issue with misspelling. For me those problem are not limited to the moroccan dialect then the choice of working with moroccan comments not arbiratry but with the challenges that we face before with english data was helpful to encouter the problems and the solutions.

2.2 Sentiment Analysis Techniques

Sentiment analysis identifies, extracts, and studies the attitude of the writer. The main task in sentiment analysis is classifying a document or a sentence based on a polarity.

The polarity of document or sentence can be positive or negative. The existing techniques for sentiment analysis can be classified into three groups, unsupervised technique, supervised technique by means of machine learning, and hybrid technique [1].

The Unsupervised technique, also called lexicon based technique [2] usually builds a lexicon by containing roots or stems of words. Every root or stem associated with polarity (e.g. −1 for negative, +1 for positive). Basically, the Lexicon based technique consists of the following steps: dataset collection, tokenization, normalization, stop words removing, and stemming.

A lexicon of stems is constructed after finishing the previous steps. Each stem in the lexicon is attached to polarity. Considering an input dataset of reviews, each review is tokenized into set of tokens. The lexicon technique searches for the token in the lexicon to get the polarity. The polarity of an inputted review is calculated by summing all polarities of emotional tokens in the review.

In contrast to the lexicon based techniques, the supervised technique machine learning method uses several learning algorithms to determine the sentiment by training on a known dataset. In this technique, a model is constructed using the extracted features based on one or more machine learning algorithms for example, Support Vector Machine (SVM) [3] or Naïve Bays (NB) [4].

A dataset of reviews is used to test and evaluate the model based on the percentage of the true classified polarities.

Finally, the hybrid technique [2] combines both of the supervised and unsupervised techniques. The output of lexicon based technique is used as input for machine leaning algorithm.

3 Related Work

Recently, deep learning has been widely used in NLP research community, including sentiment analysis in various languages, many researchers have been there to combine deep learning and machine learning concepts for the classification of sentiments,but since years deep learning considered as a powerful machine learning technique and produced state of the art results in many application domains such as computer vision, speech recognition to NLP. This section describes the several studies, related to sentiment analysis of user's opinions using the different deep learning algorithms, [5] was the first paper that applied CNN to NLP task and especially in sentiment analysis they tried to adapt a simple one convolution neural network to each sentence, they applied many models CNN-rand, CNN-static, CNN-non static and also using the pre trained word2vec with the models to gain appropriate results that attained 90. The authors [6] have proposed a convolutinal neural network for sentiments analysis with Tweets, the main task here was trained their own words embedding in a unsupervised neural network with tweets and use this representation to train a CNN, in the level of the sentence matrix they add an emoticon for positive and negative tweets which were encoded in one hot. The corpus were trained an activation function and a pooling layer and a softmax to compute the probabilities of classes the model was applied in two tasks: message level and phrase level,the proposed model atteind 80 in accuracy. In [7] the authors used an RNTN, which was trained on a constructed sentiment treebank and improved the sentence-level sentiment analysis on English datasets. After the emergence of word embeddings developed in [9], the research in NLP has taken another turn. The authors of [3] used a Recurrent Neural Networks (RNN) for fine-grained opinion analysis. In [8] an LSTM was used for predicting tweets polarity. In [16] CNN and LSTM were combined for text sentiment classification. The author of [12] defined an ensemble model based on deep neural network architectures including CNNs and LSTMs to achieve the first rank on all of the five English sub-tasks in SemEval-2017.

Despite all of the work done on English sentiment analysis using deep learning, little work has been done on Arabic data. In [13] RAE was used for Arabic text sentiment classification. The authors of [14] used an RNTN to predict the sentiment of Arabic tweets. They built a sentiment treebank [15] to be used by their model. They reported the performance of two state-of-the-art models for opinion mining in the English language when applied on Arabic data. They showed that using deep learning achieved the state-of-the-art results for the Arabic sentiment analysis.

4 Proposed Approach

In this section, we illustrate the main stages of our system. First, we collecting data from Yutube to form a dataset, then we pass through a preprocessing and cleaning phase, to remove unwanted symbols and tokens. We build a Word2Vec for this data training, finally applied The CNN with the pre-trained vectors in the data training.

4.1 Collecting Data

In the objective to applied the CNN in moroccan dialect, we were collected by hands comments from Youtube videos different in terms of categories (art, decomontary, sport). Then we apply a preprocessing and cleaning phase.

4.2 Data Preprocessing and Cleaning

In this phase, we clean the comments and remove noise and unwanted symbols As we could notice comments could consist of text, Hashtags, links, usernames, special characters, NA, emoticons and numbers etc. The goal of this phase is to maximize the number of words whose embeddings can be found in the pre-trained word embedding model.

4.3 Word Representation

Word vectors are real-number vectors that represent words from a vocabulary, in NLP those vectors are the input of our models in deep learning when we use them, it can increase the model accuracy significantly, through their presence they allow us to implicitly include external information from the world into our language understanding models.

5 Build Vocabulary for Our Comments:

In order to understand how deep learning deal with text data, think forms of data that are used as inputs in deep learning models. For example, Convolutional neural network for image use arrays of pixel values, the common thing here is that the inputs need to be scalar values, in our context of sentiment analysis thing should be going like this: There is no way for us to do common operations like dot products or backpropagation on a single string. Instead we will need to convert each word in the sentence to a vector this is called a vocabulary for our training corpus (Figs 1 and 2).

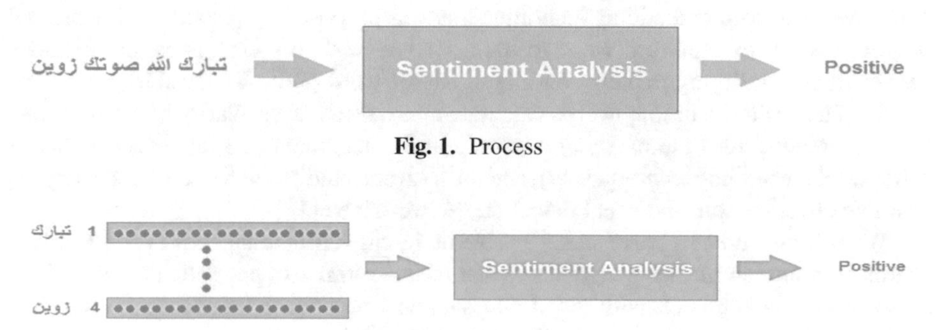

Fig. 1. Process

Fig. 2. Each word convert to a vector

The data corpus will be represented as a matrix of array which each word represented by an array of indices, the next step now is to build an embedding layer where each word will be represented by a set of feature and each sentence will be represented by a matrix of 3D.

Embedding Layer

After building the vocabulary for our sentences time to build an embedding layer each sentence ex: بصراحة صوت واعر مافيهش المؤثرات الصوتية for each word it converts to a vector of D-dimensional and then the sentence become a matrix [l, D], l is the length of the sentence and D column represent the number of embeddings vectors(word representation) Our final input to any model of deep learning taken as shown in Fig. 3.

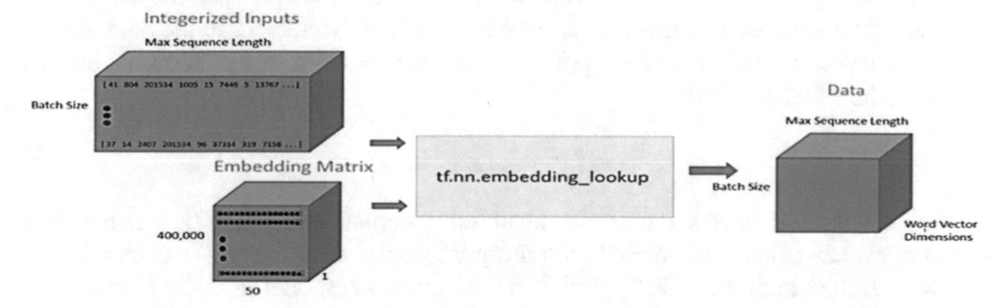

Fig. 3. Embedding vectors

In practice with tensorflow and python we can build an embedding layer for our corpus with tf.nn.embedding() which it takes as input two parameters:

W: a tensor of [|v|, D].

|v|: the size of the vocabulary.

The tensor contain the words embedding vectors (embedding Matrix) for each word in our vocabulary, this can be build by training a neural network from our corpus or use a pretrained word2vec from google or glove from amazong. In our case we used a pretrained model from google.

Morrocan Dialect with Word2Vec

Word2vec is a word embedding technique that was proposed by [9] are N-dimensional vectors that try to capture a word meaning and context in their values and includes two learning algorithms, namely continuous bag-of-word (CBOW) and skip-gram algorithms. The similarity among words calculated via the cosine similarity of word vectors in word2vec includes the meaning of words in the document and this exceeds that of other word embedding techniques [9], and thus, several studies such as emotion analysis, emotion classification, and event detection use word2vec [11].

Word2vec comes to solve many problems in english language and tried to place words with similatr meaning together so that the algorithm can perfectly genrelize from it, we tried this approach with moroccan data and we build a word2vec for our data written with arabic letters, we could in the end build a word representation for our training data and we compare in the next section the two experiments of applying the Cnn with word2vec and without pretrained word2vec.

5.1 Convolutional Neural Network

In a convolutional neural network (CNN) is an artificial neural network, is one of the main reasons why deeplearning is so popular today. That is frequently used in various fields such as image classification, face recognition, and natural language processing.

CNN Model

We first build a CNN model to predict the sentiment of the Moroccan dialect. CNNs are well-known for their good performance in classification problems. They are famous for their ability to extract important and relevant features that help with the classification task. We follow the same approach in [12]. As it is shown in Fig. 4, sentence is first fed into the embedding layer, which converts words into word vectors. Then the sentence is mapped into a matrix M of size $n \times d$. we use pretrained word2vec. In the convolution layer, convolution operations are applied on the submatrixes of M. The convolution operation here is defined as:

$$c_{k}= f_k\left(\sum_i \sum_j \omega_{ij} x_{[i:i+h-1]} + b\right) \quad (1)$$

where $b \in R$ is a bias term and f is a nonlinear function such as ReLU [17], which is used in our approach. Filters are applied with different size of windows and in each window of size h, feature matrix $c \in \mathbb{R}^{(n-h+1)\times m}$ is produced corresponding to the filters:

$$c = [c_1,\ c_2,\ \ldots,\ c_k,\ \ldots,\ c_m] \quad (2)$$

where m is the number of filters and $c_k \in \mathbb{R}^{n-h+1}$ represents the features extracted from a word sequence. In the pooling layer, we apply a max-over-time pooling operation over feature matrix and take the maximum in each column to preserve the most important features. These maximums are concatenated and then fed into a fully-connected network (L1, L2). L2 is followed by a single sigmoid neuron node to generate the prediction of the affect on the interval [0, 1].

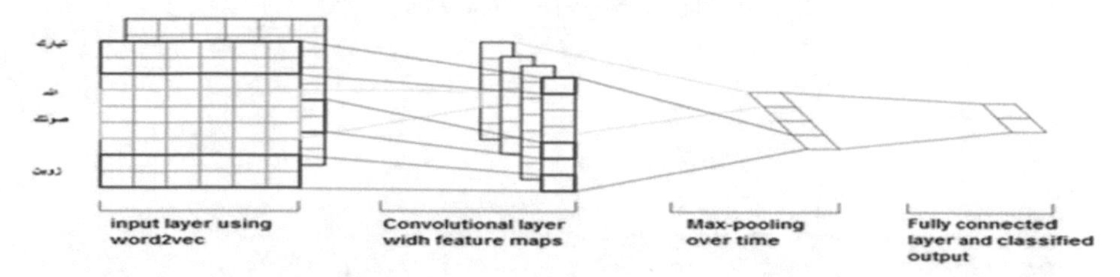

Fig. 4. CNN

6 Experimental Results

This section presents the experimental results of the proposed approach. With moroccan dialect we have some interset result using deep learning and Word2vec. Our aproach was created as following:

1. Collect Data from youtube videos from different categories (political, art, sport)
2. Build a Word2Vec for this data training those words vectors are important as input to the algorithm that we choose to work with
3. Applied The CNN with the pre-trained vectors in the data training that we collect.

Some results obtained in different phase are presented her (Figs 5, 6, 7, 8 and 9):

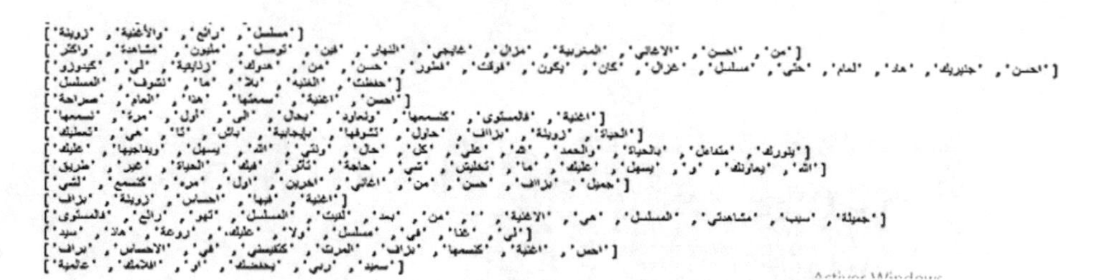

Fig. 5. Positive comments

As an observation we notice that normal sentence that describe the strong sentiment with positive words or negative words have a high accuracy with even without using a word2vec representation but when using some complex words with differents sentiments

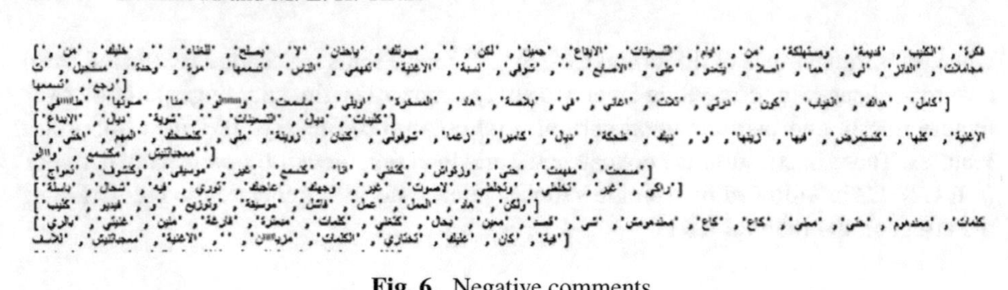

Fig. 6. Negative comments

```
import random
model.build_vocab(sentences=shuffle_corpus(corpus),update=True)

for i in range(5):
    model.train(sentences=shuffle_corpus(corpus),epochs=50,total_examples=model.corpus_count)
```

Fig. 7. Training a Word2vec

```
print(model['صوت'])

[-0.1105065   0.07059487  0.6685446   0.7933454  -0.38859662 -0.00590154
 -0.7606256  -1.1583866  -0.4720125  -1.0994427  -0.01407227  0.13592432
  0.14976397  1.0652283  -0.40751204 -0.18316865  0.14899202  0.11945751
  0.62688315  0.07266276 -0.07833085  0.31221497  0.8111801   0.8076049
 -0.21084471 -0.30218685 -0.3121586  -0.23633878 -0.66560185 -0.12840234
 -0.18638353 -0.9787066  -0.16056776 -0.13025917 -0.5878731  -0.758497
 -0.2504959  -0.468739    0.31967324  0.49697217 -0.45543998  0.04261597
  0.12917736 -0.21706252 -1.5386109  -0.5034324   0.22730258  0.75268424
  0.73258233  0.84359336 -0.71956056 -0.58518964 -0.1844755  -0.32893515
  0.00911259  0.29402563  0.4159931   0.04152976  0.50217354 -0.2750031
 -0.12986067 -0.6246914  -0.880771    0.0889065   0.21130519  0.06172145
  0.74942684 -0.34234682  0.41022336 -0.23408912 -0.44788694 -0.6063767
 -0.7932548   0.361528   -0.21933128  0.3174411  -0.8131043   0.792978
 -0.4128882   0.43468326 -0.7406508   0.64659     0.5873478  -0.17628261
 -0.38886315  0.3703026  -0.99219567 -0.5242028   0.2882508  -0.27422726
  0.03659648 -0.89389753 -0.3492958   0.23697774 -0.57943    -0.06989075
  0.08583447  0.00976471  0.5235268  -0.23026234]
```

Fig. 8. Word representation of a specific word

```
model.most_similar('زوينة')

[('بالروجولة', 0.49954551458358765),
 ('زوبيبينة', 0.48973318934440613),
 ('شابه', 0.48581668734550476),
 ('عجبتني', 0.48257160186767 58),
 ('مرزنة', 0.4818100333213806),
 ('مملة', 0.4722088575363159),
 ('عدي', 0.46502295136 45172),
 ('الهضرة', 0.46470907330513),
 ('شبابية', 0.4611663222312927),
 ('ممتازة', 0.46004408597946167)]
```

Fig. 9. Similar word representation of a specific word

in the same sentence to describe whatever negative or positive sentiment we should know the context meaning of words and thats when is important to use a vector representation as Word2vec it plays a role and how the relate together, our test accuracy do not exceed in test data that is similar to train data but is still an interest result.

We tried to applied the CNN in moroccan dialect,we were collected comments from youtube videos different in terms of categories the comments were labeled into two classes positive and negative we applied the convolutional neural network in 100 epoch the test accuracy achieve (Table 1).

Table 1. Deep learning for sentiment analysis with dialect Moroccan.

	Accuracy
CNN without pre-trained word2vec	56%
CNN with pre-trained word2vec from google	60%

The results obtained show that the proposed CNN classification model used with word2vec is better than the CNN classification model without word2vec.

7 Conclusion

In the study, we evaluated the use of word2vec in classification model via CNN based on moroccan dialect. We examined the effect of using word2vec on the results and compared the performance of word2vec. Our perspective is to continue the researchers in this field through the different algorithm from deep learning with the moroccan dialect and try to use the different approach for the word representations.

References

1. Yue, L., Chen, W., Li, X., Zuo, W., Yin, M.: A survey of sentiment analysis in social media. Knowl. Inf. Syst. **60**(2), 617–663 (2018). https://doi.org/10.1007/s10115-018-1236-4
2. Medhat, W., Hassan, A., Korashy, H.: Sentiment analysis algorithms and applications: a survey. Ain Shams Eng. J. **5**(4), 1093–1113 (2014)
3. Vapnik, C., Vladimir, C.: Support-vector networks. Mach. Learn. **20**(3), 273–297 (1995). https://doi.org/10.1007/BF00994018
4. Russell, S.J., Norvig, P.: Artificial Intelligence: A Modern Approach. Prentice Hall/Pearson Education, California (2003)
5. Kim, Y.: Convolutional neural networks for sentence classification, pp. (2, 28, 30, 37, 39, 46) (2014). https://arxiv.org/abs/1408.5882
6. Severyn, A., Moschitti, A.: Twitter sentiment analysis with deep convolutional neural networks. In: Proceedings of the 38th International ACM SIGIR Conference on Research and Development in Information Retrieval, pp. 959–962. ACM (2015)
7. Socher, R., et al.: Recursive deep models for semantic compositionality over a sentiment Treebank. In: Proceedings of the 2013 Conference on Empirical Methods in Natural Language Processing, pp. 1631–1642 (2013)

8. Wang, X., Liu, Y., Chengjie, S., Wang, B., Wang, X.: Predicting polarities of tweets by composing word embeddings with long short-term memory. In: Proceedings of the 53rd Annual Meeting of the Association for Computational Linguistics and the 7th International Joint Conference on Natural Language Processing, vol. 1, pp. 1343–1353 (2015). Long Papers
9. Mikolov, T., Sutskever, I., Chen, K., Corrado, G.S., Dean, J.: Distributed representations of words and phrases and their compositionality. In: Advances in Neural Information Processing Systems, pp. 3111–3119 (2013)
10. Peng, H., Song, Y., Roth, D.: Event detection and co-reference with minimal supervision. In: Proceedings of the 2016 Conference on Empirical Methods in Natural Language Processing, pp. 392–402 (2016)
11. Liu, H.: Sentiment analysis of citations using word2vec. https://arxiv.org/abs/1704.00177
12. Cliche, M.: BB twtr at SemEval-2017 Task 4: Twitter Sentiment Analysis with CNNs and LSTMs (2017). https://arxiv.org/abs/1704.06125
13. Al Sallab, A., Hajj, H., Badaro, G., Baly, R., El Hajj, W., Shaban, K.B.: Deep learning models for sentiment analysis in Arabic. In: Proceedings of the Second Workshop on Arabic Natural Language Processing, pp. 9–17 (2015)
14. Baly, R., et al.: A characterization study of Arabic twitter data with a benchmarking for state-of-the-art opinion mining models. In: Proceedings of the Third Arabic Natural Language Processing Workshop, pp. 110–118 (2017)a
15. Baly, R., Hajj, H., Habash, N., Shaban, K.B., El-Hajj, W.: A sentiment treebank and morphologically enriched recursive deep models for effective sentiment analysis in Arabic. ACM Trans. Asian Low-Resour. Lang. Inf. Process. **16** (2017)b
16. Zhou, C., Sun, C., Liu, Z., Lau, F.: A C-LSTM neural network for text classification (2015). https://arxiv.org/abs/1511.08630
17. Jarrett, K., Kavukcuoglu, K., Ranzato, M., LeCun, Y.: What is the best multi-stage architecture for object recognition? In: Proceedings of the International Conference on Computer Vision (ICCV 2009). IEEE (2009)

Reinforcement Learning

Policy Gradient for Arabic to English Neural Machine Translation

Mohamed Zouidine[1(✉)], Mohammed Khalil[1], and Abdelhamid Ibn El Farouk[2]

[1] LMCSA, FSTM, Hassan II University of Casablanca, Casablanca, Morocco
mohamed.zouidine-etu@etu.univh2c.ma, mohammed.khalil@univh2c.ma
[2] LLEC, FLSH, Hassan II University of Casablanca, Casablanca, Morocco
abdelhamid.farouk@univh2c.ma

Abstract. In this work, we present a strategy of using the policy gradient training method from deep reinforcement learning to train a Seq2Seq model. This strategy is based on the combination of the classical cross-entropy and the policy gradient in the training phase. To evaluate the effectiveness of this strategy, we compare two Seq2Seq models trained with two different training methods to translate from Arabic to English. The first method is the cross-entropy, and the second one is a combination of cross-entropy and policy gradient with different amounts. Experimental results show that the second training method leads to improve the performance of the sequence-to-sequence model by 0.71 BLEU points.

Keywords: Sequence-to-sequence · Machine translation · BLEU · Reinforcement learning · Policy gradient

1 Introduction

As a result of the success of deep neural networks, neural machine translation (NMT) [3,6,7,18,23] has outperformed the traditional statistical phrase-based machine translation [10]. Like for many languages (German-English, French-English, Chinese-English, ...), also for Arabic, several works [1,2,5,14,24] have shown the strong performance of the NMT against the statistical machine translation (SMT).

The idea of NMT is to use an encoder to encode a given sentence in the source language into a continuous vector and then use a decoder to generate the translation word by word in the target language. Generally, the encoders and decoders are trained by the cross-entropy (CE) training method. This training method leads to a discrepancy between training and testing phases. This discrepancy is a result of the use of the metric BLEU [15] at the testing time for the machine translation task.

Optimization methods from reinforcement learning (RL) [19], such as Policy Gradient (PG), can fill the gap between the training and the testing times, as it can directly optimize the metric BLEU at the training time. Many works

M. Lazaar et al. (Eds.): BDIoT 2021, LNNS 489, pp. 469–480, 2022.
https://doi.org/10.1007/978-3-031-07969-6_35

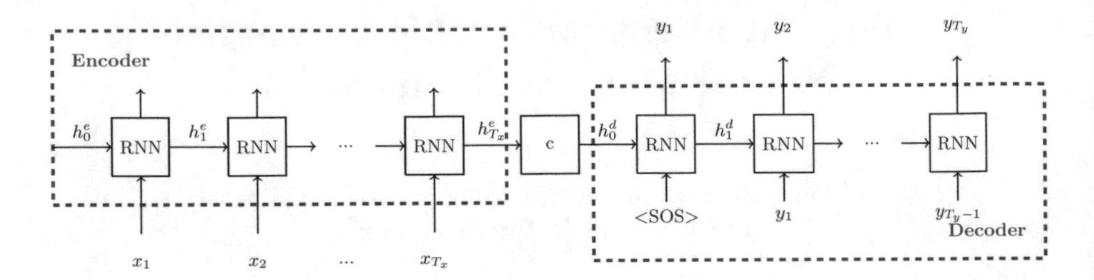

Fig. 1. Sequence-to-Sequence model for NMT. The blue side corresponds to the encoder, and the red one corresponds to the decoder. Note that we suppose that each sentence starts with a special start-of-sentence symbol "<SOS>".

[7,11,12,16,22,23] have demonstrated the effectiveness of RL techniques for training NMT models.

In this work, we compare two models of NMT using Arabic as the source language and English as the target. The first model was trained by the CE training method, whereas the second one by a combination of CE and the policy gradient (PG) training method from RL. The experiments have shown that the model trained by the combination of CE and PG performs better than the one trained only by CE.

The rest of the paper is organized as follows. In the next section, we provide the background on which this work is based. The strategy of using the PG training method to train a Seq2Seq model for translation is introduced in Sect. 3. Section 4 describes the experiments and the results obtained. Finally, in Sect. 5, the conclusion and some future works are presented.

2 Background

2.1 Neural Machine Translation

Typically, NMT models are implemented with a Recurrent Neural Network (RNN) based encoder-decoder approach. In this approach, the encoder (Fig. 1) reads a source sentence $x = \{x_1, x_2, \ldots, x_{T_x}\}$, T_x is the length of the sentence x, and generates a hidden representation for each x_i

$$h_i^e = f^e\left(h_{i-1}^e, x_i\right) \tag{1}$$

and a context vector c

$$c = h_{T_x}^e$$

f^e is a nonlinear function, the popular choices for f^e are Long Short-Term Memory (LSTM) [8] and Gated Recurrent Units (GRU) [4].

On the other side, the decoder (Fig. 1) uses the context vector to generate a target sentence $\hat{y} = \{\hat{y}_1, \hat{y}_2, \ldots, \hat{y}_{T_{\hat{y}}}\}$, $T_{\hat{y}}$ is the length of the sentence $\hat{y}$, word by word, based on the current hidden state of the decoder h_i^d and the ground-truth token y_i:

$$h_i^d = f^d\left(h_{i-1}^d, y_i\right) \tag{2}$$

as for the encoder, f^d is a nonlinear function and the popular choices for it are LSTM and GRU. Note that for the decoder $h_0^d = c$, while for the encoder generally $h_0^e = 0$.

The model is trained to predict the next token $\hat{y}_i$ by estimating the conditional probability $p\left(\hat{y}\middle|x;\, \theta\right)$:

$$p\left(\hat{y}\middle|x;\, \theta\right) = \prod_{i=1}^{T_y} p\left(\hat{y}_i\middle|\{y_1,\, \dots,\, y_{i-1}\},\, c;\, \theta\right) \tag{3}$$

where θ is the model parameters.

CE is the most used optimization method for training the model, and the objective function is defined as:

$$L_{CE} = -\sum_{i=1}^{T_y} \log p\left(\hat{y}_i\middle|\{y_1,\, \dots,\, y_{i-1}\},\, x;\, \theta\right) \tag{4}$$

where T_y is the length of the ground-truth sentence y.

Once the model is trained with this objective function, it translates a sentence by choosing the token with the highest probability:

$$\hat{y}_i = \underset{\hat{y}}{\operatorname{argmax}}\ p\left(\hat{y}\middle|\{\hat{y}_1,\, \dots,\, \hat{y}_{i-1}\},\, x;\, \theta\right) \tag{5}$$

Given y the ground-truth sentence and $\hat{y}$ the model generated sentence, the performance of the model is evaluated with a special measure.

2.2 Performance Evaluation of Machine Translation

For machine translation, the most used measure to evaluate a model is the BLEU score [15].

The BLEU metric evaluates the performance of the model by comparing n-grams of the candidate (the model generated sentence) with the n-grams of the reference (the ground-truth sentence) translation and counts the number of matches. The BLEU algorithm identifies all matches of n-grams and evaluates the performance of the match with a precision score, such that for each n-gram size, a gram in the reference sentence cannot be considered more than once. The precision score is defined as follows:

$$p_n = \frac{\#\text{ of matched n-grams}}{\#\text{ of n-grams in candidate sentence}} \tag{6}$$

A brevity penalty is used to impose that very small candidates that would achieve a precision of 1 are not supposed good translations. The brevity penalty is defined as:

$$\mathrm{BP} = e^{\min(0,\, 1-r/c)} \tag{7}$$

Algorithm 1: Training a seq2seq model.

Input: Sentences with their corresponding translation.
Result: Model optimized for translation.
Training Steps:
initialize the model at random and set N (the number of training epochs);
train the model for N epochs using Equation (4);
Testing Steps:
use the optimized model and Equation (5) to generate output $\hat{y}$;
evaluate the model using BLEU score;

where c and r are respectively the length of the candidate and the reference sentences. The BLEU score is then:

$$\text{BLEU} = \text{BP}.\exp\left(\sum_{n=1}^{N} w_n \log p_n\right) \tag{8}$$

w_n is a geometric weighting, $w_n = \frac{1}{2^n}$, and N is the maximum n-gram that the score is evaluated on, generally $N = 4$.

The BLEU score ranges from 0 to 1, and a candidate is better if their score is close to 1. The Algorithm (1) shows the steps of training and evaluating a Seq2Seq model.

2.3 Policy Gradient

Reinforcement learning (RL) [19] problems can be represented as a system consisting of an *agent* and an *environment.*

The agent and the environment interact in a series of discrete time steps (can be extended to the continuous case). At each time step t, the agent receives some information that describes the *state* s_t of the system, using this information the agent selects an *action* a_t. One time step later, $t + 1$, as a consequence of the action, the agent receives a *reward* r_{t+1} and passes into a new state s_{t+1}. The cycle (state $\rightarrow$ action $\rightarrow$ reward) repeats until the agent achieves a terminal state s_T. This process is described in Fig. 2.

The goal of the agent interacting with the environment is to maximize the sum of rewards received for each action, which means learning to choose suitable actions. In other words, the agent learns a mapping from states to probabilities of choosing each possible action. This mapping defines the *policy* π, $\pi(a|s)$ is the probability that $a_t = a$ if $s_t = s$. Learn to choose suitable actions corresponds to learn an optimal policy. There are various ways to solve this problem. In this work, we choose to use the *policy gradient* (PG) [20] method.

In the PG method, the policy is represented using a deep neural network with parameters θ (in this case we talk about deep reinforcement learning, DRL). The key idea of PG method is to learn a parameterized policy π_θ, the agent uses this policy directly to interact with the environment. During training, actions that lead to good rewards become more probable, and actions that lead to bad rewards

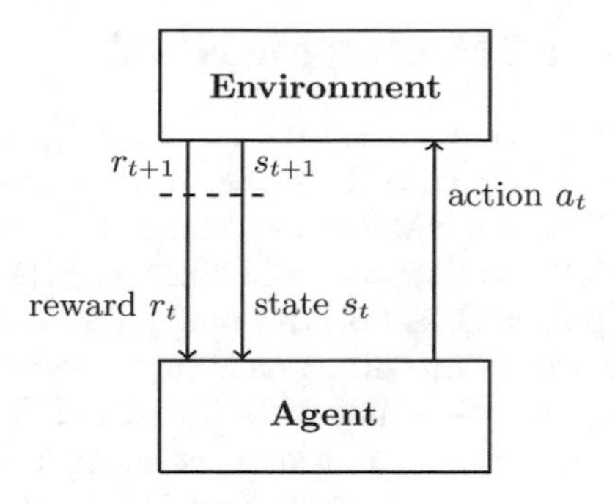

Fig. 2. Agent-environment interaction.

become less probable. Learning a good policy corresponds in this case to learn an optimal set of values for the parameters θ.

We mentioned earlier that the agent interacts with the environment in a cycle (state $\rightarrow$ action $\rightarrow$ reward). This cycle defines a trajectory denoted by:

$$\tau = s_0,\ a_0,\ r_0,\ \ldots,\ s_T,\ a_T,\ r_T$$

For each trajectory, we define the *return* $R_t(\tau)$ as the sum of discounted rewards from time step t to the end of the trajectory.

$$R_t(\tau) = \sum_{k=t+1}^{T} \gamma^{k-t-1} r_k \tag{9}$$

whit γ is the discount rate, $0 \leq \gamma \leq 1$. The return of the complete trajectory corresponds to $t = 0$, this is written as $R_0(\tau) = R(\tau)$.

Since the goal of training is to maximize the expected reward, the loss function for policy gradient is defined as follows:

$$L_{PG} = -E_{\tau \sim \pi_\theta}\left[R(\tau)\right] \tag{10}$$

The minus sign here allows us to use gradient descent instead of gradient ascent. We prove (see Supplementary Material A for the proof) that the derivative of the loss function is given as:

$$\nabla_\theta L_{PG} = -E_{\tau \sim \pi_\theta}\left[R(\tau)\sum_{t=0}^{T} \nabla_\theta \log \pi_\theta(a_t|s_t)\right] \tag{11}$$

The expectation here is calculated across many trajectories sampled from the policy π_θ, which making it impossible to exactly minimize L_{PG} if the state space is large (the same case for machine translation). One solution for this problem is the use of REINFORCE [21]. Instead of using many trajectories per policy, only one sample can be used for training at each time step:

$$\nabla_\theta L_{PG} = -R(\tau)\sum_{t=0}^{T} \nabla_\theta \log \pi_\theta(a_t|s_t) \tag{12}$$

3 Policy Gradient Training for NMT

In this section, we describe how we can use the policy gradient to train a neural machine translation model. We use the same strategy proposed by [3].

First, we suppose that the NMT model presents the agent, which interacts with the environment. The conditional probability $p\left(\hat{y}_i \middle| \{\hat{y}_1,\ \ldots,\ \hat{y}_{i-1}\},\ x;\ \theta\right)$ of the model defines the policy. Based on this policy the agent can pick an action. In machine translation, an action corresponds to generate the next word at each time step. After taking an action, the state (internal hidden state of the NMT model) is updated. Once the agent finishes generating a complete sequence $\hat{y}$, a terminal reward is observed. For machine translation, the reward is the BLEU [15] score $R(\hat{y},\ y)$, defined by comparing the generated sentence $\hat{y}$ with the ground-truth sentence y. Now, we can adopt the Eq. (12) for the NMT task as follows:

$$\nabla_\theta L_{PG} = -R(\hat{y},\ y) \sum_{t=1}^{T_{\hat{y}}} \nabla_\theta \log p\left(\hat{y}_t \middle| \{\hat{y}_1,\ \ldots,\ \hat{y}_{t-1}\},\ x;\ \theta\right) \tag{13}$$

We can observe from the Eq. (13) that the reward $R(\hat{y},\ y)$ is calculated for the entire sentence $\hat{y}$. In fact, the agent needs to pick many actions (depending on $T_{\hat{y}}$ the length of $\hat{y}$) to generate an entire sentence $\hat{y}$. One strategy to overcome this problem is the use of reward shaping [13]. With reward shaping, we define an intermediate reward at each decoding time step t as:

$$r_t(\hat{y}_t,\ y) = R\left(\{\hat{y}_1,\ \ldots,\ \hat{y}_t\},\ y\right) - R\left(\{\hat{y}_1,\ \ldots,\ \hat{y}_{t-1}\},\ y\right) \tag{14}$$

note that $R(\hat{y},\ y) = \sum_{t=1}^{T_{\hat{y}}} r_t(\hat{y}_t,\ y)$. The optimal policy does not change if we use the shaped reward instead of the the total reward [13].

Finally, we can define the loss function for the policy gradient training method as follows:

$$L_{PG} = -\sum_{t=1}^{T_{\hat{y}}} r_t(\hat{y}_t,\ y) \log p\left(\hat{y}_t \middle| \{\hat{y}_1,\ \ldots,\ \hat{y}_{t-1}\},\ x;\ \theta\right) \tag{15}$$

In general, the initial policy in reinforcement learning is random, which can be an inadequate option due to the large action space of machine translation (the target vocabulary size). Instead of starting from a random policy, we first use the CE training method to train the NMT model for N_{CE} epochs. This ensures that we start from a much better policy than a random one. After the N_{CE} epochs, we continue the training of the NMT model for N_{CE+PG} epochs, in this time we use a combination of CE and PG, such that the loss function is a linear combination of CE described in (4) and PG described in (15) as follows:

$$L_{com} = \eta L_{CE} + (1 - \eta) L_{PG} \tag{16}$$

where η is a hyperparameter that controls the passage from CE to PG loss, different values of η will be tested to evaluate its impact on the final translation performance. Algorithm (2) summarizes the process of training NMT models with policy gradient.

Algorithm 2: Policy gradient training for NMT.

Input: Sentences with their corresponding translation.
Result: Model optimized for translation.
Training Steps:
initialize the model at random and set N_{CE}, N_{CE+PG} and η;
train the model for N_{CE} epochs using Equation (4);
train the model for N_{CE+PG} epochs using Equation (16);
Testing Steps:
use the optimized model and Equation (5) to generate output $\hat{y}$;
evaluate the model using BLEU score;

4 Experiments

In this section, we experimentally compare two models. The first one is a classical NMT model [18] trained by the CE, and the second one is the same model trained with a combination of CE and PG.

4.1 Dataset

To build an Arabic-English translation data, we use the UN Parallel Corpus V1.0 publicly available[1]. This corpus is composed of manually translated documents of the United Nations from 1990 to 2014 for the six official UN languages (Arabic, Chinese, English, French, Russian, and Spanish) [25].

From this corpus, we extract a training data set comprised of 205 k sentence pairs consisting of 313,212 Arabic words and 354,292 English words. The length of each sentence in both languages ranges between 5 and 30 words. The Arabic vocabulary has 29,861 words, while the English has 15,610 words. In addition, we randomly extract a validation set and a test set each consisting of 2 k sentences.

4.2 Experimental Settings

We train two models. The first one is a sequence to sequence [18] trained with the CE, and the other one is the same model trained by the CE combined with the PG.

The encoder and decoder of the Seq2Seq consist of two layers of GRU with 512 hidden units and 256 dimensional word embeddings each, with a source vocabulary of 29,865 and a target vocabulary of 15,614. For both encoder and decoder, we apply dropout [17] to the input layer (coefficient 0.5). All the models' parameters are initialized with a unifrom distribution between −0.08 and 0.08.

The first model is trained for 100 epochs using Adam optimizer [9] with a fix learning rate ($\alpha = 10^{-3}$). While the second one is also trained for 100 epochs such that, for the first 90 epochs ($N_{CE} = 90$) we use CE as described in (4), and

[1] https://conferences.unite.un.org/UNCorpus.

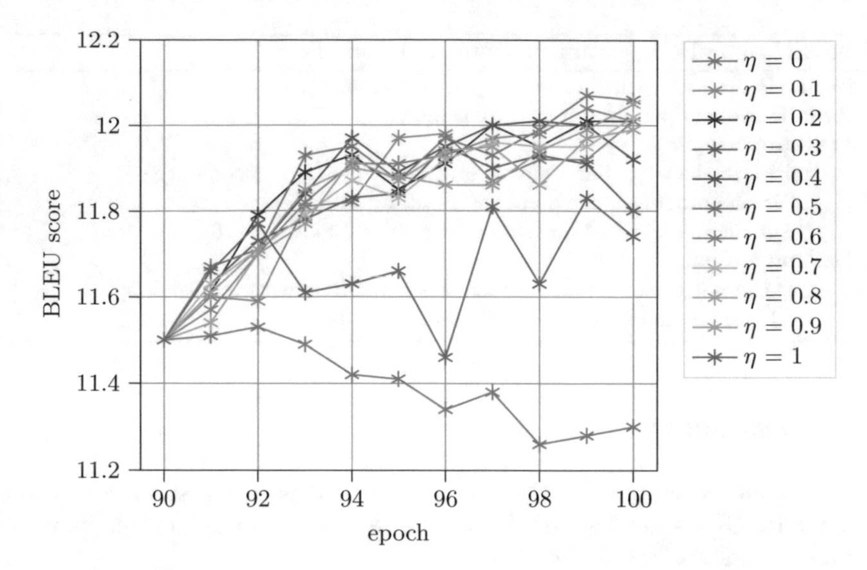

Fig. 3. BLEU scores for different η on the validation set for the last 10 epochs of model training. Note that $\eta = 1$ corresponds to the traditional Seq2Seq model [18] trained only with CE.

for the 10 epochs ($N_{CE+PG} = 10$) remaining we use CE and PG as described in (16), for both cases we use Adam optimizer ($\alpha = 10^{-5}$). For inference, we compare both greedy search and beam search.

4.3 Results

In order to compare the two training strategies, we calculate the BLEU score on the validation and test datasets.

As shown in Eq. (16), the hyperparameter η controls the passage between CE and PG loss. First, we compare different values of η by setting

$$\eta = \{0,\ 0.1,\ 0.2,\ \ldots,\ 1\}$$

to evaluate how it impacts the performance of the translation. Figure 3 presents the results on the validation set. It is clear that the combination of CE and PG is helpful to achieve better performance. We observe that $\eta = 0$ (PG without combination with CE) leads to reduce the performance of the model, and this indicates that CE is important to stabilize the training. Finally, we find that $\eta = 0.5$ is the best setting of our hyperparameter. Therefore, we set $\eta = 0.5$ for the rest of this work.

We compare two models, the first one is a traditional Seq2Seq [18] trained with the CE, and the second one is the same model trained with a combination of CE and PG as we describe in Algorithm 2. The results are presented in Table 1.

Table 1. BLEU score on the test set.

Model	BLEU score
Seq2Seq [18]	11.6
Seq2Seq+PG	**12.31**

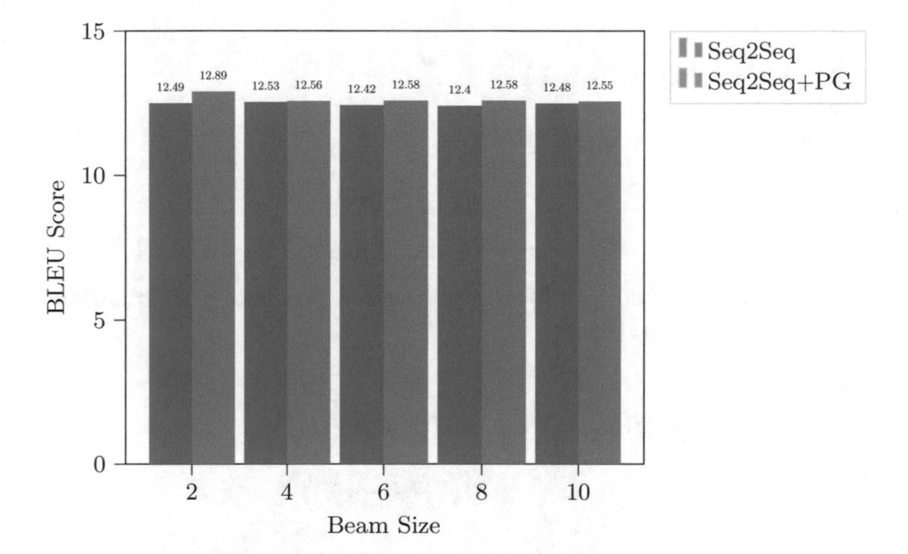

Fig. 4. BLEU scores on the test set for different beam sizes.

From Table 1, we can notice that the model trained by a combination of CE and PG has an improvement of 0.71 BLEU points against the model trained only with CE. This is due to the fact that the PG training loss is calculated using the BLEU score, which means that the model tries to optimize the BLEU score during the training phase. These results show the importance of directly optimizing the metric used at the test phase (BLEU score) during the training phase.

Additionally, we experimented with beam search. Figure 4 presents the BLEU score on the test set for different beam sizes. In both cases, we notice that the quality of generation has been improved using beam search. However, the model trained with a combination of CE and PG has outperformed the one trained only with CE in all beam sizes.

5 Conclusion

In this work, we presented the strategy of using the policy gradient training method from deep reinforcement learning to train a neural machine translation model to make a translation from Arabic to English. Two models were compared, the first one was a Seq2Seq model trained only with CE, and the second one was trained with a combination of CE and PG. We found that a model trained with a combination of CE and PG performs better than a model trained only with CE. In our future work, we aim to use other deep reinforcement learning algorithms like Actor-Critic. Furthermore, we aim to experiment with other decoding techniques besides greedy search and beam search, such as multinomial sampling.

A Proof of Equation 11

Here we demonstrate the Eq. (11).
We have the Eq. (10):

$$L_{PG} = -E_{\tau \sim \pi_\theta}\left[R(\tau)\right]$$
$$\nabla_\theta L_{PG} = -\nabla_\theta E_{\tau \sim \pi_\theta}\left[R(\tau)\right]$$

Given a function $f(x)$, $p(x|\theta)$ a parametrized probability distribution, and its expectation $E_{x \sim p(x|\theta)}\left[f(x)\right]$:

$$\begin{aligned}
\nabla_\theta E_{x \sim p(x|\theta)}\left[f(x)\right] &= \nabla_\theta \int dx\ f(x)p(x|\theta)) \\
&= \int dx\ \nabla_\theta \left(f(x)p(x|\theta)\right) \\
&= \int dx\ \left(\nabla_\theta f(x)p(x|\theta) + f(x)\nabla_\theta p(x|\theta)\right) \\
&= \int dx\ f(x)\nabla_\theta p(x|\theta) \\
&= \int dx\ f(x)p(x|\theta)\frac{\nabla_\theta p(x|\theta)}{p(x|\theta)} \\
&= \int dx f(x)p(x|\theta)\nabla_\theta \log p(x|\theta) \\
&= E_{x \sim p(x|\theta)}\left[f(x)\nabla_\theta \log p(x|\theta)\right]
\end{aligned}$$

Now, we suppose $x = \tau$, $f(x) = R(\tau)$, and $p(x|\theta) = p(\tau|\theta)$:

$$\nabla_\theta L_{PG} = -E_{\tau \sim \pi_\theta}\left[R(\tau)\nabla_\theta \log p(\tau|\theta)\right]\ (*)$$

The trajectory τ is a sequence of events a_t and s_{t+1}, respectively sampled from the agent's policy $\pi_\theta(a_t|s_t)$ and the probability of transition $p(s_{t+1}|s_t,\ a_t)$. The probability of the complete trajectory is the product of the individual probabilities:

$$p(\tau|\theta) = \prod_{t=0}^{T} p(s_{t+1}|s_t,\ a_t)\pi_\theta(a_t|s_t)$$

$$\log p(\tau|\theta) = \log \prod_{t=0}^{T} p(s_{t+1}|s_t,\ a_t)\pi_\theta(a_t|s_t)$$

$$\log p(\tau|\theta) = \sum_{t=0}^{T} (\log p(s_{t+1}|s_t,\ a_t) + \log \pi_\theta(a_t|s_t))$$

$$\nabla_\theta \log p(\tau|\theta) = \nabla_\theta \sum_{t=0}^{T} \log \pi_\theta(a_t|s_t)\ (**)$$

By putting (**) in (*), we find the Eq. (11):

$$\nabla_\theta L_{PG} = -E_{\tau\sim\pi_\theta}\left[R(\tau)\sum_{t=0}^{T} \nabla_\theta \log \pi_\theta(a_t|s_t)\right]$$

References

1. Almahairi, A., Cho, K., Habash, N., Courville, A.: First result on arabic neural machine translation (2016). ArXiv https://arxiv.org/pdf/1606.02680
2. Alrajeh, A.: A recipe for Arabic-English neural machine translation (2018). ArXiv https://arxiv.org/pdf/1808.06116
3. Bahdanau, D., Cho, K., Bengio, Y.: Neural machine translation by jointly learning to align and translate (2015). CoRR https://arxiv.org/pdf/1409.0473
4. Cho, K., et al.: Learning phrase representations using rnn encoder-decoder for statistical machine translation (2014). arXiv preprint http://arxiv.org/abs/1406.1078
5. Durrani, N., Dalvi, F., Sajjad, H., Vogel, S.: Qcri machine translation systems for iwslt 16 (2017). ArXiv https://arxiv.org/pdf/1701.03924
6. Hassan, H., et al.: Achieving human parity on automatic chinese to english news translation (2018). arXiv preprint http://arxiv.org/abs/1803.05567
7. He, D., Lu, H., Xia, Y., Qin, T., Wang, L., Liu, T.Y.: Decoding with value networks for neural machine translation. In: Advances in Neural Information Processing Systems, pp. 178–187 (2017)
8. Hochreiter, S., Schmidhuber, J.: Long short-term memory. Neural Comput. **9**(8), 1735–1780 (1997). https://doi.org/10.1162/neco.1997.9.8.1735
9. Kingma, D.P., Ba, J.: Adam: a method for stochastic optimization (2014). arXiv preprint http://arxiv.org/abs/1412.6980
10. Koehn, P., Och, F.J., Marcu, D.: Statistical phrase-based translation. In: Proceedings of the 2003 Human Language Technology Conference of the North American Chapter of the Association for Computational Linguistics, pp. 127–133 (2003). https://www.aclweb.org/anthology/N03-1017
11. Kreutzer, J., Uyheng, J., Riezler, S.: Reliability and learnability of human bandit feedback for sequence-to-sequence reinforcement learning (2018). arXiv preprint http://arxiv.org/abs/1805.10627
12. Li, J., Monroe, W., Jurafsky, D.: Learning to decode for future success (2017). arXiv preprint http://arxiv.org/abs/1701.06549
13. Ng, A.Y., Harada, D., Russell, S.: Policy invariance under reward transformations: theory and application to reward shaping. In: ICML, vol. 99, pp. 278–287 (1999)
14. Oudah, M., Almahairi, A., Habash, N.: The impact of preprocessing on Arabic-English statistical and neural machine translation (2019). arXiv preprint http://arxiv.org/abs/1906.11751

15. Papineni, K., Roukos, S., Ward, T., Zhu, W.J.: Bleu: a method for automatic evaluation of machine translation. In: Proceedings of the 40th Annual Meeting on Association for Computational Linguistics, p. 311–318. ACL 2002, Association for Computational Linguistics, USA (2002). https://doi.org/10.3115/1073083.1073135
16. Ranzato, M., Chopra, S., Auli, M., Zaremba, W.: Sequence level training with recurrent neural networks (2015). arXiv preprint http://arxiv.org/abs/1511.06732
17. Srivastava, N., Hinton, G., Krizhevsky, A., Sutskever, I., Salakhutdinov, R.: Dropout: a simple way to prevent neural networks from overfitting. J. Mach. Learn. Res. **15**(1), 1929–1958 (2014)
18. Sutskever, I., Vinyals, O., Le, Q.V.: Sequence to sequence learning with neural networks. In: Ghahramani, Z., Welling, M., Cortes, C., Lawrence, N.D., Weinberger, K.Q. (eds.) Advances in Neural Information Processing Systems, vol. 27, pp. 3104–3112. Curran Associates, Inc. (2014). http://papers.nips.cc/paper/5346-sequence-to-sequence-learning-with-neural-networks.pdf
19. Sutton, R.S., Barto, A.G., et al.: Introduction to Reinforcement Learning, vol. 135. MIT press, Cambridge (1998)
20. Sutton, R.S., McAllester, D.A., Singh, S.P., Mansour, Y.: Policy gradient methods for reinforcement learning with function approximation. In: Advances in Neural Information Processing Systems, pp. 1057–1063 (2000)
21. Williams, R.J.: Simple statistical gradient-following algorithms for connectionist reinforcement learning. Mach. Learn. **8**(3–4), 229–256 (1992). https://doi.org/10.1007/BF00992696
22. Wu, L., Tian, F., Qin, T., Lai, J., Liu, T.Y.: A study of reinforcement learning for neural machine translation (2018). arXiv preprint http://arxiv.org/abs/1808.08866
23. Wu, Y., et al.: Google's neural machine translation system: bridging the gap between human and machine translation (2016). ArXiv https://arxiv.org/abs/1609.08144
24. Zakraoui, J., Saleh, M., Al-Maadeed, S., AlJa'am, J.M.: Evaluation of Arabic to English machine translation systems. In: 2020 11th International Conference on Information and Communication Systems (ICICS), pp. 185–190. IEEE (2020)
25. Ziemski, M., Junczys-Dowmunt, M., Pouliquen, B.: The united nations parallel corpus v1. 0. In: Proceedings of the Tenth International Conference on Language Resources and Evaluation (LREC'16), pp. 3530–3534 (2016)

Dilemma Game at Unsignalized Intersection with Reinforcement Learning

Saif Islam Bouderba[1](✉) and Najem Moussa[2]

[1] LAROSERI, Department of Computer Science, Faculty of Science, University of Chouaib Doukkali, El Jadida, Morocco
bouderba.s@ucd.ac.ma
[2] Faculty of Science, University Mohammed V, Rabat, Morocco
najemmoussa@yahoo.fr

Abstract. Decision making in unsignalized traffic intersection can be challenging for autonomous vehicles. An autonomous system only depends on predefined street priorities and considering other cars as moving objects will cause the car to freeze and fail the maneuverer. Human drivers influence the cooperation of other drivers to evade such deadlock situations and convince others to change their behavior. Decision making processes should reason about the interaction with other drivers and anticipate a wide range of driver behaviors. In this paper, we present a reinforcement learning approach (Q-LEARNING) to learn how to interact with drivers with different situation. We enhanced the performance of Q-LEARNING algorithms by maintaining a belief over the level of cooperation of other drivers. We show that our drives successfully learns how to navigate with less deadlocks than without learning.

Keywords: Reinforcement learning · Q-learning · Cellular automata · Unsignalized intersection · Traffic flow · Game theory

1 Introduction

Nowadays, as a consequence of population growth, motorization and urbanization, traditional urban traffic control strategies cannot meet the requirements of the modern city. Traffic congestion takes place worldwide more frequently. It not only reduces the efficiency of transportation, but also increases air pollution and fuel consumption. This raises the issue of whether the supplied infrastructures correspond to the traffic demand. Invest in new infrastructures is more and more expensive.

As an important feature in the traffic dynamics in the cities, intersections have attracted the attention of researchers. Such intersections may include traffic circle, roundabout, T-intersection, signalized intersection and unsignalized intersections, etc. [1–5]. The unsignalized intersection, in particular, has been widely studied in the literature because, First, it plays an important role in determining road traffic capacity, particularly in rural areas. Second, car accidents are

M. Lazaar et al. (Eds.): BDIoT 2021, LNNS 489, pp. 481–491, 2022.
https://doi.org/10.1007/978-3-031-07969-6_36

generally more frequent at unsignalized intersection than at intersections with traffic lights. The crossroad rules at unsignalized intersection have a great influence on the road traffic safety, especially in the presence of aggressive drivers. Accidents caused by aggressive drivers (a driver that ignored the right-of-way rule) are becoming more frequent at unsignalized intersection, and may lead to a serious collision because it usually involves a right-angle impact type. The causes of traffic accidents at unsignalized intersections are generally complex, and involve several factors, including human factors, mechanical performance of vehicles and physical condition of the roadway.

It's essential to understand traffic dynamics in an organized environment and to detect the optimal control strategies which could solve the crossing conflict problem. When the priority rule is not respected in traffic intersection, the management of the resulting conflict is generally shared between the drivers according to their adopted strategies (Cooperator or Defector). In [6] and [7], the bottleneck problem caused by reducing lanes from double to single is studied using a game theory framework to deal with drivers decision-making processes. [8] investigated traffic flow, energy dissipation and social payoff in a traffic model with a unsignalized intersection based on the NaSch model.

Reinforcement learning has revealed its advantages in the ability to explore the environment to exploit the most appropriate actions in the dynamic situations. Researches involving reinforcement learning in the traffic signal management systems have shown significant results and thus drawn more attentions to traffic management systems' researchers. Q-Learning as one of reinforcement learning algorithm is applied in the optimization of the traffic flow in single traffic intersection [9]. Besides that, Q-learning algorithm has also been extremely valued in the researches of traffic control system as multi-agents systems [[10], [11], [12]]. In this study, Q-learning algorithm has been proposed to be studied in the unsignalized traffic intersections.

In this paper, we investigate an unsignalized traffic intersection where portion of the drivers respects the right-hand priority rule, while others ignore it. It is a very common situation in several nations with poor traffic education. In such a context, we discover the effect that the introduction of a part of "well educated" drivers which abiding by the law, and others "unwell educated" drivers which do not respect the law, engenders a prisoner's dilemma-like game at the intersection and that the game's outcomes will affect the collective performance and the efficiency of the system. Let us recall that in Ref. [13] the imitating drivers can change their strategies based on the frequency of encounters with drivers that display one or the other strategy. In our model we applied the Q-LEARNING algorithm to learn drives how to interact with others drivers with different situation.

The paper is organized as follows. In the next section, we give the definition of the model and drivers strategy-update dynamics. In Sect. 3, we present our Q-learning algorithm and the performance metrics used to study the model. In Sect. 4, we investigate the characteristics of the system by using our simulation program and where some interesting observations are analysed. In Sect. 5, we will bring our conclusion.

2 Traffic Model and Dilemma Game

2.1 Model Structure and Moving Rules

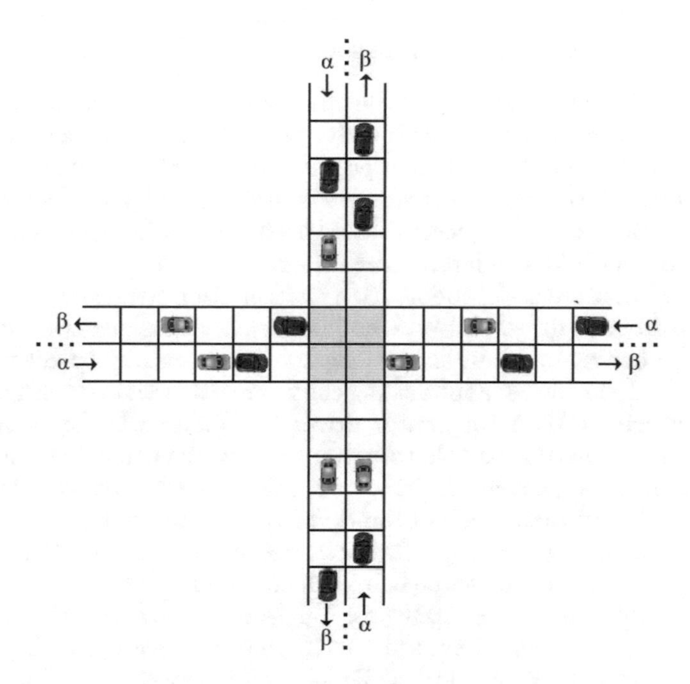

Fig. 1. Two intersecting streets. The sage cars stand for cooperators and red cars for defectors

The model studied in this paper consists of two crossing roads. Both roads are two-ways, with one lane in each direction. Each road consists of L cells of identical size and the time t is discrete. All streets cross each others at the intersection sites (see Fig. 1). Open boundary conditions are applied to all roads, with four entry/exit points. We define road 1 as the road which goes from East to West, road 2 from West to East, road 3 from South to North and road 4 from North to South. Each position can be also empty or occupied by a vehicle with the speed $v = 0, v_{max}$, where v_{max} is the maximum speed. The dynamics of vehicles in each street is controlled by the NaSch moving rules [14] are as follows:

- Acceleration: $v = min(v + 1, v_{max})$
- Slow down: $v = min(v, g)$ where g is the number of empty sites in front of the vehicle.
- Randomization: $v = max(v - 1, 0)$ with probability p_r.
- Movement: the vehicle moves v sites forward.

The boundary condition is defined as follows: at each time step and for each road i, a vehicle k is injected with probability α and then the velocity $v_k = v_{max}$ is assigned to the vehicle. When the vehicle k reaches the end of the driving-out lane j, it simply injected again in system.

2.2 Dilemma Game at the Intersection

To determine the rules that control the advancement of cars at the unsignalized intersection, it is necessary to define the priority rule once cars try to cross simultaneously the intersection. The priority rule adopted by most countries is the right-hand priority rule. In the case where this priority rule is violated, car accident (collision) will likely occur and its probability will depend on the degree of aggressiveness of the two interacting drivers.

We believe that the use of cooperation and defection strategies is a reasonable approach to describe driver attitudes and behaviors at unsignalized intersections and that the emerging steady states can be well described by dilemma game model. In our model, driver approaching the intersection may act as a cooperator (C), or a defector (D). A cooperator driver is a driver who not only respects the priority rule, but also be able to better manage the conflict situation at the intersection. However, a defector driver is the driver that crosses the intersection aggressively without taking care of the traffic on the other road. It is well known that aggressive driving is a major factor in traffic accidents [15–17].

In this paper, a dilemma game is introduced where we take into account the full set of possible interacting strategies. A driver has a strategy that determines his behavior at the intersection at the next time step. After crossing the intersection, the driver moves forward while keeping his strategy unchanged. With the aim to put into action this game, we need to define the rules of our dilemma game in order to study the evolution of traffic patterns in unsignalized traffic system.

Let V_1 and V_2 be the vehicles on road 1 (R_1) and road 2 (R_2) respectively, that attempt to cross simultaneously (see Fig. 2). Let us next specify the interaction rules at the intersection in a way that mimics what happens in real situations:

(1) If both V_1 and V_2 are cooperators then V_1 must respect the priority of V_2 and let him pass first. So V_1 must stop to give way to the oncoming vehicle V_2.

(2) If V_1 is cooperator and V_2 is defector, the situation is like (1) since V_2 has the priority to cross the intersection (see Fig. 2(b)).

(3) If V_1 is defector and V_2 is cooperator then V_1 and V_2 cross simultaneously (see Fig. 2(a)). When V_1 and V_2 cross simultaneously the intersection, a collision happens.

(4) If both V_1 and V_2 are defectors then the situation is like (3), with the difference that the collision between V_1 and V_2, when they attempt to cross simultaneously the intersection (see Fig. 2(d)). Moreover, when collision happens the stop time of both vehicles will be longer.

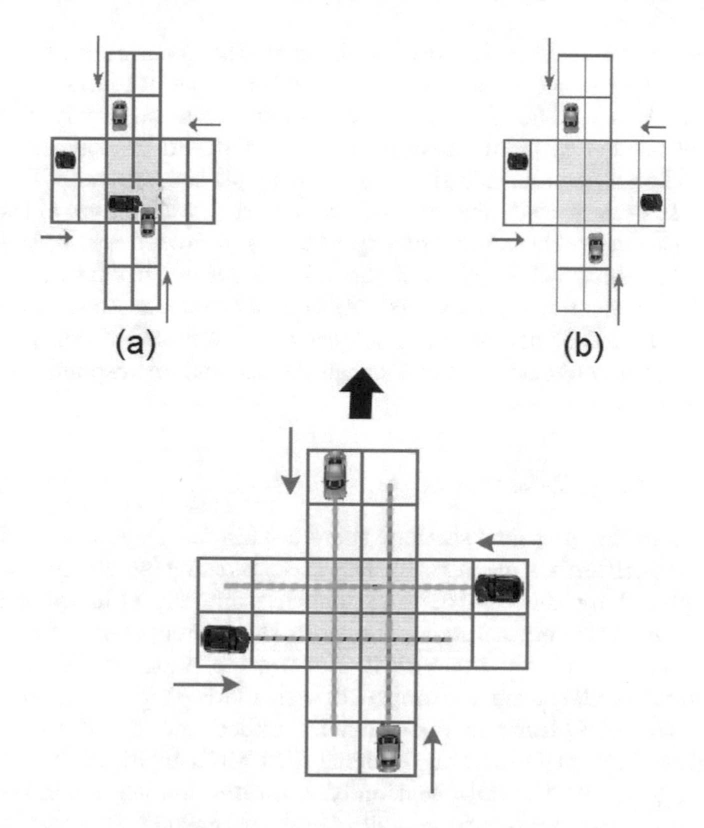

Fig. 2. Dilemma game for conflict resolution in various cases. Cars are represented by sage color for the cooperators and red color for defectors.

Situations (1) and (2) correspond to normal situations where drivers in road 1 usually respect the right-hand priority rule. In situation (3), the cooperator driver V_2 may decrease the risk of collision with the defector V_1, because he does the best he can to avoid the accident. This collision, if it will happen, usually is not fatal and the evacuation time is usually short since often the two drivers will leave the intersection to clear the road and then drawn up an accident report. In contrast, when both drivers are defectors (situation (4)), the collision is not only almost inevitable (large probability to occur), but it is usually fatal and, consequently, the cost will be much greater than in the situation (3).

3 Q-Learning Algorithm

Reinforcement learning is usually presented as Q-learning algorithm which values not only the actions taken but also the states caused by the actions. Decisions made from the past is evaluated and stored as experiences data which will provide valuable help in the future.

The most common metaphor used to illustrate the Q-learning algorithm is the relationship between a trainer and trainee, for example, a teacher and student, or parents and their child. By taking the training of a student by a teacher as example, the teacher will evaluate each actions of the student after the commands are given. When a command is given to the student, the student will respond to it, and then the teacher will observe and evaluate the performance of the student. If the student's action is within the expectations, a reward will be given to the student; and nothing will be given if the student did not behave accordingly. In this way, the student will make a connection between the commands and the actions; it will realize that only the correct command and action pairs will be rewarded. All these experiences encourage the student to respond according to the command of the teacher.

3.1 Structure of Q-Learning

As indicated in the previous section, there are several ways to describe about Q-learning algorithm's learning process. In Q-learning algorithm, the role of trainer is played by the environment model, while the Q-learning algorithm itself learns from the environment. Thus, in the development of a Q-learning algorithm, every operation includes in the process must be determined carefully. Q-Learning algorithm is composed with many steps or operations. The implementation of Q-learning operation is applied through the evaluation of (1). The flow chart of Q-learning is being illustrated in Fig. 3. The process of Q-learning starts with the initialization of the states and actions in the Q-table. After the initialization, Q-learning will identify its current state in the environment. An action will be selected from the action lists available by searching for the maximum possible rewards return by the action. Then, the actions chosen will be executed or evaluated. The rewards gained from the actions chosen will be updated in the Q-table. After the actions have been executed, Q-learning will identify the next states in the environment model. Finally, Q-learning will verify the goal accomplishment, the process will start again if the goal did not accomplished.

$$Q(s,a)_i = (1-\alpha)Q(s,a)_{i-1} + \alpha[R(s,a)_i + \gamma_a^{\prime max} Q(s',a')] \tag{1}$$

where:

s = current state
a = action taken in current state
s'= next state
a'= action taken in next state
i = iteration
α = learning rate
γ = discounting factor

Q-Learning is evaluated from quotation (1), each of evaluated Qvalue is the rewards gained from the experiences in the exploration process. Q-table is the

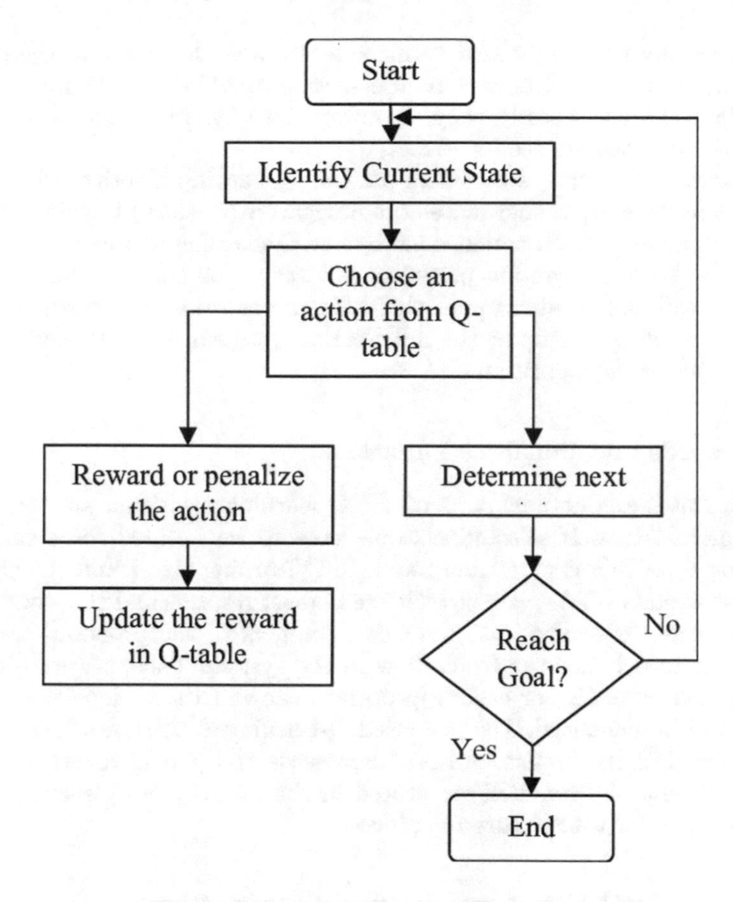

Fig. 3. Q-Learning algorithm flow chart

memory of the Q-learning algorithm, storing every single state and action pairs along with their rewards. α, is the factor that will influence the learning rate of the Q-learning algorithm. Learning rate of Q-learning is ranged from 0 to 1, and responsible for the weight of the newly learnt experience. Discounting factor γ is the variable that decides the importance of the future states. High discounting factor will make Q-learning algorithm to be too speculative, where it will focus more on the possible future rewards and neglect the importance of the current experience.

3.2 State-Action Pairs

Q-Learning algorithm gains its experience through the exploration in the modelled environment. An accurately defined environment will ease the Q-learning's exploration process. The environment of the Q-learning is formed by the states and actions [18].

In this study of unsignalized traffic intersection, the states of the designed Q-learning are the crossing cars at intersection and their strategies. There are 4 cases defined for this study, they are categorized by cars cooperation strategy crossing the intersection see (Sect. 2.2).

The actions list that is available for the Q-learning algorithm is also very important parameter, as they act as the navigators for the Q-learning algorithm in the environment. Each action done by the Q-learning will lead it to another state. If the state and action pairs are not set to be correct, then the whole Q-learning will not be able to get the optimum solution in the whole process. Therefore, crossing or stop at the intersection are defined as the actions of the proposed Q-learning algorithm.

3.3 Rewards and Penalties Functions

Although states and actions pairs of the Q-learning algorithm are set, rewards and penalties for each selected actions have to be decided for ensuring the Q-learning algorithm is performing well. In Q-learning algorithm, the basic idea is the best actions will be valued with the highest rewards and the worst actions will be assigned with the least rewards. The goal of the proposed Q-learning algorithm is to enhance the traffic flow in the system. Thus, rewards functions are computed carefully for each appropriate car and the actions that yield an accident will be penalized. The proposed Q-learning algorithm will stop after it reach the goal of the system. All of the rewards and penalties returned by the reward and penalties function are stored in the memory of Q-learning as their own experience for their future references.

3.4 Simulation Method and Performance Metrics

In this paper, simulation results for the proposed model, are carried out to investigate traffic characteristics and cooperation levels at unsignalized traffic intersection. The network size of each road is $L = 100$. In this paper, we restricted our study to the case of equivalent roads, i.e. N_1 is the number of vehicles on the system. The network density is defined as $\rho = N/L$. The following parameters are fixed throughout this paper. The maximum speed of vehicles is set as $v_{max} = 1$ and the random braking probability is $p_r = 0.1$. The initial proportion of cooperators in each road is set to 50%. The numerical results are obtained by averaging over 100 initial random configurations and 15000 time-steps after discarding 10000 initial transient steps.

Model performance metrics are developed to assess our evolutionary dilemma game behavior. We monitor the impact of car's density ρ and the traffic flow and accident probability.

- **Waiting time :** we define the waiting time of vehicles as the fraction of stopped vehicles on roads during the simulation time divided by the time interval.

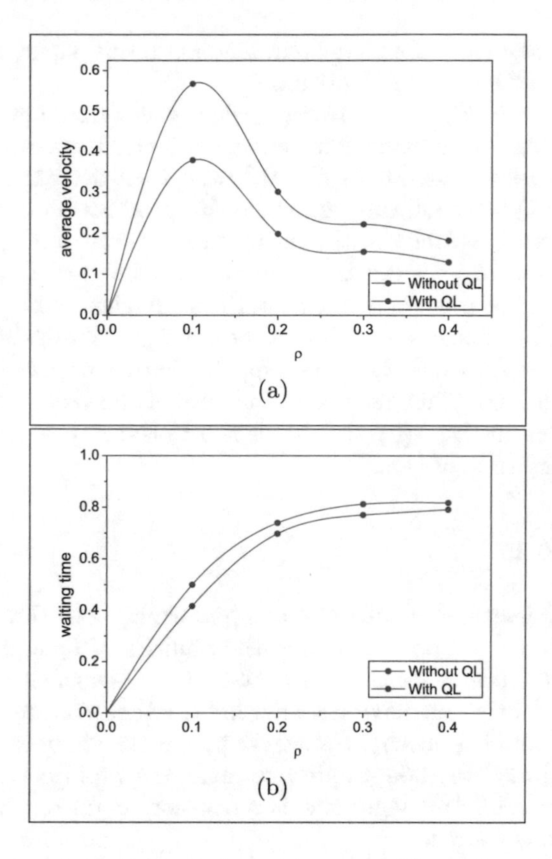

Fig. 4. The average velocity (a) and the waiting time (b) as a function of ρ for different methods.

$$\textbf{Waiting time} = \frac{1}{N \times T} \sum_{t=1}^{T} \sum_{i=1}^{N} (1 - v_i(t)) \tag{2}$$

where T is the time interval and N is the number of vehicles on roads. The variable v_i is equal to 1 if the driver i is not stopped and equal to 0 if the driver i is stopped.

- **Average velocity:** we define the average velocity of vehicles on roads as the sum velocities of all vehicles on roads during the simulation time divided by the time interval.

$$\textbf{Average velocity} = \frac{1}{N \times T} \sum_{t=1}^{T} \sum_{i=1}^{N} v_i(t) \tag{3}$$

4 Results and Discussion

In this section, we present the simulation results for our indicators of system performance in order to investigate the relationship between the states of our transportation system and the model parameters. In this paper, we are interested to the effect of the Q-learning algorithm.

Firstly, we show in Fig. 4, the average velocity as a function of the density of the system for different methods. The average velocity increases for small values of ρ until it reaches its maximum $\rho = 0.1$, and then decreases. As we can see after applied the Q-learning, the cars make the good action in the intersection, for that, the average velocity with the Q-learning (red line) give as the best result. In addition, for $\rho > 0.5$ the system is in the current phase, here, the average velocity is not more dependent on the strategies. Moreover, in Fig. 4(b), we show the waiting time as a function of the density of the system ρ for different methods. When we decrease the value ofρ, the density of cars increase and the road becomes crowded. This, results an increase of the waiting time of stopped cars, so we notice in Fig. 4(b) that with the Q-learning algorithm (red line) reduce the waiting time of cars.

5 Conclusions

The developed Q-learning traffic system is performing well throughout the simulations. This shows the potentials and the abilities of Q-learning in the traffic systems. Even without centralized control, the Q-learning of each cars able to work independently and having traffic information sharing with other cars. Q-learning algorithm has shown its strength in exploration in the dynamic traffic environment and also the adaptability towards the rapid changes of the environment by successfully manages the cars decision at unsignalized intersection within the traffic networks.

References

1. Li, X.-G., Gao, Z.-Y., Jia, B., et al.: Cellular automata model for unsignalized T-shaped intersection. Int. J. Mod. Phys. C **20**(04), 501–512 (2009)
2. Fouladvand, M.E., Sadjadi, Z., Shaebani, M.R.: Characteristics of vehicular traffic flow at a roundabout. Phys. Rev. E **70**(4), 046132 (2004)
3. Marzoug, R., Ez-Zahraouy, H., Benyoussef, A.: Simulation study of car accidents at the intersection of two roads in the mixed traffic flow. Int. J. Mod. Phys. C **26**(01), 1550007 (2015)
4. Huang, D.W.: Modeling gridlock at roundabout. Comput. Phys. Commun. **189**, 72–76 (2015)
5. Xie, D.F., Gao, Z.Y., Zhao, X.M., Li, K.P.: Characteristics of mixed traffic flow with non-motorized vehicles and motorized vehicles at an unsignalized intersection. Phys. A: Stat. Mech. Appl. **388**(10), 2041–2050 (2009)
6. Yamauchi, A., Tanimoto, J., Hagishima, A., et al.: Dilemma game structure observed in traffic flow at a 2-to-1 lane junction. Phys. Rev. E **79**(3), 036104 (2009)
7. Nakata, M., Yamauchi, A., Tanimoto, J., et al.: Dilemma game structure hidden in traffic flow at a bottleneck due to a 2 into 1 lane junction. Phys. A: Stat. Mech. Appl. **389**(23), 5353–5361 (2010)

8. Zhang, W., Chen, W.: Dilemma game in a cellular automaton model with a non-signalized intersection. Eur. Phys. J. B **85**(2), 1–8 (2012)
9. Chin, Y.K., Bolong, N., Kiring, A., Yang, S.S., Teo, K.T.K.: Q-learning based traffic optimization in management of signal timing plan. Int. J. Simul. Syst. Sci. Technol. **12**(3), 29–35 (2011)
10. Balaji, P.G., German, X., Srinivasan, D.: Urban traffic signal control using reinforcement learning agents. IET Intell. Transp. Syst. **4**(3), 177–188 (2010)
11. Abdulhai, B., Pringle, R., Karakoulas, G.J.: Reinforcement learning for true adaptive traffic signal control. J. Transp. Eng. **129**(3), 278–285 (2003)
12. Bouderba, S.I., Moussa, N.: Reinforcement learning (Q-LEARNING) traffic light controller within intersection traffic system. In: Proceedings of the 4th International Conference on Big Data and Internet of Things, pp. 1–6 (2019)
13. Paissan, G., Abramson, G.: Imitation dynamics in a game of traffic. Eur. Phys. J. B **86**(4), 1–6 (2013)
14. Nagel, K., Schreckenberg, M.: A cellular automaton model for freeway traffic. J. de physique I **2**(12), 2221–2229 (1992)
15. Osafune, T., Takahashi, T., Kiyama, N., et al.: Analysis of accident risks from driving behaviors. Int. J. Intell. Transp. Syst. Res. **15**(3), 192–202 (2007)
16. Bouderba, S.I., Moussa, N.: On the V2X velocity synchronization at unsignalized intersections: right hand priority based system. In: Belkasmi, M., Ben-Othman, J., Li, C., Essaaidi, M. (eds.) ACOSIS 2019. CCIS, vol. 1264, pp. 65–72. Springer, Cham (2020). https://doi.org/10.1007/978-3-030-61143-9_6
17. Moussa, N.: The influence of aggressive drivers on the properties of a stochastic traffic model. Eur. Phys. J. B-Condens. Matter Complex Syst. **41**(3), 421–431 (2004)
18. Watkins, C.J., Dayan, P.: Q-learning. Mach. Learn. **8**(3), 279–292 (1992)

Toward a Self-adaptive Supply Chains: L-SCOR Implementation Proposal, and Case Studies Methodology Proposal

Houda Mezouar(✉) and Abdellatif El Afia

ENSIAS, Mohammed V University of Rabat, Rabat, Morocco
houda.mezouar@gmail.com, a.elafia@um5s.net.ma

Abstract. In order to improve a business process behavior, to guarantee the service continuity, to minimalize the time wasted by automatic activities execution, L-SCOR is proposed to analyze the self-adaptability of a business process and to improve this performance. This paper gives an overall overview of the reference L-SCOR, and explains a proposal of its implementation based on reinforcement learning, and a methodology to conduct case studies based on six sigma strategies.

Keywords: Self-adaptive · Supply chain · L-SCOR · SCOR · Six sigma · Reinforcement learning · Q-learning · Web service

1 Introduction

1.1 Self-adaptive Systems

Self-adaptive systems are systems that are able to behavior according to an evaluation aiming to achieve a system goal [1]. Different research works approaching different systems has focused on improving self-adaptability by focusing on the system self-learning ability using approaches based on Hidden Markov Model [2–6]. Other works proposed frameworks for designing adaptive processes, and highlighted current research on methods and techniques for the design and engineering of adaptive software systems combining different technologies and using various tools proposing thus different architectures for adaptive software's [7–18]. other work aimed to model any system on supply chain which aims to provide a product or a service, and make this system adaptive by improving the processes that make up this supply chain and make them adaptive using performance analysis as proposed by SCOR and reinforcement learning [19–27].

This brief literature overview shows that adaptive systems in a general way are clearly a highly research issue of major importance nowadays. This work is distinguished by focusing on a way to push the process to behave in the most optimal and efficient way.

M. Lazaar et al. (Eds.): BDIoT 2021, LNNS 489, pp. 492–500, 2022.
https://doi.org/10.1007/978-3-031-07969-6_37

1.2 L-SCOR: An Extension of SCOR for a Self-adaptive Supply Chain

SCOR is a reference that represents the flows of a company allowing modeling its different structures and processes; The latest version of the SCOR (SCOR 12), considers that all supply chains are composed of the following six processes: Plan, Source, Make, Deliver, Return and Enable. Reference [1], explains L-SCOR an extension to SCOR that adds to those processes a seventh one "sL Lean" able to manage the adaptive aspect, and to its fifth performance attributes a sixth one "adaptability" with eight metrics. The Learn processes include on one hand the gathering of each process states, the gathering of the possible actions for each process and of the rewards related to each action, on the other hand they also include the establishment of the optimal policy for the process execution. At the second level of a supply chain, sL is divided into five processes, which are "sL1 - Learn Plan", "sL2 - Learn Source", "sL3 - Learn Make", "sL4 - Learn Deliver" and "sL5 - Learn Return". Concerning the third level, each of the aforementioned processes is divided into four sub-processes. For example sL1 - Learn Plan that describes the process of development and establishment of optimal policy for the self-adaptive supply chain planning is composed of the following sub-processes: "sL1.1 - Identify plan states", "sL1.2 - Identify plan actions", "sL1.3 - Identify rewards of planning actions" and "sL1.4 - Establish and communicate plan policy". At the supply chain real-time level, BPMN is used as the activities descriptive language. A control is implemented on the result of each activity execution. In the case of an anomaly, the activity calls a web-service that executes sL Learn process. This web service receives as input for the concerned activity, the execution data (as detailed in the previous levels), it reformulates the task of the activity in agent, states and actions, and it returns the optimal behavior to execute thanks to the Q-learning algorithm.

2 L-SCOR Implementation

2.1 Q-Learning

Q-Learning, a model-free algorithm of reinforcement learning, which is more in line with the self-learning mechanism of human society. Actually, it optimizes long-term benefits by rewarding positive behavior and punishing negative behavior. Watkins [28] proposes a typical description of Q-learning, the algorithm allows individuals to maintain a Q-value table for each strategy and to update the table in each round, and finally selects the strategy with the highest Q-value as the next round strategy [29]. Based on the foundation of Markov decision processes theory, Q-learning is a useful and meaningful method for solving problems of agent actions under different complicated conditions, it is very reliable because a discounted future reward is obtained when transiting from one state to another [30]. The literature offers a huge number of problematic issues that have been dealt with solutions based on Q-learning in various fields, among which we cite in the aeronautical field [31] that proposed a Q-learning based approach that stands on the experience quality of the served users, to plan the flight of stations of unmanned aerial vehicles-base. In the robotics field, [32] proposes an approach integrating Q-learning to plan the path of a robot arm. This approach has as objective to plan the path of the robot arm end-effector such that it can reach a fixed target position by getting over the minimum distance while dodging the obstacles present in the environment in an efficient manner. Within the same framework, [33] proposes an approach for the path planning of mobile robot in different environments with static obstacles in various shapes, sizes and layout. In this work, Q-learning is used in software engineer field, in order to allow process based software's to be self-adaptive, in fact the process activities will execute, when needed, the activity that implement Q-Learning algorithm, in order to obtain the optimal strategy for their operation. The principle of Q-learning algorithm is as follows:

- For each agent state, we note the action that led to this state by strengthening our existing q-value. Arrived on the state s' from state s and action a, we note the q-value as following:

$$Q(s, a) = \lambda \times (r + \gamma \times \max a' (Q(s', a'))) + (1 - \lambda) \times Q(s, a)$$

$\lambda \in [0; 1]$ is the learning rate

$\gamma \in [0; 1]$ is the discount factor

r is the reward obtained for achieving state s' from state s with action a

- The agent carries out several research cycles of rewards, from initial state to goal state and reinforces with each passage the q-value of the action which leads to rewards or leading to states leading to rewards.
- Initially, the agent does not know the states where the rewards are, does not know the arrival state of an action, so he starts by choosing random actions he explores.
- After a certain time or when it reaches a goal state, the agent resumes a search for a solution from the state initial. At each cycle, the system exhibits a behavior of less and less exploratory, and increasingly guided by quality

The Q-learning algorithm is implemented in this work as following:

```
Function Q-Learning
        ∀s ∀a Q(s, a) ← 0
        for n ← 1, nbCycles do  #nbCycles is the number of learning cycles
           λ ← 1 ;
           α ← 1 ;
           currentState ← firstState
        for i ← 1, nbMaxActions do  #the max number of actions to execute
           s ← currentState
        nb ← random(0, 1)
                  if (nb < α) then
                            a ← randomAction(s)
                  else
                            a ← argMax a' (Q(s, a'))
                  end if
      s'← a(s)
                  Q(s, a) ← α × (r + γ × max a' (Q(s ' , a' ))) + (1 − α) × Q(s, a)
                   λ ← 0.99 × λ
                  α ← 0.99 × α
                  if (s' = finalState) then
                            exit loupe i
                  end if
          end for
      end for
End Function
```

The parameters used for Q-Learning algorithm in this work are presented in Table 1 as follow.

Table 1. Q-Learning algorithm parameters.

Parameter	Value
Learning rate	0.05
Discount factor	0.7
Initial Q-Value	0
Initial value of ϵ	0.6
Decay rate	0.005

2.2 Web Services

Web services are a platform-independent technology primarily designed to connect heterogeneous applications over a network. They are widely used in different domains and are based on standard protocols that enable communication between the providers, which offer the operations, and the consumers [34]. Reusing and composing third-party

Web services have become a common practice in modern software systems engineering to provide high quality and feature rich systems. Several service ecosystems have emerged in recent years hosting a highly increasing number of published Web services by different providers on Internet [35]. The different roles involved are service provider, service registry and service requestor. The various communication/interaction between the roles are publish, discovery and bind operations. To describe a typical web service environment, we have to distinct between the service provider and service consumer roles. A service provider establishes network accessible software, and offers its description that explains the functional properties of the service and publishes it to a service repository. The service consumer utilizes the discovery operation to find the published service description from the registry, and uses the description to bind with the provider [36]. In this work web services are used to allow the communication between the implementation of a recommendation activity (that implement Q-learning algorithm) and the rest of process activities.

The implementation of sL at the real-time level of a supply chain, is done using a RESTful Web service with python Flask that calls the Q-learning algorithm implemented with python language. In this section, a RESTful API service is presented which any user, programmer or program can access to take advantage of Q-learning algorithm. In recent years, REST (REpresentational State Transfer) has emerged as the standard architectural design for web services and web APIs, due to its ability of supporting simpler programmatic access through returning either using XML or JSON [37]. RESTful Web Service is a structural mode, where the data is outlined in the terms of Uniform Resource Identifier (URI) and the behaviors are outlined as methods. Since the performance, scalability, flexibility is improved in the RESTful systems when it is compared with SOAP service, REST hypnotize the end user due to the less consumption of resources [37]. REST services are robust, use minimal bandwidth and are suitable for connecting with cloud services. Sites such as Amazon, LinkedIn and Twitter use RESTful APIs [38]. The standard REST protocol is: the client or the applicant process generates a Uniform Resource Locator (URL) that is sent to the server using a simple HTTP GET request. The Get request is received at the server and processed by the API. The HTTP RESPONSE is formatted as eXtensible Markup Language (XML) or JavaScript Object Notation (JSON) [24]. The structure of the GET request includes the resource location that is being accessed, the output format, and the Q-learning parameters, separated by "&". For web development in Python, there are many frameworks available, the most known are Django and Flask, in this project research we use Flask.

3 Case Studies Methodology

As part of this work, different case studies can be established according to a methodology that consists of three steps, which are "Supply Chain exploration" "the Supply Chain evaluation" and "the Supply Chain improvement" as shown in Fig. 1. The exploration can be done through the mapping of the first three levels of the SC using SCOR. Once the exploration is done we move on to the evaluation of the SC, this phase can be done with a proposed approach based on the data-driven quality strategy "Define, Measure, Analyze, Improve, Control" (DMAIC) which is used to drive Six Sigma projects, the define step is the mapping of the real-time level of a supply chain using BPMN. In the measure phase, we propose to apply the BPM simulation to the BPMN model; in the Analyze step, we represent the results of simulation using the root cause analysis tree diagram. in the improve phase we will take advantages of Smart supply chain characteristics to adapt the "as is" model; for the last step we propose to use a set of SCOR KPIS to keep controlling the behavior of our system. The last phase is the improvement one, the improvement is done by the implementation of the process sL of the proposed reference L-SCOR, which consist in the implementation of the web service which uses the Q-learning algorithm for general anomalies during the simulation. This web service directs the activity to work by using its optimal policy thus avoiding the situations generating the anomaly.

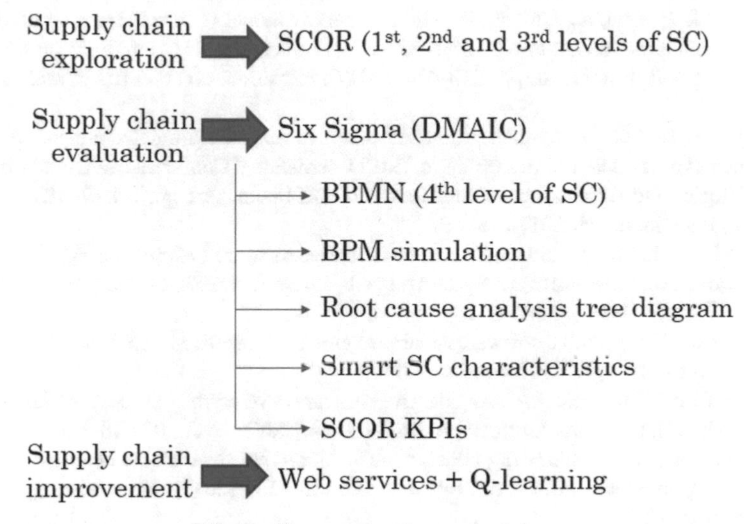

Fig. 1. The case studies methodology

4 Conclusion

This paper recalls L-SCOR goal and processes, it proposes an implementation for the sL process based on the reinforcing learning, then it proposes a methodology to adopt for the different case studies.

The proposed reference is based on SCOR, which is adapted to the different supply chain types; it extends SCOR by integrating the notion of adaptability on its both sections processes and metrics. This paper also describes how Q-learning can be integrating to process-based software in order to make its activities behaving according to the optimal strategy, and how to obtain a self-adaptive version of a traditional supply chain by combining six sigma methodology and L-SCOR.

References

1. Mezouar, H., El Afia, A.: A 4-level reference for self-adaptive processes based on SCOR and integrating Q-Learnin. In: Proceedings of the 4th International Conference On Big Data

and Internet of Things (BDIoT 2019), 23–24 October 2019. ACM, Tangier-Tetuan, Morocco (2019). https://doi.org/10.1145/3372938.3372953
2. Bouzbita, S., El Afia, A., Faizi, R.: A novel based hidden markov model approach for controlling the ACS-TSP evaporation parameter. In: International Conference on Multimedia Computing and Systems -Proceedings, pp. 633–638 (2017). https://doi.org/10.1109/ICMCS.2016.7905544
3. El Afia, A., Bouzbita, S., Faizi, R.: The effect of updating the local pheromone on ACS performance using fuzzy logic. Int. J. Electr. Comput. Eng. **7**(4), 2161–2168 (2017). https://doi.org/10.11591/ijece.v7i4.pp2161-2168
4. Kabbaj, M.M., El Afia, A.: Towards learning integral strategy of branch and bound. In: Proceedings of the 2017 International Conference on Multimedia Computing and Systems, pp. 621–626 (2017). https://doi.org/10.1109/ICMCS.2016.7905626
5. Lalaoui, M., El Afia, A., Chiheb, R.: Hidden markov model for a self-learning of simulated annealing cooling law. In: Proceedings of the 2017 International Conference on Multimedia Computing and Systems, 0, pp. 558–563 (2017). https://doi.org/10.1109/ICMCS.2016.7905557
6. Lalaoui, M., El Afia, A., Chiheb, R.: A self-adaptive very fast simulated annealing based on hidden markov model. In: Proceedings of 2017 International Conference of Cloud Computing Technologies and Applications, CloudTech 2017, 2018-January, pp. 1–8 (2018). https://doi.org/10.1109/CloudTech.2017.8284698
7. Hidaka, S., et al.: Design and engineering of adaptive software systems. In: Yu, Y., et al. (eds.) Engineering Adaptive Software Systems, pp. 1–33. Springer, Singapore (2019). https://doi.org/10.1007/978-981-13-2185-6_1
8. Mirchandani, C.: Adaptive software reliability growth. Procedia Comput. Sci. **140**(122), 132 (2018). https://doi.org/10.1016/j.procs.2018.10.309
9. Qureshi, M.R.J., Hussain, S.A.: An adaptive software development process model. Adv. Eng. Softw. **39**(654), 658 (2008). https://doi.org/10.1016/j.advengsoft.2007.08.001
10. Rodrigues, A., et al.: Enhancing context specifications for dependable adaptive systems: A data mining approach. Inf. Softw. Technol. **112**(115), 131 (2019). https://doi.org/10.1016/j.infsof.2019.04.011
11. Dalla Preda, M.: Developing correct, distributed, adaptive software. Sci Comput. Program. **97**(41), 46 (2015). https://doi.org/10.1016/j.scico.2013.11.019
12. Chen, L., Huang, L., Li, C., Wu, X.: Self-adaptive architecture evolution with model checking: A software cybernetics approach. J. Syst. Softw. **124**(228), 246 (2016). https://doi.org/10.1016/j.jss.2016.03.010
13. Ding, Z., Zhou, Y., Zhou, M.: Modeling self-adaptive software systems with learning petri nets. IEEE Trans. Syst. Man Cybern. Syst. **46**(4), 483–498 (2016). https://doi.org/10.1109/TSMC.2015.2433892
14. Han, D., et al.: FAME: A UML-based framework for modeling fuzzy self-adaptive software. Inf. Softw. Technol. **76**(118), 134 (2016). https://doi.org/10.1016/j.infsof.2016.04.014
15. Lee, E., et al.: RINGA: Design and verification of finite state machine for self-adaptive software at runtime. Inf. Softw. Technol. **93**(200), 222 (2018). https://doi.org/10.1016/j.infsof.2017.09.008
16. Shevtsov, S.: Developing a reusable control-based approach to build self-adaptive software systems with formal guarantees. In: Proceedings of the 24th ACM SIGSOFT International Symposium on Foundations of Software Engineering, FSE 2016, 13–18 November 2016. ACM, Seattle. https://doi.org/10.1145/2950290.2983949
17. Wang, Q.: Towards a rule model for self-adaptive software. ACM Sigsoft Softw. Eng. Notes **30**(1), 8–12 (2005). https://doi.org/10.1145/1039174.1039198

18. Georgas, J.C., Taylor, R.N.: Policy-based self-adaptive architectures: A feasibility study in the robotics domain. In: Proceedings of the International Workshop on Software Engineering for Adaptive and Self-managing Systems SEAMS 2008 Leipzig, Germany, 12–13 May 2008. ACM, Leipzig. https://doi.org/10.1145/1370018.1370038
19. Mezouar, H., El Afia, A.: Proposal of an approach to improve business processes of a service supply chain. International Journal of Mechanical Engineering and Technology (IJMET) **10**(03), 978–989 (2019). http://www.iaeme.com/ijmet/issues.asp?JType=IJMET&VType=10&IType=3
20. Mezouar, H., El Afia, A.: A process simulation model for a proposed moroccan supply chain of electricity. In: Proceedings of the International Renewable and Sustainable Energy Conference (IRSEC), 14–17 November 2016. IEEE, Marrakech (2017). https://doi.org/10.1109/IRSEC.2016.7983999
21. El Afia, A., Mezouar, H.: A global mapping of the moroccan supply chain of hospital drugs and a simulation of the dispensation process. In: Proceedings of the 2nd International Conference on Big Data, Cloud and Applications (BDCA 2017), 29–30 March 2017. ACM, Tetouan (2017). https://doi.org/10.1145/3090354.3090465
22. Mezouar, H., El Afia, A., Chiheb, R., Ouzayd, F.: Proposal of a modeling approach and a set of KPI to the drug supply chain within the hospital. In: Proceedings of the 3rd International Conference on Logistics Operations Management (GOL), 23–25 May 2016. IEEE, Fez (2016). https://doi.org/10.1109/GOL.2016.7731691
23. Mezouar, H., El Afia, A., Chiheb, R.: A new concept of intelligence in the electric power management. In: Proceedings of the 2nd International Conference on Electrical and Information Technologies (ICEIT 2016), 4–7 May 2016. IEEE, Tangier (2016). https://doi.org/10.1109/EITech.2016.7519596
24. Mezouar, H., El Afia, A., Chiheb, R., Ouzayd, F.: Toward a process model of moroccan electric supply chain. In: Proceedings of the International Conference on Electrical and Information Technologies (ICEIT 2015), 25–27 March 2015. IEEE, Marrakech (2015). https://doi.org/10.1109/EITech.2015.7162990
25. Mezouar, H., El Afia, A.: Proposal for an approach to evaluate continuity in service supply chains: Case of the Moroccan electricity supply chain. Int. J. Electr. Comput. Eng. (IJECE) **9**(6), 5552–5559 (2019). https://doi.org/10.11591/ijece.v9i6.pp5552-5559
26. Mezouar, H., El Afia, A.: A retirement pension from a supply chain side: Case of the moroccan retirement pension. In: Ezziyyani, M., Bahaj, M., Khoukhi, F. (eds.) AIT2S 2017. LNNS, vol. 25, pp. 103–115. Springer, Cham (2018). https://doi.org/10.1007/978-3-319-69137-4_11
27. Mezouar, H., El Afia, A.: Performance analysis model for service supply chains: Case of the retirement supply chain. Int. J. Eng. Technol. (IJET) **7**(3), 1429–1438 (2018)
28. Watkins, C.J., Dayan, P.: Technical note: Q-learning. Mach. Learn. **8**, 279–292 (1992). https://doi.org/10.1023/A:1022676722315
29. Ding, H., Zhang, G., Wang, S., Li, J., Wang, Z.: Q-learning boosts the evolution of cooperation in structured population by involving extortion. Pysica A **536**, 122551 (2019). https://doi.org/10.1016/j.physa.2019.122551
30. Tong, Z., Chen, H., Deng, X., Li, K, Li, K: A scheduling scheme in the cloud computing environment using deep Q-learning. Inf. Sci. (2019). https://doi.org/10.1016/j.ins.2019.10.035
31. Colonnese, S., Cuomo, F., Pagliari, G., Chiaraviglio, L.: Q-SQUARE: A Q-learning approach to provide a QoE aware UAV flight path in cellular networks. Ad Hoc Netw. **91**, 101872 (2019). https://doi.org/10.1016/j.adhoc.2019.101872
32. Sadhu, A.K., Konar, A., Bhattacharjee, T., Das, S.: Synergism of firefly algorithm and Q-learning for robot arm path planning Swarm Evol. Comput. **43**(50), 68 (2017). https://doi.org/10.1016/j.swevo.2018.03.014

33. Soong, L.E., Pauline, O., Chun, C.K.: Solving the optimal path planning of a mobile robot using improved Q-learning. Robot. Auton. Syst. **115**, 143–161 (2019). https://doi.org/10.1016/j.robot.2019.02.013
34. Oliveira, R.A., et al.: An approach for benchmarking the security of web service frameworks. Future Generation Comput. Syst. (2019, in press). https://doi.org/10.1016/j.future.2019.10.027
35. Almarimi, N., et al.: Web service API recommendation for automated mashup creation using multi-objective evolutionary search. Appl. Soft Comput. **85**, 105830 (2019). https://doi.org/10.1016/j.asoc.2019.105830
36. Sambasivam, G., et al.: An QoS based multifaceted matchmaking framework for web services discovery. Future Comput. Inf. J. **3**(371), 383 (2018). https://doi.org/10.1016/j.fcij.2018.10.007
37. Padmanaban, R., Thirumaran, M., Anitha, P., Moshika, A.: Computability evaluation of RESTful API using primitive recursive. Inform. Sci. (2018, in press). https://doi.org/10.1016/j.jksuci.2018.11.014
38. Mcgrath, H., Kotsollaris, M., Stefanakis, E., Nastev, M.: Flood damage calculations via a RESTful API. Int. J. Disaster Risk Reduction **35**, 101071 (2019). https://doi.org/10.1016/j.ijdrr.2019.101071

The Retirement Supply Chain Improvement Using L-SCOR

Houda Mezouar(✉) and Abdellatif El Afia

ENSIAS, Mohammed V University of Rabat, Rabat, Morocco
houda.mezouar@gmail.com, a.elafia@um5s.net.ma

Abstract. In order to design an adaptable business process, the SCOR model was extended to L-SCOR model integrating the concept of adaptability at the two sections "Process" and "Metrics". This work represents a case study that gives the chance to apply L-SCOR on the retirement supply chain. This paper started by giving an overall overview about L-SCOR model then detailed the studied supply chain models at two levels of L-SCOR. The second level model expressed the flows between the planning processes and the business process sL3 "Learn Make" that describes the management of adaptability for rights liquidation. The fourth level model explains how the recommendation web service will be used for this business process. The paper also detailed the prerequisites for the recommendation web service which implements the Q-Learning algorithm, and discussed the returning Q-table and the optimal strategy for the called activity.

Keywords: SCOR · L-SCOR · Q-Learning · Web service · Adaptability · BPMN · Business process · Supply chain

1 Introduction

1.1 L-SCOR: An Extension of SCOR for a Self-adaptive Supply Chain

SCOR is a reference that represents the flows of a company allowing modeling its different structures and processes [1]; SCOR makes it possible to create for the supply chain actors a shared language and to harmonize their practices. It thus allows the improvement of the supply chain's performance and builds standardized indicators [2]. Different research works approaching different fields had used the reference SCOR for the process improvement and its performance analysis [3–12]. The latest version of the SCOR (SCOR 12), considers that all supply chains are composed of the following six processes: Plan, Source, Make, Deliver, Return and Enable. Reference [13], suggests an extension to SCOR by adding to those processes a seventh one "sL Lean", in order to manage the adaptive aspect of a supply chain processes, and to its fifth performance attributes, a sixth one "adaptability" with eight metrics. It also offers a sL implementation approach at the fourth level of a supply chain. L-SCOR suggests in the supply chain strategic level to add the process sL. "sL Learn" process describes the activities associated with developing policies to be adopted in order to provide the ideal functioning of the process. The Learn

M. Lazaar et al. (Eds.): BDIoT 2021, LNNS 489, pp. 501–514, 2022.
https://doi.org/10.1007/978-3-031-07969-6_38

processes include on one hand the gathering of each process states, the gathering of the possible actions for each process and of the rewards related to each action, on the other hand they also include the establishment of the optimal policy for the process execution. In the configuration level, sL is divided into processes related to establishing the ideal policies for self-adaptive supply chains, which are “sL1 - Learn Plan”, “sL2 - Learn Source”, “sL3 - Learn Make”, “sL4 - Learn Deliver” and “sL5 - Learn Return”. Concerning the supply chain operational level, each of the second level processes are divided into sub-processes. sL1 - Learn Plan describes the process of development and establishment of optimal policy for the self-adaptive supply chain planning. It consists of the following sub-processes: “sL1.1 - Identify plan states”, “sL1.2 - Identify plan actions”, “sL1.3 - Identify rewards of planning actions” and “sL1.4 - Establish and communicate plan policy”. It receives as input the supply chain requirements and resources from the planning processes to give the ideal supply chain planning policy as an output. sL2 - Learn Source describes the process of development and establishment of optimal policy to meet the sourcing conditions of a self-adaptive supply chain. It's composed of the following activities: “sL2.1 - Identify sourcing states”, “sL1.2 - Identify sourcing actions”, “sL1.3 - Identify rewards of sourcing action” and “sL1.4 - Establish and communicate sourcing policy”. It receives as input the products requirements and sources from the planning processes, to return as output the optimal supply chain sourcing policy. sL3 - Learn Make describes the process of development and establishment of optimal policy to meet the production requirements of a self-adaptive supply chain. It's composed of the following activities: “sL3.1 - Identify production states”, “sL3.2 - Identify production actions”, “sL3.3 - Identify rewards of production actions” and “sL3.4 - Establish and communicate production policy”. It gets from the planning processes, the production conditions and resources as input, to give as output the ideal supply chain production policy. sL4 - Learn Deliver outlines the development process of the optimal policy in order to meet the delivery requirements of a self-adaptive supply chain. It's composed of the following activities: “sL4.1 - Identify delivery states”, “sL4.2 - Identify delivery actions”, “sL4.3 - Identify rewards of Delivery actions” and “sL4.4 - Establish and communicate delivery policy”. It receives as input the delivery requirements, resources and capabilities from the planning processes, to return as output the optimal policy for supply chain delivery. sL5 - Learn Return describes the process of development and establishment of optimal policy to satisfy anticipated and unanticipated return conditions of a self-adaptive supply chain. It consists of the following sub-processes: “sL5.1 - Identify Return states”, “sL5.2 - Identify Return actions”, “sL5.3 - Identify rewards of return actions” and “sL5.4 - Establish and communicate return policy”. It receives as input, the return requirements and resources from the planning processes, to give as output the ideal supply chain return policy.

At the fourth level of the supply chain, the use of the BPMN as a descriptive language of an activity is proposed. With the integration of a control, on the execution result of each activity. In the event of a processing anomaly, the process calls a web-service that executes sL Learn process. This web service receives as input for the concerned activity, the execution data (as detailed in the previous levels), it reformulates the task of the activity in agent, states and actions, and it returns the optimal behavior to execute thanks to the Q-learning algorithm. The activity runs a second time with the data received, and if its behavior is normal, the process goes to the next activity, otherwise we call the web-service with the new data as shown in Fig. 1.

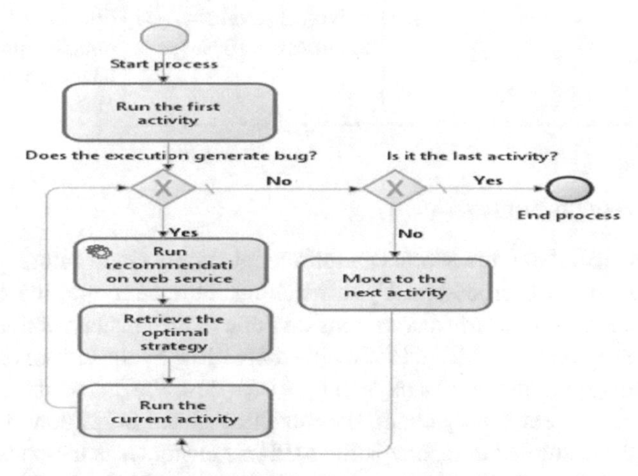

Fig. 1. sL implemented at the fourth level of the supply chain

There are over two hundred and fifty SCOR metrics that are codified and structured in a hierarchical framework. Those metrics are grouped in five performance attributes, which are 'asset management efficiency', 'reliability', 'costs', 'responsiveness' and 'agility'. In addition to those five performance attributes, L-SCOR add a sixth one: "Adaptability". The "Adaptability" performance attributes outlines the supply chain ability to deal with changes and unexpected malfunction. The tactical level metrics perform as diagnostics for the strategic level metrics. This implies that by examining the tactical level performances metrics, performance gaps or improvements for the strategic level metrics are explained. Likewise, operational level metrics perform as diagnostics for tactical level metrics. Table 1 presents the eight metrics related to the proposed performance attribute.

Table 1. The "Adaptability" L-SCOR metrics.

	Strategic metrics	Tactical metrics	Operational metrics
Adaptability	perfect change adaptation AD1.1:[Total Perfect adapted changes]/[Total Number of changes] × 100%	% changed activities AD2.1:[total changed activities]/[total activities] × 100%	# of activity requirements AD3.1 # of activity resources AD3.2 # of activity states AD3.3 # of activity actions AD3.4
		% adapted activities AD2.2:[total adapted activities]/[total changed activities] × 100%	% adapted activities use AD3.4:[number of used optimal policies]/[number of adapted activities] × 100%

1.2 The Retirement Supply Chain

References [14, 15], detail the Moroccan retirement supply chain actors, processes and the flows between these processes. Upon reaching retirement age, the citizen begins receiving a pension from the Moroccan Pension Fund (MPF) instead of a salary from his recruiter. This pension is established following a file filing by the citizen administration, whose processing gives in output a payment [14]. To describe and explain the processes of the studied retirement supply chain, the modeling of the SC is done via SCOR. As shown in Fig. 2 the studied company is the MPF, its customer is the pensioner, and its supplier is the affiliate. The affiliate manages the process Deliver (sD) that describes the management rules for its contribution to the pension plan. The MPF manages three processes, Source (sS) that describes the management of the affiliation and the contribution; Make (sM) that describes the management of the rights liquidation, Deliver (sD) that describes the management of the pensioner's payment. The pensioner manages the Source (sS) process that describes its benefits of a pension. The process Plan (sP) is the one that balance aggregate demand and supply to develop a course of action, which best meets sourcing, production, and delivery requirements.

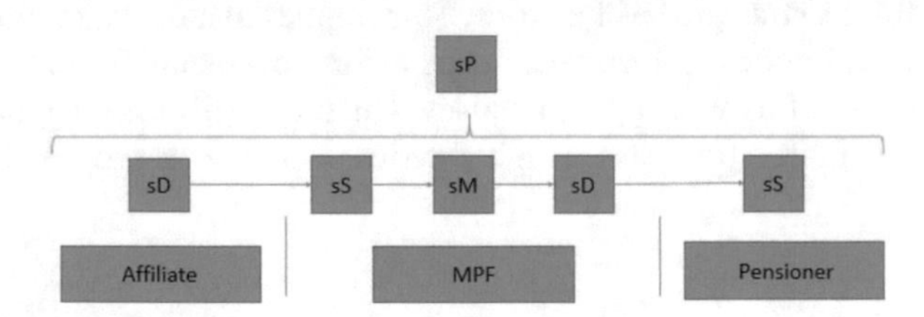

Fig. 2. sL the Strategic level model of the Moroccan retirement SC

The studied product (the retirement pension) is a make-to-order product, so the tactical level processes are: sS2 (Source Make-to-Order Product), sM2 (Make-to-Order), sD2 (Deliver Make-to-Order Product). We note that the processes Plan shown in Fig. 2.2

are sP1 (Plan Supply Chain), sP2 (Plan Source), sP3 (Plan Make), sP4 (Plan Deliver) as show in Fig. 3.

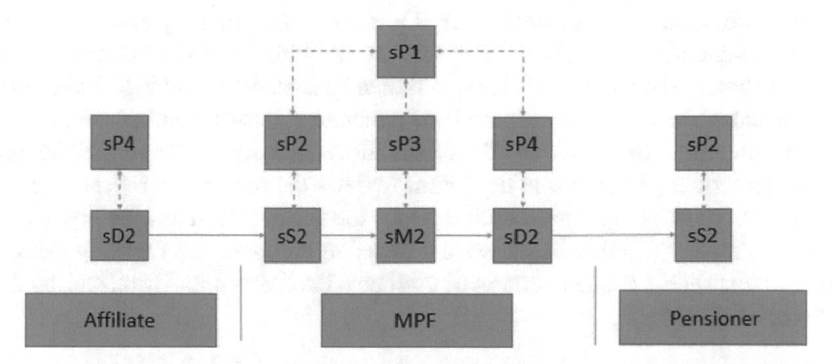

Fig. 3. sL the Tactical level model of the Moroccan retirement SC

At the operational level, organizations can specify the activities of sub-processes, best practices, the functionality of the software and existing tools. For this case, the MPF is the studied company, [14] detail at this level the process sM2, sS2, sD2.

At the real-time level, the focus is on the operational study of the retirement records circulation processes within the MPF, so the itinerary of each file from its reception by the department concerned until its final archiving is traced. This study also concerns the intermediate stages of the each file processing as well as the interveners in each step, which allowed to define the number of intervening in the processing of files and to stop all the documents generated following the liquidation of records. Among the pension two types (civil and military), this study is based on 'the management of civil pension rights'. According to Fig. 4, 'the management of civil pension rights' process is composed of six lanes: "head of career tracking service", "Dispatching", "Affiliation", "Contribution", "Liquidation", and "Concession". The process is initiated by the start event "receive folders", and it ends by the execution of one of the three end events: "folder sent to the organization", "Non-compliant folder", or "conceded folder". Once the folders are received from the different affiliate's administrations, the head of career service assign these folders to the affiliation officers, and then he designates a monitoring agent who will be responsible for the follow-up of the processing of folders, he starts by dispatching the folders to the assigned officers. The affiliation officer studies the folder, if the folder is not complete (all papers are received) he rejects it, if not he validates it, and then he certifies the affiliation data before printing the certification sheet. Then he sends the certified folder with its certification sheet to the concerned contribution officer and a copy of it to the monitoring agent. Once the contribution officer receives the folder, he verifies the affiliation certification, if it is valid, he certifies the contribution, if not he rejects the certification. The liquidator studies the certified folder (affiliation and contribution), if it is not valid he reject the folder. If a folder is Rejected before the certification and/or before the liquidation, the monitoring agent establishes a rejection letter that describes the rejection causes; this letter is signed by the head of career service be-fore being send to the concerned organization. For each rejected folder, once it is sent to the organization, the management of civil pension rights ends by executing the error end event "folder sent to the organization". For the valid certified folder, the liquidator verifies the compliance with the rights, if this compliance is not valid he rejects the folder, in this case the management of civil pension rights ends by executing the error end event "Non-compliant folder". For the folders with valid rights compliance, the liquidator validates the legal conditions, and then he runs the liquidation via the system (which will calculate the pension amount and generate a pension number) before printing the liquidation report, then he sends the folder for verification to the verifier. In case the verifier rejects the liquidation, the liquidator reviews the folder and repeats the verifier validates the liquidation operation until it. Once the verifier validates the liquidation, he concedes the folder (which will generate a decision number) and finally generates the decision. In this case, the management of civil pension rights ends by executing the end event "conceded folder".

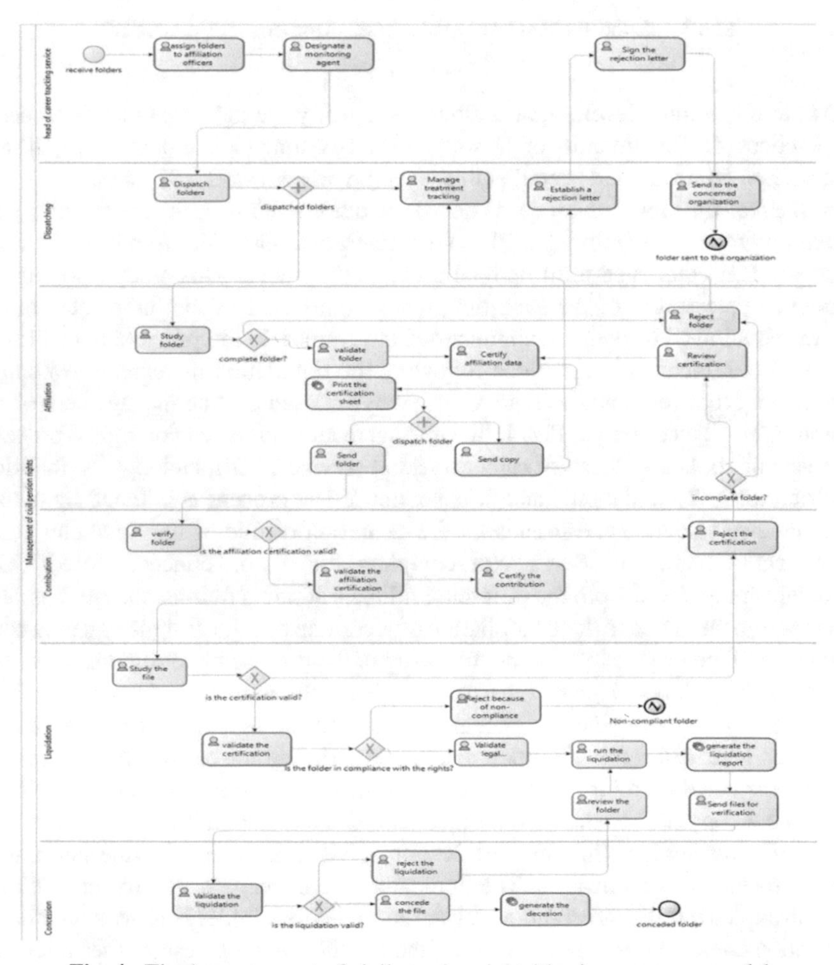

Fig. 4. The 'management of civil pension rights' business process model

2 The Retirement Supply Chain Modelling with L-SCOR

For a self-adaptive retirement SC, L-SCOR is adopted At the second level of L-SCOR, Fig. 5 presents the self-adaptive retirement Supply chain (SC) modelling. We note that the planning processes interact with the learning process sL3 "Learn Make" that describes the management of adaptability for rights liquidation.

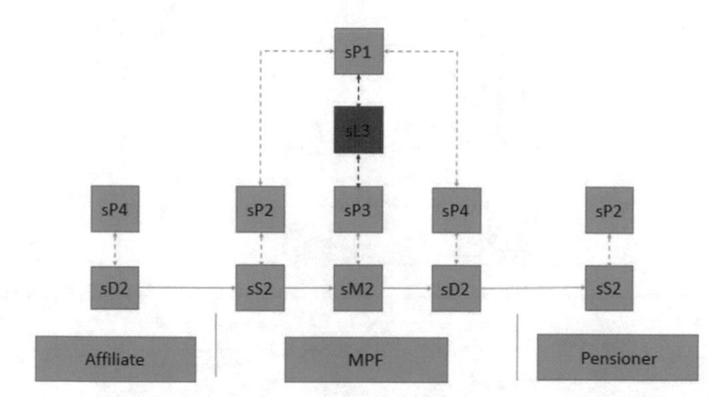

Fig. 5. The Tactical level model of the self-adaptive Moroccan retirement SC

Concerning the studied business process, which is in this case "the civil pension rights management" of the self-adaptive retirement SC presented by Fig. 6, we notice on the BPMN model the addition of the script activity "Execute recommendation system". This one applies the Q-learning algorithm, and gets the necessary data for its execution from the activity "Study the file", returning thus the optimal combination (state, action) for the treatment of the current pension file.

3 Implementation and Results Analysis

3.1 Q-Learning as a Reinforcement Learning Algorithm

In this work Q-learning is chosen as a reinforcement learning (RL) algorithm from a range of other RL algorithms. This choice is justified on one hand by its low computational footprint and being a 'model-free' algorithm, it combines learning and execution, simultaneously in any given environment having states and possible actions; Which is a very important point knowing that the proposed web-service would be used by different processes (different agents). On the other hand the proposed web-service learns by experience replay, so we need an off-policy learning method, that's why the on-policy methods have been discarded, like SARSA that expects that the actions in every state are chosen based on the current policy of the agent that usually tends to exploit rewards. So among the free-model algorithms based on off-policy learning we choose Q-Learning as a standard RL algorithm to first evaluate the proposed implementation of sL.

Watkins [16] proposes a typical description of Q-learning, the algorithm allows individuals to maintain a Q-value table for each strategy and to update the table in

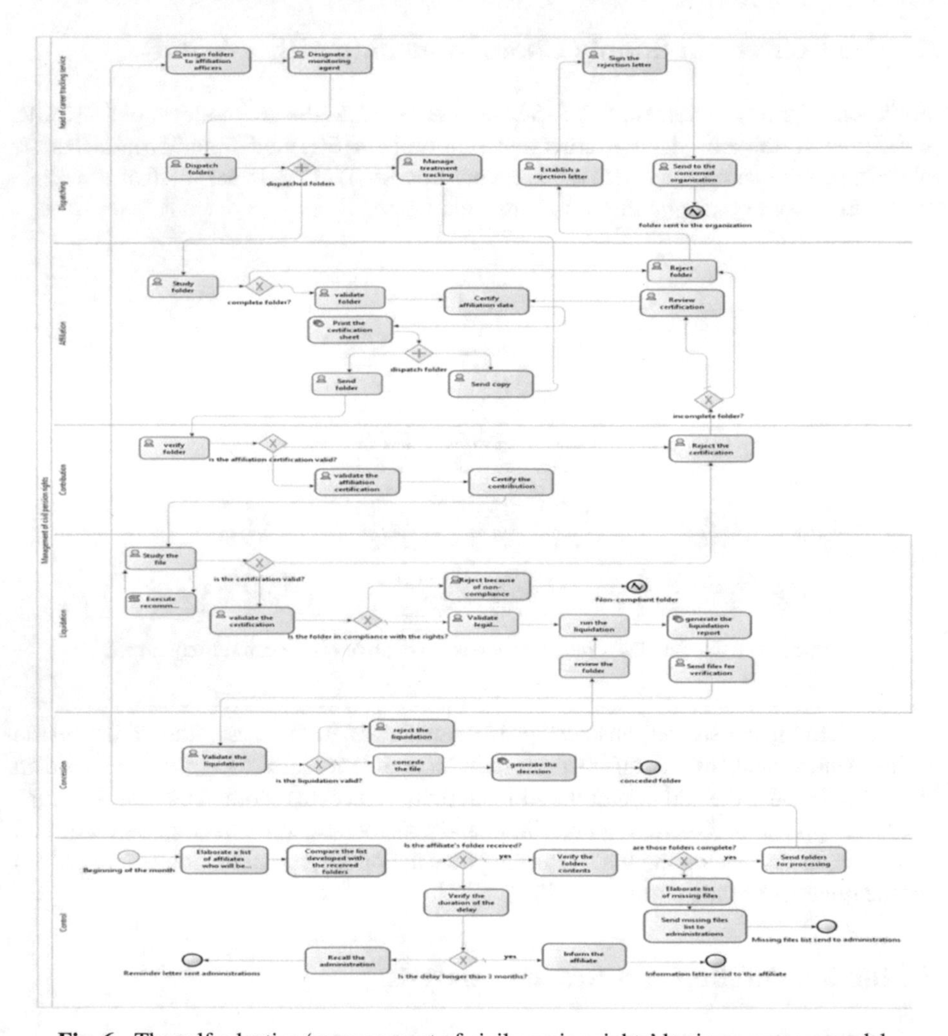

Fig. 6. The self-adaptive 'management of civil pension rights' business process model.

each round, and finally selects the strategy with the highest Q-value as the next round strategy [17]. Based on the foundation of Markov decision processes theory, Q-learning is a useful and meaningful method for solving problems of agent actions under different complicated conditions, it is very reliable because a discounted future reward is obtained when transiting from one state to another [18]. The principle of Q-learning algorithm is as follows:

– For each agent state, we note the action that led to this state by strengthening our existing q-value. Arrived on the state s' from state s and action a, we note the q-value as following:

$$Q(s,\ a) = \lambda \times \left(r + \gamma \times \max a'\left(Q\left(s',\ a'\right)\right)\right) + (1 - \lambda) \times Q(s,\ a)$$

$\lambda \in [0; 1]$ is the learning rate
$\gamma \in [0; 1]$ is the discount factor
r is the reward obtained for achieving state s' from state s with action a

- The agent carries out several research cycles of rewards, from initial state to goal state and reinforces with each passage the q-value of the action which leads to rewards or leading to states leading to rewards.
- Initially, the agent does not know the states where the rewards are, does not know the arrival state of an action, so he starts by choosing random actions he explores.
- After a certain time or when it reaches a goal state, the agent resumes a search for a solution from the state initial. At each cycle, the system exhibits a behavior of less and less exploratory, and increasingly guided by quality.

The Q-learning algorithm is implemented in this work as following:
Function Q-Learning

```
∀ s ∀ a Q(s, a) ← 0
for n ← 1, nbCycles do  #nbCycles is the number of learning cycles
   λ ← 1 ;
   α ← 1 ;
   currentState ← firstState
        for i ← 1, nbMaxActions do  #the max number of actions to execute
             s ← currentState
             nb ← random(0, 1)
        if (nb < α) then
        a ← randomAction(s)
        else
        a ← argMax a' (Q(s, a'))
        end if
                  s'← a(s)
        Q(s, a) ← α × (r + γ × max a' (Q(s ' , a' ))) + (1 − α) × Q(s, a)
        λ ← 0.99 × λ
        α ← 0.99 × α
        if (s' = finalState) then
        exit loupe i
        end if
         end for
        end for
        End Function
```

3.2 State and Action Spaces

The first step to apply Q-Learning algorithm is defining the state space. In real cases, we are focusing to obtain results in a limited period. If the state space is very large, the algorithm will take too long to converge. The treatment steps through which a retirement folder passes, trace all the possible states for this agent. The possible states of a retirement folder are as follows: 'Received', 'Certified affiliation', 'Certified contribution', 'Liquidated', 'Revised', 'Conceded', 'Rejected', and 'Incomplete'. An identifier is associated with each state respectively as follows: 'S0: Received', 'S1: Certified affiliation', 'S2: Certified contribution', 'S3: Liquidated', 'S4: Revised', 'S5: Conceded', 'S6: Rejected', 'S7: Incomplete', 'S8: Stuck in a state', 'S9: Returned to a previous state', and 'S10: Skipped a treatment step'. S the space of states is the set of these identifiers, S = {S0, S1, S2, S3, S4, S5, S6, S7, S8, S9, S10}. The possible operations on the retirement folder treatment present all the agent actions as follows: 'Reject', 'Certify Affiliation of received folder', 'Certify contribution for certified affiliation folder', 'Review certified contribution folder', 'Liquidate certified contribution folder', 'Concede liquidated folder', 'Stay in the same folder state' and 'Return to a previous state'. An identifier is associated with each action respectively as follows: 'A0: Reject', 'A1: Certify Affiliation of received folder', 'A2: Certify contribution for certified affiliation folder', 'A3: Review certified contribution folder', 'A4: Liquidate certified contribution folder', and 'A5: Concede liquidated folder'. The actions space A is the set of these identifiers A = {A0, A1, A2, A3, A4, A5}.

3.3 The Reward Strategy

The purpose of choosing an action transitioning to some state is illustrated by the reward strategy that plays a key role in learning speed. The reward value can be negative or positive. A negative reward value is used to express a penalty for an undesirable action. The rewards that are generated after a state transition are a short-term outcome. Because of including future rewards in the evaluation of algorithm actions, it may return a low immediate reword for an action linked to a high Q-value. A fixed reward function is implemented generating static rewards. A reward of +1 is given when the agent reaches the target. A reward of 0,5 is given when the agent takes action that can get him close to the target to encourage agents searching for defined targets. A reward of 0,1 is given when the agent takes action that can get him close to the target but repeating a task (a revision). A reward of 0 is given when the agent takes an action that does not get him close to the target and does not make him back down. A penalty of −0,5 is given when the agent takes action that make him back down. Table 2 describes the used reward strategy.

3.4 The Value Function

The Q values of state-action pairs is stored in Q-table. All values stored in the table are initially set to zero. The rows in the table represent the retirement folder's states, and the columns represent the action corresponding to the given state. During training, each entry in the table is populated by updating the Q values. After sufficient training, the Q values converges. The updating equation is shown below:

$$Q(s, a) \leftarrow \alpha \times \left(r + \gamma \times \max a'\left(Q\left(s', a'\right)\right)\right) + (1 - \alpha) \times Q(s, a)$$

Table 2. Reward strategy for the 'retirement folder' agent.

State	Reward
Certify affiliation	0,5
Certify contribution	0,5
Liquidate folder	0,5
Revise treatment	0,1
Concede folder	1
Reject folder	0
Folder Stuck in a state	0
Return folder to a previous state	−0,5
Skip a treatment step	0

where 's' is the current joint state, and 'a' is the selecting action. 's' is the next state that the retirement folder transfers to from the current state after performed by the action 'a'. 'α' is the learning rate, it specifies to what extent recently learned information supersedes old information. 'γ' is the discount factor, it determines the importance of future rewards. After training, the retirement folder is able to execute the best action for each state.

3.5 The Results

The search table where the agent determines the maximum anticipated future rewards for action at each state is called a Q-table. This table will lead the agent to the best action at each state. In the Q-table, the columns represent the available actions and the rows represent the states. Every Q-table outcome represents the utmost envisaged a future reward that the agent will obtain if it executes that action at that state. In order to learn and improve each value of the Q-table in every repetition of the iterative process, the algorithm "Q-learning" is executed. Before exploring the environment by the agent, the Q-table accords identical arbitrarily fixed values. While the agent still in the exploring step, the Q-table attributes a better and better estimation by repeatedly updating Q(s, a). Exploration and exploitation are the two strategies of the training process. Exploration is finding more information about the environment and exploitation is exploiting known information to maximize the reward. The agent aim is to maximize the expected cumulative reward. However, the agent may trap in local optimum and fail to find a feasible solution. That being the case, Q-learning uses what is called "the epsilon greedy strategy". An exploration rate ϵ is specified as the rate of steps that the agent takes actions randomly. At the beginning of the mission, ϵ rate have to be at its maximum value, since the agent ignores the Q-table values, it needs to do a lot of exploration by arbitrarily choosing its actions. Therefore, a random number N is generated. If $N > \epsilon$, the agent will then do the exploitation, meaning that the agent uses what is already known to select the best action in each step. Else, the agent will do the explorations. The idea is that the agent must have a big ϵ rate at the beginning of the Q-function training. As the agent explores the environment, the ϵ rate reduces and the agent begins the environment exploitation.

Then, reduce ϵ rate gradually since the agent starts being more confident at estimated Q-values. The Q-Learning used parameters are defined in Table 3 as follow.

Table 3. Q-Learning used parameters for "the retirement folder" agent

Parameter	Value
Learning rate	0.05
Discount factor	0.7
Initial Q-value	0
Initial value of ϵ	0.6
ϵ Decay rate	0.005

The Q-table returned by the recommendation web service for the 'retirement folder' agent is shown in Table 4.

Table 4. The returned Q-Table for "the retirement folder" agent

	A0	A1	A2	A3	A4	A5
S0	0	0.7999	0	0	0	0
S1	0	0	0.7999	0	0	0
S2	0	−0.499997	0	0.47934	0.7999	0
S3	0	−0.499997	−0.499997	0.47934	0	0.99999
S4	0	−0.499997	−0.499997	0.47934	0.7999	0.99999
S5	−0.4999	−0.499997	−0.499997	−0.499997	−0.499997	0
S6	0	0	0	0.47934	0	0
S7	0	0	0	0.47934	0	0
S8	0	0	0	0.47934	0	0
S9	0	0	0	0.47934	0	0
S10	0	0	0	0.47934	0	0

The optimal strategy imposed by the recommendation web service for the "retirement folder" agent according to the returned Q-Table is as follows:

OS = {(S0, A1), (S1, A2), (S2, A4), (S3, A5), (S4, A5), (S5, A5), (S6, A3), (S7, A3), (S8, A3), (S9, A3), (S10, A3)}.

4 Conclusion

The reference L-SCOR is adopted for the improvement of the Moroccan retirement SC. this paper started by expressing the SC at the second level of L-SCOR, in order to clarify the flows between the planning processes and sL3 "Learn Make" that describes the management of adaptability for rights liquidation. Then it established at the fourth level of L-SCOR the business process "the management of civil pension rights", thus explaining, how the recommendation web service will be used for this business process. Last and not least, it explained, the communication protocol between the calling activity and the recommendation web service, the prerequisites for the implementation of the Q-Learning algorithm, by defining the environment adapted to this case agent, the action space, the state space, and the reward table. Finally, the paper discussed the returning Q-table and the adopted strategy used by the called activity.

References

1. Mezouar, H., El Afia, A.: Proposal of an approach to improve business processes of a service supply chain. Int. J. Mech. Eng. Technol. (IJMET) **10**(03), 978–989 (2019). http://www.iaeme.com/ijmet/issues.asp?JType=IJMET&VType=10&IType=3
2. Mezouar, H., El Afia, A.: A 4-level reference for self-adaptive processes based on SCOR and integrating Q-learnin. In: The 4th International Conference on Big Data and Internet of Things (BDIoT 2019), 23–24 October 2019. ACM, Tangier-Tetuan, Morocco (2019). https://doi.org/10.1145/3372938.3372953
3. Mezouar, H., EL Afia, A.: A process simulation model for a proposed Moroccan supply chain of electricity. In: International Renewable and Sustainable Energy Conference (IRSEC), 14–17 November 2016. IEEE, Marrakech, Morocco (2017). https://doi.org/10.1109/IRSEC.2016.7983999
4. El Afia, A., Mezouar, H.: A global mapping of the Moroccan supply chain of hospital drugs and a simulation of the dispensation process. In: The 2nd International Conference on Big Data, Cloud and Applications (BDCA 2017), 29–30 March 2017. ACM, Tetouan, Morocco (2017). https://doi.org/10.1145/3090354.3090465
5. Mezouar, H., El Afia, A., Chiheb, R., Ouzayd, F.: Proposal of a modeling approach and a set of KPI to the drug supply chain within the hospital. In: The 3rd International Conference on Logistics Operations Management (GOL), 23–25 May 2016. IEEE, Fez, Morocco (2016). https://doi.org/10.1109/GOL.2016.7731691
6. Mezouar, H., El Afia, A., Chiheb, R.: A new concept of intelligence in the electric power management. In: The 2nd International Conference on Electrical and Information Technologies (ICEIT 16), 4–7 May 2016. IEEE, Tangier, Morocco (2016). https://doi.org/10.1109/EITech.2016.7519596
7. Mezouar, H., El Afia, A., Chiheb, R., Ouzayd, F.: Toward a process model of Moroccan electric supply chain. In: The International Conference on Electrical and Information Technologies (ICEIT 15), 25–27 March 2015. IEEE, Marrakech, Morocco (2015). https://doi.org/10.1109/EITech.2015.7162990
8. Bouzbita, S., El Afia, A., Faizi, R.: A novel based Hidden Markov model approach for controlling the ACS-TSP evaporation parameter. In: International Conference on Multimedia Computing and Systems -Proceedings, pp. 633–638 (2017). https://doi.org/10.1109/ICMCS.2016.7905544
9. El Afia, A., Bouzbita, S., Faizi, R.: The effect of updating the local pheromone on ACS performance using fuzzy logic. Int. J. Electr. Comput. Eng. **7**(4), 2161–2168 (2017). https://doi.org/10.11591/ijece.v7i4.pp2161-2168
10. Kabbaj, M.M., El Afia, A.: Towards learning integral strategy of branch and bound. In: International Conference on Multimedia Computing and Systems -Proceedings, pp. 621–626 (2017). https://doi.org/10.1109/ICMCS.2016.7905626

11. Lalaoui, M., El Afia, A., Chiheb, R.: Hidden Markov model for a self-learning of simulated annealing cooling law. In: International Conference on Multimedia Computing and Systems -Proceedings, pp. 558–563 (2017). https://doi.org/10.1109/ICMCS.2016.7905557
12. Lalaoui, M., El Afia, A., Chiheb, R.: A self-adaptive very fast simulated annealing based on Hidden Markov model. In: Proceedings of 2017 International Conference of Cloud Computing Technologies and Applications, CloudTech 2017, January 2018, pp. 1–8 (2018). https://doi.org/10.1109/CloudTech.2017.8284698
13. Mezouar, H., El Afia, A.: Proposal for an approach to evaluate continuity in service supply chains: case of the Moroccan electricity supply chain. Int. J. Electr. Comput. Eng. (IJECE) **9**(6), 5552–5559 (2019). https://doi.org/10.11591/ijece.v9i6.pp5552-5559
14. Mezouar, H., El Afia, A.: A retirement pension from a supply chain side: case of the moroccan retirement pension. In: Ezziyyani, M., Bahaj, M., Khoukhi, F. (eds.) AIT2S 2017. LNNS, vol. 25, pp. 103–115. Springer, Cham (2018). https://doi.org/10.1007/978-3-319-69137-4_11
15. Mezouar, H., El Afia, A.: Performance analysis model for service supply chains: case of the retirement supply chain. Int. J. Eng. Technol. (IJET) **7**(3), 1429–1438 (2018). https://doi.org/10.14419/ijet.v7i3.13929
16. Watkins, C.J., Dayan, P.: Technical note: Q-learning. Mach. Learn. **8**, 279–292 (1992). https://doi.org/10.1023/A:1022676722315
17. Ding, H., Zhang, G., Wang, S., Li, J., Wang, Z.: Q-learning boosts the evolution of cooperation in structured population by involving extortion. Pysica A **536**, 122551 (2019). https://doi.org/10.1016/j.physa.2019.122551
18. Tong, Z., Chen, H., Deng, X., Li, K., Li, K.: A scheduling scheme in the cloud computing environment using deep Q-learning. Inf. Sci. **512**, 1170–1191 https://doi.org/10.1016/j.ins.2019.10.035.LNCS. Accessed 21 Nov 2016

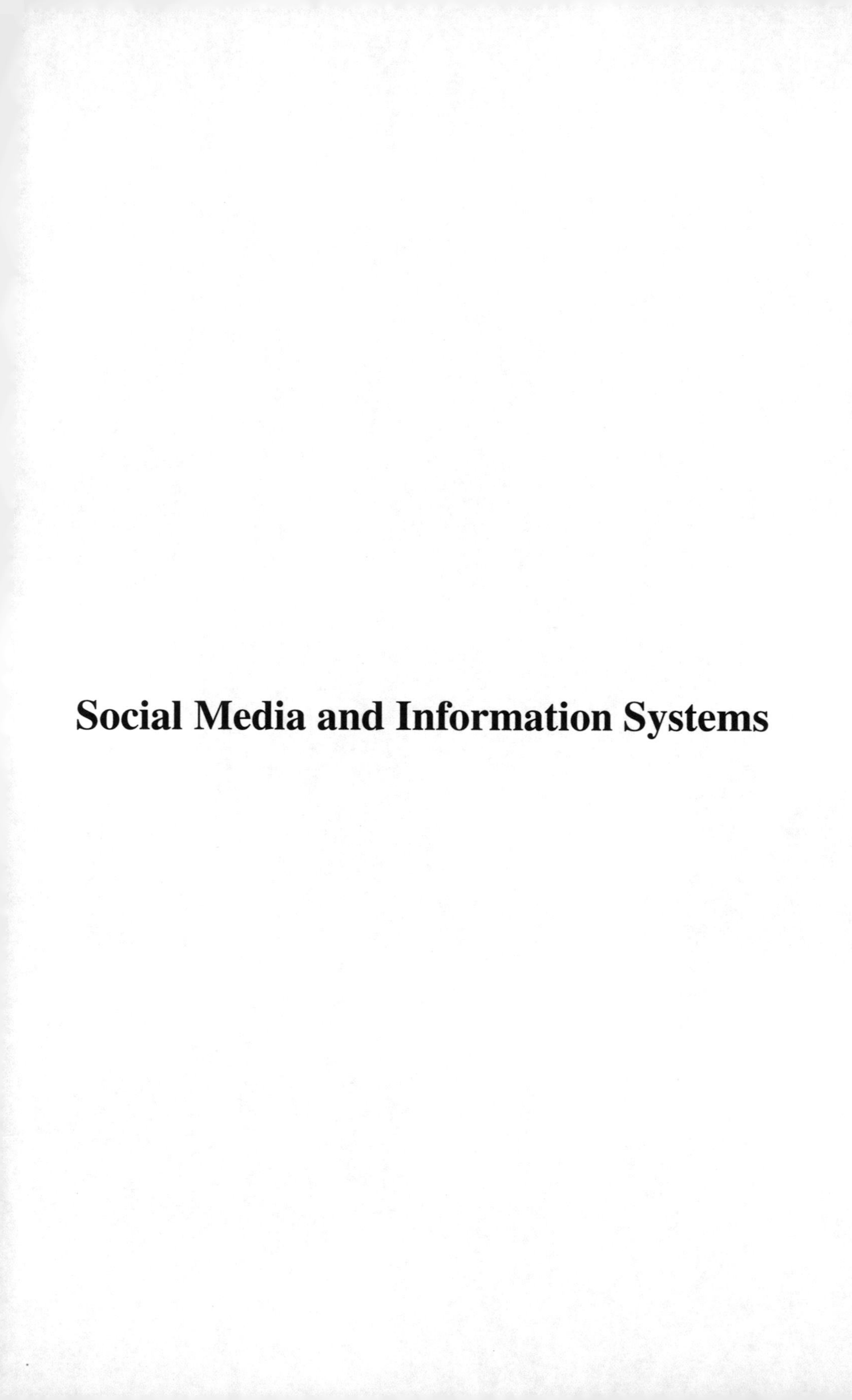

Social Media and Information Systems

A Hybrid Machine Learning Method for Movies Recommendation

Redwane Nesmaoui(✉), Mouad Louhichi, and Mohamed Lazaar

ENSIAS, Mohamed V University, Rabat, Morocco
redwane.nesmaoui@um5r.ac.ma

Abstract. Recently, the application of machine learning algorithms is very useful in marketing by companies nowadays. Overall, it has become a big factor on the companies success and growth in term of the number of users or revenues, since it helps to suggest the right content to the right people in an easy way without going through a long complicated process to choose an element in a list of millions elements. This research has a goal of evaluating several recommending mining algorithms in machine learning by adopting a model that combines the content-based (constrained system to people) and collaborative approach and compares it with a paralleled algorithm, and we assume that can help to get the right recommendations to users. The model's results show that it can positively solve this issue and help users to find the right content that they want to watch, and also predict if they like the new trending content.

Keywords: Artificial intelligence · Machine learning · Recommendation system · Movie recommendation · Collaborative filtering · Content based filtering · Parallel computing

1 Introduction

Artificial Intelligence systems are becoming ubiquitous and very important to have smart industries such as healthcare, robotics, transportation, manufacturing, retail, banking, and energy. However, to make Artificial Intelligence systems deployable in social environments, industry, and business-critical applications, several challenges, one of these challenges is the Recommender Systems.

Recommender Systems are a specific form of information filtering aimed at presenting information elements (films, music, books, news, images, web pages, etc.) that might be interesting to the user [1].

As the internet has become an important part of our generation, Recommendation Systems are used on online sales websites. They allow e-merchants to automatically highlight products likely to interest visitors and maximize profits to them [2]. The selection of products displayed is then personalized according to different criteria to increase the turnover generated by sales. Amazon is known for using a powerful recommendation system, but the practice is spreading and is

M. Lazaar et al. (Eds.): BDIoT 2021, LNNS 489, pp. 517–528, 2022.
https://doi.org/10.1007/978-3-031-07969-6_39

no longer reserved for large e-commerce sites. Recommendation systems can also be used to enrich the user experience, by offering articles on a news site likely to interest the reader, or on a music site, a musical selection likely to please the user.

Content recommending is one of the biggest challenging areas for the companies and as an area in machine learning applications because there are a huge number of data nature on it. Movies Netflix [8] as an example needs a big process and requires so many additional pre-processing steps especially that the data is unstructured.

Here we have a main huge issue, and it is the accuracy that can measure the variety of data, so we have two main issues Accuracy and performance because to have a good accuracy we need to build a good strategy that can help us to get a big performance to solve this issue. To solve these challenges, we proposed to focus on parallel processing in distributed environments and cloud as a high-performance computing resource (HPC), by varying the number of processing GPU and applied them on the SVD (Singular Value Decomposition) and compare it to the different choosing models and algorithms.

The main goal of this research paper is to present a study that compares three algorithms with a distributed algorithm using machine learning algorithms and focusing on several models with accuracy in the goal of evaluating each model and compare it with the others.

To solve this issue there are different types of recommender systems with different approaches some of them are classified as below:

1.1 Content-Based Filtering Systems (CBF Based Systems)

The main principle of the Content-based Filtering Systems is recommending items based on a comparison between item profile (Content founded to be relevant to the user) and user profile (Collection of assigned keywords collect by algorithm from items found relevant (or interesting) by the user). In general, it is like to find an item who has similar characteristics. For example, a person goes to watch his favorite series 'X' on a website. Unfortunately, series 'X' has no more episodes and because of this, a friend with same taste recommends the person to watch series 'Y' which have similar gender to series 'X'. This is an instance of content-based filtering.

Advantages of Content-Based Filtering

- he capability to recommend unrated items.
- Easy to explain the work of the recommender system by listing the Content features of an item.
- It needs only the rating of the concerned user and not any other user of the system.

Disadvantages of Content-Based Filtering

- It doesn't work for new users because there isn't a previous rated item by him, and it is required to evaluate the user preferences to provide him new recommendations.
- No recommendation of serendipitous items.
- This approach is limited by what the user likes or does not like.

1.2 Collaborative Filtering-Based Systems (CF-Based Systems)

A collaborative filtering system recommends items based on similarity measures between users and/or items. The system recommends items preferred by similar users. This scenario is based on the concept of asking someone, who has similar taste, to suggest a Movie.

Advantages of Collaborative Filtering-Based Systems

- It is dependent on the relation between users which implies that it is content independent.
- They can suggest serendipitous items by observing the user's similarity rather than the item's similarity.
- They can suggest real quality assessment of items based on other people's experience.

Disadvantages of Collaborative Filtering-Based Systems

- It can't provide recommendations for new items because there are no prerating from users.
- Existing of users (Gray sheep) who may neither agree nor disagree with the majority of the users. They may introduce difficulties to produce accurate collaborative recommendations [4]
- Difficulties of finding enough people's rates due to the exceeding number of users by a great margin (Data Sparsity Problem) [5]

2 Related Works

Several research efforts cover Movies Recommendation Systems using different types of features and classifiers algorithms like collaborative approach, content-based approach, hybrid approach, etc.

2.1 Movie Recommender System Using Single Value Decomposition And K-means Clustering (2020)

Based on the MoviesLens dataset M. Sandeep Kumar and J. Prabhu [7] proposed a hybrid model collaborative movie recommendation system that performs a combination of K-means clustering with ant colony optimization technique (ACO-KM).

2.2 Movie Recommender System Using K-Means Clustering AND K Nearest Neighbor (2019)

Ahuja, Solanki and Nayyar [8] used the collaborative filtering approach (the most commonly used method was the K-nearest-neighbor approach between users [9]), this approach based on computing the user's item rating with the other user and rate the similarity, it can also use the k most similar users to this user and predict items for him.

2.3 Fully Content-Based Movie Recommender System With Feature Extraction Using Neural Network (2017)

For the content-based approach H. Chen, Y. Wu, M. Hor, and C. Tang [10], used the most useful way (computing the Cosine similarity (Metric used to measure how similar the movies are irrespective of their size. Mathematically, it measures the cosine of the angle between two vectors projected in a multi-dimensional space)).

3 Methodology

The resolution of this subject issue requires the analyze of several types of algorithms (Random Forest, KNN, SVD and SVD++ algorithm.) and try to understand their characteristics and performance, because each recommendation algorithm needs different condition to provide results based on the same existed data by computing (MAE) Mean Absolute Error and (RMSE) Root Mean Square Error value.

For Random Forest and KNN Content we will use the following Approach:

For User KNN, Item KNN, SVD and SVD+ we will use the following Approach:

3.1 Random Forest

Random forest is basically a set of decision trees. Each tree classifies (often linearly) the dataset using a subset of variables. The number of trees in the forest and the number of variables in the subset are hyper-parameters and must be chosen a prior. The number of trees is the order of hundreds, while the subset of variables is quite small compared with the total number of variables. Random forests provide also a natural way of assessing the importance of input variables (predictors). This is achieved by removing one variable at a time and assessing whether the out-of-bag error changes or not. If it does, the variable is a must for the decision.

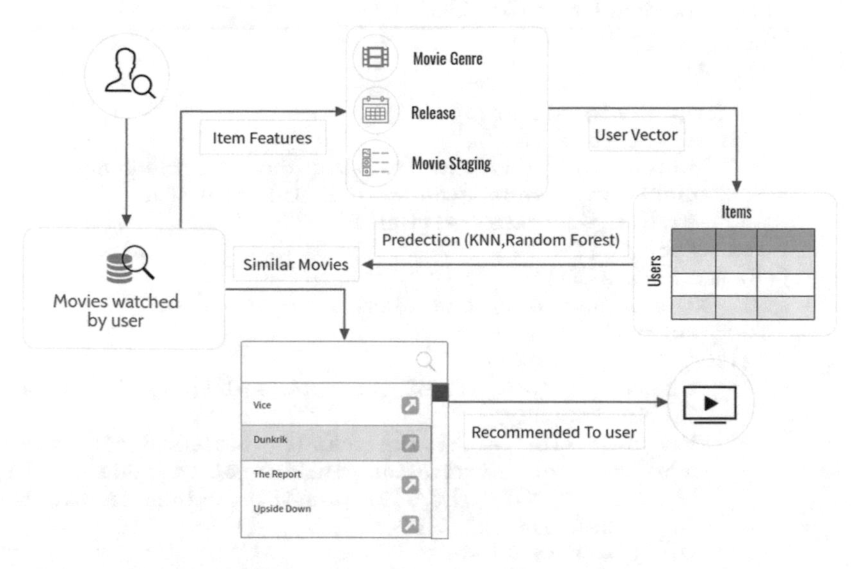

Fig. 1. This approach is based on looking for the movies that user watched it before by focus on their gender, release, or staging like that we can compare this list of movies with the other movies exists in the database and try to provide him new similar content recommendations.

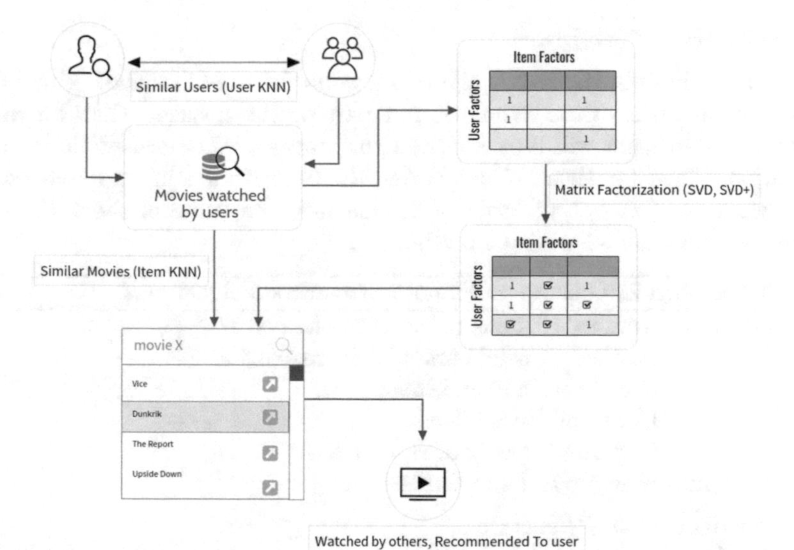

Fig. 2. This approach is based on comparing the user watched and rated movies with another user's movies, and try to find similar users, then provide recommendations based on the same users records

Algorithm 1 Random Forest Classifier

Input Training data(Td)
Output class label

```
1) To form t classifiers:
   for i = 1 to t do
       Select Tdi from the training data Td randomly
       Generate a root node called Rni with Tdi
       Invoke GenerateTree(Rni )
   end for
2) GenerateTree(Ni)
   if Ni consists only one class instance then
       return
   else
       Choose possible p% of the node(splitting features)
       in Ni randomly
       For splitting elect the feature called F with the
       more gain of information Build f child nodes of Ni,
       Ni1 ,..., Nif , here Fi possible values in can be
       in F that are Fi1 Fif
       for j = 1 to Fi do
           Replace Nij the contents of to TDj , here
           TDj is instances of Ni which is a match to Fi
       Invoke GenerateTree(Ni)
       end for
   end if
```

3.2 K-Nearest Neighbors

To make a prediction, the K-NN algorithm would be based on the entire dataset. Indeed, for an observation, which is not part of the dataset, that we want to predict, the algorithm will look for the K instances of the dataset closest to our observation. Then for these K neighbors, the algorithm will be based on their output variables (output variable) y to calculate the value of the variable y of the observation that we want to predict.

Algorithm 2 : General outline of neighborhood algorithms

input : Number of items to be recommended $N \in \mathbb{N}$,
Number of neighbors used for ranking $k \in \mathbb{N}$,
User to recommend items to u,
List of all items *Items*,
User-Item matrix of ratings R
output: N items to be recommended

foreach *item* $\in$ *Items* **do**
  if *item* $\notin$ *u.rated_items* **then**
    item.rank $\leftarrow$ rank_according_to_nearest_neighbors(k, u, *item*)

descending_rank_sort(*Items*)

return top(N, *Items*)

3.3 Singular Value Decomposition (SVD)

The Singular-Value Decomposition or SVD is a matrix decomposition method for reducing a matrix to its constituent parts in order to make certain subsequent matrix calculations simpler.

Algorithm 3: Matrix Factorization based Recommendation

input : Set of users U, items Y,
Matrix of users-items rankings R,
Number of latent features l,
Number of items to be recommended $N \in \mathbb{N}$,
output: Top-N recommendations $R(U) \in \mathcal{I}^N$

```
for i ← 1 to numIterations do
    foreach user ∈ U and item ∈ Y with rating R[u, i] do
        predicted_rating = U[u, ] * M[i, ]^t
        err = R[u, i] − predicted_rating
        U[u, ] = adjust_by_gradient_descent(U[u, ])
        M[i, ] = adjust_by_gradient_descent(M[i, ])
return N items with highest predicted ranking
```

3.4 Similarity Computing

To find the similarities between past user behaviors, we measure the similarity as a critical component of user-based collaborative filtering recommendation, so that we can present it as a table of N users vector with item scores.

Table 1. presents a fictitious example of a binary matrix containing information of the type " user u liked/do not like item i". This information can also be " bought/did not buy", " consulted/did not consult", etc. They can also be measured on a higher number of classes: "set 1/2/3/4/5 stars" etc. Another important point is how time influences the user's profile. The interests of users, generally, evolve the data in the user model should therefore constantly be readjusted to remain in line with the user's new interests.

	Item 1	Item 2	Item 3	Item 4	Item 5	...
User 1	5			4		...
User 2		1	4			...
User 3		5				...
User 4			1	5	1	...
...	...	...	...	...	...	...

3.5 CosSim Similarity

Cosine similarity is the measure of similarity between two vectors, if we take two movies that we want to compute the cosine similarity between. We will call these two movies X and Y. So, the cosine similarity of X and Y is given by:

- **User KNN :**

$$CosSim(x,y) = \frac{\sum_i x_i y_i}{\sqrt{\sum_i x_i^2}\sqrt{\sum_i y_i^2}}$$

- **Item KNN :**

$$CosSim(x,y) = \frac{\sum_i((x_i - \bar{(i)})(y_i - \bar{(i)}))}{\sqrt{\sum_i(x_i - \bar{(i)})^2}\sqrt{\sum_i(y_i - \bar{(i)})^2}}$$

3.6 Mean Absolute Error (MAE)

Mean Absolute Error is a metric used to compute the average of all the absolute value differences between the true and the predicted rating. f all the absolute value differences between the true and the predicted rating. The lower the MAE [2] the better the accuracy.

$$MAE = \frac{\sum_{i=1}^{n} \mid y_i - x_i \mid}{n}$$

3.7 Root Mean Square Error (RMSE)

Root Mean Square Error computes the mean value of all the differences squared between the true and the predicted ratings and then proceeds to calculate the square root out of the result. Consequently, large errors may dramatically affect the RMSE rating, rendering the RMSE metric most valuable when significantly large errors are unwanted. The root mean square error between the true ratings and predicted ratings is given by:

$$RMAE = \sqrt{\frac{\sum_{i=1}^{n}(y_i - x_i)^2}{n}}$$

4 Experiments And Results

4.1 Data Set

To implement our algorithms, we have used the MovieLens dataset of 100.000 [11] rates and for the parallel approach 20.000.000 rates [12].

4.2 Results

In this section, we present and discuss the results obtained from the application of the sequential approaches which is based on the 100.000 rates dataset where we used:

- KNN Content and Random Forest: those two algorithms are based on predicted recommendations by using previous movies watched characteristics for example Gender, Release year etc.

Table 2. The data is represented as the following format, where the dataset is represented as a 2D Matrix in a python numpy array, Mi,j (Mi represents the movies and the values are centered on zero by subtracting the mean from the respective elements). Which is divided as a ratio of 70% (Trainset) and 30% (Test-set).

User id	Movie id	Rating	Timestamp
1	17	2.5	835355681
2	592	5.0	1298923247
3	2903	3.0	949896309
4	31	1.0	1260759144
...	...	...	...

- KNN Item, KNN User, SVD (with 2 elements of ranking), SVD++ (with 3 elements of ranking): those algorithms are using user's records to compare them with other users, then provide recommendations based on the similarity between them.

Table 3. Compares the result of the proposed system technique. These tables show a comparison of RMSE and MAE among all the proposed techniques. It is seen from the tables that for the existing technique the best value of RMSE 0.8959 and MAE is 0.6871 of SVD++ because we have more criteria on the movies.

Techniques	RMSE	MAE
Random	1,4385	1,1478
Content KNN	0,9375	0,7263
User KNN	0,9961	0,7711
Item KNN	0,9995	0,7798
SVD	0,9039	0,6984
SVD++	0,8959	0,6871

We tried to implement the previous algorithms on the 20 M MoviesLens rate, but we did not have the chance to get results because of the limitation of physical resources to work on this data, which makes us think about finding a solution which is the parallel Approach.

Parallel Approach. The main context of this Approach is paralyzing the algorithms processing and making each process work on a part of the algorithms. For example if we have an addition of numbers from 1 to 100 and we have 4 processors each one of them will calculate 25 additions, then we calculate the addition of the 4 obtained results. We base our approach in this logic, so we use Apache Spark which gives us the possibility to paralyze our algorithms and we decide

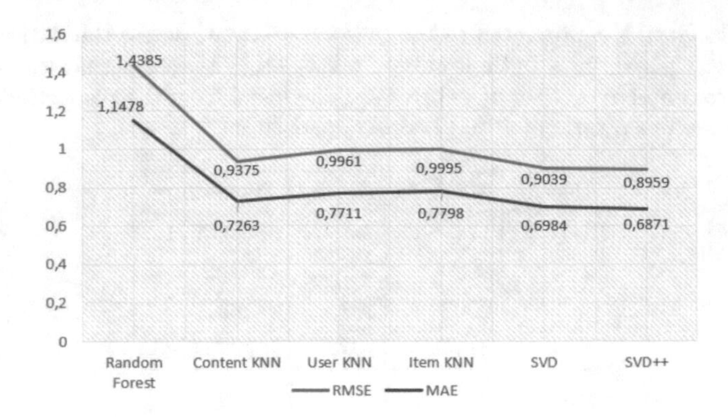

Fig. 3. In general, collaborative filtering gives better results as an MAE and RMSE than the Content Based Filtering, because CF (collaborative filtering) uses more parameters than the CBF (Content-Based Filtering).

to work on the more efficient one SVD (Singular Value Decomposition) and we try to variate the number of CPUs to get the same running time as the SVD applied on the 100 k dataset using a computer with the following equipment:

- Processor: Intel Core i7-8750H $8 \times 2.2 - 4.1$ GHz
- Memory: 16384 MB, DDR4-2666.

And we variate the number of cores from 1 to 8 and we find out that the best practice is using 6 cores, and we tried to find a way that can allow us to automate the process of choosing the number of cores but didn't find it.

And we obtain the following results:

Table 4. Compares the results of the proposed parallel algorithm. It shows the different values of RMSE and MAE for SVD Parallel (100 k) and SVD Parallel (20 M).

Techniques	RMSE	MAE
SVD Parallel (100 k rating)	0,9039	0,6984
SVD Parallel (20 M rating)	0,8959	0,6871

Limitation. For the theoretical side, there is a limitation in the part of choosing the number of cores that we should use to get the best results as the term of time. For the processing limitation, we have the problem of using other algorithms with large data such as (20 M), and it would be solved in the future works.

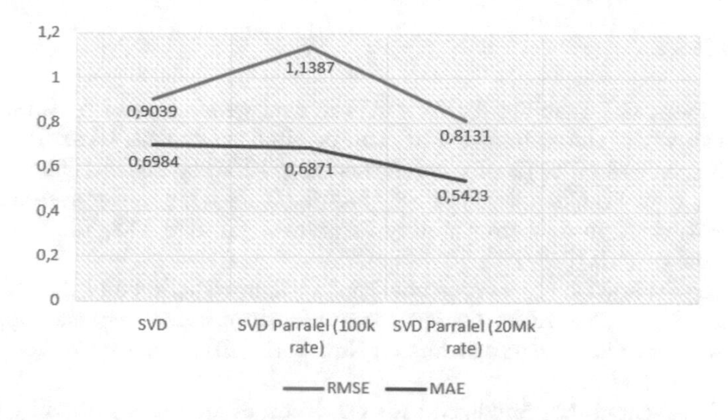

Fig. 4. Compares the values of RMSE and MAE of SVD, SVD Parallel (100 k), and SVD Parallel (20 M), and it shows that the best values are of SVD Parallel (20 M) and it is possibly due to the algorithm has more capabilities to understand and learn data more than we 100 k that give us the bad results as RMSE and MAE.

5 Comparison

This Research paper's method was compared to similar work made by Bao and Xia [13], by using the same dataset and similar techniques. Through this paper and the proposed methods, there is an improvement of two methods SVD (RMSE $\simeq$ 0.9375) and SVD++ (RMSE $\simeq$ 0.9250) against SVD (RMSE $\simeq$ 0.9039) and SVD++ (RMSE $\simeq$ 0.8959) for the proposed method. (Table 4). It is a small gain as an RMSE, It is a small gain as an RMSE,but it would have a big impact when the number of the item in the dataset increases.

6 Conclusion

In our Research Paper, various prediction techniques are used based on the MoviesLens dataset. The SVD (Singular value decomposition) technique which based on the matrix factorization perform better than the others, and it performs further better by adding new block value matrix, and for the paralleled techniques it performs better in a huge number of rating, and further by changing the number of processes. We count on implementing innovative approaches to improve the quality of the recommendation, automate the work process and improve the user's experience as well as increasing the efficiency of the suggestion because it represents a mix between the IT and marketing industries. In future work we expect to run all the algorithms of our model using parallel approaches, for instance, we can compare all the algorithms using a huge number of data, and how they will behave, this later concept can improve the quality of recommendations in term of time and RMSE and MAE and can let us build a parallel hybrid recommendation system.

References

1. Adomavicius, G., Tuzhilin, A.: Toward the next generation of recommender systems: a survey of the state-of-the-art and possible extensions. IEEE Trans. Knowl. Data Eng. **17**, 734–749 (2005). https://doi.org/10.1109/TKDE.2005.99
2. Chen, L., Hsu, F., Chen, M., Hsu, Y.: Developing recommender systems with the consideration of product profitability for sellers. Inf. Sci. **178**, 1032–1048 (2008). https://doi.org/10.1016/j.ins.2007.09.027
3. Steck, H.: Evaluation of recommendations: rating-prediction and ranking. In: Proceedings of the 7th ACM conference on Recommender systems. pp. 213-220. Association for Computing Machinery, New York (2013). https://doi.org/10.1145/2507157.2507160
4. Zheng, Y., Agnani, M., Singh, M.: Identification of grey sheep users by histogram intersection in recommender systems. In: Cong, G., Peng, W.-C., Zhang, W.E., Li, C., Sun, A. (eds.) ADMA 2017. LNCS (LNAI), vol. 10604, pp. 148–161. Springer, Cham (2017). https://doi.org/10.1007/978-3-319-69179-4_11
5. Grčar, M., Mladenič, D., Fortuna, B., Grobelnik, M.: Data sparsity issues in the collaborative filtering framework. In: Nasraoui, O., Zaïane, O., Spiliopoulou, M., Mobasher, B., Masand, B., Yu, P.S. (eds.) WebKDD 2005. LNCS (LNAI), vol. 4198, pp. 58–76. Springer, Heidelberg (2006). https://doi.org/10.1007/11891321_4
6. Rahul, M., Kumar, V., Yadav, V., Rishabh: Movie recommender system using single value decomposition and K-means Clustering. In: IOP Conference Series: Materials Science and Engineering, vol. 1022, p. 012100 (2021). https://doi.org/10.1088/1757-899X/1022/1/012100
7. Kumar, M.S., Prabhu, J.: A hybrid model collaborative movie recommendation system using K-means clustering with ant colony optimisation. Int. J. Internet Technol. Secured Trans. **10**, 337–354 (2020). https://doi.org/10.1504/IJITST.2020.107079
8. Ahuja, R., Solanki, A., Nayyar, A.: Movie recommender system using K-means clustering AND K-nearest neighbor. In: 2019 9th International Conference on Cloud Computing, Data Science Engineering (Confluence), pp. 263–268 (2019). https://doi.org/10.1109/CONFLUENCE.2019.8776969
9. Croft, W.B., Metzler, D., Strohman, T.: Search Engines: Information Retrieval in Practice. Addison-Wesley, Boston (2010)
10. Chen, H.-W., Wu, Y.-L., Hor, M.-K., Tang, C.-Y.: Fully content-based movie recommender system with feature extraction using neural network. In: 2017 International Conference on Machine Learning and Cybernetics (ICMLC), pp. 504–509 (2017). https://doi.org/10.1109/ICMLC.2017.8108968
11. MovieLens 100K Dataset. https://grouplens.org/datasets/movielens/100k/. Accessed 20 Jan 2021
12. MovieLens 20M Dataset. https://grouplens.org/datasets/movielens/20m/. Accessed 20 Jan 2021
13. Zhouxiao, B., Haiying, X.: Movie Rating Estimation and Recommendation (2012)

On the Sensitivity of LSTMs to Hyperparameters and Word Embeddings in the Context of Sentiment Analysis

Bousselham El Haddaoui(✉), Raddouane Chiheb, Rdouan Faizi, and Abdellatif El Afia

ENSIAS, Mohammed V University in Rabat, Rabat, Morocco
{bousselham.haddaoui,r.chiheb,r.faizi,a.elafia}@um5s.net.ma

Abstract. Recurrent neural networks are still providing excellent results in sentiment analysis tasks, variants such as LSTM and Bidirectional LSTM have become a reference for building fast and accurate predictive models. However, such performance is difficult to obtain due to the complexity of the models and the hyperparameters choice. LSTM based models can easily overfit to the studied domain, and tuning the hyperparameters to get the desired model is the keystone of the training process. In this work, we provide a study on the sensitivity of a selection of LSTM based models to various hyperparameters and we highlight important aspects to consider while using similar models in the context of sentiment analysis.

Keywords: Sentiment analysis · Neural networks · LSTM · Bidirectional LSTM · GloVe · Word2vec

1 Introduction

The digital world is growing in a surprisingly fast way, more than 4.9 billion internet users[1] and 3.6 billion active social media users[2] are experiencing the freedom to exchange about their interests. The ability to express opinions and sentiments towards topics and entities is even easier with the diversity of social media, this allows us to reach various targets and strengthen the bonds of communities with shared interests. The considerable amount generated from this exchange results in massive data and a hidden value that academics, business and government entities are willing to understand in order to improve the quality of their services. In this respect, research focused on various mechanisms to mine value from this massive data. Sentiment Analysis (SA), as a research field, is intended to extract opinions and sentiments from social media and other sources in an automated way considering time and cost constraints. It's also seen as a complex opinion mining problem [1].

SA has evolved from a complex natural language processing NLP task focusing on text processing and applied learning algorithms, to a more complete and industrial

[1] https://www.internetworldstats.com/stats.htm.
[2] https://www.statista.com.

M. Lazaar et al. (Eds.): BDIoT 2021, LNNS 489, pp. 529–542, 2022.
https://doi.org/10.1007/978-3-031-07969-6_40

approach where it can be defined as a framework with several components such as data acquisition, preprocessing, classification and visualization [2]. The classification task still the main focus of this field of research, a wide range of techniques were proposed covering machine learning and deep learning algorithms in different learning modes: supervised, semi-supervised and unsupervised. Another important task consists of word embeddings, a set of techniques allowing a better representation of words. Various aspects are learned from raw text and encoded into word embeddings such syntactic, lexical and semantic attributes. The tasks are subject to continuous improvements and studies, the sensitivity of one to another can impact heavily the quality of the results.

The study of the sensitivity of models helps define optimal hyperparameters in order to create robust and general-purpose combinations of those techniques. In our study, we will present results of experiments on the sensitivity of long short-term memory neural networks (LSTMs) [3] that have good performance for sequence modeling [4] with two trending and popular word embeddings techniques. The rest of this paper is organized as follows. Section 2 provides theoretical foundations and key findings from existing studies about LSTMs and Word Embeddings. Section 3 presents a detailed description of the proposed system, the experimentation process and the test environment. Section 4 will be dedicated for the study's results and discussion. Conclusions are summarized in Sect. 5.

2 Background Information

SA studies usually focus on social media as a primary source of data, and open platform such as Twitter[3] provides open access public data according to their policy for research purposes. In this context, most user opinions and sentiments in social medias are expressed via short or medium size sentences. The sequential aspect of words in these sentences make it a suitable use case for recurrent neural networks (RNNs) which have a proven performance in this area [5], and make better alternatives for traditional machine learning algorithms for SA related tasks. RNN based models suffer from a known vanishing gradient problem which is linked to error vanishing as it propagates back [6]. This issue is present when learning long data sequences resulting in smaller gradients, thus insignificant learning. A variant of RNN is proposed to handle this phenomenon which is LSTM.

2.1 Long Short-Term Memory Neural Networks

LSTMs come to address the problem of long-term dependencies in data sequences, and gave the success to RNNs in this field due to their ability to connect information in larger contexts. Their ability to control the data flow provides new mechanisms to keep and forget information [7] during the learning process in order to keep long-term dependencies and prevent RNNs vanishing gradient problems.

LSTM Network Architecture

The architecture of an LSTM neural network is similar to a regular RNN, it follows the

[3] https://twitter.com.

same chaining structure of cells, the main difference is the internal architecture. The structure of the cell illustrated in Fig. 1 contains regulators that interact in a very specific way.

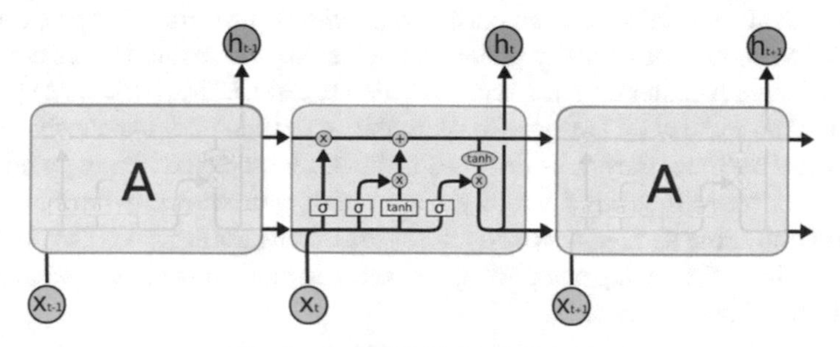

Fig. 1. An LSTM cell architecture

A basic LSTM cell is characterized by 3 main gates: input, forget and output gate. Each gate is a sigmoid (Eqs. 1, 2 and 3) that outputs a value of either 0 or 1. Depending on this value, the gate will throw away the information or it will keep it. Other variants of LSTM cells architectures can be found such as LSTM with and without forget gate, LSTM cell with peephole connection [8] that adds connections from the cell state to the gates in order to consider also the previous internal state, and the Gated Recurrent Unit (GRU) [9] that reduce the LSTM complexity by integrating the existing gates into two gates: the update and reset gates.

$$i_t = \sigma(w_i[h_{t-1}, x_t] + b_i) \tag{1}$$

$$f_i = \sigma(w_f[h_{t-1}, x_t] + b_f) \tag{2}$$

$$o_i = \sigma(w_o[h_{t-1}, x_t] + b_o) \tag{3}$$

where:

i_t: the input gate
f_i: the forget gate
o_i: the output gate
σ: the sigmoid function
w_x: the weights for the respective gates' neurons
h_{t-1}: the output of the previous LSTM block at (t–1)
x_t: the input at the current timestamp
b_x: the biases for the respective gates

The standard LSTM neural network architecture is usually composed of a single LSTM layer with a feedforward output layer, and this should resolve mid-complexity problems. However, modern issues such as sentiment analysis require advanced network architecture in order to get reasonable performance. Popular architectures include stacked and bidirectional LSTMs architecture that are widely used, more architectures were presented in [10]. Stacked LSTMs architectures focus on the depth of the network, they use multiple LSTM layers to learn high-level features. The issue of the accumulated errors between layers resulting in a vanishing gradient was addressed in [11] by adding a residual connection that centers layers gradients and errors. The proposed model was applied to sentiment intensity prediction achieving promising results compared to existing methods. Another trending type is Bidirectional LSTMs (BiLSTM) [12] which learn faster than regular LSTMs due to their ability to train simultaneously in the positive and negative time, they consider both right and left context at any position of the input. Authors stated that this flexibility comes with approximately the same parameters and training time as regular RNNs, the value of integrating BiLSTMs in SA tasks was explored in [13] showing notable improvement in performance compared to regular RNN and LSTM based models.

LSTM Network Hyperparameters

LSTM neural networks have several parameters to configure in order to obtain good results, each parameter addresses a special aspect of the model such as: the learning rate, the complexity and the training process and duration. Those parameters are called hyperparameters and their tuning task is mandatory to have performant and stable models, below is a list of a neural network hyperparameters:

- **The learning rate:** An important hyperparameter that reflects how fast the model should adapt to the problem, it ranges between 0.0 and 1.
- **The optimizer:** The optimizer choice defines the model's training speed and final predictive performance.
- **The network size:** The number of hidden neurons for each neural network layer.
- **The network depth:** The number of neural network layers used to define the model.
- **The batch size:** the number of training samples (N) used for one iteration, there are three commonly used modes: batch mode (N = total dataset size), mini-batch mode (1 < N < total dataset size) and the stochastic mode (N = 1). The network internal states are updated after each batch.
- **The number of epochs:** Refers to how many times the entire datasets is passed once through the model, it can start from one epoch to a reasonable number where the model achieves expected results.
- **The weights initialization:** the initial values of the neural network weights, they are updated during the training process.
- **The loss function:** a method to evaluate the cost of training, it becomes high when predicted values derivate too much from expected results.
- **The drop out:** A simple yet effective neural network regularization technique that prevents model overfitting. It works by disabling temporarily a set of network layer units along with their input and output connections in a way that results in a different layer size and architecture making the model less possible to overfit to training data.

As listed above, those are a set of hyperparameters to consider when training a neural network. Through the literature there is no predefined configuration recommended to work for most cases, but neural networks present a sensitivity to each of these hyperparameters to a certain extent which is the aim of this work.

LSTM Network Optimization
In the SA context achieving high performance is related to the model's quality, and tuning neural networks to obtain the desired outputs usually becomes an optimization problem to solve. Many optimization techniques were proposed to address this issue such the Stochastic Gradient Descent (SGD) [14], RMSProp, Adam [15], etc.

SGD is the foundation of optimization techniques, due to its simplest iterative nature and ability to find optimal configuration of neural networks. The learning rate can be configured to allow slow or fast learning, the performance of the algorithm is proven and the results are applicable to unseen data benefiting from the generalization characteristic of SGD [16], [17]. There are also other techniques that enhance the SGD performance and lead to faster learning and enhanced performance such Momentum [18], Nesterov [19]. Another recent enhancement to SGD performance was introduced by the Adaptive Moment Estimation (Adam) [20] algorithm, a method that computes adaptive learning rates for each parameter. Authors stated that $\alpha = 0.001$, $\beta_1 = 0.9$, $\beta_2 = 0.999$ and $\varepsilon = 10^{-8}$ are good default settings for the algorithm for testing machine learning related problems. The choice of a model's optimizer can be very challenging, especially when common findings show special trends for every optimizer. In [21], results of an empirical comparison show that recent algorithms such as RMSProp and Adam should never underperform SGD, Nesterov or Momentum if their parameters are chosen carefully. They also studied the effect of other hyperparameters on the training time and performance of optimizers, the batch size for example can increase the gaps between the training times of optimizers.

Another common phenomenon to consider while optimizing neural networks is overfitting, it occurs naturally when the model fits too much to the training data obtaining near perfect predictions but fails to apply on real world data. Tuning the previous hyperparameters helps prevent this issue at some extent but would require other techniques to get better results. In this respect, there is another important hyperparameter to consider which is the dropout [22], it allows to disable randomly neural network units preventing them from co-adapting too much. The technique significantly reduces overfitting and improves the model's performance.

2.2 Word Embeddings

Neural networks usually deal with numerical values as inputs, and in the SA field most inputs are text entries to process. The task of mapping textual information to numerical values while encoding domain knowledge with low dimensionality was the focus of many research studies. To address this area, Word Embeddings [23] were introduced as a new approach that encodes syntactic and semantic information in denser feature spaces. They can be categorized into two types [24]: prediction-based models which are context aware and leverage local data, and count-based models that tend to use global information like word count, frequencies, etc. In this section we will focus on two frequently used word embeddings which are Word2vec [25] and Global Vectors (Glove) [26].

Word2vec

Learning word representations from huge datasets with high quality was a big challenge as most studies used to train models on a few hundreds of millions of words with low dimensionality 50–100. In [25], Mikolov et al. provided a new model with two variances to address those limitations: Continuous Bag-of-Words Model (CBOW) and Continuous Skip-Gram Model (Skip-gram). CBOW is intended to predict a word representation from its context while the Skip-gram model maximizes classification of a word based on another which can be defined as predicting the surrounding words given the current word w(t) (see Fig. 2). The techniques achieved high performance compared to state-of-the-art baselines in the same study and unlocked new discoveries as they allow the learning of syntactic regularities and even semantic ones.

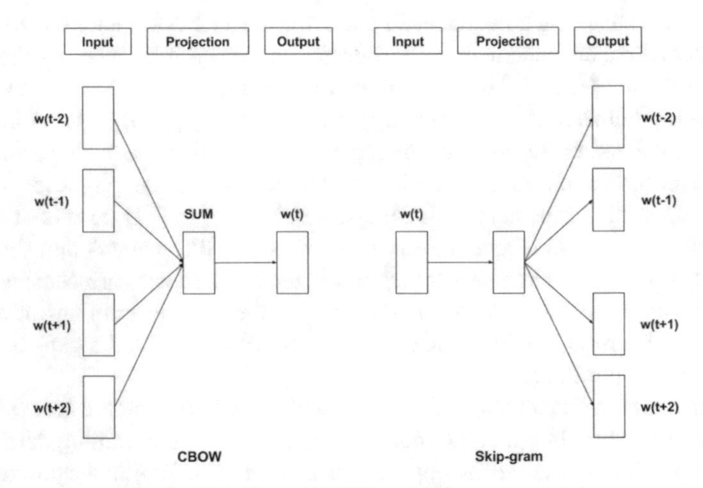

Fig. 2. Word2vec CBOW and Skip-gram models

Word2vec is widely used with LSTM neural networks for sentiment analysis, and can be combined with other techniques to enhance the model's performance. In [27], authors trained a CBOW Word2vec model and used the TF-IDF algorithm to capture word importance in the corpus which is not encoded in the word vector representation. The study showed promising results using word2vec representation along with the LSTM model applied to emotional text orientation detection.

Global Vectors - GloVe

Global Vectors [26] is a new unsupervised approach for word representation that was introduced due to the lack of visibility related to how old methods had success in capturing syntactic and semantic regularities. In GloVe, Pennington et al. tried to combine both global matrix factorization that generates low-dimensional word representations, and local context window models. This process allowed the authors to introduce a global log-bilinear regression technique for learning word representations that encode more features into denser spaces. As the previous techniques, GloVe relies on statistics of word occurrence in a corpus to build a word-word co-occurrence matrix, and it trains on nonzero elements in this matrix which bring more efficiency.

The model performs better than similar techniques in reference tasks for work analogy, similarly and named entity recognition. It's still a reference word representation used with deep learning methods for sentiment analysis related tasks [28].

3 System Design

The proposed framework to study the sensitivity of LSTMs to their hyperparameters and word embeddings relies on three key components: the use of well-known provided word embeddings adapted with the context of sentiment analysis, various models of LSTM based architectures and a set of scenarios that tests the models against different values of the hyperparameters.

3.1 The Dataset

The dataset used for the experiment was constructed from a dataset used in the SemEval competition for sentiment analysis between the edition 2013 and 2016 [29–32]. The data was extracted from twitter and provided with sentiment labels, each tweet was assigned a sentiment value of either positive, neural or negative. A total of 50 133 tweets were gathered for our analysis with the following ratios: 44.9% neural, 39.5% positive and 15.6% negative. The data annotations were performed using both Amazon's Mechanical Turk[4] and CrowdFlower, each tweet was labeled by five persons followed by a labeling consolidation which chooses a tweet label based on a rule of 3 out of 5 meaning 3 annotators agreed on the same label. Otherwise, an average value was calculated and approximated to the nearest labeling category. As for our study we chose to perform a binary classification and removed all neutral tweets from the datasets keeping only positive and negative samples to reduce model complexity and training time.

3.2 Data Preprocessing and Word Embedding

Before the training phase on the gathered dataset, preprocessing and text transformation tasks are needed in order to clean our dataset and feed appropriate data representations to our LSTM-based models. Since the focus in this study doesn't cover the effect of preprocessing techniques on the model, only few recommended [33] tasks were performed such as: removing stop words (we used nltk[5] stop words), URLS, mentions and hashtags using a python library named tweet-preprocessor[6].

For the word representation transformation, we used a general purpose pre-trained word vectors using GloVe[7] with a word-word co-occurrence statistic from a 2 billion

[4] https://www.mturk.com.

[5] https://www.nltk.org.

[6] https://pypi.org/project/tweet-preprocessor.

[7] https://nlp.stanford.edu/projects/glove.

tweets corpus. The pre-trained word vectors are intended to be twitter general purpose and not domain customized, and they are provided with various vector dimensions 25,50,100 and 200. The learned embedding matrix from the training dataset and word embeddings is used to initialize the embedding layer in our LSTM models.

3.3 The Models Architecture

The system's core model is designed as stacked layers of LSTM and BiLSTM neural networks, each layer feeding its output to the next one. The hyperparameters values will be specified later in the test scenarios along with the optimizer type. Each model's layer uses a hyperbolic tangent (tanh) as an activation function, a hard sigmoid for recurrent activation and we also activate the dropout mechanism. The hard sigmoid is a linear approximation of the sigmoid function which is faster to compute. The output layer uses a sigmoid function as an activation, and returns a single output value as our problem is a binary classification.

The models are divided into two categories; LSTM and BiLSTM models, each category providing three variants with respectively 1,2 and 3 layers. We name our models for the rest of the paper as LSTM1, LSTM2, LSTM3, BiLSTM1, BiLSTM2 and BiLSTM3 denoting the model type and the number of used layers. They were implemented using open-source python libraries such as Keras[8] for deep learning sequential models, Numpy[9] for scientific computing and Pandas[10] for data analysis and manipulation. The experimentation and test runs were performed using Google Colab[11] which is a Google platform for data science.

3.4 Test Scenarios and Performance Metrics

The sensitivity study of our models against the identified hyperparameters can be very hard to perform with all the possible combinations, we identified key test scenarios that focus on important LSTM parameters. The list of scenarios is presented below:

- S01: The Sensitivity of LSTM and BiLSTM to Batch Size and Epochs
- S02: The Sensitivity of LSTM and BiLSTM to Dropout
- S03: The Sensitivity of LSTM and BiLSTM to Word Embeddings Dimensions
- S04: The Sensitivity of LSTM and BiLSTM to Optimizers
- S05: The Sensitivity of LSTM and BiLSTM to Word Embeddings Type

Our models were trained with a simple strategy, we use the best parameters values results from a test scenario as default in the next scenario. The evaluation process is performed on validation data with a ratio of 20% the size of the dataset, the monitored metric to compare models is the best validation accuracy.

---

[8] https://keras.io.

[9] https://numpy.org.

[10] https://pandas.pydata.org.

[11] https://colab.research.google.com.

4 Results and Discussion

In this section we present and discuss the results of our study, for each test scenario the used parameters are mentioned along with the results. It's important to note that we didn't provide an exhaustive list of values for each hyperparameter but only usually used values in previous research that may be with notable impact on the results, also the first three tests (S01, S02 and S03) were conducted with less training data (~60%) than the last two ones due to computation time and complexity constraints. The loss calculation is performed with the binary cross-entropy suitable for binary classification tasks.

4.1 The Sensitivity of LSTM and BiLSTM to Batch Size and Epochs

The test scenario was conducted using GloVe twitter embeddings with the following parameters: dimension = 100, dropout = 0.2, layers hidden units = 100 and optimizer = Adam. Results presented in Table 1 shows that the overall performance didn't seem to be affected by batch size or epochs variations. A stable behavior is shown using higher epochs even when we change the number of layers, while low epochs slightly change the performance depending on the used batch size. Good results were provided using 100–200 epochs with small medium batch size 200–500. Another finding is that BiLSTM performed better than LSTM by slight margin except for a few test cases. Results showed that the highest obtained accuracy was 0.8451 with BiLSTM2 for the couple batch size = 500 and epochs = 200.

4.2 The Sensitivity of LSTM and BiLSTM to Dropout

The drop sensitivity test is conducted using the following values: dimension = 100, batch size = 500, epochs = 200, layers hidden units = 100 and optimizer = Adam. These values rely on results from the S01 test, and we use various dropout values denoted as D(N) where N is the percentage of used dropout. Results presented in Table 2 showed no notable improvement using various values from dropout, a slight improvement is observed compared to the baseline from the previous test scenario using a dropout of 0.25 (D25). Another surprising observation is that LSTM based models are more likely to benefit from dropout variations than the BiLSTM based ones.

4.3 The Sensitivity of LSTM and BiLSTM Word Embeddings Dimension

Word embeddings are a key component of our system, the dimension size is reflecting the length of word vectors. In this test we use GloVe word embeddings with various dimensions available for general purpose sentiment analysis with the following values: batch size = 500, dropout = 0.25, epochs = 200, layers hidden units = 100 and optimizer = Adam. Outputs in Table 3 shows that high dimensions 100–200 provide good results. Another observation is that LSTM can perform slightly better than BiLSTM but with an unstable accuracy while increasing the number of layers. On the other hand, BiLSTM models have a stable performance while increasing the number of layers with high dimensions 100–200.

Table 1. S01 The sensitivity of LSTM and BiLSTM to batch size and epochs

	LSTM1	LSTM2	LSTM3	BiLSTM1	BiLSTLM2	BiLSTM3
B100E50	0.8353	0.8324	0.8304	0.8324	0.8333	0.8294
B200E50	0.8353	0.8245	0.8294	0.8373	0.8324	0.8265
B500E50	0.8353	0.8343	0.8294	0.8373	0.8304	0.8314
B1000E50	0.8363	0.8284	0.8265	0.8314	0.8275	0.8294
B100E100	0.8324	0.8382	0.8304	0.8343	0.8265	0.8324
B200E100	0.8343	0.8402	0.8304	0.8275	0.8431	0.8343
B500E100	0.8392	0.8333	0.8304	0.8363	0.8363	0.8353
B1000E100	0.8373	0.8353	0.8324	0.8382	0.8382	0.8353
B100E200	0.8324	0.8382	0.8402	0.8373	0.8333	0.8343
B200E200	0.8333	0.8284	0.8333	0.8392	0.8314	0.8382
B500E200	0.8382	0.8392	0.8373	0.8392	0.8451	0.8343
B1000E200	0.8284	0.8343	0.8333	0.8324	0.8353	0.8343
B100E300	0.8353	0.8353	0.8363	0.8353	0.8324	0.8363
B200E300	0.8343	0.8343	0.8343	0.8392	0.8353	0.8373
B500E300	0.8353	0.8353	0.8265	0.8402	0.8353	0.8343
B1000E300	0.8333	0.8343	0.8333	0.8392	0.8412	0.8333

Table 2. S02 The sensitivity of LSTM and BiLSTM to dropout

	LSTM1	LSTM2	LSTM3	BiLSTM1	BiLSTLM2	BiLSTM3
D15	0.8333	0.8353	0.8353	0.8314	0.8402	0.8343
D20	0.8382	0.8392	0.8373	0.8392	0.8451	0.8343
D25	0.8461	0.8373	0.8353	0.8353	0.8353	0.8373
D30	0.8382	0.8314	0.8363	0.8441	0.8373	0.8373
D35	0.8353	0.8451	0.8392	0.8363	0.8382	0.8343
D40	0.8343	0.8412	0.8343	0.8392	0.8422	0.8363
D45	0.8441	0.8451	0.8392	0.8392	0.8353	0.8363
D50	0.8382	0.8382	0.8422	0.8373	0.8431	0.8412

4.4 Sensitivity of LSTM and BiLSTM to Optimizers

The optimizer type affects the model's performance, tuning the learning rate the model is perhaps the most important task. In this test scenario we study the behavior of our models with three popular optimizers: SGD with Momentum, RMSProp and Adam. We assigned the following values for our models: dimension = 100, batch size = 500,

Table 3. S03 The Sensitivity of LSTM and BiLSTM word embeddings dimension

	LSTM1	LSTM2	LSTM3	BiLSTM1	BiLSTLM2	BiLSTM3
WE25	0.8069	0.8059	0.8078	0.8108	0.8098	0.8088
WE50	0.8275	0.8275	0.8324	0.8304	0.8324	0.8245
WE100	0.8441	0.8314	0.8431	0.8343	0.8382	0.8431
WE200	0.8431	0.8402	0.8343	0.8422	0.8382	0.8392

dropout = 0.25, epochs = 200, layers hidden units = 100. The results illustrated in Table 4 shows that adaptive gradient optimizer Adam is providing notable improvements to all our models, RMSProp results are compared to those of Adam. We observe that models give different performance depending on the optimizer which shows a sensitivity towards the optimization technique. We also note that the optimizers were used with their default values, a tuning of the parameters of each optimizer may lead to better results.

Table 4. S04 Sensitivity of LSTM and BiLSTM to optimizers

	LSTM1	LSTM2	LSTM3	BiLSTM1	BiLSTLM2	BiLSTM3
SGD	0.8406	0.8345	0.8204	0.8403	0.8399	0.8423
RMSPROP	0.8629	0.8667	0.8702	0.8691	0.8638	0.8707
ADAM	0.866	0.8712	0.8714	0.8729	0.8696	0.8712

4.5 The Sensitivity of LSTM to Word Embeddings Type

Word embeddings play a key role in our model since we don't focus on the feature engineering part and rely on unsupervised proposed techniques GloVe and Word2vec for word representation. In this test, we compare our models initialized with 3 different embeddings weights, GloVe (100 and 200 dimensions) and Google News pre-trained vectors (300 dimensions) word embeddings. The rest of the hyperparameters were assigned the following values: batch size = 500, dropout = 0.25, epochs = 200, layers hidden units = 100 and optimizer = Adam. Results in Table 5 shows that the models performed better with GloVe for the two variants compared to word2vec with higher dimensions. We note that GloVe word embeddings were trained on a twitter corpus while word2vec was trained on Google News which is more suitable for formal language.

5 Conclusion

In this paper we studied the sensitivity of LSTM based models to their hyperparameters and word embeddings algorithms. The results of our study show that the LSTM based models are very sensitive to their parameters variations but with different degrees, the

Table 5. S05 The sensitivity of LSTM and BiLSTM to word embeddings type

	LSTM1	LSTM2	LSTM3	BiLSTM1	BiLSTLM2	BiLSTM3
GloVe_100	0.8682	0.8682	0.8703	0.8638	0.8687	0.8709
GloVe_200	0.8743	0.8741	0.8718	0.8716	0.8734	0.87
W2V_300	0.8645	0.8645	0.8653	0.8665	0.8658	0.8662

number of the model's layers adds limited improvements when using high epochs. Other parameters and components should be chosen carefully such as the dropout, the optimizer and the word embeddings type and dimension, a low dropout value or word embedding dimension reduces the performance of the model. We note also that GloVe performs better with the LSTM model than other similar approaches, and BiLSTM can be good alternatives to LSTM for SA tasks. At the end of our experiments, final models performed way better than top ranked submissions on the SemEval task 2013 to 2016.

Future work will investigate the effect of domain trained GloVe word embeddings on our models, in addition we will conduct an in-depth study of the learning rate and model's architectures tuning to achieve high performance and stable models.

References

1. Liu, B.: Sentiment analysis and opinion mining. Synth. Lect. Hum. Lang. Technol. **5**(1), 1–167 (2012). https://doi.org/10.2200/S00416ED1V01Y201204HLT016
2. El Haddaoui, B., Chiheb, R., Faizi, R., El Afia, A.: Toward a sentiment analysis framework for social media, 1–6 (2018). https://doi.org/10.1145/3230905.3230919
3. Hochreiter, S., Schmidhuber, J.: Long short-term memory. Neural Comput. **9**(8), 1735–1780 (1997). https://doi.org/10.1162/neco.1997.9.8.1735
4. Purnamasari, P.D., Taqiyuddin, M., Ratna, A.A.P.: Performance comparison of text-based sentiment analysis using recurrent neural network and convolutional neural network. In: Proceedings of the 3rd International Conference on Communication and Information Processing - ICCIP 2017, Tokyo, Japan, pp. 19–23 (2017). https://doi.org/10.1145/3162957.3163012
5. Yin, W., Kann, K., Yu, M., Schütze, H.: Comparative Study of CNN and RNN for Natural Language Processing. ArXiv170201923 Cs, Feb 2017. http://arxiv.org/abs/1702.01923. Accessed: 14 Nov 2020
6. Hochreiter, S.: The vanishing gradient problem during learning recurrent neural nets and problem solutions. Int. J. Uncertain. Fuzziness Knowl. Based Syst. **06**(02), 107–116 (1998). https://doi.org/10.1142/S0218488598000094
7. Schmidhuber, J., Cummins, F.: Learning to forget: Continual prediction with LSTM, 6 (1999)
8. Gers, F.A., Schmidhuber, J.: Recurrent nets that time and count. In: Proceedings of the IEEE-INNS-ENNS International Joint Conference on Neural Networks. IJCNN 2000. Neural Computing: New Challenges and Perspectives for the New Millennium, Como, Italy, pp. 189–194. vol. 3 (2000). https://doi.org/10.1109/IJCNN.2000.861302
9. Chung, J., Gulcehre, C., Cho, K., Bengio, Y.: Gated feedback recurrent neural networks, 9 (2015).
10. Yu, Y., Si, X., Hu, C., Zhang, J.: A review of recurrent neural networks: LSTM cells and network architectures. Neural Comput. **31**(7), 1235–1270 (2019). https://doi.org/10.1162/neco_a_01199

11. Wang, J., Peng, B., Zhang, X.: Using a stacked residual LSTM model for sentiment intensity prediction. Neurocomputing **322**, 93–101 (2018). https://doi.org/10.1016/j.neucom.2018.09.049
12. Schuster, M., Paliwal, K.K.: Bidirectional recurrent neural networks. IEEE Trans. Signal Process. **45**(11), 2673–2681 (1997). https://doi.org/10.1109/78.650093
13. Zhou, P., Qi, Z., Zheng, S., Xu, J., Bao, H., Xu, B.: Text classification improved by integrating bidirectional LSTM with two-dimensional max pooling. ArXiv161106639 Cs, Nov 2016. http://arxiv.org/abs/1611.06639 (2020). Accessed 18 Nov 2020
14. Robbins, H., Monro, S.: A stochastic approximation method. Ann. Math. Stat. **22**(3), 400–407 (1951). https://doi.org/10.1214/aoms/1177729586
15. Dozat, T.: Incorporating Nesterov Momentum into Adam, p. 4 (2016)
16. Allen-Zhu, Z., Li, Y.: Can SGD learn recurrent neural networks with provable generalization? ArXiv190201028 Cs Math Stat, May 2019. http://arxiv.org/abs/1902.01028 (2020). Accessed 21 Nov 2020
17. Keskar, N.S., Socher, R.: Improving generalization performance by switching from adam to SGD. ArXiv171207628 Cs Math, Dec 2017. http://arxiv.org/abs/1712.07628 (2020). Accessed 21 Nov 2020
18. Polyak, B.T.: Some methods of speeding up the convergence of iteration methods. USSR Comput. Math. Math. Phys. **4**(5), 1–17 (1964). https://doi.org/10.1016/0041-5553(64)90137-5
19. Nesterov, Y.E.: A method for solving the convex programming problem with convergence rate O (1/k^ 2). Dokl. Akad. Nauk SSSR **269**, 543–547 (1983)
20. Kingma, D.P., Ba, J.: Adam: A method for stochastic optimization. ArXiv14126980 Cs, Jan 2017. http://arxiv.org/abs/1412.6980 (2020). Accessed 21 Nov 2020
21. Choi, D., Shallue, C.J., Nado, Z., Lee, J., Maddison, C.J., Dahl, G.E.: On empirical comparisons of optimizers for deep learning. ArXiv191005446 Cs Stat, Jun 2020. http://arxiv.org/abs/1910.05446 (2020). Accessed 18 Nov 2020
22. Srivastava, N., Hinton, G., Krizhevsky, A., Sutskever, I., Salakhutdinov, R.: Dropout: A simple way to prevent neural networks from overfitting, p. 30 (2014)
23. Turian, J., Ratinov, L.-A., Bengio, Y.: Word representations: A simple and general method for semi-supervised learning, p. 11 (2010)
24. Almeida, F., Xexéo, G.: Word embeddings: A survey. ArXiv190109069 Cs Stat, Jan 2019. http://arxiv.org/abs/1901.09069 (2020). Accessed 21 Nov 2020
25. Mikolov, T., Chen, K., Corrado, G., Dean, J.: Efficient estimation of word representations in vector space. ArXiv13013781 Cs, Sep 2013. http://arxiv.org/abs/1301.3781 (2020). Accessed 14 Nov 2020
26. Pennington, J., Socher, R., Manning, C.: Glove: Global vectors for word representation. In: Proceedings of the 2014 Conference on Empirical Methods in Natural Language Processing (EMNLP), Doha, Qatar, pp. 1532–1543 (2014). https://doi.org/10.3115/v1/D14-1162
27. Yuan, H., Wang, Y., Feng, X., Sun, S.: Sentiment analysis based on weighted word2vec and att-LSTM. In: Proceedings of the 2018 2nd International Conference on Computer Science and Artificial Intelligence - CSAI 2018, Shenzhen, China, pp. 420–424. (2018). https://doi.org/10.1145/3297156.3297228
28. Lee, J.Y., Dernoncourt, F.: Sequential short-text classification with recurrent and convolutional neural networks. In: Proceedings of the 2016 Conference of the North American Chapter of the Association for Computational Linguistics: Human Language Technologies, San Diego, California, pp. 515–520 (2016). https://doi.org/10.18653/v1/N16-1062
29. Nakov, P., Rosenthal, S., Kozareva, Z., Stoyanov, V., Ritter, A., Wilson, T.: SemEval-2013 task 2: Sentiment analysis in twitter. In: Proceedings of the Seventh International Workshop on Semantic Evaluation (SemEval 2013), Atlanta, Georgia, USA, pp. 312–320 (June 2013). https://www.aclweb.org/anthology/S13-2052. Accessed 23 Nov 2020

30. Rosenthal, S., Ritter, A., Nakov, P., Stoyanov, V.: SemEval-2014 task 9: Sentiment analysis in twitter. In: Proceedings of the 8th International Workshop on Semantic Evaluation (SemEval 2014), Dublin, Ireland, pp. 73–80 (August 2014). https://doi.org/10.3115/v1/S14-2009
31. Rosenthal, S., Mohammad, S.M., Nakov, P., Ritter, A., Kiritchenko, S., Stoyanov, V.: SemEval-2015 task 10: Sentiment analysis in twitter. ArXiv191202387 Cs, Dec 2019. http://arxiv.org/abs/1912.02387 (2020). Accessed 23 Nov 2020
32. Nakov, P., Ritter, A., Rosenthal, S., Sebastiani, F., Stoyanov, V.: SemEval-2016 task 4: Sentiment analysis in twitter. In: Proceedings of the 10th International Workshop on Semantic Evaluation (SemEval-2016), San Diego, California, pp. 1–18 (June 2016). https://doi.org/10.18653/v1/S16-1001
33. Angiani, G., et al.: A Comparison between Preprocessing Techniques for Sentiment Analysis in Twitter (2016)

Improved Content Based Filtering Using Unsupervised Machine Learning on Movie Recommendation

Yassine Afoudi[1(✉)], Mohamed Lazaar[1], Mohammed Al Achhab[2], and Hicham Omara[2]

[1] ENSIAS, Mohammed V University in Rabat, Rabat, Morocco
yassine_afoudi@um5.ac.ma, yassine.afoudi@gmail.com
[2] Abdelmalek Essaâdi University, Tétouan, Morocco

Abstract. In our world of massive entertainment options and with thousands of choices on every movie platform, the user found himself in the circle of confusion over which movie to choose. Here the solution is using the recommender systems to predict user's interests and recommend items most likely to interest them. Recommender systems are utilized in a variety of areas and are most commonly recognized as playlist generators for video and music services, product recommenders for online stores as AliExpress and Amazon..., or content recommenders for social media platforms and open web content recommenders.

In this paper, we propose a new powerful recommender system that combines Content Based Filtering (CBF) with the popular unsupervised machine learning algorithm K-means clustering. To recommend items to an active user, K-means is then applied to the movie data to give each movie a specific cluster and after founding the cluster to which the user belongs, the content-based approach applies to all movies with the same cluster. The experimentation of well-known movies, we show that the proposed system satisfies the predictability of the Content-Based algorithm in GroupLens. In addition, our proposed system improves the performance and temporal response speed of the traditional collaborative filtering technique and the content-based technique.

Keywords: Content based filtering · K-means clustering · Movie recommendation · Recommender system

1 Introduction

During the last few decades, with the rise of Netflix, Amazon, YouTube, and many other such web services and the amount of online information available is increasing day by day, it leaves users in new trouble, they find it too difficult to choose what they really want to see or buy. That's why recommender systems today are unavoidable in our daily online journeys, they help users by suggesting relevant items that may interest them (books, movies, restaurants, music...

M. Lazaar et al. (Eds.): BDIoT 2021, LNNS 489, pp. 543–555, 2022.
https://doi.org/10.1007/978-3-031-07969-6_41

etc.), from the large number of choices available on the web or other electronic information sources.

Many companies like Amazon, Netflix and Spotify use recommendation systems which allow them to generate a huge amount of income and revenue. The reason these companies and others are seeing increased revenue is that they are bringing real value to their customers, by suggesting items and making their lives easier, also by knowing what a user wants the threat of losing a customer to a competitor decreases.

There are many approaches that are used in recommendation systems and we can distinguish between them by analyzing the type of data used to generate recommendations. In general, there are three well known techniques used in the recommendation field, Collaborative Filtering (CF), Content-basedFiltering (CBF) and Hybrid Models.

Collaborative Filtering (CF) is a method for making automatic predictions about a user's interests by gathering preferences or taste information from multiple users. The basic assumption of the collaborative filtering approach is that if Person X has the same opinion as Person Y in an issue, then it is more likely that X has a Y's opinion in a case different from that of the randomly chosen person. CF generally expressed based on two categories of rating data, explicit rating, is a rate given by a user to an item on a sliding scale, such as 5 stars for The Joker movie as most direct feedback from users to show how much they like an item, or an implicit rating, indirectly suggests user preference, such as page views, clicks, buy records, whether or not to listen to a music track, etc. Content-based filtering methods are based on a description of the item and a profile of the user's preferences then uses item features as title, genre, actors, description, etc. to recommend other items similar to what the user likes, based on their previous actions or explicit feedback. To improve performance, most recommendation systems now use a hybrid approach, combining collaborative filtering, content-based filtering, and other approaches in different ways to gain their complementary benefits.

In this article, we propose a powerful recommender system based on a Content filtering approach and combined with the K-means clustering, the well-known unsupervised machine learning technique used to identify clusters of data objects in a dataset.

2 Outline

In the first section, we will briefly describe the recommendation system as well as the most well-known approaches and some related work in the context of movie recommendation. The next two sections will then describe the proposed architecture and how to evaluate a recommender system. The following section will be dedicated to evaluate our system with other techniques in the area of recommendation and discuss the results. Finally, we will summarize the conclusion.

3 Related Work

Several research efforts have been made to create powerful movie recommendation systems using different recommendation approaches. Z. Wang et al. [9] proposed a hybrid model-based movie recommendation system which utilizes the improved K-means clustering coupled with genetic algorithms (GAs) to partition transformed user space. They use principal component analysis (PCA) data reduction technique to dense the movie population space which could reduce the computation complexity in movie recommendation as well. RamniHarbir Singh et al. [8] illustrated a movie recommendation system using content-based filtering and the KNN algorithm with the principle of cosine similarity to make recommendations. N. Pradeep et al. [7] builds a movie recommendation system based on content on cast, keywords, team and genres, then a single column is created as the sum of the 4 attributes, and it acts as a dominant factor for that system recommendation of films. Putra Pandu Adikara et al. [6] focus on the hybrid approach by combining content-based filtering and collaborative filtering using a graph-based model. Yadav Vikash et al. develop a movie recommender system with the help of clustering using K-means clustering technique and data pre-processing using Principal Component Analysis. In our previous paper [4] we proposed a new intelligent recommender system that combines collaborative filtering (CF) with the popular unsupervised machine learning algorithm K-means clustering. Also, we use certain user demographic attributes such as the gender and age to create segmented user profiles, when movies are clustered by genre attributes using K-means and users are classified based on the preference of items and the genres they prefer to watch, then to recommend items to an active user, Collaborative Filtering approach then is applied to the cluster where the user belongs.

In our system, a new content recommendation system is represented, based on three parts, namely K-means clustering, vector space modelling and content based model. Implicit user ratings are used to assign each user in a specific cluster, we also use the textual characteristics of the items to represent the dataset in a vector space model and build our proposed content-based model to recommend Top-N movies similar to the active user.

4 Proposed Work

This paper proposes K-means clustering algorithm based on content-based approach in the movies recommendation system, Fig. 1 represent the proposed architecture, it mainly consists of the following three functional modules: k-means clustering module, vector space module with user profile module and finally recommendation module.

4.1 K-means Clustering Module

We are testing our system with a movielens dataset, with each movie represented with features like title, movie id, and 19 genre attribute (if a movie has a specific

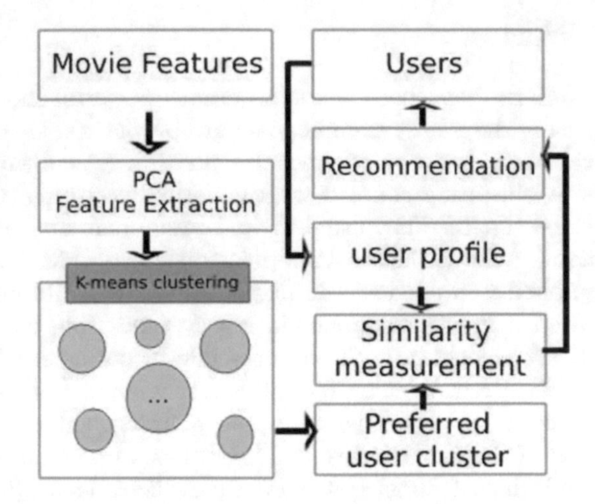

Fig. 1. Adopted architecture.

genre, we found 1 as a value otherwise 0), for this we plan to use a feature selection or feature extraction technique to reduce the number of attributes, in general during our previous work we found that there was an impact on using the Principal Component Analysis (PCA) technique on the results of movie recommendation, this technique is a dimensionality reduction technique used when we have a slow algorithm because the dimension of input is too high and we want to speed it up. After applying the PCA technique, we continue our work with only 10 features of PCA movie to minimize the memory occupied by our system.

K-means clustering is a module used to group movies into a specific number of classes, this part describes the standard k-means algorithm.

K-means clustering is one of the simplest and most popular unsupervised machine learning algorithms. As a rule, unsupervised algorithms make inferences from sets of data using only vectors of entered without reference to known or labelled results.

Simply, K-means looks for a fixed number (k) of clusters in a dataset, in other words, the K-means algorithm identifies k number of centroids and calculates the distance between each object and each cluster centre, then assign it to the nearest cluster, update the averages of all clusters, repeat this process until the criterion function converged.

The K-Means algorithm needs a way to compare the degree of similarity between the different features. Thus, two data which are similar, will have a reduced dissimilarity distance, while two different objects will have a greater separation distance, in this work, we use the k-means clustering algorithm based on the Euclidean similarity approach.

The Euclidean distance is the geometric distance, for example, if we have a matrix X with i quantitative variables. The Euclidean distance d between two features x1 and x2 is calculated as follows:

$$d(x_1, x_2) = \sqrt{\sum_{i=1}^{n}(x_{1n}, x_{2n})^2} \tag{1}$$

In this work, we give our 10 PCA components to the K-means approach and we obtain as output our dataset of movies classified into a specific number of clusters.

4.2 Vector Space Model and User Profile

As we all know, the collaborative filtering algorithm is vulnerable to the data sparsity of the rating matrix, because it measures the similarity between users based on the similarity between users' ratings, and ignores the impact of movies on the calculation of similarity. In this paper, we introduce movie attributes in the calculation of the similarity, which aims to provide more information for the measurement of similarity and relieve the influences caused by data sparsity.

Here, we get the movie attributes by analyzing some the textual features of each movie as the title, because they are easy to be accessed, read and understood. These attributes are used as recommendation factors and also used to describe documents in a vector space model. Term frequency and inverse document frequency (TF-IDF) can recognize the important words or phrases of movies. It is the most common weighting method used to describe documents in the vector space model.

TF-IDF consists of TF and IDF, which are Term Frequency and Inverse Document Frequency respectively. TF represents the frequency a word appears in the document. The main idea of IDF is if a term appears more times in other documents, the term will be less important.

Term Frequency. Term Frequency is the number of times the term Ti appears in document Dj, which can be represented by TF (Tij). After removing the stop words, the more Ti appears in the document, the more important the term for the document is. This can be defined as follows:

$$TF(T_{ij}) = N(T_i, D_j)/N(D_j) \tag{2}$$

N(Ti, Dj) is the number of times term Ti appears in document Dj and N(Dj) is the total number of terms in document Dj.

Inverse Document Frequency. Document frequency is the number of documents U containing a particular term Ti, which is represented by N(U,Ti). If the term Ti appears more in all U documents, the term Ti can be the weakest for represent document Dj.

Inverse document frequency means that the represent ability of term Ti for document Dj and its amount in all documents N(U,Ti) is inverse proportion, for example we know that the term "the" is so common in all documents. The term "the" is not a good keyword to distinguish relevant and irrelevant documents and terms. Hence, an inverse document frequency factor is incorporated which diminishes the weight of terms that occur very frequently in the document set and increases the weight of terms that rarely occur. Inverse document frequency can be represented by IDF(Ti):

$$IDF(T_i) = \log(N(U)/N(U,T_i)) \tag{3}$$

N(U) is documents total number, IDF(Ti) decreases with the increase of N(U,Ti). The less N(U,Ti) is, the more representative Ti is for Dj .

Normalization. In order to reduce the inhibition of stop words, we will normalize each variable. After normalization, the calculation of TF-IDF is:

$$TF - IDF(T_{i,j}) = TF(T_{i,j}) \times \log(N(U)/N(U,T_i)) \tag{4}$$

Equation 4 based on the principle of taking the term that is more representative for a document is the word that appears in the document more often and less often in other documents.

The main mission of user profile module is to build for each user a profile based on a set of parameters for recommendation purposes, in our work we take the rating data from each user, and we only use movies with a rating greater than or equal to 2.5. After getting the highest rated movies for an active user, we assign for each movie in the list its cluster class, after that we calculate the sum of each movie class in the list and we use the maximum class result as a representative class of the most liked user movie group. For example, we have a user X who likes movie classes [1,1,1,2,3,3,4,1,1], here, we can assign that our user liked movies classified in class 1 more than the other classes.

4.3 Recommendation Module

After finding the user's preferred cluster, we apply TF-IDF on the title of his viewed movies, then we calculate the cosine distance between the user profile vector and all the TF-IDF movie vectors with his same cluster, then we take the top N results as a recommendation after removing previously viewed movies to avoid double recommendation.

The cosine similarity between two vectors (or two movies on the Vector Space) is a measure that calculates the cosine of the angle that separates them. This metric is a measurement of orientation and not magnitude, it can be seen as a comparison between movies (document) on a normalized space because we do not only consider the magnitude of the number of words (TF-IDF) of each movie, but the angle between the movies. What we need to do is to build the cosine similarity equation, which is to solve Eq. 5 below:

$$cos(\theta) = \frac{A \cdot B}{||A|| \cdot ||B||} \tag{5}$$

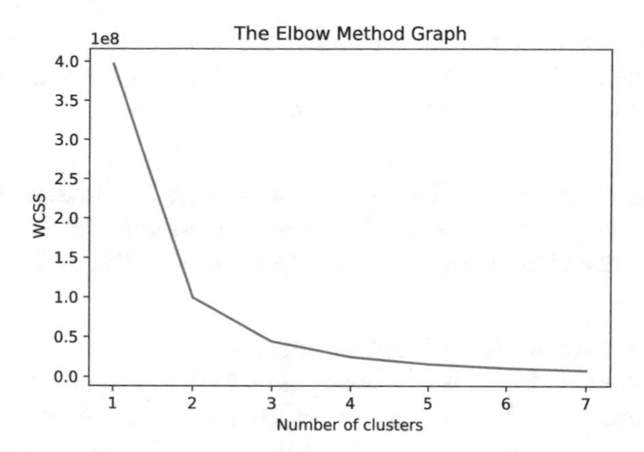

Fig. 2. The variance of 7 numbers of clusters on our dataset (Elbow approach)

5 Result and Discussion

5.1 The Dataset

We chose the Movielens 100 k dataset to evaluate our experiences because it is widely used and publicly available. MovieLens datasets were collected by the GroupLens research project at the University of Minnesota. We are using this dataset for a study aimed at generating movie recommendations for users. This dataset contains 100,000 ratings (scale of 1 to 5) of 943 users out of 1,682 movies and we're only taking users who rated 20 or more movies.

5.2 Implementation

Experimental Steps. First step is to import the dataset in our project, and then we use file ua.base as the training set and ua.test as the test set. The second step is to use PCA feature extraction to minimize the genre attributes in the movie dataset from 19 genres to 10 components, after that we use K-means clustering to group the whole dataset into a specific number of clusters, by plotting the different 7 numbers of clusters as a function of the variance, based on the approach of Elbow and by observing the Fig. 2 we choose the k=6 the number of our k-means clustering.

Once the movies are grouped into 6 clusters, we assign each movie its cluster class number, then we look for a user's preferred cluster based on the most viewed class, then we create a user profile for all users applying the TF-IDF approach on the title of their movies viewed on the train set, finally we take an active user profile and we apply the content-based approach by calculating the cosine similarity between his profile and all movies TF-IDF vectors in its preferred class to make recommendations and give the result to the user after deleting all previously viewed movies.

We will test and evaluate our recommendations by investigating whether the movies recommended to a user based on the training set are in the list of items the user has seen and rated in the test set.

Experimental Platform. All our experiments were implemented using Python and compiled using Jupyter notebook. We ran all our experiments on a MacBook Pro with Intel core i7 processor having a speed of 2.7 GHz and 8 GB of RAM.

Experiment Results. After building our system, we tested it for all users in the dataset and evaluate the results using time recommendation speed and other offline techniques with the state of the art approaches in recommendation filed. The result of the experiment is presented in Table 1 for user ID 100.

Table 1. Result experiment for user ID 100

Movie id	Title	Strenght
35	**Free willy2 the adventure home**	0.698535
39	**Strange days**	0.698535
248	**Grosse pointe blank**	0.310460
23	**taxi driver**	0.310460
62	**Stargate**	0.310460

5.3 Evaluation

In recommender systems, we are most likely interested in recommending top-N items to the user. In this way, the user will receive a list of recommendations, ranked from the best to the worst. As a matter of fact, there are some cases when the user does not care about the exact order of the list, some good recommendations are enough.

As a result, many evaluation techniques are involved in this area, classified as online and offline techniques. For those online, we need real users to give their opinion and feedback on the recommendation given, in this work we will use the offline techniques based on the movies cited for each user in the test set.

In this study, we used the Precision-Recall at k method to evaluate our result where k is a user definable integer that is set by the user to match the top-N recommendations objective.

The Precision-Recall at k method also is an offline way to evaluate a recommender system where k is a definable integer defined by the user to match the goal of the Top-N recommendations. For this method, we test every model by giving a list of movie recommendations for all users and we evaluate results by

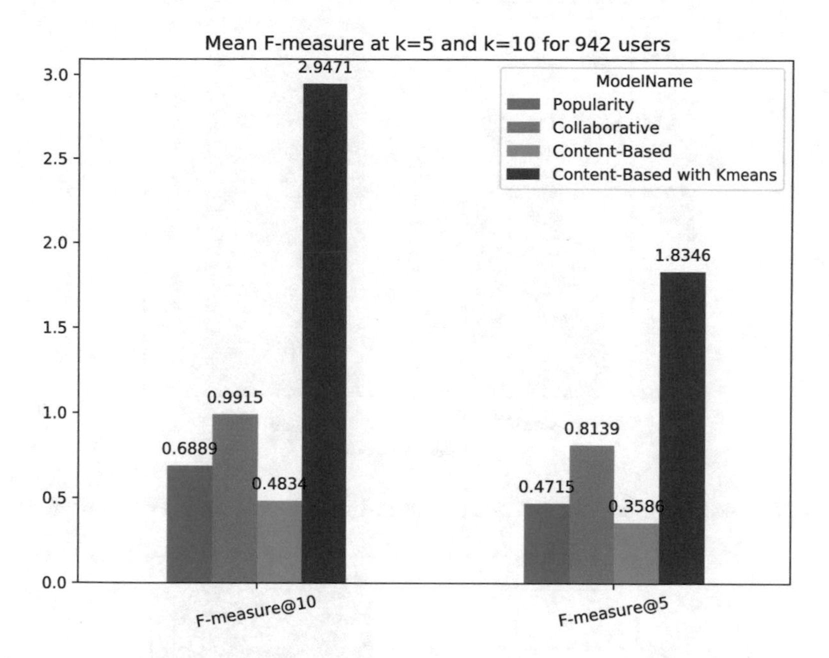

Fig. 3. The mean F-measure plot of four models at K = 5 and K = 10 using Movielens dataset

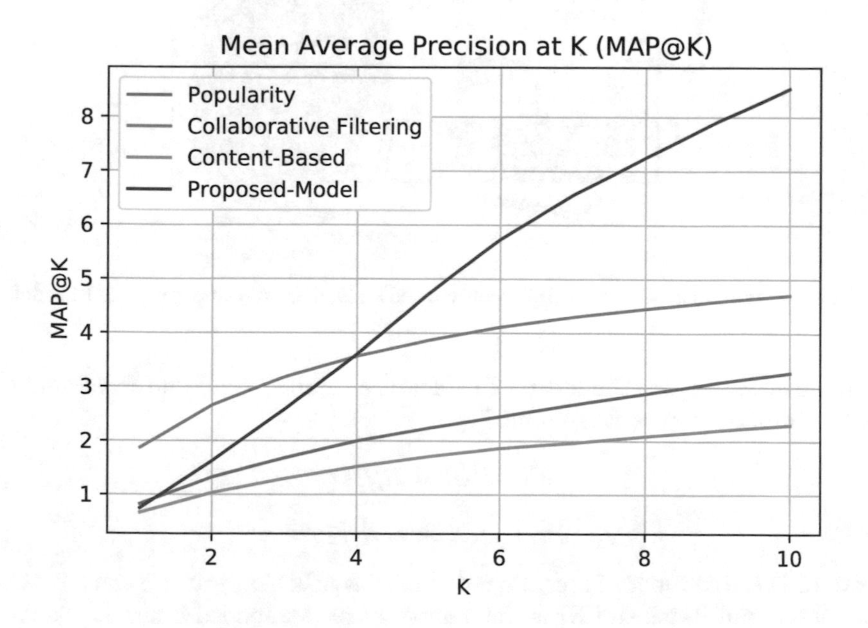

Fig. 4. The mean average precision at K

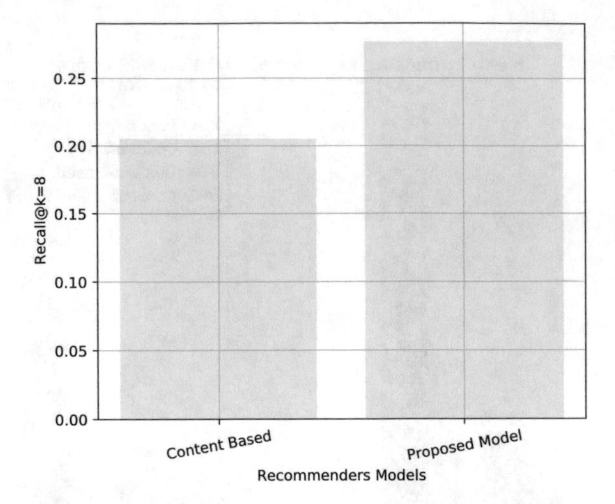

Fig. 5. Recall comparison between content-based model and proposed model

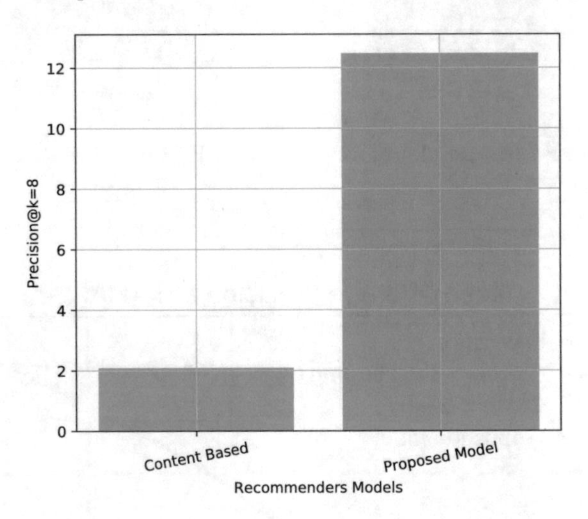

Fig. 6. Precision comparison between content-based model and proposed model

calculate and compare the mean of F-Measure result. Recall and Precision at k are then determined as Eqs. 6 and 7.

$$Recall@k = R_i/Nr \tag{6}$$

$$Precision@k = R_i/Tr \tag{7}$$

where Ri is the number of recommended movies at k that are relevant, Tr is the total of relevant items and Nr is the number of recommended items at @k. After calculating The Recall and Precision at k we should normalize the result, then we introduce F-Measure, which is:

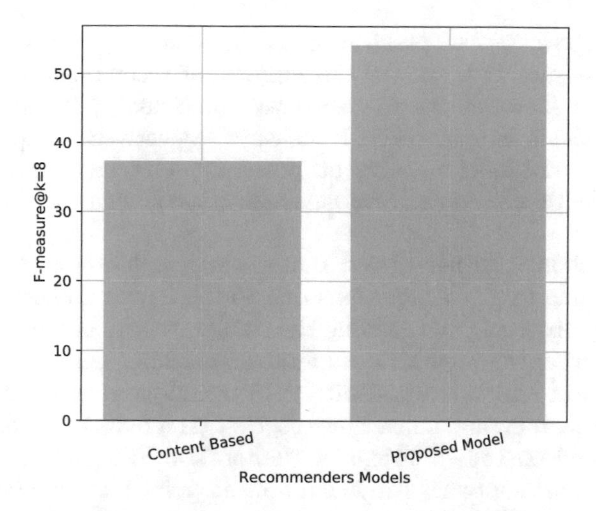

Fig. 7. F-measure comparison between content-based model and proposed model

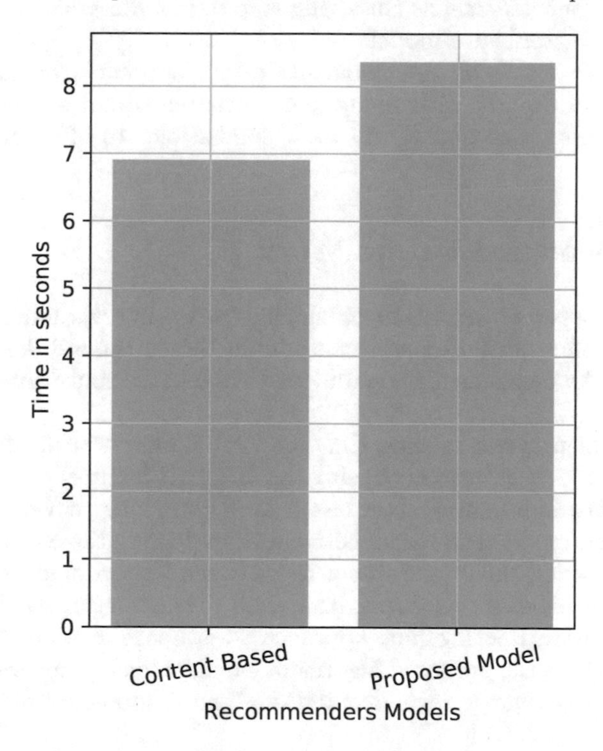

Fig. 8. Time comparison between content-based model and proposed model

$$F - measure@k = (2 \times P@k \times R@k)/(P@k + R@k) \tag{8}$$

After building our system, we want to know and evaluate which model gives us powerful recommendation and best accuracy of our Movielens 100 k database.

As explained above we use precision-recall method, Fig. 3 shows us a plot of the mean F-measure at 5 and 10 the number of recommended movies for all users, the Fig. 4 shows us the mean average precision at k and Fig. 5,6 and 7 shows us the Recall, Precision and F-measure respectively comparison between the Content-Based Model and our proposed model, when the Fig. 8 shows us the time in seconds taken by the two models to recommend 20 movies to the all users.

Finally, it should be mentioned that recommendation systems take a bit space and runtime to give suggestions due to the large number of features and parameters. By observing and reading the results, we can assume that the use of K-means clustering the popular unsupervised machine learning technique with Content based Filtering associated with PCA technique gives us much better performance and accuracy, also a fast time response on movies recommendation than the models based on the traditional Collaborative Filtering or Content-Based approaches but our approach still has a limited point like the attributes associated with the items. Obviously, if the item does not have descriptive attributes, it cannot be used in the K-means clustering step to find out which cluster contains this item or in vector space model.

To give more details on the generosity of our approach, we can say that, if we have an item dataset with many genre attributes and a user dataset, our approach can work well and it will be a good choice to offer customers what they want.

6 Conclusion and Future Work

This article proposes an algorithm which uses the unsupervised machine learning algorithm K-means clustering with the content based the well known approach in the field of the recommendation and we take into account using PCA feature extraction.

The experimental results show that the RMSE was reduced when we use the PCA feature extraction instead of using the 19 movie features, after that we take these features to build our system based on 6 clustering movies, then we take the most viewed movie class for an active user and apply the content-based approach by calculating the cosine similarity between the selected user profile and all movies with his same preferred cluster to give powerful recommendations. Using the Precision Recall offline approach to evaluate a recommendation system, we found that the proposed algorithm can effectively improve the accuracy and performance of movie recommendation. The proposed method is therefore feasible.

In another study, we will continue to explore and improve other techniques used in recommendation systems and try to use advanced machine learning algorithms and deep learning models.

Acknowledgment. The authors would like to thank the Smart System Lab our research laboratory and Al Borchers for cleaning up this data.

References

1. Afoudi, Y., Lazaar, M., Al Achhab, M.: Impact of feature selection on content-based recommendation system. In: 2019 International Conference on Wireless Technologies, Embedded and Intelligent Systems (WITS), pp. 1–6 (2019)
2. Feng, J., Xia, Z., Feng, X., Peng, J.: RBPR: a hybrid model for the new user cold start problem in recommender systems. Knowl.-Based Syst. **214**, 106732 (2021)
3. Wang, J., Lv, J.: Tag-informed collaborative topic modeling for cross domain recommendations. Knowl.-Based Syst. **203**, 106119 (2020)
4. Afoudi, Y., Lazaar, M., Al Achhab, M.: Intelligent recommender system based on unsupervised machine learning and demographic attributes. Simul. Model. Pract. Theor. **107**, 102198 (2021)
5. Zhu, Y., Lin, Q., Lu, H., Shi, K., Qiu, P., Niu, Z.: Recommending scientific paper via heterogeneous knowledge embedding based attentive recurrent neural networks. Knowl.-Based Syst. **215**, 106744 (2021)
6. Adikara, P.P., Sari, Y.A., Sigit, A., Setiawan, B.D.: Movie recommender systems using hybrid model based on graphs with Co-rated, genre, and closed caption features. Register: Jurnal Ilmiah Teknologi Sistem Informasi **7**(1), 31–42 (2021)
7. Reddy, S.R.S., Nalluri, S., Kunisetti, S., Ashok, S., Venkatesh, B.: Content-based movie recommendation system using genre correlation. In: Satapathy, S.C., Bhateja, V., Das, S. (eds.) Smart Intelligent Computing and Applications. SIST, vol. 105, pp. 391–397. Springer, Singapore (2019). https://doi.org/10.1007/978-981-13-1927-3_42
8. Singh, R., Maurya, S., Tripathi, T., Narula, T., Srivastav, G.: Movie recommendation system using cosine similarity and KNN, **9**, 2249–8958 (2020)
9. Wang, Z., Yu, X., Feng, N., Wang, Z.: An improved collaborative movie recommendation system using computational intelligence. J. Vis. Lang. Comput. Distrib. Multimedia Syst. DMS2014 Part I **25**(6), 667–675 (2014)
10. Cami, B.R., Hassanpour, H., Mashayekhi, H.: A content-based movie recommender system based on temporal user preferences. In: 2017 3rd Iranian Conference on Intelligent Systems and Signal Processing (ICSPIS), pp. 121–25. IEEE, Shahrood (2017)
11. Duwairi, R., Abu-Rahmeh, M.: A novel approach for initializing the spherical K-means clustering algorithm. Simul. Model. Pract. Theor. **54**, 49–63 (2015)
12. Harper, F.M., Konstan, J.A.: The MovieLens datasets: history and context. ACM Trans. Interact. Intell. Syst. **5**(4), 1–19 (2016)
13. Katarya, R.: Movie recommender system with metaheuristic artificial bee. Neural Comput. Appl. **30**(6), 1983–1990 (2018). https://doi.org/10.1007/s00521-017-3338-4

The Impact of the k-Nearest Neighbor Parameters in Collaborative Filtering Recommender Systems

Kawtar Najmani[1(✉)], Lahbib Ajallouda[2], El Habib Benlahmar[1], Nawal Sael[1], and Ahmed Zellou[2]

[1] Laboratory Information Technology and Modeling, Faculty of Sciences Ben M'sik, Hassan II University, Casablanca, Morocco
kawtarnajmani@gmail.com

[2] Laboratory Software Project Management, Mohammed V University in Rabat, ENSIAS, Rabat, Morocco
{lahbib_ajallouda,ahmed.zellou}@um5.ac.ma

Abstract. Recommender Systems (RS) have become very important recently, they are a main component of many applications in different fields. They aim to give beneficial information according to the profile of each user among the huge existing online information. RS are based on several approaches to provide the best results and give satisfaction to the active user.

Collaborative filtering is one of these approaches. It helps to choose a product according to the consumer's preference from many and various choices. It uses the k-Nearest Neighbor (kNN) technique for the extraction of similar users from the group of users.

In this paper, we will study the effect of the parameters of the kNN algorithm on the obtained results. For that, we have varied the value of k, then we have measured for each value the prediction accuracy, using the Root Mean Square Error (RMSE), and the Mean Absolute Error (MAE) metrics. The experiments are carried out also to find the value of k which gives good results in both metrics. Then we have calculated also the RMSE and MAE metrics for different similarities in order to find the similarity which gives good results compared to others.

Keywords: Recommender Systems · Collaborative filtering · k-Nearest Neighbor · Neighbor parameter · Similarity measures

1 Introduction

One of the main areas of research relating to the problem of information overload, is the recommender systems field (RS) which play an important role in modern online services nowadays. The principle of these systems is to provide recommendations adapted to the preferences and needs of users. They are based on two main notions, the user, who is the target individual to whom the recommendation is given, and the item, which is the

M. Lazaar et al. (Eds.): BDIoT 2021, LNNS 489, pp. 556–567, 2022.
https://doi.org/10.1007/978-3-031-07969-6_42

recommended element to this user, it can be of various natures (videos, product, movie, restaurant, hotel, etc.).

Recommender systems have become increasingly popular and are today a main component of many applications in several domains, such as, Netflix and Allocine for movie recommendation, it's also used in e-commerce domain, for instance, Amazon and E-bay. In addition, RS is applied in video games field like Steam and Microsoft Xbox Live and in many applications in social networks like Facebook, Linkedin and Twitter. In order to improve the tourist experience, there are many applications which use recommender systems, among them, Expedia and TripAdvisor. Even though these domains are different, but they share the same goal to propose to a target user the subset of useful items that interests him from a huge information. It reduces considerably the effort that the user can make to access what interests him and thus contributes to his satisfaction and confidence towards the system.

These systems are principally based on three approaches, the content-based filtering which is based on the description of the element in order to give recommendations to the current user. The second approach is the collaborative based filtering, which is based on the user profile history that contains much information like, the ratings given to the items, then search the user closer to the active user and finally provide recommendations. Ratings can be explicit, the system gives to customers directly a questionnaire to know their satisfactions, or implicit from the studying patterns or click-stream behavior of the user [1]. And the hybrid technique, which combines different techniques in order to overcome the limitations of the traditional approaches.

The CF technique uses the kNN technique to extract similar users from a group of users. In this paper, we will analyse the influence of the parameter k of this algorithm on the accuracy of the results using the Root Mean Square Error (RMSE), and the Mean Absolute Error (MAE) metrics. In addition, the experiments aim to find the value of k which gives good results.

The rest of this paper is organized as follows. In Sect. 2 we explain the principle of collaborative filtering recommendation systems. Section 3 briefly describes the algorithm k-Nearest Neighbor (kNN). Section 4 presents our experiments on MovieLens dataset and discussion of the results. The final section contains the conclusion.

2 Collaborative Filtering Recommender Systems

The purpose of the collaborative algorithm in recommender systems, is to use the ratings that the active user has made of certain items, in order to recommend these same items to the users that are close to him. The idea is that, if a user likes an item, then users who have the same preferences to this user will also like this item. For instance, if user X has the same tastes as user Y, and X buys a product from an e-commerce site and gives it a good rating, based on collaborative technique, this product will be recommended to the user Y.

Collaborative technique is applied in some applications in different domains, for example, Grouplens [2], Netflix for the recommendation of movies. Amazon [3] and eBay [4] in e-commerce domain. It is also applied in social networks, like Facebook,

LinkedIn, and Twitter. Moreover, the collaborative filtering is used in Last.fm [3] application for music recommendation. This approach is divided into two main types, Memory Based Approach and Model Based Approach.

Memory Based Approach: This type is more popular and very used in practice. It is called also nearest-neighbor or user-based collaborative filtering [5]. It uses a vote matrix which contains user preferences to items that can interest a new user. The goal of memory-based collaborative filtering is to predict the importance of items for a target user based on the user vote. So, user ratings stored by the system are directly used to predict scores for new items.

Model Based Approach: Memory Based Approach directly uses the stored ratings in the prediction. However, Model Based Approach uses these ratings to learn a predictive model. The principal characteristics of users and items are captured by a set of model parameters, which are learned from training data and used to predict new ratings [6]. The process of creating a model is performed by various machine learning algorithms like the Bayesian Network, Clustering and Association Rule approaches.

3 K-Nearest Neighbor Algorithm

The k-nearest neighbor model (kNN) is a supervised machine learning algorithm. It is the most popular model in collaborative filtering approach [5].

The principle of this algorithm is the prediction of the class of a new input, by providing it with a set of training data, a function of distance and an integer k, for any new test point A, for which it must make a decision, the algorithm searches in the set of training data for the K closest neighbors to A using the Euclidian distance or other distance, and assigns A to the class which is the most frequent among these k neighbors.

k-Nearest Neighbor algorithm has many advantages, it's very simple to understand. In addition, the kNN classifier is considered as a lazy learner, so it does not need to learn and maintain a given model, as a result, the system can adapt to quick changes in the user ratings matrix [6].

Even if the algorithm has many advantages, it has also several limitations. The big challenge in k-Nearest Neighbor is the selection of the parameter k, if it's too small, the model will be sensitive to noise points. However, if the value of k is very large, the neighborhood might include too many points from other classes [6]. Moreover, the algorithm keeps the training data in memory in order to make predictions. For that, we have to pay attention to the size of the data used while its implementation. Below the pseudo code of the KNN algorithm.

```
D = {(x ', c), c ∈ C} is the learning set
And x be the example whose class we want to determine
Algorithm
Start
for each ((x ', c) ∈ D) do
Compute the distance dist (x, x ')
end
for each {x ' ∈ KNN (x)} do
Count the number of occurrences of each class
end
Assign to x the most frequent class;
end
```

4 Experiments

In this section we experimentally evaluate the performance of the parameters of the k-Nearest Neighbor algorithm using the different similarities. We have used different metrics to see if the accuracy will be modified by changing the parameter k, and then we will choose the optimal value of k which gives more precision and good results.

In our experiments, we have used the well-known Movie-Lens dataset which is developed by the GroupLens research group at the University of Minnesota. It contains the user ratings for movies 100,000 ratings, 9,000 movies and 600 users [7].

4.1 Evaluation Metrics

In order to evaluate the quality of a recommender system, there are various types of metrics or measures. They are divided into two classes, statistical accuracy metrics which evaluate the accuracy of a technique by comparing the predicted ratings with the actual rating and decision support accuracy metrics which evaluate the effectiveness of a prediction engine in helping a user to select items that have high quality from the set of all items [5].

In our experiments we have used the most popular statistical accuracy metrics, the Root Mean Square Error (RMSE) and the Mean Absolute Error (MAE). The formula of each metric as follows:

$$MAE = \frac{\sum_{i \in N} |p_i - q_i|}{|N|}$$

$$RMSE = \sqrt{\frac{\sum_{i \in N} (p_i - q_i)^2}{|N|}}$$

with pi is the actual rating, qi is the predicted rating, and N is the set of items which have the rating predictions.

4.2 Experimental Evaluation

4.2.1 The Impact of the Similarity Measures

In our experimental evaluation, we have used surprise which is a python scikit for building and analyzing recommender systems that deal with explicit rating data [8].

In order to know the impact of the similarity measures on the kNN algorithm we have calculated the MAE and RMSE metrics for cosine, pearson_baseline, pearson and the Mean Squared Differencc (MSD) similarities.

Cosine

We have started the calculation by the cosine similarity, we have calculated the RMSE and the MAE metrics for it in order to know the value of k which gives the optimal results in both metrics.

The Fig. 1 shows the calculation of RMSE for different k values for cosine similarity, we can conclude that the Min RMSE is 0.8948499820521632 when the parameter k is equal to 40.

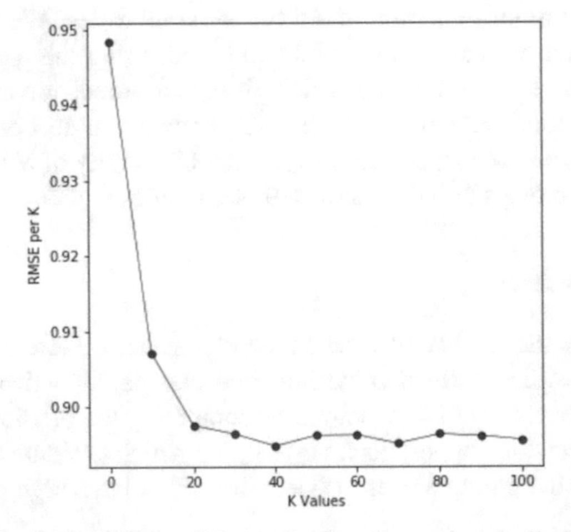

Fig. 1. RMSE for different k values for cosine similarity

The Fig. 2 shows the calculation of the MAE metric for different k values for cosine similarity, we can conclude that the Min MAE is 0.6843631697909892 when the parameter k is equal to 70.

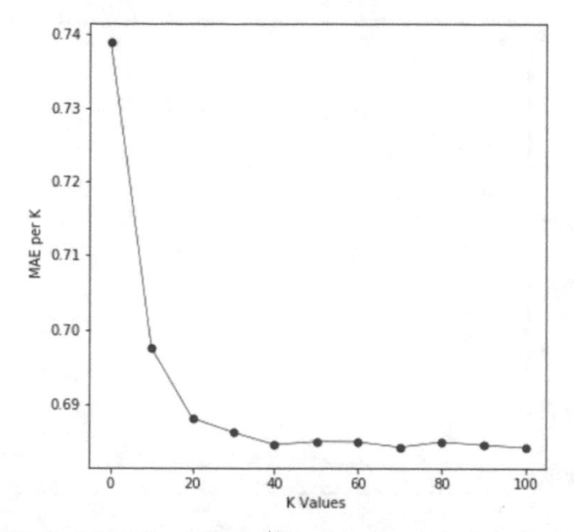

Fig. 2. MAE for different k values for cosine similarity

Here is the result found in the calculation of the both metrics RMSE and MAE for cosine similarity:

- Min RMSE: 0.8948499820521632 when k = 40
- Min MAE: 0.6843631697909892 when k = 70

Pearson_baseline

We have calculated the RMSE and the MAE metrics for Pearson_baseline similarity as shown respectively in Figs. 3 and 4.

We have found from the Fig. 3 that the Min RMSE is 0.8851022038814824 when the parameter k is equal to 20.

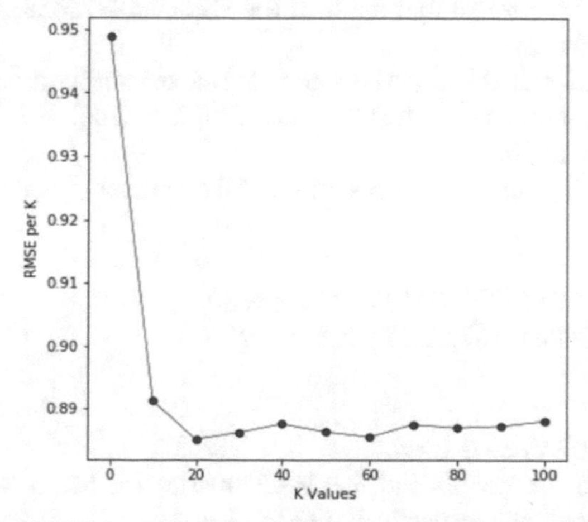

Fig. 3. RMSE for different k values for pearson_baseline similarity

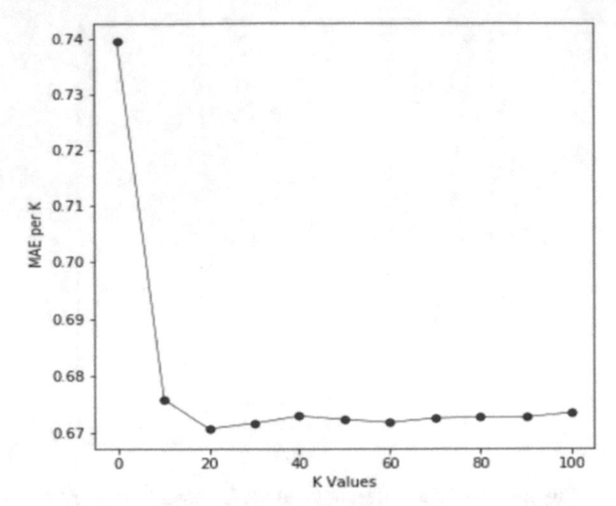

Fig. 4. MAE for different k values for pearson_baseline similarity

The Fig. 4 shows the calculation of the MAE metric for different k values for pearson_baseline similarity. We have found that the Min MAE is 0.6706869718700585 when the parameter k is equal to 20.

Here is the result found in the calculation of the both metrics RMSE and MAE for pearson_baseline similarity:

- Min RMSE: 0.8851022038814824 when k = 20
- Min MAE: 0.6706869718700585 when k = 20

Pearson

The Fig. 5 shows the calculation of the RMSE metric for different k values for pearson similarity. We have found that the Min RMSE is 0.8887545874694105 when the parameter k is equal to 30.

We have also calculated the MAE metric for pearson similarity as shown in Fig. 6. We can conclude from the figure that the Min MAE is 0.6706869718700585 when the parameter k is equal to 20.

Here is the result found in the calculation of the both metrics RMSE and MAE for pearson similarity:

- Min RMSE: 0.8887545874694105 when k = 30
- Min MAE: 0.6706869718700585 when k = 20

Mean Squared Difference (MSD)

We have calculated the RMSE and the MAE metrics for Mean Squared Difference (MSD) similarity as shown respectively in Figs. 3 and 4.

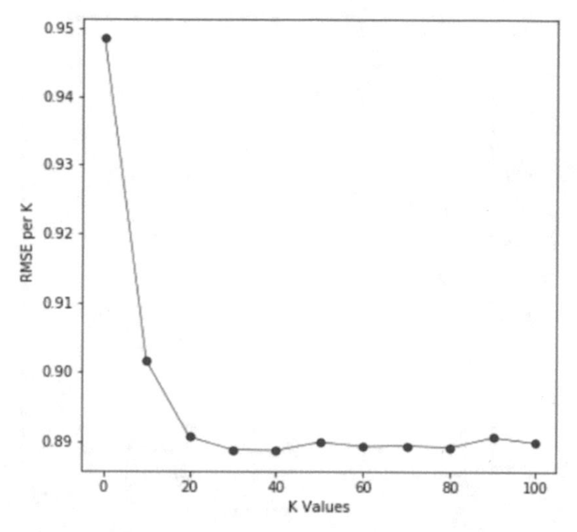

Fig. 5. RMSE for different k values for pearson similarity

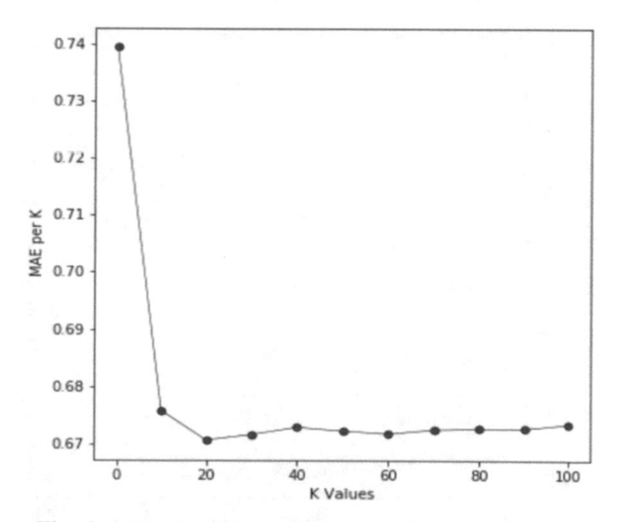

Fig. 6. MAE for different k values for pearson similarity

We have found from the Fig. 7 that the Min RMSE is 0.8913532423185575 when the parameter k is equal to 50.

The Fig. 8 shows the calculation of the MAE metric for different k values for the MSD similarity. We have found that the Min MAE is 0.6807832987122796 when the parameter k is equal to 50.

Here is the result found in the calculation of the both metrics RMSE and MAE for Mean Squared Difference (MSD) similarity:

- Min RMSE: 0.8913532423185575 when $k = 50$

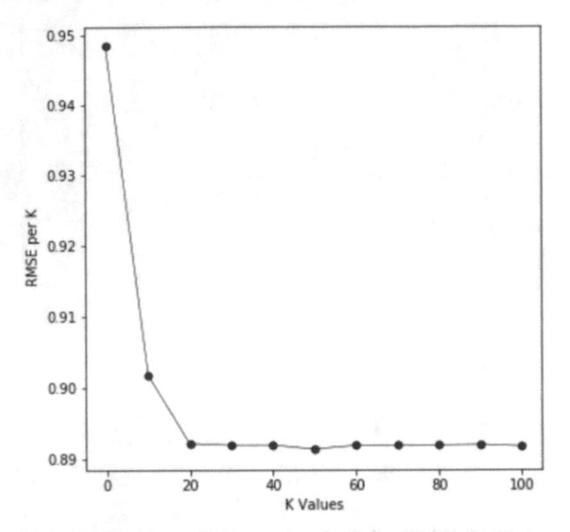

Fig. 7. RMSE for different k values for MSD similarity

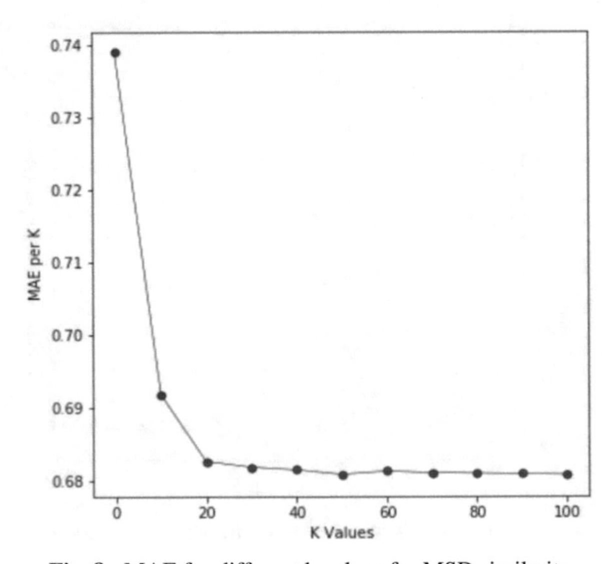

Fig. 8. MAE for different k values for MSD similarity

- Min MAE: 0.6807832987122796 when k = 50

Results:
The table below shows the results of the calculation result of the RMSE and the MAE metrics of each similarity to know the value of k which gives the optimal results in both metrics.

Table 1. Evaluation metrics for each similarity

	Evaluation metrics	
Similarity measures	Min RMSE	Min MAE
Cosine	0.8948499820521632	0.6843631697909892
Pearson_baseline	0.8851022038814824	0.6706869718700585
Pearson	0.8887545874694105	0.6706869718700585
MSD	0.8913532423185575	0.6807832987122796

The results of the pearson_baseline similarity is more optimal than the cosine similarity, for example, in RMSE metric we have in cosine the value 0.8948499820521632 and in pearson_baseline we have 0.8851022038814824 which is less than the value of cosine similarity, and the same thing for MAE metric, in cosine similarity we have 0.6843631697909892 and in pearson_baseline we have the value 0.6706869718700585 which is less than it.

Compared to Pearson similarity, we have found that the minimum value of the calculation of RMSE metrics in pearson is 0.8887545874694105 which is greater than the minimum value of the pearson_baseline and for MAE metric we have the same result, the minimum value of pearson similarity is greater than the value of the similarity pearson, which confirm that the pearson_baseline gives better results compared to cosine and peason similarities.

From the Table 1, we have the minimum RMSE of the Mean Squared Difference (MSD) similarity is equal to 0.8913532423185575 and the value of minimum MAE is 0.6807832987122796 and these values are greater than the values of the pearson_baseline similarity.

If we compared the calculation of both metrics of the four similarities, we observe that the values of the similarity pearson_baseline are the most minimal compared to the other values of the other similarities. Therefore, we conclude from the results that the pearson_baseline similarity gives good and optimal results compared to others.

4.2.2 The Impact of the Neighbor Parameter k

The k parameter in the kNN algorithm has an important impact on the obtained results. For that, we have to choose the right value for better accuracy and optimal results, but it's not obvious to find it and we cannot choose an arbitrary value, we need to study the results for different k values to find the optimal value which gives a good result.

To study the effect of the neighbor parameter k on the obtained results, we have varied its value and for each one, we measure the prediction accuracy in terms of the Root Mean Square Error (RMSE), and the Mean Absolute Error (MAE), like shown respectively in Figs. 1 and 2 for cosine similarity, 3 and 4 for pearson_baseline similarity, 5 and 6 for pearson similarity, 7 and 8 for MSD similarity. The value of k is varied from 0 to 100.

The figures show that the RMSE metric vary according to k value. For k between 0 and 20, the result of this metric has high values so, we can conclude that in this interval, the system will be ineffective, it will not give a good recommendation and will give a bad accuracy. After the value 20 of k, or more precisely between 20 and 100, the metric values start to decrease, so we have better predictions. In this case, the system will provide optimal results and satisfy the customer.

From the graphs, we observe that the minimum obtained value of the measure of RMSE is when k = 40 for cosine similarity, k = 20 for pearson_baseline similarity, k = 30 for pearson similarity and k = 50 for MSD similarity.

The figures also show the variation of the results of the MAE metric for each value of k between 0 and 100. Like the RMSE metric, we have found that for each value of k lower than 20, it is difficult for the system to give recommendations that can help user, and which is adapted to his profile. After the value 20, the metric values start to decrease, so they indicate that the model will predict the more accurate responses than the k value lower than 20. From the obtained graphs from the measure of MAE, we have found that k = 70 gives the lowest value for cosine similarity, for pearson_baseline and pearson similarities, k = 20 gives the lowest value and k = 50 for the MSD similarity.

We can conclude that the minimum value of RMSE which gives the optimal results when the value of k is between 20 and 50 and for MAE, when the value of k is between 20 and 70 we find good results in comparison to other values.

5 Conclusion

The collaborative filtering approach is one of the prime approaches used in recommender systems which aim to give the best recommendations and better predictions to a target user. This technique uses the kNN algorithm for the selection of similar users or the neighbors near to a specific user.

In this work, we have analyzed the impact of the parameters of the kNN algorithm on the obtained results. For this reason, we have started with the k parameter to find the value of k for which we can get the optimal accuracy for the algorithm, we have varied the value of k and for each one, we have measured the prediction accuracy in terms of the Root Mean Square Error (RMSE), and the Mean Absolute Error (MAE) metrics. In addition, we have studied the impact of the similarity measures on the results, the similarities that we have used are, cosine, pearson_baseline, pearson and MSD.

Table 2. Evaluation metrics for each similarity with the values of k

	Evaluation metrics			
Similarity measures	Min RMSE	k	Min MAE	k
Cosine	0.8948499820521632	40	0.6843631697909892	70
Pearson_baseline	0.8851022038814824	20	0.6706869718700585	20
Pearson	0.8887545874694105	30	0.6706869718700585	20
MSD	0.8913532423185575	50	0.6807832987122796	50

Based on the obtained graphs from the experimental results and from the Table 2, we have found that the minimum value of RMSE which gives the optimal results depending on each similarity measure when k is between 20 and 50 and for MAE when k is between 20 and 70 we find good results in comparison to other values. We have compared the minimum values of both metrics of each similarity and we have concluded that the pearson_baseline similarity has the minimum value of the calculation of the MAE and the RMSE metrics and gives the optimal results compared to other similarities. In the future, we are working on a new recommender system in the domain of videos, which gives to users an interesting recommendation depending on their profiles and preferences.

References

1. Klašnja-Milićević, A., Ivanović, M., Nanopoulos, A.: Recommender systems in e-learning environments: A survey of the state-of-the-art and possible extensions. Artif. Intell. Rev. **44**(4), 571–604 (2015). https://doi.org/10.1007/s10462-015-9440-z
2. Konstan, J.A., Miller, B.N., Maltz, D., Herlocker, J.L., Gordon, L.R., Riedl, J.: GroupLens: Applying collaborative filtering to usenet news. Commun. ACM **40**(3), 77–87 (1997)
3. Ha, T., Lee, S.: Item-network-based collaborative filtering: A personalized recommendation method based on a user's item network. Inform. Proc. Manag. **53**(5), 1171–1184 (2017)
4. Haviv, A.: Recommendation Systems in eBay: One of the Largest Semi-Unstructured Marketplace. Newell-Simon, 30 Nov
5. Sarwar, B., Karypis, G., Konstan, J., Riedl, J.: Item-based collaborative filtering recommendation algorithms. In: Proceedings of the Tenth International Conference on World Wide Web–WWW 2001, pp. 285–295. Hong Kong, China (2001)
6. Ricci, F., Rokach, L., Shapira, B. (eds.): Recommender Systems Handbook. Springer, US (2011)
7. https://grouplens.org/datasets/movielens/
8. http://surpriselib.com/

Automatic Detection of Fake News on Twitter by Using a New Feature: User Credibility

Yasmine Lahlou(✉), Sanaa El Fkihi, and Rdouan Faizi

IRDA Group, ADMIR Laboratory, Rabat IT center, ENSIAS, Mohamed V University of Rabat, Rabat, Morocco
yasmine-emi@hotmail.fr

Abstract. Nowadays, searching news on social media like Twitter or Facebook is something usual. Any internet users can create a lot of content: posts, comments, and they can also redistribute information with retweet option for example. Nevertheless, a large portion of these pieces of news is fake and its main aim is simply to mislead people. In this case, information credibility on social networks is an increasing important issue. This article develops a method to automatically detect fake news on Twitter by calculating a user credibility. Many approaches uses NLP techniques to analyse the content of tweets to predict the credibility of news. Our approach is based on social context feature; we propose a new feature user credibility.

Keywords: Fake news · Social networks · Twitter · User-credibility · Machine learning

1 Introduction

Knowing the right information at the right time is very important. However, with the increasing proliferation of social media platforms, any internet user can generate information or news. Reports shows that over a trillion posts are made on social media per second mostly, in this fact they make traditional facts checking impossible. The traditional fact checking by journalist became difficult with the big volume of information that is generated online [1]. In general, networks use social media for conversation or chat. Now users use Twitter or Facebook to consume information or news. It has been confirmed that nearly six-in-ten twitter users (59%) get news on Twitter [1]. One of the social media services that are usually used for sharing information and building rapport is Twitter, it is also used for the delivery of newsflashes or headlines, this characteristic create a best tool for political propaganda [1]. However, the ease of information sharing, and the absence of evaluation allowed the spread of malicious information. For instance, Twitter is an international network with 152 million daily active users worldwide. It is most popular in the United States, where it counts on 59.35 million 2 users as of January 2020, followed by Japan and the United Kingdom. In China, Iran, and North Korea: the government [2] has blocked the platform. On Twitter, 140-character messages, called Tweets, are shared, and users can post messages (Tweet) and re post (Retweet) or like

M. Lazaar et al. (Eds.): BDIoT 2021, LNNS 489, pp. 568–580, 2022.
https://doi.org/10.1007/978-3-031-07969-6_43

a Tweet. Users can keep track of the posts of others (follow) and are tracked by other users (followers). These features make this network highly interactive and allow rapid and broad dissemination of information. As well, the circulation of false information is often associated with novelty, time-critical events, and emergencies like the recent crisis COVID 19 [2]. Our model aims to detect fake news on twitter by using the information of the \Author' Source of information, which is an important feature in Twitter, and then analyse the relation between user credibility and his tweet. The contributions of this study are as follows: a novel fake news analysis modelling system is built by analyzing Author's Tweets and Content of tweet together. This paper is organized as follows: The background and features of Twitter described in Sect. 2; related work is outlined in Sect. 3. The collection and preprocessing is described in Sect. 4; the statistical analysis, visualization and discussion of characteristics of fake news based on the analyzed results is in Sect. 5; and the conclusion is in Sect. 6.

2 Background

2.1 Twitter

Twitter is a social networking tool that allows people to share information, in a real time news feed through posting brief comments about their experiences and thoughts. Public messages sent and received via Twitter | or 'tweets' | are limited to no more than 140 characters and can include links to blogs, web pages, images, videos and all other material online [3]. In Twitter, we can share information, follow, retweet the tweets, Fiona Maclean et al. regroup the main functions of twitter in this Table 1:

Table 1. Useful Twitter terminology.

Term	Definition
Follow	Following another user means that all their tweets will appear in your feed
Who to follow list	This is a list of Twitter's suggestions of people or organizations that you might want to follow, based on points of similarity with your profile
Unfollow	To stop seeing someone else's tweets,
Block	From time to time a spammer or other unsavory character may appear in your Followers list. Click the icon next to the unwanted follower's name so that the 'Block [their name]'
Retweet or RT	To share somebody else's tweet that you have seen in your feed,
@	Used in tweets when you want to mention another user

(*continued*)

Table 1. (*continued*)

Term	Definition
#	Hashtag – used to categorize tweets. Popular topics are referred to as trending topics and are sometimes accompanied by hashtags
Direct Message or DM	These are private messages that you can send to other Twitter users
Shortened URLs	Given that a typical web address is rather long and clumsy, which you can paste into tweets

2.2 News Features

The process of fake news detection depends on the analysis of news characteristics. These functions are mainly divided into two categories: functions related to news content are called: content-based functions, and functions related to communication context are called social-based functions [4]. The content-based-features includes the analysis of content: Language function, writing style function, semantic function, emotional function and visual-based function, but the function based on social context is related to the analysis of news and communication through different social networks. The purpose of the news function is to analyze the way fake news is spread and how different users interact with the news and how they share news among them [5]. In our article, we use social context feature, because all the recent research of Fake news detection "FND" usually only depends on the analysis of features based on news content. Using social context-based features analysis is important in this case, which the size of tweet is slow 140 characters [3], and then this approach is an important complement. There are two basic categories of functions based on social context: user-based functions and network-based functions [4]: User based features: The Objectif of using this feature is to analyse the online users, which create, share and diffuse fake news on social network. By using the characteristics of the authors (user verified, profile, followers, following). The user-based features analysis approach can be divided into user profiling features analysis, posting behavior and temporal features analysis and credibility features analysis [5]. 4 User profiling features represents the fundamental user data involves the language used in his account, the location, the profile creation time, verification of the account, how many posts have been authored, etc. Through the interaction of online users with the news, the credibility of the news can be detected and predicted. For example, if a piece of news is widely "liked" or "commented" by unreliable users and abnormal accounts, the news is likely to be fake news [6]. It is recognized that the credibility of news can be specified by the authenticity of users interacting with the news [6]. Network-based features analysis: Two different networks can represent the interaction between online users and news releases: homogeneous and heterogeneous. Each of them has its own characteristics and behaviors. Homogeneous networks have the same node and link types. There are three types of homogeneous networks: friendship networks, diffusion networks, and credibility networks. Each of these types is potentially useful in detecting and mitigating fake news. However, heterogeneous networks have a different set of node and link types, which the common types of heterogeneous networks for analyzing fake news: knowledge networks, stance networks, and interaction networks [7].

3 Related Work

Most of the work done to detect fake news are based on supervised methods. They present the problem as a binary classification by using feature extraction and machine learning algorithms. Generally, they are two categories of features: news content and social context. For news content-based approaches, linguistic features are extracted such as lexical and syntactic features.

Carlos Castillo et al. [8] did one of the first studies on identifying fake news on Twitter. They extract some features related to the content or the author of the tweet. They trained a supervised classifier to predict credibility levels on Twitter events including SVM, decision trees, decision rules, and Bayes networks. The results shows that the best results achieved by a J48 decision tree method with an accuracy of 89%. In this paper, the authors conclude that text features are not enough by themselves for this task. The study indicated that the best features are related to the users.

In the same way, Noha Y et al. [8] develop a model for automatic detecting fake news based on text analysis using word N-grams.

They obtained results indicated that word N-gram features are more relevant compared with content and source-based features. The best results achieved 85% using a combination of unigrams and bigrams with LSVM classifier.

Both of these authors consider that content features are not sufficient to get the best the model to detect fake news on twitter, extracting social or network features is being more important.The next research try to combine content and network features: Maria Nefeli et al. [9] define a method for detection fake news by using linguistic and network features.

The model is tested on a new original dataset consisting of 2,366 tweets in English, regarding the Hong Kong events (August 2019). The naïve bayes model achieves a best accuracy of 99%.

Hamdi et al. confirmed that consider extracting some useful features of users from Twitter social graph (followers/following) by using graph embedding which captures the irregular behavior of SOFNs. It takes into consideration the relationship between Twitter users. In this paper, we propose an efficient automated SOFNs detection model by combining user features and social features. As social feature contain important clues, our proposed model is expected to improve the performance of SOFNs detection.

In the same case of social features, Lin Tian et al. [10] Propose to analyse the comments of user to learn his attitude and then combine it with content analysis for early detection of rumours. They propose convolutional neural network (CNN) CNN and BERT neural network language models to learn attitude representation for user comments Methodology and proposal approach.

4 Our Approach

4.1 New Feature: User Credibility

In twitter, user can view news feed on their homepage, which are according to who they follow. Moreover, user can also view trending news. There are many fake news on Twitter, so finding incredible users who tweet lies is an important task.

In the model proposed, we target the detection of fake news on twitter; in particular, we use hybrid approach to combine the extraction of two different features: social context based features and the content based features. Our objective is to define the importance of analyzing the relation between user or the author and his post: “tweet”.

Then we will calculate a new feature: ‘user credibility’, by using a model of probability. After that we will analyse the principal words using by the credible or incredible user.

We employ the following general definition of social context [1]:

$$\text{Social Context} = (\mathrm{C}, \mathrm{U}, \mathrm{R}, \mathrm{I})$$

where: U is the set of content generated; C is the set of users; I is the set of interactions between users, and of users with content; R is the set of relations between users, between pieces of content, and between users and content. Our objectif is to calcul the credibility of users U by using two variables C & R.

To calcul user credibility, we use Bayes’ theorem, it works on conditional probability, which is the probability that an event will happen, given that a certain event has already occurred.

Bayes ‘theorem asserts that

$$\mathrm{P(H|E)} = \frac{\mathrm{P(E|H)P(H)}}{\mathrm{P(E)}} \tag{1}$$

where,

- P(H) is the prior probability of event H. Prior probability means that it does not consider any information about the occurrence of event E;
- P(E) is the prior probability of event E;
- P(H|E) is the conditional probability of event H given that E is true;
- P(E|H) is the conditional p

To classify fake news, the following concept is used, From (1), the conditional probability that a news is a fake is based on the conditional probability that user C publish fake F or True news T, it’s calculated by (2)

$$\mathrm{P(F|C)} = \frac{\mathrm{P(C|F)P(F)}}{\mathrm{P(C|F)P(F) + P(C|T)P(T)}} \tag{2}$$

where,

- P(F|C) is the conditional probability that a news article is fake given that users C publish it is;

- P(C|F) is the conditional probability of finding a user C publishing fake news articles;
- P(F) is the probability that given news article is fake news article;
- P(C|T) is the conditional probability of finding a user C publishing true news article;
- P (T) is the probability that given news article is true.

4.2 Data Preprocessing

Given the need for social context feature (followers, following) to calculate the user's credibility score, we choose a dataset that contains this features, for this reason we use an original dataset consisting of 2,366 tweets in English, regarding the Hong Kong events.

Before exploring the dataset, cleaning and preprocessing functions are required before extracting features: content and social features. We clean the dataset by removing tweet id, hyperlinks, emoticons, punctuations and non-letter characters. To reduce the size of the dataset, we use some preprocessing functions like stop word removal, stemming and tokenizing.

Stop Word Removal:
Stop words are insignificant words which occur commonly across all the tweets but actually they are insignificant like; a, an, the, will, was, were, is, are, to, of, that, these, what, when etc. These are words commonly used a lot in sentences to help connect thought or to assist in the sentence structure.

We remove these words, because they are not discriminant when used as features in the classification task.

Stemming:
Stemming is the process of removing suffixes and reduce words to their word stem. For example, words like (connects, connected, connecting, connection) all have the same meaning. Removing the suffixes (-ed, -s, -ion, -ing) and leaving the single word (connect) will reduce the number of unique words and make classification more efficient.

Stemming simply is changing the words into their original form, and decreasing the number of word types or classes in the data. For example, the words "Running", "Ran" and "Runner" will be reduced to the word "run." We use stemming to make classification faster and efficient. Furthermore, we use Porter stemmer, which is the most commonly used stemming algorithms due to its accuracy.

Figure 1 shows an example of a tweet before and after preprocessing.

text	Stemming
I woke up this morning to find a variation of ...	i woke up this morn to find a variat of this h...

Fig. 1. Example of stemming tweet

4.3 Features Extraction

Content Based Features:
Term Frequency (TF) Term Frequency is an approach that utilizes the counts of words appearing in the documents to figure out the similarity between documents. An equal length vector that contains the words counts represents each document.

Next, each vector is normalized in a way that the sum of its elements will add to one. Each word count is then converted into the probability of such word existing in the documents. For example, if a word is in a certain document it will be represented as one, and if it is not in the document, it will be set to zero.

A term importance increases with the number of times a word appears in the document; however, this is counteracted by the frequency of the word in the corpus. One of the main characteristics of IDF is it weights down the term frequency while scaling up the rare ones. For example, words such as "the" and "then" often appear in the text, and if we only use TF, terms such as these will dominate the frequency count. However, using IDF scales down the impact of these terms.

User Based Features:
To find user credibility we use a 7 points based method:

Points corresponding to each factor are:

- Verified Profile: 2 points
- Find Followers/Following ratio and use a threshold. Here if ratio > 0.8, assign 2 points to the user. If the ratio > 0.4, assign 1 point to the user.
- Linked to some other social networking site: - 1 point
- User has listed his website - 1 point
- User has added his/her description - 1 pointe credible.

So extract the full URL from TinyURLs used in each of the tweet and if there exists a corresponding full URL which opens, assign a point.

4.4 Classification Process

The implementation of the project is done in Python language. We have used a general process of machine learning: data collection, data preprocessing, feature extraction, feature selection on our dataset and modeled our classifier using LSTM.

We implemented LSTM on our own and compared our accuracy with other classification algorithms like KNN, Naïve Bayes, Decision Tree and neural networks.

We split the dataset into training and testing sets. For instance, in the experiments presented subsequently, we use 5-fold cross validation, so in each validation around 80% of the dataset is used for training and 20% for testing (Fig. 2).

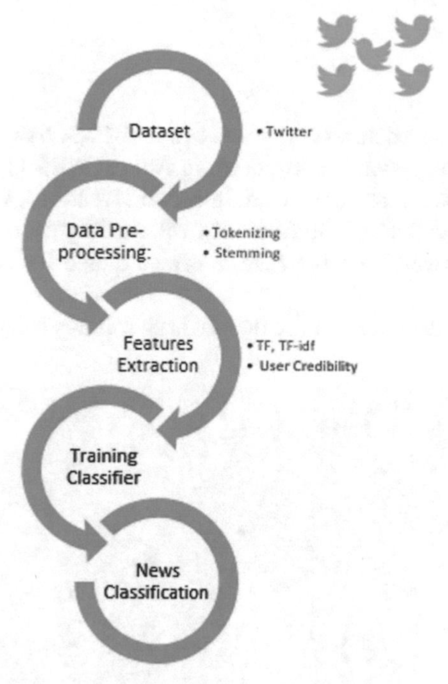

Fig. 2. Classification process

4.5 Models

Naïve-Bayes

Naïve Bayes is a straightforward machine-learning model that can be applied for classification. It adopts the Bayes Theorem and it is a collection of estimated probabilities.

Naive Bayes learners and classifiers can be extremely fast compared to methods that are more sophisticated [11].

Logistic Regression

Logistic Regression is a Machine Learning classification algorithm that is used to predict the probability of a categorical dependent variable. In logistic regression, the dependent variable is a binary variable that contains data coded as 1 (yes, success, etc.) or 0 (no, failure, etc.). In other words, the logistic regression model predicts $P(Y = 1)$ as a function of X [12].

Random Forest

A random forest is a meta estimator that fits a number of decision tree classifiers on various sub-samples of the dataset and uses averaging to improve the predictive accuracy and control over-fitting.

5 Experiments

5.1 Dataset

In this article, we will use a dataset composed of a set of 2,366 tweets written in English, regarding the Hong Kong events, and posted in August 2019 [10]. Our approach uses both linguistic and network features. Certain network features, which are available for these tweets, are also collected. Additionally, certain linguistic features are extracted from the text of each tweet. A label feature is also added for each tweet, with either value "fake", or "real".

The distribution of data label in the dataset is as follows (Fig. 3):

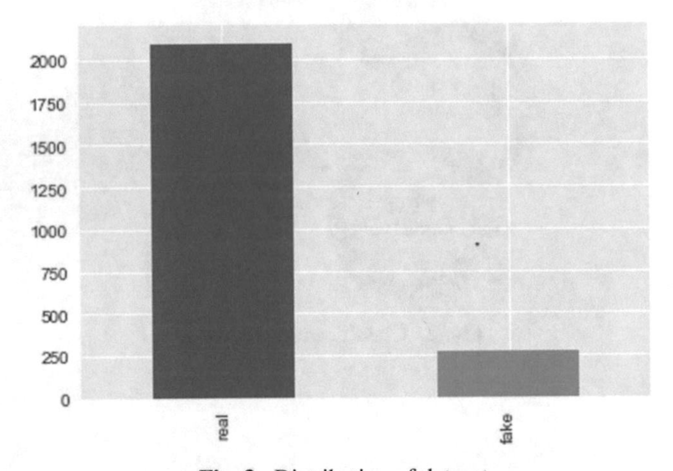

Fig. 3. Distribution of dataset

In this dataset the following information is available:

- tweet_text
- user_id
- user_display_name
- user_screen_name
- follower_count
- following_count
- account_creation_date
- tweet_time
- in_reply_to_userid
- like_count
- retweet_count

5.2 Performance Evaluation of Proposed Model

Most widely used procedure for evaluating the performance of the classification model (in term of accuracy, specificity, and sensitivity) is by using confusion matrix. This paper uses the most widely accepted metrics with the following classification:

- True Positive (TP): When the news is predicted as fake and it is actually annotated as fake.
- True Negative (TN): When the news is predicted as real and it is actually annotated as real.
- False Positive (FP): When the news is predicted as fake and it is actually annotated as real. • False Negative (FN): When the news is predicted as real and it is actually annotated as fake.

5.3 Data Exploratory

In this analysis, we will use two classes of features: Content based features that we have extract from the tweets, number of words in both fake and real tweets, and then we will calcul the score of user credibility as described in section of feature.

Content based features:
Figures 4, 5 shows the distributions of numbers of words in fake and real tweets:

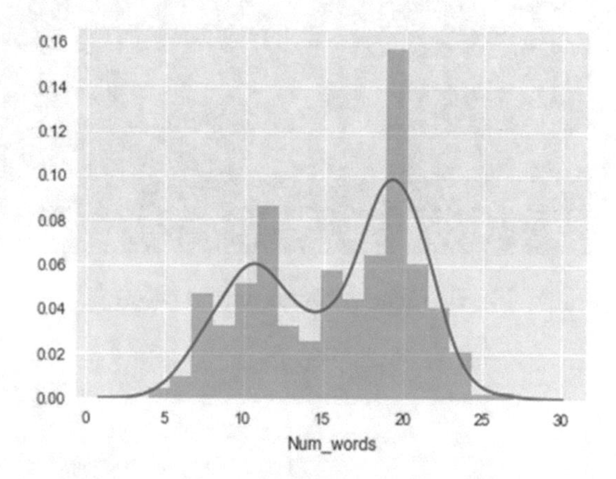

The text_tweet analysis give us a simple conclusion about the words, number of words using in fake & real tweets.

Social Context Features
To complete our analysis, we need to calcul the user credibility to understand the relation between user and the tweet.

Based on a 7 points method we get the distributions of user based on her score: Credible, highly credible, and not credible as figures in this figure (Fig. 6):

By this figure, we understand that they are an important relation between the user_credibility and the class of tweets.

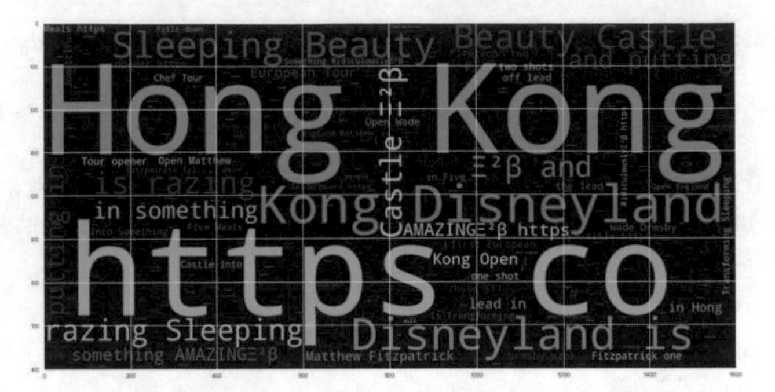

Fig. 4. Frequent words on fake tweets

Fig. 5. Frequent words on real tweets

	tweet_text	user_credibility	class
0	I must stick to the post, says a police office...	Highly Credible	real
1	President Xi Jinping says he believes that the...	Highly Credible	real
2	President Xi Jinping says that China will not ...	Highly Credible	real
3	President Xi said the central government fully...	Highly Credible	real
4	#HongKong police arrest 4 people for money lau...	Highly Credible	real
5	U.S. 'pouring oil over fire' in Hong Kong to u...	Highly Credible	real
6	Facing fierce and long-lasting social violence...	Highly Credible	real
7	Double-decker bus crashes into tree in Hong Ko...	Highly Credible	real
8	Hong Kong, Macao join hands to tap opportuniti...	Highly Credible	real
9	While Hong Kong police are dealing with violen...	Highly Credible	real

Fig. 6. User credibility

6 Experiments Results

To be more exhaustive in our analysis, we run the model with the new feature -user credibility- and without it, by this method we can understand the value of this score, and in general confirm that the combination between content feature and social feature is very important to perform the detection of fake news on twitter.

For this reason, we decide to run four algorithms of machine learning and compare the results with the paper of.

Models Using User Credibility:
The table offer the results obtained by combining 2 features: Linguistic and social features specially the score of user credibility.

Classifier	Accuracy	Precision	F1-score
Naïve Bayes	96	96	95
Logistic Regression	98	98	96
Random Forest	99	99	96

- ***Models without user credibility***

Classifier	Accuracy	Precision	F1-score
Naïve Bayes	84	84	88
Logistic Regression	85	85	89
Random Forest	85	85	88

7 Discussion

In this work, we have shown a practical approach for treating the identification of fake news on Twitter as a binary machine-learning problem. While that translation to a machine-learning problem is rather straight forward, the main challenge is to define a news feature based on user or source of information.

By adding this new feature we get a high accuracy of 93% with Random forest; then we see the importance of the user to detect fake news on twitter.

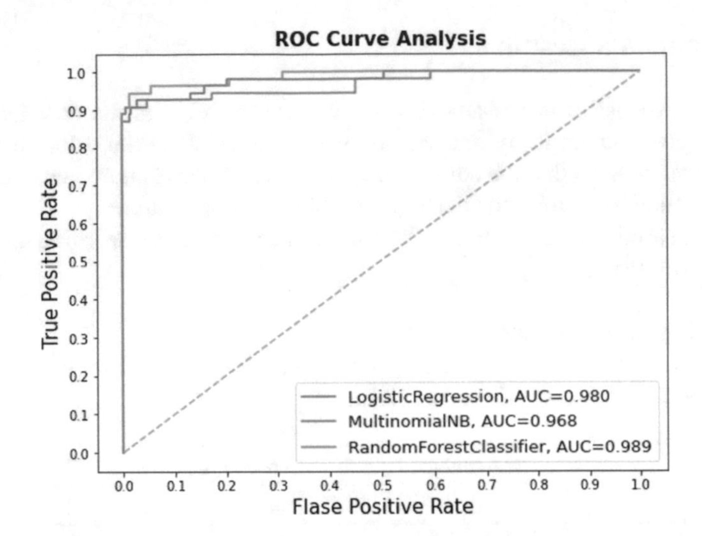

Our future challenge is to determine the graph analysis of all users, to get the original sources. We will see by using the graph embedding the history of the chain of sharing tweets.

References

1. Shearer: News Use Across Social Media Platforms 2016 (2016)
2. Pulido, R.: False News Around COVID-19 Circulated Less On Sina Weibo. International Journal of Multidisciplinary Research Review, pp. 1–22 (2020)
3. Fiona M.: Understanding Twitter. British Journal of Occupational Therapy, p. 76 (2013)
4. Okoro: A hybrid approach to fake news detection on social media. Nigerian Journal of Technology (NIJOTECH), **37**(12) (2018)
5. Content-Social Based Features for Fake News Detection Model from Twitter. International Journal of Advanced Trends in Computer Science and Engineering, 8(16), 1–6 (2019)
6. Carlos Castillo, «Information Credibility on Twitter,» international conference on World wide web, 2011
7. Kai, S.: Studying Fake News via Network Analysis: (2018)
8. Hassan, N.Y.: Credibility Detection in Twitter Using Word N-gram Analysis and Supervised. Int. J. Intell. Eng. Syst. **13**(11), 10 (2020)
9. Nikiforos, M.N., Vergis, S., Stylidou, A., Augoustis, N., Kermanidis, K.L., Maragoudakis, M.: Fake news detection regarding the Hong Kong Events from tweets. In: Maglogiannis, I., Iliadis, L., Pimenidis, E. (eds.) AIAI 2020. IAICT, vol. 585, pp. 177–186. Springer, Cham (2020). https://doi.org/10.1007/978-3-030-49190-1_16
10. Lin, T.: Early Detection of Rumours on Twitter via Stance Transfer Learning. European Conference on Information Retrieval (2020)
11. Scikit_learn. Naive Bayes. Scikit_learn, [En ligne]. Available: https://scikitlearn.org/stable/modules/naive_bayes.html. [Accès le 2020]
12. Li, S.: Building A Logistic Regression in Python, Step by Step. Towards data science, 2017 09 29. [En ligne]. Available: https://towardsdatascience.com/building-a-logistic-regression-in-python-stepby-step-becd4d56c9c8. [Accès le 2020]
13. Safety, T.: Twitter Safety. Information operations directed at Hong Kong, 19 08 2019. [En ligne]. https://blog.twitter.com/en_us/topics/company/2019/information_operations_directed_at_Hong_Kong.html. [Accès le 20 5 2020]

User-Enriched Embedding for Fake News Detection on Social Media

Oussama Hebroune(✉) and Lamia Benhiba

ENSIAS, Mohammed V University in Rabat, Rabat, Morocco
{oussama_hebroune,lamia.benhiba}@um5.ac.ma

Abstract. Recent political, pandemic, and social turmoil events have led to an increase in the popularity and spread of misinformation. As demonstrated by the widespread effects of the large onset of fake news, humans are inconsistent if not outright poor detectors of fake news. Thereby, many efforts are being made to automate the process of fake news detection. The most popular of these approaches include blacklisting sources and authors that are unreliable. While these tools are useful, in order to create a more complete end to end solution, we need to account for more difficult cases where reliable sources and authors release fake news. As such, the goal of this paper is to propose an approach for detecting the language and behavioral patterns that characterize fake and real news through the use of social network analysis and natural language processing techniques. We have built a model that catches many intuitive indications of real and fake news using users and submissions attributes, and thus laid the foundation for an approach that concatenates multiple embeddings for better fake news detection on social media.

Keywords: User representation · Social Network Analysis · Fake news detection · Community detection · NLP

1 Introduction

Fake news may be a relatively new term but it is not necessarily a new phenomenon. Fake news has technically been around at least since the appearance and popularity of one sided, partisan newspaper in the 19th century. However, advances in technology and the spread of news through different types of media have increased the spread of fake news today.

Defined as "fabricated information that mimics news media content in form but not in organizational process or intent" [1], fake news is often used to propagate false information in order to change people's behavior. As such, their effects are increasingly felt as it has been shown in their non-negligible influence on the 2016 US presidential elections [2].

Fake news is thereby becoming a societal issue and scientists are unanimous on the necessity of finding solutions to fight their proliferation. Early detection approaches that relied on blacklisting unreliable sources and authors proved to be insufficient. Technologies such as Artificial Intelligence (AI) and Machine Learning (ML) tools offer great

M. Lazaar et al. (Eds.): BDIoT 2021, LNNS 489, pp. 581–599, 2022.
https://doi.org/10.1007/978-3-031-07969-6_44

promise for researchers to build systems which could automatically detect fake news. However, detecting fake news on social media is a challenging task to accomplish as it requires models to summarize the posts and compare them to actual news or against debunking websites such as snopes.com, in order to classify them. Classic methods are often content-based, and rely on textual and/or image representation. Recent embedding approaches however take into account content authors and their immediate social context. This paper aims to explore how a combination of content and social embeddings can help detect fake news on social media.

By using social network and textual analysis, we investigate intrinsic properties of fake and real posts in a social media platform and propose a user-enriched representation approach to distinguish them. In particular, we aim to answer the following research questions:

- Do users' attributes help identify fake posts?
- Does concatenating vectors embeddings give better classification results?
- Does our model succeed in detecting fake news?

The remainder of the study will be structured as follows: Sect. 2 lays the background by over viewing related research on fake news detection on social media, Sect. 3 will describe the methodology used during different stages of our study. Section 4 will cover our implementation, results and discuss different interpretations and limitations. The last section will provide a summary of contributions and suggestions on future work opportunities.

2 Introduction

In this section, we first overview prior works on fake news detection on online social platforms. Then, we present embedding approaches for user representation on social media, as they offer a general context for our proposed approach.

2.1 Fake News Detection on Social Media

Fake news definition relies on two characteristics: authenticity and intent. Authenticity means that fake news contains false information that can be verified as such. This means that conspiracy theory is not included in fake news as there are difficult to be proven true or false in most cases. The second part, intent, means that the false information has been written with the goal of misleading the reader.

Current research focuses mostly on combining multimodal features or using social features and author information in order to improve the quality of classifications. Multiple approaches can be used in order to extract features and use them in models as will be described in the next sections.

Image and Text-based Approach. Yang et al. [3] focused on content to classify fake news. They proposed a CNN architecture that relies on the text and images contained in articles in order to make the classification. Authors used Kaggle fake news dataset, and additionally scrapped real news from trusted sources such as New York Times and Washington Post. Their network was made of two branches: one text branch and one image branch. The textual branch was then divided into two subbranch: textual explicit: derived information from text such as length of the news and the text latent subbranch, which is the embedding of the text, limited to 1000 words.

The image branch is also made of two subbranch, one containing information such as image resolution or the number of people present on the image, the second subbranch use a CNN on the image itself.

User-engagement based approach. Tacchini et al. [4] focus on using social network features in order to improve the reliability of their fake news detector. Based on a dataset collected using Facebook Graph API, the authors considered user interactions with posts as features (e.g. likes) within pages from two main categories: scientific news and conspiracy news. They used logistic regression and harmonic algorithm [5] to classify news in categories hoax and non-hoax. Harmonic Algorithm is a method that allows transferring information across users who liked some common posts. Their approach led to good results (accuracy of 99%) where harmonic algorithm outperformed logistic regression.

These approaches consider different aspects of the fake news detection problem: content vs. authors. Our premise is that combining the two might offer a more comprehensive user representation and could thus provide better classification performance. Our approach is thus based on jointly learning the graph and textual embeddings in order to feed them to our classification algorithm. To do so, we make use of recent approaches in attributed graph representations.

2.2 User Representation on Social Media

User representation learning helps create low-dimensional embeddings of users that capture users' characteristics and take into account the context of their contributions as well. On social media, social actors are associated with rich attributes, so it is crucial to capture those attributes and combine them with network structure to learn a comprehensive representation that can be later used for machine learning tasks (Fig. 1).

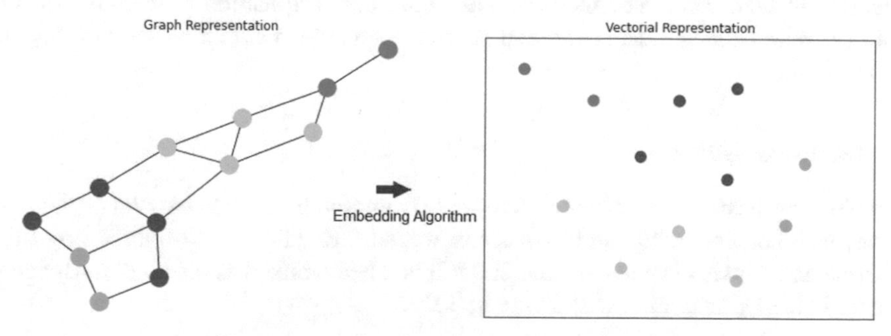

Fig. 1. Illustration of social network embedding.

Existing works are increasingly interested in learning representations that focus on the relational context of users within the social network.

Deepwalk [6] uses random walks to capture the structure of graphs starting at every vertex of the graph. Each of these random walks forms a sequence of vertices. Therefore, the algorithms for word representation that uses sequences of words (sentences) as the input can thereby be exploited for this graph representation. Word representation or word embedding is an important tool in language modeling [7], which helps algorithms extract similar words. The idea is that given a set, similar nodes would appear within similar context. A context of a node is the set of surrounding words in the same sequence.

Other approaches such as LINE [8] proposes to preserve the first- and second-order proximity when learning node representations. While Node2Vec [9] suggests designing a biased random walk on the flexible node's neighborhood.

However, all of these methods only utilize the topological structure, ignoring the useful attribute of nodes.

LANE [10] proposes to incorporate the label information into the attributed network embedding. Unlike the previous network embedding methods, LANE is mainly based on spectral techniques [11]. It adopts the cosine similarity to construct the corresponding affinity matrices of the node attributes, network structure, and labels. Then, based on the corresponding Laplacian matrices, LANE is able to map the three different sources into different latent representations, respectively. In order to build the relationship among those three representations, LANE projects all these latent representations into a new common space by leveraging the variance of the projected matrix as the correlation metric. The learned representations of nodes are able to capture the structure proximities as well as the correlations in the label informed attributed network.

Another approach is Tri-Party Deep Network Representation (TriDNR) [12]. It uses a coupled neural network framework, TriDNR learns vertex representations from three information sources: network structure, vertex content and vertex labels. To capture the vertex content and label information.

3 Methodology

In this section, we will describe our proposed approach and explain each step from data collection to fake news classification. The entire data acquisition process is based on publicly available data. Users content that abide to privacy restrictions was not included in our dataset.

3.1 Data Acquisition

Data Source: Reddit. We have collected data from Reddit, an online social news aggregation, web content rating, and discussion website. Reddit was co-founded by Steve Huffman and Alexis Ohanian in June 2005. It had been ranked as the sixth most visited website in the US and ninth worldwide in 2017.

The main goal of the study is to identify the similarities and the differences between the fake and the true posts on both a textual content level and a social network level. Reddit was the most reliable platform for this research for the following reasons:

- Existence of forums (subreddits) that address specific topics.
- Heavily uncensored content
- Popularity of the platform
- Availability of an API which helps us extract posts and comments from every subreddit
- Rich metadata

Data Collection. Data was gathered from 5 subreddits: r\theonion, r\nottheonion, r\misleadingthumbs, r\usnews and r\photoshopbattles. The five subreddits were created to state the credibility of posts.

To collect the posts and comments from each subreddit, we used Praw API [13] in order to get specific distinguished metadata about the posts and comments shared by the users for this study and for any potential future work ahead. The API helps provide enhanced functionality and search capabilities for searching Reddit comments and submissions. We eventually gathered posts and comments from the selected subreddits. The final dataset contains a total of 13467 users for real posts and 3898 for fake ones, with a total post of 44911 and total comments of 1071263 (Table 1).

Table 1. Subreddits, fake, real posts and total comments.

Subreddits	Fake posts	Real posts	Total comments
r\theonion, r\nottheonion, r\misleadingthumbs r\photoshopbattles	10274	34637	1071263
r\usnews			

Our proposed approach can be summarized in three main steps: We first prepare the extracted data through data pre-processing and annotation processes. Using embedding techniques, we learn embeddings for each author, including their comments and submission. These embeddings are then concatenated and fed to a classification model as an input (Fig. 2).

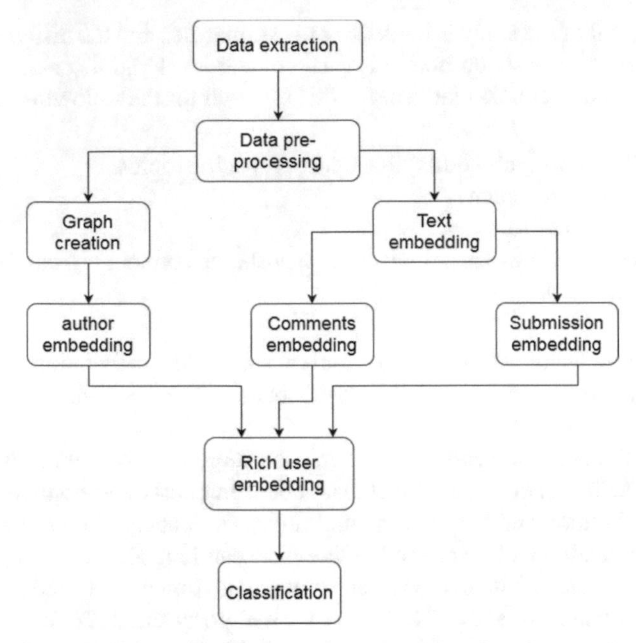

Fig. 2. The proposed user-enriched embedding-based approach.

Data Pre-processing. The data pre-processing step is dedicated to transforming our raw data into an exploitable one. It also helps prune data in order to keep only the data needed for analysis. Preprocessing data for our approach entails cleaning, labeling, and formatting it into the right format for each branch of the embedding process.

Based on the data extracted using the Praw API, we chose the following fields for the users, the submissions, and the comments respectively.

Metadata on the users were defined as follows:

- author: the account name of the poster.
- comment_karma: The comment karma for the Redditor.
- link_karma: The link karma for the Redditor.
- verified_mail: Whether or not the Redditor has verified their email.
- account_verified: Whether or not the Redditor has verified his email.
- reddit_employee: Whether or not the Redditor is a Reddit employee.
- moderator: Whether or not the Redditor mods any subreddits.
- gold: Whether or not the Redditor has active Reddit Premium status.

Data on submissions is defined as follows:

- id: the id of the post.
- author: the account name of the poster.
- title: the title of the link.
- num_comments: the number of comments that belong to this post. Includes removed comments.

- score: the net-score of the link. A submission's score is simply the number of upvotes minus the number of down votes.
- created_utc: the time of creation in UTC epoch-second format.
- subreddit: subreddit on which the post is submitted excluding the /r/prefix.

For the comments, we identified the following attributes:

- id: the id of the comment.
- author: the account name of the poster.
- created_utc: the time of creation in UTC epoch-second format.
- link_id: subreddit on which the comment is submitted excluding the /r/prefix.
- parent_id: ID of the submission (post/ comment) this comment is a reply to, either the link or a comment in it
- body: the raw text. This is the unformatted text which includes the raw markup characters.
- subreddit: subreddit of thing excluding the /r/ prefix.

In order to increase the performance of textual models, especially since we're dealing with text coming from social media, we apply NLP pre-processing techniques. The cleaning includes:

- Removing Non-alphanumeric data such as HTML tags, links.
- Tokenization: separating each word of text.
- Removing Stop Words.
- Lemmatization.

Data Labeling. The dataset annotation was made manually on a data sample where '0' labels fake posts and '1' real ones. We then applied a pseudo-labeling method to annotate the entirety of the dataset [14]. This method uses a small set of labeled data along with a large amount of unlabeled data following these steps:

- Train model on a batch of labeled data with a 3 fully connected layer.
- Use the trained model to predict labels on a batch of unlabeled data.
- Use the predicted labels to calculate the loss on unlabeled data.
- Combine labeled loss with unlabeled loss and back propagate (Fig. 3).

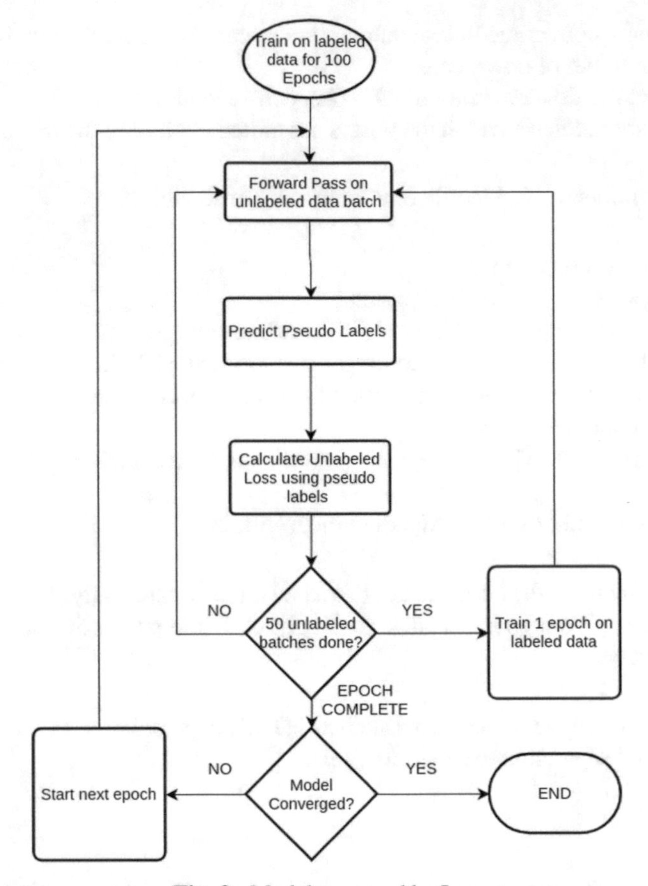

Fig. 3. Model proposed by Lee.

Graph Creation. Social network analysis (SNA) is an interdisciplinary descriptive, conceptual and empirical framework that represents complex systems as networks, expressed in terms of relationships among actors. SNA helps uncover the structure and behaviors within a social construct through a set of metrics on three levels: node (micro), network (macro) and substructures levels (meso-scale).

We model our social network as a social interaction graph G = (N, E) where N represent the authors and commenters within the subreddit and E the set of interactions among them (an interaction is a reply or a comment on a comment). Hence, for every pair of nodes i and j, a link is drawn from i to j when i replies to j. Links can be directed or undirected depending on the conducted analysis and self-loops are removed (Fig. 4).

Fig. 4. From Reddit to network.

Authorembedding: Graph Convolutional Network. Neural Networks have gained massive success in the last decade. However, a lot of data in the real world have underlying graph structures which are non-Euclidean. The non-regularity of data structures has led to recent advancements in Graph Neural Networks. In the past few years, different variants of Graph Neural Networks are being developed with Graph Convolutional Networks (GCN) [15] being one of them. GCNs are also considered as one of the basic Graph Neural Networks variants.

GCNs are a very powerful neural network architecture for machine learning on graphs. In fact, they are so powerful that even a randomly initiated 2-layer GCN can produce useful feature representations of nodes in networks.

A graph convolutional network (GCN) is a neural network that operates on graphs. We used the same graph obtained from the authors and commenters G = (N, E), which takes as input:

- an N × F feature matrix, X, where N is the number of nodes and F is the number of input features for each node.
- an N × N matrix representation of the graph structure such as the adjacency matrix A of G.

To implement our GCN we used Spektral [17], which is a Python library for graph deep learning based on TensorFlow 2.

Textual Embedding. One of the most important techniques in natural language processing is word embedding, where words are mapped to real number vectors. Word embedding is able to capture a word's context, semantic and syntactic similarity, relationship with other terms, in a document. The most known word embedding model is Word2Vec [16], it's a two-layer neural network that takes text corpus as its input and outputs a set of vectors representing each word of the corpus.

Doc2Vec is an extension of Word2vec that encodes entire documents as opposed to individual words. Doc2Vec vectors represent the theme or overall meaning of a document.Doc2Vec [18] has two main training algorithms, the first one is Distributed Memory Model of Paragraph Vectors (PV-DM) and the Distributed Bag of Words version of Paragraph Vector (PV-DBOW). The main difference between the two algorithms is that (PV-DBOW) uses the document to draw the word, while the (PV-DM) attempts to guess the output from its neighboring words (context words) with the addition of a paragraph ID (Figs. 5 and 6).

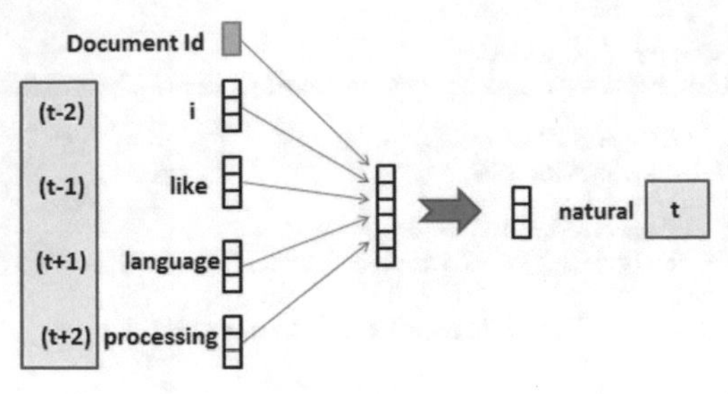

Fig. 5. Distributed memory model.

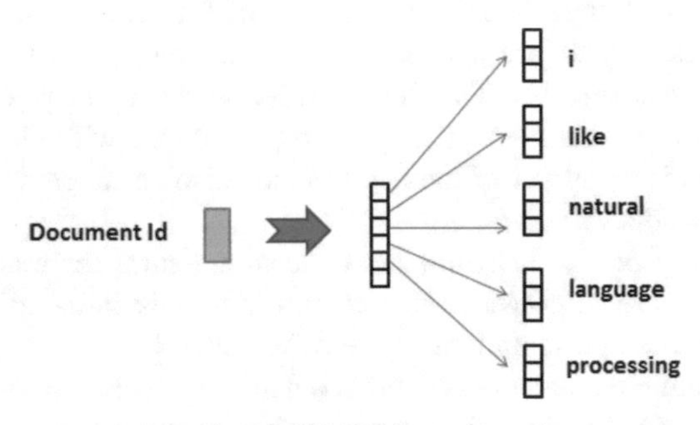

Fig. 6. Distributed bag of words.

For this study, we will use PV-DBOW to embed submissions as it has shown its efficiency with small training data corpuses and PV-DM for comments in order to preserve more the context and the semantics of the comments.

Classification Task. After we get each embedding, we concatenate them into one vector of length of 137. The dataset has been split be choosing 80% of the posts for training, and 20% for testing. 6 classification models have been chosen to be trained on our dataset: SVM, Logistic Regression, Naïve-Bayes, Stochastic Gradient, Deep Neural Network and XGBoost.

In order to evaluate each model, two evaluation metrics have been used, accuracy andf1-score. The f1-score metric is used because it combines the recall and the precision and clarifies cases where accuracy can be misleading when classifying rare cases.

4 Implementation and Results

In this section, we will describe the data through an exploratory data analysis, and examine the results of our classification model. Then we will present the findings and debate the results of the analysis we performed and explore the future works.

4.1 Exploratory Analysis

Exploratory Data Analysis refers to the critical process of performing initial investigations on data so as to discover patterns, spot anomalies, test hypothesis and check assumptions with the help of summary statistics and graphical representations.

A good starting point for the analysis is to make some data exploration of the data set. The first thing to be done is a statistical analysis such as counting the number of texts per class or counting the number of words per sentence. Then it is possible to try to get an insight of the data distribution by making dimensionality reduction and plotting data in 2D.

The first analysis shows that the number of fake posts is smaller with respect to the number of reliable posts as it's shown in.

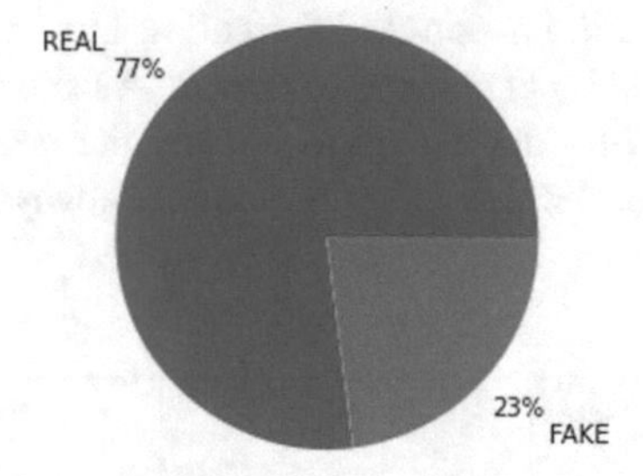

Fig. 7. Proportion of fake and real posts in our dataset.

When investigating how many users we have per class, we found that 14% of our users are responsible for misinformation Fig. 7 (Fig. 8).

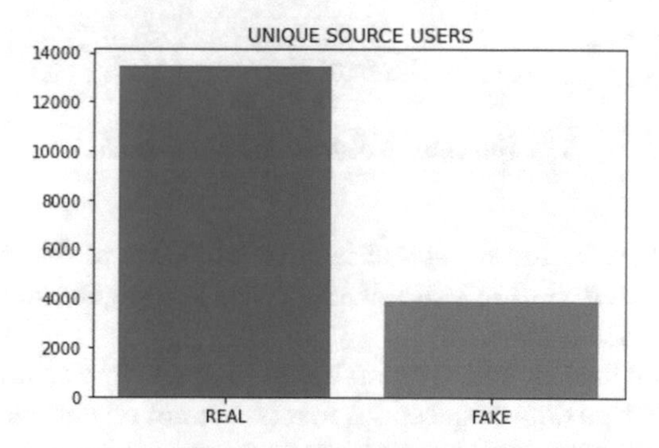

Fig. 8. Users per class.

Another interesting feature, is the account creation time of the posts. We found that almost all authors of the fake posts are new accounts, their life span (from their creation until they posted the fake post) is smaller than the authors of real posts (Fig. 9).

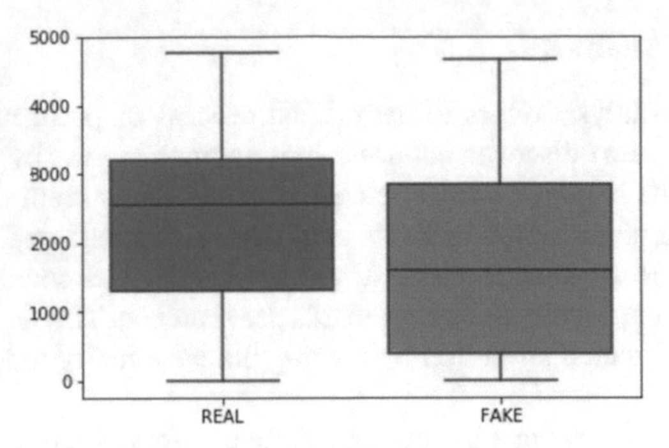

Fig. 9. Box plot of the life of users.

Besides the life span of the account, the creation time of the posts give us a little insight on the time preferred by each class to post in. As shown in the figure below (the x-axis represents the hour of a day from0 to 24, and the y-axis represents the number of posts in respective hour), the fake posts are almost always posted at night (between 20 h-24 h) (Fig. 10).

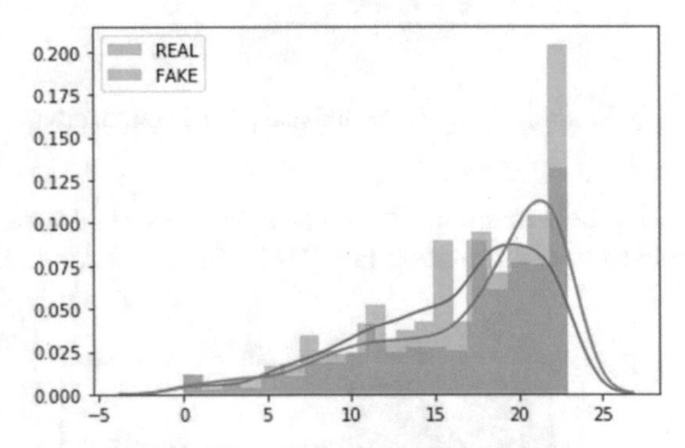

Fig. 10. Hourly distribution of the posts.

Another important feature to look at is the distribution of the number of words in the text. It can be seen in Fig. 11 that reliable posts have slightly more words than fake posts, but the difference is minimal.

To compute word distribution, we start by splitting text into an array of sentences on stop punctuation such as dots or questions marks, but not on commas. The second step consists of filtering words that are contained in these sentences, to do so, stop words (words such as 'a', 'an', 'the'), punctuation, numeric values and tags (such as html tags) are removed. The number of remaining words is used for the metric's computation (Fig. 11).

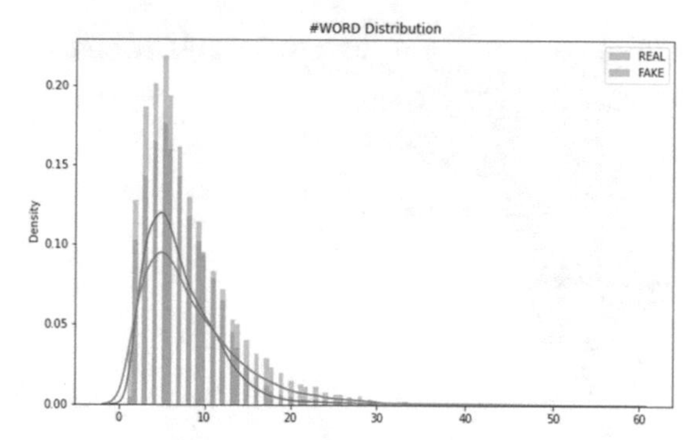

Fig. 11. Word distribution.

Fake and real posts' word cloud representation helps us recognize their most overused words (Figs. 12 and 13).

Fig. 12. Fake posts' word cloud.

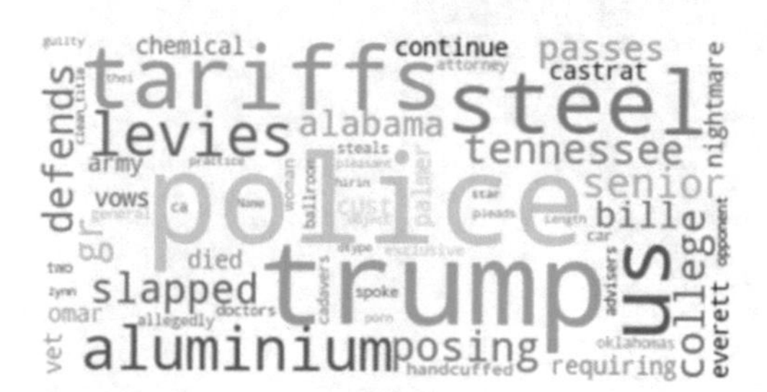

Fig. 13. Real posts' word cloud.

We observed that in the majority of fake posts, their authors are using words that are more related to uncovering illegal stuffs such as "footage" and "leaked", which attract most of the reader's attention.

Word cloud relay a general overview of the content but don't delve into the details. In order to get more insight and better understanding, we use the LDA method to identify the discussed topics in each category (Fig. 14).

We begin to see discursive patterns in each of the groups when looking at their most relevant words.

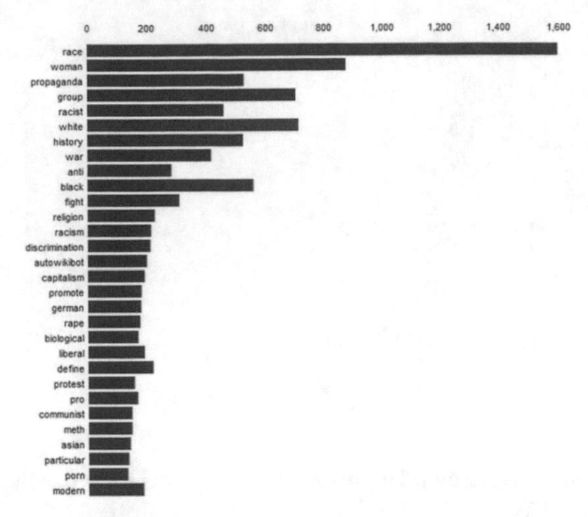

Fig. 14. Most relevant terms for topics in fake submissions.

The most notable observation regarding words in fake posts is the presence of a wealth of concepts related to aggressive discourse and racism. Which complements the assumption that fake posts exploit the weak point of the situation to mislead the readers (Fig. 15).

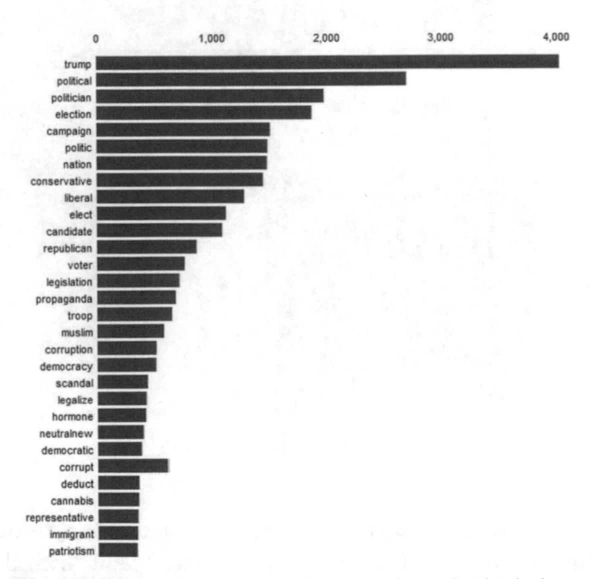

Fig. 15. Most relevant terms for topics in real submissions.

On the other hand, judging from the most relevant terms in the real posts, we can say that real posts' authors rely on logic and facts to argument their posts. Their posts are more characterized by the presence of themes such as "law", "politic" and "government".

To better defend the assumptions that fake posts use aggressive discourse, we conducted a sentiment analysis on the submissions. Figures 16 and 17 show our that negative sentiments are more prevalent in fake posts:

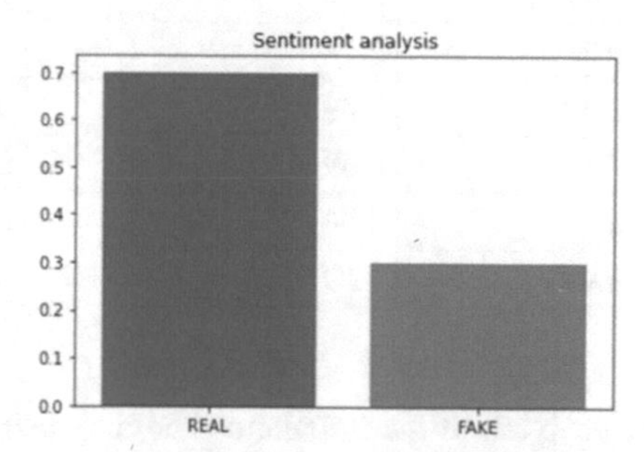

Fig. 16. Ratio of positive sentiments.

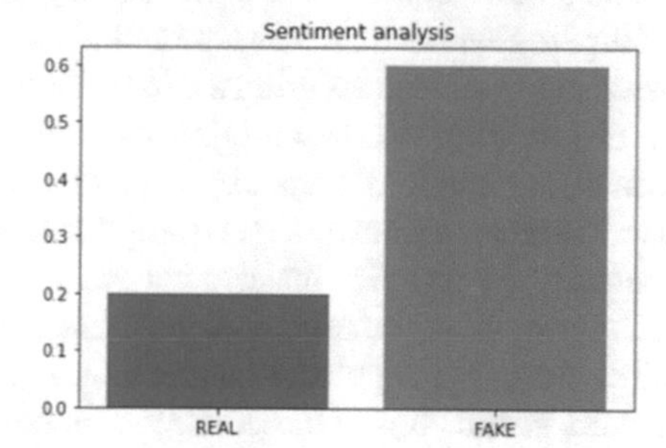

Fig. 17. Ratio of negative sentiments.

4.2 Model Implementation and Results

Model Architecture. The studies stated in the literature review influenced our choice of the model architecture. For our supervised classification model, we're using the following main models:

- GCN embedder: the GCN model will take 2 inputs, the Node Features Matrix (X) and Adjacency Matrix (A). We are going to implement 2-layer GCN with Dropout layers and L2 regularization. We are also going to set the maximum training epochs to be 200 which was showen a better performance.
- 2 words embedders: the Doc2Vec model takes as input the padded sequences and we instantiate it with a vector size with 20 dimensions for the submissions, and 100 for the comments.
- A classification algorithm: after concatenating all the embeddings into one matrix, we feed it to 6 classifications models. Results are state in the next subsection.

Results. The concatenation resulted into a 44911×137 matrix, that we fed to multiple classification models. We present here the results, and mainly the accuracy and the f1-score (Table 2).

Table 2. Classification results for imbalanced data.

Models	Accuracy	F1-score real posts	F1-score fake posts
SVM	0.94	0.96	0.86
logistic regression	0.77	0.78	0.06
Naïve Bayes	0.30	0.39	0.18
stochastic gradient	0.30	0.39	0.28
neural network	0.89	0.94	0.70
XGBoost	0.98	0.99	0.96

We tried to solve the problem of the imbalanced data by using two techniques. The first is under-sampling which uses a subset of the majority class to train the classifier, then the training set becomes more balanced and the training process becomes faster. The second one is over-sampling which achieves a more balanced class distribution by duplicating minority class instances and preserving the information. But both approaches suffer some drawbacks. The random under-sampling method can potentially remove certain important data points, and random oversampling can lead to overfitting.

So, we decided to use Synthetic Minority Over-sampling Technique [19]. SMOTE has been designed to generate new samples that are coherent with the minor class distribution. The main idea is to consider the relationships that exist between samples and create new synthetic points along the segments connecting a group of neighbors. The table below shows the results of the chosen classification models on the balanced data (Table 3):

Table 3. Classification results for balanced data.

Models	Accuracy	F1-score real posts	F1-score fake posts
SVM	0.92	0.95	0.85
logistic regression	0.61	0.72	0.32
Naïve Bayes	0.30	0.17	0.39
stochastic gradient	0.77	0.85	0.12
neural network	0.75	0.80	0.65
XGBoost	0.98	0.99	0.98

By comparing the two tables, we observed that overall, the neural network performances decreased in the balanced dataset. The SMOTE technique helped the Naïve Bayes and Stochastic Gradient outperform their previous results in our initiate dataset. The model that achieved better results in our case was XGBoost. Both tables show a stability of results in both balanced and imbalanced datasets, yet there is a slight difference in recall results for the fake class. In the balanced dataset the recall was 0.99, compared to imbalanced dataset which was 0.92.This, means that there are more misclassified fake posts as real posts, which can be sensitive in credibility classification (Fig. 18).

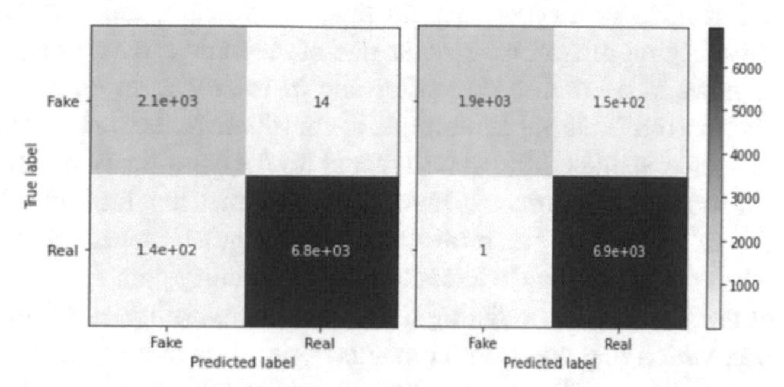

Fig. 18. Confusion matrix for XGBoost model (balanced data on the left, imbalanced data on the right).

Discussion. Focusing on the findings of our research questions, we can affirm that the fake and the real posts share many lexical features yet talk about the same topics in slightly identifiable different ways, they both share concerns about controversial events. But what distinguishes the real comments and posts from the fake ones is the presence of an argumentative and radical discourse. The exploratory data analysis made that clear with the frequent presence of political and law terms within the real posts. Looking further into each group's unique words, the presence of racism terms "black", "white", "asian", "discrimination" suggest that the fake posts are generally more aggressive, superficial and manipulative. The sentiment analysis complemented this idea by asserting that negative sentiments prevailed for more than 60% of the fake posts.

Moreover, post attributes helped us get more insight on each group. The Fake posts has less "comment link" and "karma link" than the real posts which gives us an indication on users' interaction with the posts. The "upvote" and "score" of each post describes the appreciation of the posts by the communities, real posts have a big number of upvotes than fake posts. This could be interpreted as a dislike of the posts by other subreddits members. It's clear that attributes could play a big role into distinguishing the fake posts from the real posts.

In addition, we made minimal assumptions in our social network model and started with a directed graph in order to preliminarily depict the existence of interactions among community members, the flow of these interactions and identify patterns of connections in the communities overall.

Nevertheless, future works can address the use of sentiment analysis with our approach of text and network analysis, allowing to glean insight on the polarity of opinions in each community and include other types of users' data like images. Furthermore, we want to better label our data, taking in consideration the noise, but also the possibility multilabeling. Rather than only using "fake" and "real" labels, we want to identify manipulative posts, misleading posts, imposter posts and satire posts.

5 Conclusion

Fake news has become omnipresent in our online social platforms. Their potential of harm on the other hand is far-reaching. In fact, the World Economic Forum (WEF) considers massive digital misinformation as one of the main threats to our society [19]. Detecting fake news is becoming thereafter one of the most important challenges of today's internet research. This paper attempts to contribute to the fight against fake news by, proposing a user-rich embedding based model for the classification of fake news on a social media platform. The approach is based on the premise that multiple characteristics can help identify fake news: origin, proliferation and linguistic tone. An experiment was conducted on data collected via Reddit API and exploratory data analysis unveiled the contribution of users' features in distinguishing posts credibility and dissimilarities on the lexicon level, which that was complemented by sentiment analysis that showed the linguistic tone of each class. We hence built a model that takes as input a vector obtained by concatenating: comments embeddings, user embedding and submission embedding and fed it to a classification baseline algorithm. The results of these analyses suggest the following: 1) fake news are proliferated more by unverified users compared to verified users; and 2) fake news stories are written in a specific linguistic tone, though further analysis needs to be conducted to specify its characteristics. The results expand the understanding of fake news as a phenomenon and motivate future work, which includes: 1) expanding the study on additional attributes, 2) designing and developing a multi classification model to better detect the different levels of fakeness or legitimacy.

References

1. Lazer, D.M.J., et al.: The science of fake news. Science **359**(6380), 1094–1096 (2018)
2. Allcott, H., Gentzkow, M.: Social media and fake news in the 2016 election. J. Econ. Perspect. **31**(2), 211–236 (2017)
3. Yang, Y., Zheng, L., Zhang, J., Cui, Q., Li, Z., Yu, P.S.: TI-CNN: convolutional neural networks for fake news detection (2018). http://arxiv.org/abs/1806.00749
4. Tacchini, E., Ballarin, G., Della Vedova, M.L., et al.: Some like it hoax: automated fake news detection in social networks (2017). http://arxiv.org/abs/1704.07506
5. Karger, D., Oh, S., Shah, D.: Iterative learning for reliable crowdsourcing systems. Adv. Neural. Inf. Process. Syst. **24**, 1953–1961 (2011)
6. Perozzi, B., Al-Rfou, R., Skiena, S.: Deepwalk: online learning of social representations. In: Proceedings of the 20th ACM SIGKDD International Conference on Knowledge Discovery and Data Mining, pp. 701–710, August 2014
7. Bengio, Y., Ducharme, R., Vincent, P., Jauvin, C.: A neural probabilistic language model. J. Mach. Learn. Res. **3**(Feb), 1137–1155 (2003)
8. Tang, J., Qu, M., Zhang, M., Mei, Q.: Large-scale information network embedding. In: Proceedings of the 24th International Conference on World Wide Web International World Wide Web Conferences Steering Committee, pp. 1067–77 (2003)
9. Grover, A., Leskovec, J.: Node2vec: scalable feature learning for networks. In: Proceedings of the 22nd ACM SIGKDD International Conference on Knowledge Discovery and Data Mining, pp. 855–864, August 2016
10. Huang, X., Li, J., Hu, X.: Label informed attributed network embedding. In: Proceedings of the Tenth ACM International Conference on Web Search and Data Mining, pp. 731–739, February 2017

11. Chung, F.R., Graham, F.C.: Spectral graph theory (No. 92). American Mathematical Soc. (1997)
12. Pan, S., Wu, J., Zhu, X., Zhang, C., Wang, Y.: Tri-party deep network representation. Network **11**(9), 12 (2016)
13. Boe, B.: PRAW: The Python Reddit API Wrapper (2012). https://github.com/praw-dev/praw/. Accessed 29 Sep 2017

Author Index

M. Lazaar et al. (Eds.): BDIoT 2021, LNNS 489, pp. 601–602, 2022.
https://doi.org/10.1007/978-3-031-07969-6

Printed in the United States
by Baker & Taylor Publisher Services